T0140646

Theodosius
Sphaerica
Arabic and Medieval Latin Translations
Edited by Paul Kunitzsch and Richard Lorch

BOETHIUS

Texte und Abhandlungen zur Geschichte der Mathematik und der
Naturwissenschaften

Begründet von Joseph Ehrenfried
Hofmann, Friedrich Klemm und
Bernhard Sticker

Herausgegeben von
Menso Folkerts

Band 62

Theodosius
Sphaerica

Arabic and Medieval Latin Translations
Edited by Paul Kunitzsch and Richard Lorch

Franz Steiner Verlag Stuttgart 2010

Bibliografische Information der Deutschen Nationalbibliothek:
Die Deutsche Nationalbibliothek verzeichnet diese
Publikation in der Deutschen Nationalbibliografie;
detaillierte bibliografische Daten sind im Internet über
<http://dnb.d-nb.de> abrufbar.

ISBN 978-3-515-09288-3

© 2010 Franz Steiner Verlag, Stuttgart
Gedruckt auf säurefreiem, alterungsbeständigem Papier.
Druck: AZ Druck und Datentechnik Kempten
Printed in Germany

To Menso Folkerts

in grateful recognition

of many years of friendship, support and

scholarly encouragement

Preface

Theodosius' *Sphaerica* is on the geometry of the sphere, a subject of great importance for astronomy. Like many other such books in Greek, it was translated into Arabic in the ninth century and from Arabic into Latin in the twelfth. Much has been said of the medieval Arabic-Latin translations, but there has been little detailed study of them. In this book we present a critical edition of an Arabic text and, on facing pages, a parallel text of the Latin derived from it. The Latin was almost certainly written by Gerard of Cremona, the most prolific of the Arabic-Latin translators.

Financial support from the Deutsche Forschungsgemeinschaft, Bonn, and from the Kurt-Vogel-Stiftung, Munich, is gratefully acknowledged.

It is a pleasure to thank Menso Folkerts, University of Munich, for making the facilities of the Institut für Geschichte der Naturwissenschaften available to us – particularly for allowing access to its library of microfilms.

We are especially grateful to the late Dr. Anton Heinen for giving us a copy of the Theodosius works in the Lahore manuscript.

Munich, September 2009   Paul Kunitzsch and Richard Lorch

Contents

Introduction

The Sphaerica

Theodosius is known as the author of the *Sphaerica*, the *De habita-tionibus* and the *De diebus et noctibus*. Little is known of his life, but from citations by Strabo (d. ca. 25 AD)[1] he seems to have lived ca. 100 BC. His three works all belonged to the "little astronomy" – "little" in contrast to the great Astronomy of Claudius Ptolemy (2c. AD), later known as the *Almagest*. Arabic translations were considered to belong to the similar collection of mathematical and astronomical works called in the Arabic literature the "middle books" ("middle", that is, between the *Elements* of Euclid and Ptolemy's *Almagest*).

Unlike most of the "little astronomy", including Theodosius' *Opera minora*, the *Sphaerica* is presented as a purely geometric trea-tise, though it is clear that many of the propositions have an astro-nomical application[2]. There are occasional lapses into astronomical language, e.g. "the visible pole" in some propositions of Books II and III. The style of the *Sphaerica* is that of Euclid's *Elements*: for each proposition there is an enunciation, a display of this by reference to a lettered diagram, and a proof. There are, too, correspondences of some of the early theorems (on spheres) and the third book of the *Elements* (on circles).

There are numerous versions and redactions of the *Sphaerica* in Arabic, Latin and Hebrew.

The Arabic Tradition

In Arabic there are two known translations, represented by **ANH** (edited here) and **FC**, and several redactions, those in **L**, **K**, the well-known *Taḥrīr* by Naṣīr al-Dīn al-Ṭūsī (d. 1274) and the redactions by Ibn Abī [al-]Shukr (d. between 1281 and 1291) and Taqī al-Dīn ibn Maʿrūf (d. 1585)[3].

---

[1]On Theodosius' life and works see R. Fecht, "De Theodosii vita et scriptis", in his edition of *De habitationibus* and *De diebus et noctibus*; also Czinczenheim, 8–17, Ziegler, Bulmer-Thomas, the introduction to Ver Eecke's French translation of the *Sphaerica*, and Heath [1921], II, 245-246.

[2]See Czinczenheim, 957-984. For a description of the mathematical content of the work, see also Heath [1921], II, 246–252, and Ver Eecke, *op. cit.*, xii–xix.

[3]For Taqī al-Dīn, see Ḥājjī Khalīfa, col. 142.

In the colophon of **L** it is said that the manuscript was copied from a copy which was copied from a copy in the hand of Ibn al-Sarī (i.e. the famous commentator of Arabic translations of Greek mathematical and philosophical works, also known as Ibn al-Ṣalāḥ, d. 1154).

The Arabic title of the *Sphaerica* was not literally translated from the Greek title, Σφαιρικά. It appears in various forms: *Kitāb al-ukar*[4], *Kitāb al-kurāt*[5] and *Kitāb al-kura*[6] all occur, the first being the most frequent.

No translator is mentioned in the bio-bibliographical literature: the *Fihrist* of Ibn al-Nadīm and Ibn al-Qifṭī's biographical work. They mention Theodosius' three works, but give no translator for any of them[7]. Of **ANH**, only **N** mentions that the text was revised by Thābit ibn Qurra (*iṣlāḥ*); but at the beginning of Book II it is said, instead, that it is a translation (*tarjama*) by Thābit. In **FC** it is stated that Qusṭā b. Lūqā translated it (*tarjama*). At the beginning of the text in **L** it is said that it was the translation (*naql*) of Abū Zayd Ḥunayn b. Isḥāq. However, this ascription is unlikely, because Ḥunayn is mostly known for his translations of medical and philosophical works. At the beginning of **K**, which is similar to Ṭūsī (see below), the scribe writes that Aḥmad, son of Caliph al-Muʿtaṣim[8], ordered the *Sphaerica* to be brought from Greek into Arabic (*ikhrāj*) and that Qusṭā b. Lūqā translated (*tarjama*) it.

Most detailed is the description in the introduction to Ṭūsī's *Taḥrīr* of the *Sphaerica* (completed in 1253), where he says: "There has ordered its translation (*naql*) from Greek into Arabic Abū 'l-ʿAbbās Aḥmad ibn al-Muʿtaṣim bi-llāh, and Qusṭā ibn Lūqā al-Baʿlabakkī undertook its translation (*naql*) until the fifth proposition of the third book; then somebody else (*ghayruhu*) undertook the translation of the rest, and Thābit ibn Qurra revised it (*aṣlaḥahu*)"[9].

Ṭūsī's text was later repeated almost *verbatim* by Ḥājjī Khalīfa (d. 1657) in his description of the *Sphaerica*, *Kitāb al-ukar*: "There has ordered its translation (*naql*) from Greek into Arabic al-Mustaʿīn bi-

---

[4]Thus in **A**, **K** and Ṭūsī.

[5]**N** (at the end of Book III).

[6]**H** (at the beginning) and **FC**.

[7]Ibn al-Nadīm, 269, lines 5–7, the name being wrongly given as *Thywdwrs*; Ibn al-Qifṭī, 108, lines 1–5 and 11–14.

[8]*Sic*; in reality Aḥmad was the grandson of al-Muʿtaṣim and later became caliph himself with the honorary title of al-Mustaʿīn (r. 862–866). Al-Muʿtaṣim reigned 833–842.

[9]Ṭūsī, 2.

llāh Abū 'l-ʿAbbās Aḥmad ibn al-Muʿtaṣim during his caliphate, and Qusṭā ibn Lūqā al-Baʿlabakkī undertook its translation (*naql*) until the fifth proposition of the second (*sic*)[10] around the year 250[11]; then somebody else (*ghayruhu*) undertook the translation of the rest, and Thābit ibn Qurra revised it (*aṣlaḥahu*). Then the learned Naṣīr al-Dīn Muḥammad ibn Muḥammad al-Ṭūsī, who died in the year 672 (= 1274 AD), and the excellent Taqī al-Dīn Muḥammad ibn Maʿrūf al-Rāṣid, who died in the year 993 (= 1585 AD), made recensions of it"[12].

The Arabic Manuscripts

A[13]: Istanbul, Seray, Ahmet III 3464, ff. 20v–53v. This codex contains altogether seventeen treatises – *inter alia* most of the middle books – in several hands[14]. Seven of them, among them the *Sphaerica*, are in the same hand, three of which (but not the *Sphaerica*) are dated to August or September 1228 and written by Muḥammad b. Abī Bakr b. Muḥammad. Three other texts, in different hands, are dated to 1219, 1233 and 1290.

N: Lahore, private library M. Nabī Khān, pp. 185–281. This codex contains, besides the *Sphaerica*, also a copy of the *De habitationibus* (pp. 282–294)[15]. The first two pages of the *Sphaerica* (pp. 185, 186) are lost and have been supplied, in another hand, by two pages from Ṭūsī's *Taḥrīr*. The genuine text begins at the top of p. 187, with *al-mustaqīma* (Prop. I 1, line 5, in the present edition). In the colophon the scribe explains that the text he copied was in the hand of a direct descendant of Thābit ibn Qurra. From his further report it is clear that, at some stage of the transmission, the diagrams were corrected by al-Ḥasan ibn Saʿīd:

> Finished is the third chapter of Theodosius' book on the spheres, and with its ending the entire book is finished with the praise of God. It is fourteen theorems and the number of the theorems of

---

[10]*al-thāniya*, the second, is an easy miswriting in Arabic script for *al-thālitha*, the third; the word for "Book" (*al-maqāla*) has been omitted here.

[11]I.e. 864 AD. This detail is not in Ṭūsī. The given year falls in the reign of Caliph al-Mustaʿīn.

[12]Ḥājjī Khalīfa, col. 142.

[13]From **A** the text has been edited in the unpublished dissertation by T. J. Martin, University of St. Andrews, 1975.

[14]For a detailed list of the items, see Lorch [2001], 22-23.

[15]We are very grateful to the late Dr. Anton Heinen for giving us copies of the Theodosius texts in this manuscript. Unfortunately, we have no access to the rest of this manuscript and cannot describe it.

the three chapters is 59, [in] the correction by Thābit b. Qurra
al-Ḥarrānī al-Ṣābiʾ.

I have copied this book from the handwriting of Qurra b. Sīnān b.
Manṣūr b. Saʿīd b. Thābit b. Sinān b. Thābit b. Qurra al-Ḥarrānī
al-Ṣābiʾ in the city of Mosul (God protect it!) in the Niẓāmīya
Madrasa (God give it long life!), when six nights remained of
Jumādā I of the year 554 H [= 13 June 1158] (upon its patron
be the finest *salām*!).

I found written at the end of the book: "al-Ḥasan b. Saʿīd has
finished devising the diagrams [*tashkīl*] of this book, but the vol-
ume from which he copied the figures [*ashkāl*] was not reliable.
Moreover there was corruption in it, so it was necessary to collate
it with the figures [*ashkāl*] in another copy. That was on the eve
of Tuesday, eight nights remaining of twelve [i.e. Dhū ʾl-Ḥijja] of
the year 421 [20 December 1030]. Praise be to God richly and
His blessings upon Muḥammad and all his family!"

Some notes on the text by al-Ḥasan ibn Saʿīd are edited below, after
the text itself.

H: Paris, BnF hebr. 1101, ff. 1–53r, 86r–87r, in Hebrew script. This
contains, besides the *Sphaerica*, only one other text[16]: the treatise on
the use of the astrolabe by Abū [al-]Ṣalt Umayya b. ʿAbd al-ʿAzīz b.
Abī [al-]Ṣalt (d. 1034) in 88 chapters[17]. At the beginning this is entitled
Risāla (sic) al-asṭurlāb li-Abī Ṣalt. At the end it is called *Kitāb al-ʿamal
bi-l-asṭurlāb*.

These three manuscripts, **ANH**, represent the translation that
is edited here. Another translation is represented by **FC**. Further, **L**
and **K** appear to be two different reworkings of these translations. The
details of these manuscripts are as follows:

F: Florence, Laur. Med. 124, 76ff., and **C**, Cambridge, University Library, Add. 1220, ff. 1r–50r, are both in Hebrew script. They seem
to represent another translation, which both manuscripts attribute to
Qusṭā ibn Lūqā.

L: Leiden, Or. 1031, pp. 22–72. According to the colophon this text
seems to be a reworking by Ibn al-Ṣalāḥ (see above), despite the ques-

---

[16]Ff. 56r–85r, the intervening pages being blank and ff. 86r–87r containing Prop.
III 7, which is omitted in the main copy.

[17]See Steinschneider, 364.

tionable ascription of the translation in the manuscript to Ḥunayn b. Isḥāq.

K: Private library (formerly in the possession of H. P. Kraus), ff. 33v–64r, 7/13c., is also apparently a reworking. It is striking that the preface has a wording that is almost identical to Ṭūsī's preface in his *Taḥrīr*. This seems to point to **K**'s being written after the *Taḥrīr*. The author of **K** here used some terms different from Ṭūsī's and omitted the second part of Ṭūsī's preface on the details of Qusṭā's translation. Another possibility is that Ṭūsī and the author of **K** used the same source.

The Latin Tradition

Two Latin texts of the *Sphaerica* circulated in the Middle Ages, the one edited here and a longer version that has been ascribed to Campanus[18]. Our text is clearly a translation of the Arabic of **ANH**. Since it is of the translation style of Gerard of Cremona – literalism and some characteristic translations (e.g. *cum* for *idhā*, *si* for *in*, etc.)[19] – and since the *Sphaerica* appears in the well-known list of Gerard's translations compiled by his students[20], we may safely attribute the translation to Gerard.

The Latin Manuscripts

P: Paris, BnF, lat. 9335, ff. 1r-19v, ca. 1200[21].
R: Vatican, lat. 1548, ff. 25r-50v, 14c.[22]
V: Vatican, Ottob. lat. 2234, ff. 54ra-64rb, 14c.[23]
M: Madrid, Biblioteca Nacional, lat. 10010, ff. 1v-13r, 14c.[24]
Kg: Cracow, Jagiellonian Univ. Library, 1924, pp. 223-257, 13-14c.[25]
O: Oxford, Bodleian Library, Auct. F.5.28, ff. 29v-51v, 13c.[26]
Fi: Florence, Bibl. Nazionale Centrale, c.s. J.I.32, ff. 135v-165v, 13c.
Z: Venice, Bibl. Nazionale Marciana, 1647 (f.a. 332), ff. 261r–289r, 13c.

---

[18]Lorch [1996], 169–171.

[19]For Gerard's style, see Kunitzsch [1974], 104–110 and 214–217.

[20]See the recent edition by Burnett, 276: *Liber Theodosii de speris tractatus .III.*

[21]160 ff. See Björnbo. He here dates the codex to the 14th century, but Bernhard Bischoff dated it to the late 12th century (private communication, 1.9.89). The items are all or mostly translations by Gerard of Cremona.

[22]76ff. See Nogara, 59–60.

[23]See Daly and Ermatinger, 22-23.

[24]Formerly Toledo 98–24; 86 ff. See Millás Vallicrosa, 208–211.

[25]318 pp. See Wisłocki, 461.

[26]For a full description of **O**, **Fi**, **Z** and **B**, see Busard and Folkerts, 64–67, 49–51, 78–80, 36–38, resp.

B: Berlin, Staatsbibliothek, lat. qu. 510, ff. 94v-112, 13c.
Ps: Paris, BnF, lat. 7399, ff. 139v-173v, 14c.[27]
Va: Vatican, Reg. lat. 1069, ff. 1r-44r[28].

The above manuscripts have been collated for the edition. The following manuscripts have not been included in the collation:

Db: Dresden, Sächsische Landesbibliothek, Db 86, ff. 128r–158v, 13c.[29]
Pt: Paris, BnF, lat. 3359, ff. 89r–114v, mid-14c.[30]
Od: Oxford, Bodleian Library, Digby 178, ff. 107r–111r, 15c.[31]
Bf: Berlin, Staatsbibliothek, lat. fol. 633, ff. 1r–47v, 15c.[32]
G: Glasgow, University Library, Hunt. 394, pp. 1–83, late 15c.[33]
Mq: Milan, Biblioteca Ambrosiana, Q 69 sup., ff. 41r–68v[34].

The following "mixed" manuscripts, which carry the Campanus text until III 10 and then go to the end with the Gerard translation, have also not been used in the collation:

Cracow, Jagiellonian Univ. Library, 1924, pp. 207–222, 13–14c.
Schweinfurt, Stadtbibliothek, H 81 (not paginated; 2nd part), 16c.
Vatican, lat. 3380, ff. 1r–24r, 16c.

The Edition

Arabic

The Arabic text was established by comparing **A** with **N**: in cases of disagreement the reading was decided by reference to the Greek, in the new edition by Claire Czinczenheim. The entire readings of **A** and **N** are to be found either in the text or in the apparatus. **H**, containing as it does numerous faults of orthography and other trivial mistakes, is not regularly recorded, but only in cases of doubt in the **AN** text. In general, trivial grammatical mistakes in pointing and in such matters as the orthography of the *hamza* have been silently corrected. As for the readings reported in the apparatus, a quotation from one manuscript

---

[27]See *Cat. Paris*, 351-352.

[28]This is the only item in the codex. There is no printed description.

[29]Described in Busard and Folkerts, 41–43. Not used because of physical damage. The manuscript has readings similar to **O** in the *Sphaerica* and also in Robert of Chester's version of Euclid (see *ibid.*, 43).

[30]173 ff. See *Cat. BN*, 279–287.

[31]115 ff. See Macray, 190–192.

[32]There is no printed description of this codex. It contains only Theodosius' *Sphaerica* and *De habitationibus*.

[33]158 ff. See Young and Aitkin, 314–315.

[34]70 ff. See Gabriel, 324–325.

is reproduced exactly (or as exactly as possible) as it appears in the manuscript, but when text from two or more manuscripts is quoted, correct pointing and orthography are imposed on it.

Angle brackets, < >, are used to include material added by the editors and not directly supported by the manuscripts. Square brackets, [ ], enclose readings that are uncertain because of physical damage.

Latin

The Latin text was established by transcribing **P**, which appears to be, of all the Latin manuscripts, the closest to the Arabic; the transcription was then compared with ten other manuscripts, taken in approximate order of agreement with the Arabic (see the order of the list of manuscripts, above). In every case the reading was decided by comparison with the Arabic and with Gerard's known style. The titles of the three books and the colophons are edited only from **P**.

The equivalences of the diagram letters may be seen in Table 1. In the edition of the Latin we have kept to Gerard's lettering, with the exception of his *thel*, 7, ʜ and *Ge*, which we render by Ť, Ź, Ď and Ǵ, respectively.

General Remarks

The numbering of the propositions is not constant in the various texts. Throughout, the Latin agrees with **A**, and this is the numbering we have adopted for the edition. Table 2 is a conspectus of the proposition numbers in Books I and II. In Book III the numbering is constant – though in **H** III 7 is omitted in its place and added at the end of the codex.

The mathematical summary, printed after the text, is a translation of the mathematical argument of the Arabic text. It also contains, in the footnotes, remarks about the presence or absence of some elements of meaning in the Arabic and Latin; but no attempt has been made to make such remarks complete.

It is our intention to provide future scholars with reliable information rather than to draw conclusions ourselves. But we note that right at the start there is extra material in the Arabic-Latin texts not to be found in the Greek edited by Czinczenheim: definitions 6 and 7, on the distances of circles from the centre of the sphere, and much of definitions 8–11, on the inclination of planes. Further obvious differences appear in the early theorems of Book I.

Table 1

Greek	Arabic	Latin
A	ا	A
B	ب	B
Γ	ج	G
Δ	د	D
E	ه	E
Z	ز	Z
H	ح	H
Θ	ط	T
I	ى	(I)
K	ك	K
Λ	ل	L
M	م	M
N	ن	N
Ξ	س	S
O	ع	Q
Π	ف	F
P	ق	C
Σ	ر	R
T	ش ت	O
Υ	ث	P
Φ	ث	Y
X	خ	X
Ψ	ذ	*thel* (T́)
Ω	ظ	7 (Ź)
ꝯ	ض	ʮ (D́)
५	غ	*Ge* (Ǵ)
⅄ / ↑	و	U

Table 2

In the following table "id." means that the respective number is the same as in our present edition shown here in the first column. Square brackets, "[ ]", mean that the number they enclose was cut off in the photocopies of **N**.

Proposition Numbers in Book I

Present edition	Arabic manuscripts			Greek edition
	A	**N**	**H**	
1–8	id.	id.	id.	id.
9	id.	[9]	—[a]	–
10	id.	[10]	9	9
11	id.	id.	[b]	10
12	id.	id.	10	11
13	id.	[13]	11	12
14	id.	[14]	12	13
15	id.	[15]	13	14
16	id.	[16]	14	15
17	id.	id.	15	16
18	id.	id.	16	17
19	id.	id.	17	18
20	id.	[20]	18	19
21	id.	id.	19	20
22	id.	2[2]	20	21[c]

[a] not counted separately – incorporated in Prop. 8.

[b] added at the end of Book I, without number.

[c] Props. 22–23 of the Greek edition do not exist in the Arabic versions.

Proposition Numbers in Book II

Present edition	Arabic manuscripts			Greek edition
	A	**N**	**H**	
1–10	id.	id.	id.	id.
11	id.	id.	id.	id.[a]
12	id.	13	id.	14
13	id.	[14]	id.	13
14	id.	15	id.	15
15	id.	16	id.	16
16	id.	17	id.	17
17	id.	18	id.	18
18	id.	19	id.	19
19	id.	[20]	id.	20
20	id.	21	id.	21
21	id.	[22]	id.	22
22	id.	23	id.	23

[a] In the Greek, Prop. 12 begins in Prop. 11, line 33, of the present edition.

Theodosius

Sphaerica

Arabic and Latin Texts

بسم الله الرحمن الرحيم

المقالة الأولى من كتاب ثاوذوسيوس فى الأكر

الكرة هى شكل مجسم يحيط به سطح واحد جميع الخطوط المستقيمة التى
تخرج من نقطة واحدة من النقط التى فى داخله وتلقى السطح مساوٍ
بعضها لبعض ، ومركز الكرة هى تلك النقطة ،

ومحور الكرة هو خط ما مستقيم يمر بالمركز وينتهى فى كلتى الجهتين إلى
سطح الكرة إذا أثبت الخط وأديرت الكرة عليه ، وقطبا الكرة طرفا
المحور ،

الشىء الذى يقال له فى الكرة قطب دائرة هو نقطة تكون على سطح الكرة
جميع الخطوط المستقيمة التى تخرج منها إلى الخط المحيط بالدائرة مساوٍ
بعضها لبعض ،

يقال فى الكرة إن بعد الدوائر من مركزها بعد متساوٍ إذا كانت الأعمدة التى
تخرج من مركز الكرة إلى سطوح الدوائر مساوٍ بعضها لبعض ، والدائرة
التى هى أبعد هى التى يقع عليها عمود أطول ،

---

1 الرحيم [بسم ... *om.* H 2 ثاوذوسيوس [*corr. ex* ثاודوس: A, האודוסיוס H 2 الأكر [אלכרה H
4 النقط [הנקטה A القطه [وتلقى :H ملعى A 4 مساوٍ [מסאוי H 5 بعضها [בעצה H 6 ما [om. H
6 بالمركز [במרכזהא H 6 إلى [illeg. A 7 الخط [כט ואחד H 7 طرفا [טרפי H 9 هو [הי
H 10 إلى [*supra* A 10 مساوٍ [מסאוי H 12 متساوٍ [مساوٍ A 13 مركز [אלכבה *add. et del.* H
13 مساوٍ [מסאוי H 13 والدائرة [ואלדואיר H 14 هى [הוא H

Pars prima libri theodosii de speris incipit.

Spera est figura corporea una quidem superficie contenta intra quam unum punctorum ipsius existit a quo omnes linee recte protracte, que illi superficiei occurrunt, sunt ad invicem equales. Et punctum illud spere est centrum. 5

Meguar vero spere est quelibet linea recta per centrum transiens et ab utraque parte ad superficiem spere perveniens, cum ipsa figitur et spera circa ipsam volvitur. Spere autem poli sunt meguar extremitates.

Res que in spera polus circuli dicitur est punctum super superficiem spere consistens, a quo omnes linee recte ad circumferentiam circuli 10 protracte ad invicem sunt equales.

Circulorum in spera a centro elongatio equalis dicitur cum perpendiculares que a centro spere ad circulorum superficies protrahuntur ad invicem sunt equales. Circulus vero qui magis est remotus est super quem longior cadit perpendicularis. 15

---

2 Spera est] *marg.* **B**   2 corporea] *om.* **R**   3 punctorum] punctum **KgBPsVa**   3 ipsius] *om.* **Z**   3 linee recte] *tr.* **PRVMKgOFiZ**   4 occurrunt] occurrant **BPs**   4 ad invicem] *om.* **PRVMKgOFiZ**   5 spere est] *tr.* **KgBPsVa**   6 Meguar] Mengar **BOFiZBPs**, *corr. ex* Megar **Va**   6 linea recta] *tr.* **PRVOZ**   6 centrum] spere *add.* **OFiZ**   7 parte] spere *add.* **R**   7 superficiem spere] *tr.* **PRVOFiZ**   7 perveniens] veniens **Kg**, *corr. ex* protracte **BPs**   7 cum] et **Va**   7 et] quod **Va**   8 ipsam] *supra* **B**   8 volvitur] involvitur **BPsVa**   8 meguar] mengar **FiBPs**, *corr. ex* megar **Va**   9 Res que] Quod **KgBPsVa**   9 polus circuli] *tr.* **PRVOFiZ**   9 est] *supra* **BPs**   9–10 superficiem spere] *tr.* **PRVMKgOFiZ**   10 linee recte] *tr.* **MKg**   10 recte] *om.* **OFiZ**   10 circumferentiam circuli] *tr.* **PRVOFiZ**   11 ad invicem sunt] *permut.* **MKg**   12 a] *supra* **B**, *marg.* **Ps**, e **Va**   12 elongatio equalis] *tr.* **Va**   12 equalis] equales **Kg**, *om.* **B**, *supra* **Ps**   12–13 perpendiculares] perpendicularis **Va**   13 spere] *om.* **Va**   13 circulorum superficies] superficiem circulorum **Kg**   13–14 ad invicem sunt equales] *permut.* **Va**   15 quem] quam **OKgPs**   15 longior] *corr. ex* longitudo **R**

يقال إن السطح مائل على سطح آخر إذا تعلم على الفصل المشترك للسطحين   15

نقطة ما وأخرج منها فى كل واحد من السطحين خط مستقيم قائم على

الفصل المشترك على زوايا قائمة فأحاط الخطان المخرجان بزاوية حادة ،

والميل هو الزاوية التى يحيط بها ذانك الخطان المستقيمان ،

ويقال إن ميل السطح عن السطح مثل ميل سطح آخر عن سطح آخر إذا

كانت الخطوط المستقيمة التى تخرج من الفصول المشتركة للسطوح على زوايا   20

قائمة فى كل واحد من السطوح من نقط بأعيانها محيطة بزوايا متساوية ،

والتى تكون زواياها أصغر فهى أكبر ميلاً ·

ا   إذا قُطع بسيط كرى بسطح ما فإن القطع الحادث خط محيط بدائرة ·

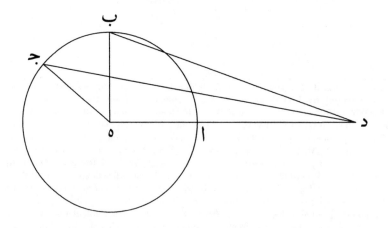

עלי [1عن 19 H אלמכרנה [المخرجان 17 H .om [على الفصل المشترك 17-16 H סטח [السطح 15

יכון [تكون 22 H בזאויה [بزوايا 21 H محيطة [محيط 21 H זאויה [زوايا 20 H עלי [2عن 19 H

אלכרי [فإن 1 H מיל [ميلاً 22 H אכתר [أكبر 22 H הי [فى 22 H זאויאהא [زواياها 22 H

add. et del. H 1 אלכט אלחדאת [القطع الحادث H

Superficies super aliam superficiem inclinata dicitur cum super communem differentiam duarum superficierum quodlibet punctum signatur, a
quo in utraque duarum superficierum linea recta super communem differentiam orthogonaliter erigitur, et due linee protracte angulum continent acutum. Inclinatio autem est angulus qui ab illis duabus rectis 20
lineis continetur.

Inclinatio superficiei a superficie dicitur equalis inclinationi superficiei
alterius ab altera superficie cum linee recte que protrahuntur a communibus differentiis superficierum orthogonaliter in unaquaque superficierum existentes ab eisdem punctis angulos equales sunt continentes. 25
Quarum vero anguli sunt minores, maioris sunt inclinationis.

1 Cum spere superficiem aliqua secat superficies, sector proveniens in
superficie spere est linea circulum continens.

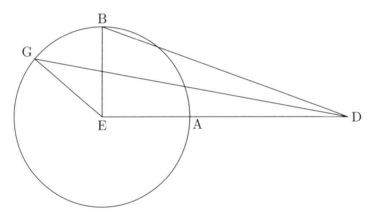

---

16 inclinata] erecta **OFiZ**; sive erecta *supra* **B** 16 inclinata dicitur] *tr.* **BPsVa** 16–
20 super communem ... continent acutum]] ab uno puncto communis septionis [*sic*] earum linee ipsi perpendiculares per utramque superficiem de[...]cte angulum non rectum content **Va**
17–19 duarum superficierum quodlibet ... differentiam] *om.* **OFiZBPs** 18 in] *supra* **PR** 18–
19 differentiam] distantiam **Kg** 19 et] inclinata cum **FiZO** (*corr. ex* et **O**) 19 linee] in
illis duabus *supra* **B** 19–20 continent] continente **Kg** 20 Inclinatio autem] Inclinacionis est
Va 20 autem] *om.* **Ps** 20 illis] istis **V** 20 rectis] *om.* **MFiZ** 21 continetur] continentur **Fi**
22 Inclinatio] autem *add.* **OFiZBPs** 22–26 Inclinacio superficiei ... sunt inclinationis]] Superficierum unius super alteram inclinacio cl'is [*sic*] dicitur alius supra aliam inclinacioni cum
a singulis punctis communium sectionum suarum linee perpendiculares per omnes superficies illas ducte angulos illos continent. Inter t' quas superficies est angulus acutior vel minor earum alterius super alteram inclinacio est maior **Va** 22 inclinationi superficiei] *tr.* **V**
23 recte que protrahuntur] protracte in rectum **BPs** 23–24 communibus differentiis] quibus
distantiis **Kg** 24 superficierum] superficie **FiZ**, superficies **Kg**, *corr. ex* super **B** 24–
25 superficierum] superficie **ZBPs** 25 existentes] existens **R** 25 continentes] continuentes
Fi 26 Quarum] quantum **Z** 26 vero] *om.* **BPs** 1 secat superficies] *tr.* **BPsVa**
1 proveniens] perveniens **KgBPs** 2 superficie spere] *tr.* **OFiBPsVa**, spere superficiem **Z**
2 continens] *corr. ex* contingens **Ps**

فليُقطع بسيط كرى بسطح ما وليحدث فى بسيط الكرة قطعاً وهو

خط اب‍ج ، فأقول إن اب‍ج خط محيط بدائرة فإن كان السطح القاطع يمر

بمركز الكرة فقد تبين أن خط اب‍ج محيط بدائرة وذلك أن الخطوط

المستقيمة التى تخرج من المركز إلى خط اب‍ج مساوٍ بعضها لبعض فإن كان

الأمر كذلك فقد تبين أن مركز الكرة ومركز الدائرة واحد بعينه ، وإن لم

يمر السطح القاطع بمركز الكرة فلنتوهم مركز الكرة نقطة د وليخرج من

نقطة د إلى السطح الذى يمر بخط اب‍ج عمود ده وليلق السطح على نقطة

ه وليخرج خطا ه‍ب ه‍ج وليوصل خطا دب دج فلأن نقطة د مركز الكرة

يكون خط دب مساوياً لخط دج فالمربع الكائن من خط دب مساوٍ للمربع

الكائن من خط دج والمربعان الكائنان من خطى ده ه‍ب مساويان للمربع

الكائن من خط دب وذلك أن الزاوية التى يحيط بها خطا ده ه‍ب قائمة

والمربعان الكائنان من خطى ده ه‍ج مساويان للمربع الكائن من خط دج

وذلك أن الزاوية التى يحيط بها خطا ده ه‍ج قائمة فالمربعان الكائنان من

خطى ده ه‍ب مساويان للمربعين الكائنين من خطى ده ه‍ج ويسقط مربع

خط ده المشترك فيبقى المربع الكائن من خط ه‍ب مساوياً للمربع الكائن من

2 فليُقطع] פלחקטע H 2 قطعاً] قطع (supra) A 3 فأقول ... بدائرة] marg. A 3 خط [2 om.

واحده] واحد A 5 المستقيمة] hic incipit N 6 واحد] واحده H دائرة] דאירה H 4 اب‍ج] خط add. A

ولیخرج] ול"כרג H 7 يمر] A يكن A القاطع] يمر add. A 7 الكرة [2] supra A 7 القاطع] يمر

add. et من نقطة] منه A 8 السطح الذى يمر بخط] סטח H 8 ده] דהו H 9 وليخرج] עלי

del. H 9 خطا [1] כט H 9 ه‍ج] ה ו הג כיף מא וקעא H 9 وليوصل] ולנצל H 9 خطا [2] כט H

9 د] הו add. H 10 مساوياً] מסאויה H 10 مساو ... لخط] marg. A 11 ده هب] דה והב H

11 ه‍ب] supra A 11 مساويان] מסאוין H 12 ده هب] דה והב H 13 ه‍ج] وهج H 13 خط] om.

H 14 يحيط] תחיט H 14 خطا] om. H 15 مساويان] מסאוין H 15 للمربعين الكائنين] ללמרבעאן

add. H איךא ;A ه‍ج [2] 16 add. et del. A ده هب مساوبان من حطين [2] خطى 15 אלכאינאן

8 اب‍ج] س نا نا س supra A 12 ه‍ب] س مر ا supra A

Superficiem igitur spere superficies aliqua secet et proveniat in
superficie spere sector, qui sit linea *ABG*. Dico igitur quod linea *ABG*
continet circulum. Quod sic probatur: si enim superficies secans fuerit   5
transiens per centrum spere, tunc iam manifestum est quod linea *ABG*
est circumferencia circuli. Quod ideo est quoniam omnes linee recte que
a centro spere ad lineam *ABG* protrahuntur ad invicem sunt equales.
Si ergo res ita fuerit, tunc iam manifestum erit quod centrum spere et
centrum circuli unum et idem existit. Quod si superficies secans per   10
centrum spere non transierit, imaginabor spere centrum punctum *D*, a
quo ad superficiem que transit per lineam *ABG* protraham perpendicularem *DE*, que superficiei occurrat super punctum *E*; et protraham
duas lineas *EB EG*, et producam duas lineas *DB DG*. Et quia punctum
D est centrum spere, ergo linea *DB* quocumque modo cadat est equalis   15
linee *DG*. Ergo quadratum factum ex linea *DB* est equale quadrato
facto ex linea *DG*. Sed duo quadrata facta ex duabus lineis *DE EB*
sunt equalia quadrato facto ex linea *DB*. Quod ideo est quoniam angulus qui continetur a duabus lineis *DE EB* est rectus. Et duo quadrata
facta ex duabus lineis *DE EG* sunt equalia quadrato facto ex linea *DG*.   20
Quod ideo est quoniam angulus qui a duabus continetur lineis *DE EG*
est rectus. Duo igitur quadrata ex duabus lineis *DE EB* facta duobus
quadratis ex duabus lineis *DE EG* factis sunt equalia. Demam autem
quadratum commune quod fit ex linea *DE*; remanet ergo quadratum

3 igitur] *supra* **M**, ergo **BPsVa**   3 superficies aliqua] *tr.* **MKgFiZ**   3 proveniat] *corr. ex*
proveniet **Z**, perveniat **BPs**   4 spere sector] *tr.* **M**   4 *ABG*] *AGB* **Kg**   4 igitur] *om.*
BPsVa   4 quod linea] *om.* **V**   5 probatur] comprobatur **OZ**   5 enim] *om.* **V**
5 superficies] superficiens **Z**   5–6 fuerit transiens] transierit **BPsVa**   6 iam] *om.* **BPsVa**
6–7 quod ... circuli] propositum **BPsVa**   7–8 Quod ... equales] quia omnes linee ducte
a centro ad *ABG* sunt equales sibi invicem **BPsVa**   9–10 Si ergo ... existit] *om.* **BPsVa**
10 existit] consistit **Fi**   10 Quod] Sed **BPs**, Et **Va**   10–11 per centrum spere] *om.* **B**, *su-*
pra **Ps**   10–11 per centrum ... transierit] *permut.* **Va**   12 lineam *ABG*] *tr.* **BPsVa**; circumductam *supra* **Ps**, *add. in textu* **B**   13 superficiei] secanti *supra* **B**   14 producam] ponam
OFiZ   15 est[1]] positum *add.* **PRVMKgOFiZ**   15 centrum spere] *tr.* **BPsVa**   16 Ergo] *corr. ex*
quia **Ps**   16 factum ex linea] *om.* **BPsVa**   17 facto ex] *om.* **BPsVa**   17 linea *DG*] *DG* linee
BPsVa   17 duabus] *om.* **BPsVa**   17 lineis *DE EB*] *tr.* **BPsVa**   18–19 sunt ... rectus] *marg.*
R   18 linea] *om.* **B**   18–20 Quod ideo ... *DG*] *marg.* **BPs**, *om.* **Kg**   18 ideo est] *tr.*
PRVMOFi   18 est] *om.* **Va**   18 quoniam] quia **OFiZ**   19 a duabus] *om.* **B**   20 facto] ex
duabus lineis *DE EG* sunt equalia quadrato facto *add.* **R**   21 ideo est] *tr.* **Kg**
21 quoniam] quia **BPsVa**   21 a duabus] ab his duabus **PRVMKg**, ab his **OFiZ**, *om.*
BPsVa   22 Duo] dico **FiZ**   22 ex duabus] *om.* **BPsVa**   22 lineis *DE EB*] *tr.* **BPsVa**
23 Demam] demo **V**   23–24 Demam ... commune] dempto ergo quadrato communi **BPsVa**
24 quadratum commune] *tr.* **FiZ**   24 fit] sit **FiV**   24 ergo] *om.* **BPsVa**

خط ه‍ ج‍ فخط ب‍ ه‍ إذاً مساوٍ لخط ج‍ ه‍ وكذلك ايضاً نبين أن جميع الخطوط

المستقيمة التى تخرج من نقطة ه‍ إلى خط ا‍ ب‍ ج‍ مساوٍ بعضها لبعض فا‍ ب‍ ج‍

خط محيط بدائرة ونقطة ه‍ مركز الدائرة .

وقد تبين من ذلك أنه إذا أخرج من مركز الكرة إلى دائرة من الدوائر 20

التى فى الكرة عمود فهو يقع على مركز الدائرة ، وذلك ما أردنا أن نبين .

ب‍   كيف نجد مركز كرة معلومة .

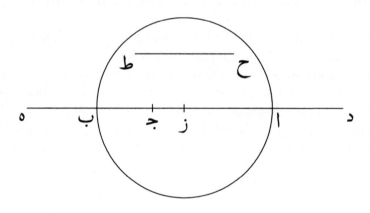

فلنتوهم كرة معلومة نريد أن نجد مركزها فلنقطعها بسطح ما

فيكون القطع الحادث دائرة ولتكن الدائرة التى تحدث دائرة ا‍ ب‍ فإن كان

السطح القاطع يمر بمركز الكرة فقد تبين أن مركز الكرة والدائرة مركز واحد

وقد علمنا كيف نجد مركز دائرة معلومة ، وإن لم يمر السطح القاطع بالمركز 5

H هَذ‍] ج‍ ه‍ 17   N اضا] إذاً 17   פ‍אדא כט בֹה] לֹּּغط ب‍ ه‍ إذاً 17   خط 17

H תقע] يقع 21   N ان] إذا 20   H אנהא] أنه 20   A يساوى] مساوٍ 18   H תכרנאן] تخرج 18

N 2 أن ... دائرة 4-5   illeg. N   H פֹליכֹן N, نكون] فيكون 4-5   H וליקטעֹהא סטֹח] فلنقطعها بسطح 2

add. H   אֹבֹ] دائرة 5   A مركزا واحدا] مركز واحد 4

A supra   ا‍ ج‍ س‍] دائرة 5

factum ex linea *EB* equale quadrato facto ex linea *EG*. Linea igitur   25
EB equalis etiam existit linee *GE*. Similiter quoque ostenditur quod
omnes recte linee, que a puncto *E* egrediuntur ad lineam *ABG*, ad
invicem sunt equales. Ergo linea *ABG* est linea circuli circumducta et
punctum *E* est centrum circuli.

Ex hoc igitur manifestum est quod cum a centro spere ad unum   30
circulorum qui sunt in spera protrahitur perpendicularis, super centrum circuli cadit. Et illud est quod demonstrare voluimus.

2 Quomodo spere date centrum reperiatur.

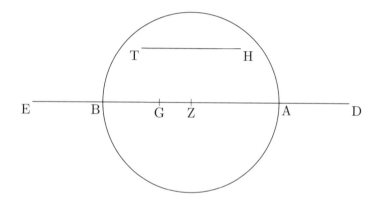

    Imaginabor igitur speram datam cuius centrum volo reperire,
quam secet aliqua superficies, sitque sector proveniens circulus et sit
circulus qui provenit circulus *AB*. Si ergo superficies secans fuerit transiens per centrum spere, tunc iam manifestum est quod centrum spere   5
et circuli est unum. Iam autem scivimus qualiter circuli dati centrum

25 factum ex linea] *om.* **BPsVa**   25 facto ex linea] *om.* **BPsVa**   25 Linea igitur] ergo linea
BPsVa   26 equalis etiam existit] equatur etiam **MKg**, est equalis **BPsVa**   26 *GE*] *in corr.*
B   26 Similiter quoque] Simili modo **BPsVa**   26 ostenditur] ostendetur **OZBPsVa**, ostende
FiZ   27 omnes] omnis **Fi**   27 recte] *om.* **BPsVa**   27 recte linee] *marg.* **R**   27 *E*] *om.*
Fi   27 egrediuntur] progrediuntur **BPsVa**   27–28 *ABG*, ad invicem ... linea[2]] *marg.*
R   27–28 ad invicem] *om.* **BPsVa**   28 linea *ABG*] *tr.* **BPsVa**   28–
29 est linea ... centrum circuli] continet circulum **BPsVa**   28 linea[2]] *om.* **V**   29 circuli] *om.*
Fi   30 *ante* Ex] Corollarium *add.* **PRVMOBPs** (Coroll' **PV**, Correll' **R**, Coroll'r **M**,
Correll'm **O**, corell' *marg.* **B**, Correl' *marg.* **Ps**)   30 igitur] *om.* **BPsVa**   30 est] *om.*
Z   31–32 centrum circuli] *tr.* **PRVMKgOFiZ**   32 Et ... voluimus] *om.* **FiZ**, et ita
est quod dico **KgBPsVa** (dico: d'ico **Kg**)   32 demonstrare] monstrare, de *supra* **R**
32 voluimus] volumus **O**   2 igitur] *om.* **BPsVa**   2 datam] *om.* **Z**   3 secet aliqua] *tr.*
PRVMKgOFiZ   3–4 et sit circulus] *om.* **FiZ**   3–4 et sit circulus ... circulus[2]] *om.* **BPsVa**
3–4 sit circulus] *tr.* **Kg**   4 ergo] igitur **Fi**   4 ergo superficies] superficies igitur **BPsVa**
4–5 secans fuerit transiens] *AB* transeat **B**   5 spere[1]] *supra* **BPs**   5 iam] *om.* **BPsVa**
5 est] *om.* **OFiZ**   6 unum] idem **BPsVa**   6 Iam] *om.* **BPsVa**   6 autem scivimus] *tr.* **BVa**,
scivimus itaque **Ps**   6 scivimus] scimus **FiZ**   6 qualiter] *om.* **BPsVa**   6 circuli dati] *tr.*
PRVMKgOFiZ

فليكن مركز دائرة اب نقطة جـ وليخرج من نقطة جـ خط قائم على سطح

دائرة اب على زوايا قائمة وهو خط جد ولينفذ فى كلتى الناحيتين وليلق

بسيط الكرة على نقطتى د ه وليُقطع خط ده بنفصين على نقطة ز ، فأقول

إن نقطة ز مركز الكرة فإن لم يكن الأمر كذلك وأمكن أن يكون المركز نقطة

غيرها فلتكن نقطة حـ وليخرج من نقطة حـ خط يلقى سطح الدائرة على

نقطة طـ على زوايا قائمة وإذا أخرج من مركز الكرة إلى دائرة من الدوائر

التى فى الكرة خط مستقيم يكون عموداً عليها فإنه يمر بمركز الدائرة فنقطة

طـ مركز الدائرة وقد كانت نقطة جـ أيضاً مركزها وذلك ممتنع فإن وقع العمود

المخرج على نقطة جـ فقد خرج من نقطة واحدة بعينها على سطح واحد

بعينه فى جهة واحدة خطان مستقيمان على زوايا قائمة وذلك غير ممكن

فليس نقطة حـ مركز الكرة وكذلك أيضاً نبين أنه لا يمكن أن يكون مركز

الكرة نقطة أخرى غير نقطة ز فنقطة ز مركز الكرة ·

reperiatur. Ergo iam manifestum est nobis quomodo reperiamus centrum spere. Quod si superficies secans per centrum non transierit, sit
igitur circuli *AB* centrum punctum *G*; a quo protraham lineam super
superficiem circuli *AB* orthogonaliter insistentem, que sit linea *GD*, et   10
protrahatur in duas partes quousque superficiei spere occurrat super
duo puncta *D* et *E*; deinde secetur linea *DE* in duo media in puncto
Z. Dico igitur quod punctum *Z* est centrum spere. Quod si res non
ita fuerit et fuerit possibile ut centrum sit punctum preter ipsum, sit
igitur punctum *H*, protraham itaque a puncto *H* lineam perpendicula-   15
rem occurrentem superficiei circuli *AB* super punctum *T*. Cum autem
a centro spere ad unum circulorum qui sunt in spera linea recta protrahitur et est super ipsum perpendicularis, tunc ipsa per circuli centrum
transit. Ergo punctum *T* est circuli centrum. Sed punctum *G* iam eius
etiam fuerat centrum; quod est impossibile. Si igitur perpendicularis   20
protracta ceciderit super punctum *G*, tunc ab uno et eodem puncto
ad unam et eandem superficiem in parte una due recte linee orthogonaliter protrahuntur; quod est impossibile. Punctum igitur *H* non est
centrum spere. Simili quoque modo ostenditur quod non est possibile
ut centrum spere sit aliud punctum preter punctum *Z*. Punctum igitur   25
Z est spere centrum.

7 reperiatur] reperire **BPsVa**   7 iam … reperiamus] scimus   reperire   **BPsVa**
7 reperiamus] reperiatur **RFiZ**   9 igitur] *om.* **BPsVa**   9 punctum] *om.* **BPsVa**
9 super] *om.* **Kg**   10 *AB*] v *add.* **Fi**   10 insistentem] insistente **V**, existentem
MKg   10 insistentem, que] *om.* **BPsVa**   10 sit] sitque **Va**   10 *GD*] *BG* **OPsVa**
11 superficiei] superficies **Fi**   11 superficiei spere] *tr.* **PRVMKg FiZ**   12 *E*] *corr. ex A* **BPs**,
A **Va**   12 secetur] resecetur **BPsVa**   12 media] per medium **FiZ**   13 igitur] ergo **MBPsVa**
13 punctum] *om.* **BPsVa**   13 centrum spere] *tr.* **BPsVa**   13–16 Quod si … circuli *AB*] si non
sit centrum [*supra* **Ps**, *om.* **B**] preter *Z* et sit puncum *H* protraham itaque [ita **Va**] a puncto *H* perpendicularem super superficiem *AB* **BPsVa**   13 res] *om.* **Fi**, ista *add.* **MKg**   13–14 non ita] *tr.*
OFiZ   15 igitur] ergo **MKg**   15 a puncto] punctum **R**   17 a centro … spera] super
unum circulorum a centro spere **BPsVa** (er' *pro* super **Va**)   17 unum] unam **R**   17–
18 protrahitur] ducatur **BPsVa**   18 est] *om.* **Kg**, sit **BPsVa**   18 per] *supra* **V**   19–
20 Sed … centrum] *om.* **Va**   19 punctum[2]] *om.* **BPs**   19 iam … fuerat] est circuli
BPs   19 eius etiam fuerat] *permut.* **M,Kg**   20 etiam] *om.* **Fi**   20 centrum] ut dic-
tum est *add.* **BPsVa**   20 igitur] ergo **M**, autem **BPsVa**   21–22 ab uno … due] *om.* **Z**
21 et] ab *add.* **R**   22 ad unam … una] eiusdem superficiei **BPsVa**   22 recte] *om.* **BPsVa**
23 protrahuntur] eriguntur **BPsVa**   23 *H*] *B* **Fi**   24 Simili quoque modo] Simili quo
modo **P**, Similiter quomodo **V**, similiter **BPsVa**   24–26 non est … spere centrum] nullum
punctum preter *Z* erit centrum **BPsVa**   25 preter] quam **M**, *om.* **Kg**   25 punctum[2]] *om.* **M**
25 igitur] *supra* **R**   26 est] *supra* **P**

فقد تبين من ذلك أنه إذا كانت دائرة فى كرة وأخرج من مركز الدائرة

إلى سطح الكرة خط مستقيم قائم عليه على زوايا قائمة فإن مركز الكرة

يكون على ذلك الخط القائم ، وذلك ما أردنا أن نبين .

20

ج   إذا ماست كرة سطحاً من غير أن يقطعها فإنها تماسه على نقطة واحدة

فقط .

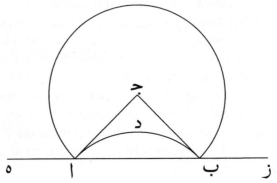

فإن أمكن فلتماس كرة سطحاً من غير أن يقطعها على أكثر من نقطة

واحدة فلتماسه على نقطتين وهما نقطتا آ ب فليكن مركز الكرة نقطة جـ

وليوصل خطا اجـ جـب وليخرج السطح الذى يمر بخطى اجـ جـب ويحدث   5

قطعاً يكون إما فى بسيط الكرة فدائرةً وإما فى السطح فخطاً مستقيماً ولتكن

الدائرة التى تحدث فى بسيط الكرة دائرة داب والخط المستقيم الذى يحدث

فى السطح خط هابز فلأن السطح لا يقطع الكرة لا يقطع أيضاً خط

الدائرة [AH: supra N, الكرة in textu N   18 [من 2] repet. N   18 [وأخرج من A   18 [وأخرج من
إلى سطح الكرة [om. A; على سطح الدائره H   19 [الكرة 1] supra N الدائرة 19 [قائم ... قائمة
al_سطح add. et del. N; ألسطح N. om] ذلك 1   20 [زوايا 19 H ازايه A على سطح الدايره يكون عمودا
H   20 [الخط] وذلك add. et del. A   20 [ما] illeg. N   3 [يقطعها] سطعه N فلتماسها
4 [فليكن] ولكن N   4 [ج] supra A   6 [قطعاً] illeg. N   6 [يكون] ميكون N   6 [سطح] add. N
6 [مستقيماً خطاً] فقط N   7 [الدائرة] illeg. N   7 [داب] داب ح A   8 [لا] supra A   8 [خط أيضاً] om. H

Ex hoc manifestum est quod cum in spera fuerit circulus, ad cuius superficiem linea recta orthogonaliter erecta protrahetur a centro circuli, tunc centrum spere super illam lineam erectam consistit. Et illud est quod demonstrare voluimus.

30

3 Cum speram superficies tetigerit ita quod eam non secet, non tanget eam nisi tantum super punctum unum.

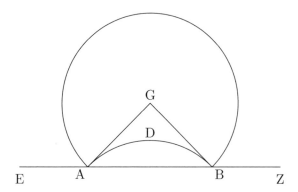

Quod si fuerit possibile, tangat speram superficies ita quod eam non secet in punctis pluribus uno. Contingat itaque eam super duo puncta, que sint puncta *A* et *B*. Sitque centrum spere punctum *G*,   5 et protraham duas lineas *AG GB* et inveniam superficiem que transit per duas lineas *AG GB* et provenient duo sectores, in superficie spere circulus et in superficie linea. Sitque circulus qui in spere superficie provenit circulus *DAB* et linea recta que provenit in superficie linea

27 *ante* Ex] Corollarium *add.* **PRVMOBPs** (Coroll' **PV**, Corroll' **R**, Corll' **M**, Correll'm **O**, Coroll' *marg.* **B**, Correl' *marg.* **Ps**); unde *add.* **PRVMKgOFiZ**   27 Ex] *supra* **R** 27 manifestum est] patet **BPsVa**   27 in] *supra* **BPs**   27–28 ad cuius superficiem] a cuius superficie **RV**, *corr. in* a cuius superficie **P**   28 linea recta orthogonaliter erecta] *permut.* **BPsVa**   28 erecta] erectam, *corr. ex* ereactan **R**, *supra* **M**, *om.* **Kg**, centra **Fi**, erigitur **BPsVa**   28 protrahetur] protrahitur **Kg**, *om.* **BPsVa**   28 centro] scilicet *add.* **PR**   29 circuli] spere **Kg**   29 centrum spere] *tr.* **PRVMKgOFiZ**   29 lineam] *om.* **BPsVa** 29 consistit] consistet **PBPsVa**   29 Et illud … voluimus] *om.* **BPsVa**   30 voluimus] volumus **OZ**   1 superficies] sup*era* **Fi**   1 tetigerit] tangit **OFiZ**   1 eam] *om.* **R**   1 eam non] *tr.* **MKgBPs**   1 eam non secet] *om.* **Va**   1 tanget] tangit **FiZ**   2 tantum] *om.* **BPsVa** 2 super] sunt **Kg**   2 punctum unum] *tr.* **PVMKgFi**   3 si] *om.* **Kg**, *supra* **B**   3 fuerit] fuit **Kg**   3 possibile] ut *add.* **MKgBPsVa**   3 speram] spera **Z**   3 speram superficies] *tr.* **Kg** 3–4 eam non] *tr.* **MKgBPsVa**   4 punctis pluribus] *tr.* **BPsVa**; nl' *add.* **Fi**   4 uno] immo **BVa**   4 itaque] *om.* **BPsVa**   4 itaque eam] *tr.* **MFi**   4 itaque eam super] *permut.* **Z** 5 que sint puncta] que sunt **Fi**, *om.* **OBPsVa**   5 Sitque] sit **BPsVa**   6 *AG*] et *add. supra* **Ps**   6 et[2]] *om.* **Kg**   6 inveniam] invenies **OBPsVa**   7 et] *repet* **Fi**   7 provenient] provenerint **KgFiZBP**, proveniunt **O**, perveniant **Va**   7 sectores] scilicet *add.* **PRVMKgFi**, scilicet *supra* **O** 8 superficie[1]] proposita *add.* **FiZ**, *add. supra* **B**; *illeg. supra* **O**   8 qui] que **Kg**, *corr. ex* que **Ps** 8 spere] *om.* **B**, *supra* **Ps**   8 spere superficie] *tr.* **VOVa**   9 linea recta] *tr.* **M**   9 linea[2]] *om.* **Va**

ه‍ا ب‍ز دائرة د‍ا‍ب فإذ قد تُعلم على الخط المحيط بالدائرة نقطتان كيف ما

وقعتا وهما نقطتا آ ب‍ فالخط الذى يصل بين نقطة آ ونقطة ب‍ يقع داخل   10

دائرة د‍ا‍ب وقد وقع أيضاً خارجها وذلك ممتنع فليس تماس كرة سطحاً من

غير أن يقطعها على أكثر من نقطة واحدة ·

د   إذا ماست كرة سطحاً ما من غير أن يقطعها فإن الخط المستقيم الذى يصل

بين المركز وبين نقطة التماس عمود على السطح المماس ·

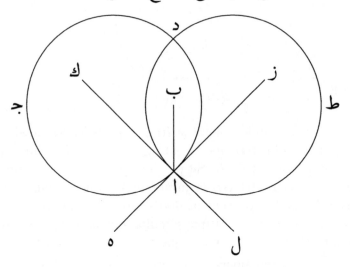

فلتماس كرة سطحاً ما على نقطة واحدة وهى نقطة آ من غير أن

يقطعها وليكن مركز الكرة نقطة ب‍ ونصل بين نقطتى آ ب‍ بخط ا‍ب ،

فأقول إن خط ا‍ب عمود على ذلك السطح وذلك أنه إذا أخرج سطح يمر   5

بخط ا‍ب أحدث فى بسيط الكرة دائرة ا‍ج‍د وفى السطح خط ه‍ا‍ز المستقيم

ونصل ... H: om. AN   4 بخط ا‍ب [سطحاً 11 corr. ex سطحان A   داب‍ ح [داب‍ ح A   11 داب‍ [11

نקطה H [نقطتى 4

حٰٓاس‍ [المستقيم supra A   6 ب‍ حٰٓ س‍ [داب‍ [11 supra A

EABZ. Et quia superficies non secat speram neque etiam linea *EABZ*   10
secat circulum *DAB*, et quia super circumferenciam circuli duo puncta
iam sunt signata quocumque evenerit modo, que sunt puncta *A B*,
ergo linea que coniungit quod est inter punctum *A* et punctum *B* cadit
intra circulum *DAB*. Iam autem ceciderat etiam extra ipsum; quod est
impossibile. Non igitur speram contingit superficies ita quod non secet   15
eam super puncta plura uno. Et illud est quod demonstrare voluimus.

4 Cum superficies aliqua contigerit speram ita quod eam non secet, linea
recta que coniungit quod est inter centrum spere et punctum contactus
erit perpendicularis super superficiem contingentem.

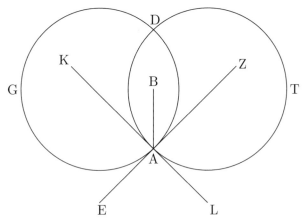

Contingat igitur speram superficies aliqua super punctum unum,
quod sit punctum *A*, ita quod eam non secet, et sit centrum spere   5
punctum *B* et copulabo duo puncta *A B* linea recta. Dico igitur quod
linea *AB* est perpendicularis super illam superficiem. Quod ideo est
quoniam, cum protrahetur superficies transiens per lineam *AB*, proveniet in superficie spere circulus *ADG* et in superficie linea *EAZ* recta;

10 *EABZ*[1]] ea *BZ* **Kg**, *EBAZ* **OBVa**, *corr. ex EBAZ* **Ps**   10 *EABZ*[2]] ea *BZ* **Kg**   11 quia] que
Fi   11 circuli] sunt semper *add.* **OBPsVa**   12 iam sunt] *om.* **OBPsVa**   12 sunt] inde
FiZ   12 evenerit] eant **OBPsVa**   12 evenerit modo] *tr.* **FiZ**   13 ergo linea que] *repet.* **V**
13 inter] intra **PsVa**   14 intra] inter **KgOPsVa**   14 Iam autem] linea etiam **OBPsVa**; eadem
supra **OB**   14 ceciderat etiam] *tr.* **MKgOBPsVa**   15–16 Non ... uno] patet ergo primum **OB**
PsVa   16 puncta] predictum **Kg**   16 Et illud ... voluimus] *om.* **OFiZBPsVa**   1 aliqua] *om.*
PRVMKgOZBPsVa, *supra* **Fi**   1 contigerit] *mut. ex* contingerit **R**, contingerit **PVM**, contingat **OBPsVa**, contingit **Kg**   1 speram] *om.* **Va**   1 eam non secet] *permut.* **VOBPsVa**
2 coniungit] contingit **Z**   3 erit] *mut. ex* evenerit **R**   4 speram superficies aliqua] superficies
speram **OBPsVa**; contingentem *add. et del.* **Ps**, *add.* **Va**   4 punctum unum] *tr.* **O**
5 punctum] *om.* **OBPsVa** 5 eam] eadem **Fi** 5 eam non] *tr.* **OBPsVa** 6 copulabo] conpleabo
Fi   6 linea recta] *om.* **BPsVa**   6 igitur] ergo **ROBPsVa**   7 illam] *corr. ex* aliam **R**
7 est[2]] cum **Va** 8 cum] *supra* **BPs** 8 protrahetur] pertrahetur **Z** 8–9 proveniet] provenit
Z   9 *ADG*] *ABG* **PRVM**   9 et in superficie] proposita **Fi**   9 superficie] proposita *add.* **Z**,
add. supra **B**   9 linea] linee **V**   9 *EAZ* recta] *tr.* **FiZ**

وليمر أيضاً بخط اب سطح آخر وليحدث فى بسيط الكرة دائرة اط وفى

السطح خط كال ولأن السطح يماس الكرة يكون خط هاز أيضاً مماساً

لدائرة ادج فلأن خط هاز المستقيم يماس دائرة ادج على نقطة ا وقد أخرج

من نقطة ا إلى مركز الدائرة خط اب يكون خط اب عموداً على خط هاز   10

ومن البيّن أن نقطة ب مركز دائرة اجد لأن سطح دائرة اجد يمر بخط با

الذى يخرج من مركز الكرة وكذلك أيضاً نبين أن خط با عمود على خط

كال فلأن خط با المستقيم عمود على الفصل المشترك لتقاطع خطى هز

كل يكون خط اب عموداً على السطح الذى يمر بهما والسطح الذى يمر

بخطى هز كل مماس للكرة ،   وذلك ما أردنا أن نبين ·   15

ه   إذا ماست كرة سطحاً ما من غير أن يقطعها وأخرج من موضع المماسة من

السطح خط قائم عليه على زوايا قائمة فإن مركز الكرة يكون على ذلك

الخط القائم ·

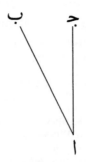

7 بخط [H: خط A, *illeg.* N   8 خط [*illeg.* N   8 supra A   15 N مماس [مماس   1 ما [*illeg.* N   ولأن [فلان N   A *om.*] ما

9 ادج [1 لأنه إن قطعها يلزم أن يكون […] لها فى أكثر من موضع واحد وحينئذ يكون السطح قاطعاً للكرة

10 supra A   خط هاز [رح س   *marg.* A   15 بخطى [supra A س نا س   د نا س   وقد وضعناه مماساً لها هذا خلف

et transeat etiam per lineam AB superficies alia et proveniat in super- 10
ficie spere circulus AT et in superficie linea KAL. Et quia superficies
contingit speram, ergo linea EAZ etiam contingit circulum ADG. Et
quia linea recta EAZ contingit circulum ADG super punctum A, a quo
iam protracta est linea AB ad centrum circuli, ergo linea AB est perpendicularis super lineam EAZ. Manifestum autem est quod punctum 15
B est centrum circuli AGD, quoniam superficies circuli AGD transit
per lineam BA que a spere centro protrahitur. Similiter quoque demonstratur quod linea BA est perpendicularis super lineam KAL. Et
quia linea recta BA est perpendicularis super communem differentiam
sectionis duarum linearum EZ KL, ergo linea AB est perpendicula- 20
ris super superficiem que transit per eas et superficies que transit per
duas lineas EZ KL contingit speram. Et illud est quod demonstrare
voluimus.

5 Cum superficies aliqua speram contigerit ita quod eam non secet et a
loco contactus fuerit a superficie linea orthogonaliter super eam erecta,
centrum spere super illam erectam lineam consistet.

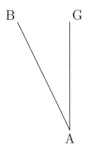

10 transeat] transiat *corr. ex* transiet **Ps**   11 *KAL*] *KL* **Fi**, *AKL* **Z**, *KLA* **BVa**, *corr. ex*
KLA **Ps**   12 linea *EAZ* etiam] *permut.* **FiZ**   13 *ADG*] *corr. ex DG* **Kg**   13 *A*] *supra*
Ps   14 iam] *om.* **Kg**   14–16 ergo linea … transit] transit circulus *ABG* **BPsVa** (*ABG*
om. **B**)   15 autem est] *tr.* **MKgOZ**, est nunc **Fi**   16 quoniam] transit per lineam quia
non **Fi**   17 centro] centrum **R**   17 protrahitur] propterea erit *AB* perpendicularis super lineam *EAZ* *add. marg.* **Ps**, *add. in textu* **Va**   17 Similiter quoque] Similiterque
BPsVa (que *supra* **Ps**)   17–18 demonstratur] demonstretur **B**, demonstraretur **PsVa**
18 quod] quoniam **Kg**   18 *BA*] *corr. ex GA* **Ps**   19 *BA*] *corr. ex BOA* **Ps**
19 est] *om.* **R**   19–20 differentiam sectionis] *tr.* **BPsVa**   20 sectionis] *corr. in* sectoris **R**   20 *EZ KL*] *EAZ KAL* **OFiZ**   20–21 perpendicularis] supra lineam *add.* **Kg**
21 superficies] quidem *add.* **OFiZ**   22 *EZ KL*] *EAZ KAL* **FiZ**   22 contingit] *om.* **BVa**, *su-*
pra **Ps**   22 speram] contingit *add. supra* **B**   22–23 Et … voluimus] *om.* **Va**   22 illud] hoc
OBPs   22–23 quod demonstrare voluimus] *om.* **BPs**   23 voluimus] volumus
OZ   1 speram contigerit] contingerit speram **M**, contingat speram **BPsVa**   1 contigerit] *corr.*
ex contingerit **PR**, contingerit **V**, contingit **Kg**, contingat **O**   1 eam non] *tr.* **MKgOBPsVa**
2 a superficie] ad superficiem **PRVMOFiZ**   2 linea] *om.* **Fi**; protracta *add.* **PRVMKgFiZ**
3 super] *supra* **P**, *om.* **Z**   3 illam] *om.* **O**   3 illam erectam] *tr.* **Kg**   3 lineam] *om.* **BPsVa**

فلتماس كرة سطحاً ما على نقطة آ من غير أن يقطعها وليخرج من

نقطة آ عمود على السطح وهو خط اٰب ، فأقول إن مركز الكرة يكون

على خط اٰب فإن لم يمكن ذلك فأمكن غيره فليكن مركز الكرة نقطة جـ

وليوصل خط جآ فلأنه قد ماست كرة على نقطة آ سطحاً من غير أن يقطعها

وقد أخرج من مركز الكرة إلى موضع الماسة خط جآ يكون خط جآ عموداً

على السطح وقد كان خط بآ أيضاً عموداً عليه فقد خرج من نقطة واحدة

بعينها على سطح واحد بعينه خطان مستقيمان على زوايا قائمة وهما خطا

اٰب اٰجـ فى جهة واحدة بعينها وذلك ممتنع فليست نقطة جـ مركز الكرة

وكذلك أيضاً نبين أنه لا يمكن أن يكون المركز نقطة أخرى ليست على خط

بآ فمركز الكرة على خط بآ ، وذلك ما أردنا أن نبين ·

و ما كان من الدوائر التى تكون فى الكرة ماراً بمركز الكرة فهو أعظمها وما كان

من الدوائر الباقية بعده من المركز بعداً متساوياً فهى متساوية وما كان بعده

من المركز أكثر فهو أصغر ·

فلتكن فى كرة دوائر اٰب جد هز ولتكن دائرة جد مارة بمركز الكرة

وليكن بعد دائرتى اٰب هز أولاً من المركز بعداً متساوياً ، فأقول إن أعظم

هذه الدوائر دائرة جد وإن دائرتى اٰب هز متساويتان وذلك أنا نصيّر مركز

عمـودا علـه ابضا [أيضاً عموداً عليه 9 الدائرة [مركز 6 add. et del. A وامكن [فأمكن 6 A

متساويان [متساويتان 6 A مساويا [متساوياً 2 om. A فمركز الكرة على خط بآ [13

على[1س اٰس بٮ supra A 11 ممتنع [اٰس مح supra A 5

Superficies itaque aliqua contingat speram super punctum *A* ita
quod non secet eam; et protrahatur a puncto *A* perpendicularis super   5
superficiem, que sit linea *AB*. Dico igitur quod centrum spere existit
super lineam *AB*. Quod si hoc non fuerit ita, sit preter ipsum, si est
possibile; sit itaque centrum spere punctum *G*, et producam lineam
GA. Et quia superficies contingit speram super punctum *A* ita quod
eam non secat et a centro spere iam protracta est linea *GA* ad locum   10
contactus, est ergo linea *GA* perpendicularis super superficiem. Sed
linea *BA* iam etiam fuit perpendicularis super eam: ergo ab uno et
eodem puncto iam protracte sunt ad unam et eandem superficiem due
recte linee ei orthogonaliter insistentes, que sunt due linee *AB AG*, in
una et eadem parte, quod est impossibile. Punctum igitur *G* non est   15
centrum spere. Simili quoque modo monstratur quod non est possibile
ut punctum aliud quod non sit super lineam *BA* sit centrum. Centrum
igitur spere super lineam *BA* consistit. Et illud est quod demonstrare
voluimus.

6 Circulorum in spera existencium qui per centrum spere transit eis est
maior; reliqui vero circuli quorum longitudo a centro equalis est sunt
equales, sed cuius longitudo a centro est maior minor existit.

Sint ergo in spera circuli *AB GD EZ* et sit circulus *GD* per
centrum spere transiens. Sitque primum longitudo duorum circulorum   5
AB EZ a centro equalis. Dico igitur quod maior his circulis est circulus
GD et quod duo circuli *AB EZ* sunt equales. Quod ideo est quoniam

---

4 itaque] *supra* **Kg**, igitur **Va**   4 super punctum *A*] *om.* **OBPsVa**   5 non secet eam] *permut.*
PRVMKg, FiZ   5 protrahatur] *in corr.* **Fi**   5 perpendicularis] perpendiculari **V** (e *supra*)
5 super] *supra* **R**   6 igitur] itaque **OBPsVa**   7 non fuerit ita] *permut.* **OBPsVa**, ita non fuerit ita **Kg**, ita *om.* **FiZ**   7 preter] *corr. ex* punctum **Ps**   7 preter ipsum] supra alteram **B**
7 si^2] *supra* **R**   8 si est possibile] *marg.* **P**   8 sit itaque] et sit **OBPsVa**   8 producam] *corr.*
ex productum **Ps**   9 quia] quod quidem **Fi**, quod quia **OZ**   9 contingit] continet **Va**   9 *A*] *om.*
V   10 eam non] *tr.* **Kg**   10 non] *supra* **B**   10 secat] *corr. ex* secet **P**, secet **BPsVa**
10 spere] *om.* **OBPsVa** 10 protracta] producta **Va** 10 *GA*] *GD* **Fi** 10–11 ad ... *GA*] *marg.*
R   11 contactus] *A add.* **OBVa**, *supra (in textu add. et del. E)* **B**   11 est ergo linea] est
igitur linea **O**, linea igitur **BPsVa**   12 *BA*] *corr. ex BOA* **Ps**   12 iam] tam **R**, *om.* **OBPsVa**
12 eam] ipsam **OBPsVa**   12 uno] una **Fi**   12 et] ab *add.* **Va**   13 iam] *om.* **O**   13 sunt] super
unam et eandem superficiem uero [?] deb'e dici^{sa} *add.* **Kg**   13–14 ad unam ... insistentes] due
linee perpendiculares super unam et eandem superficiem **OBPsVa**   14 due linee] *om.* **BPsVa**
14 *AB*] *AD* **Va**   14 *AG*] *A* **Fi**   14 in] ex **OBPsVa**   15 non] *om.* **Z**   16 centrum spere] *tr.*
PRVMKgFiZ   16 Simili quoque modo] Similiter **OBPsVa**   16 monstratur] ostenditur
MKgOBPsVa   16 possibile] inpossibile **Fi**   17 ut] quod **O**   17 punctum aliud] *tr.* **OBPsVa**
17 *BA*] et *add.* **BVa**, *add. et del.* **Ps**   17 centrum] spere centrum **OBPsVa**, centrum spere **FiZ**
17–18 Centrum ... consistit] *om.* **BPsVa**   18 igitur] ergo **OFiZ**   18 Et illud est] *om.* **FiZ**   18–
19 Et illud ... voluimus] *om.* **OBPsVa**   19 voluimus] volumus **Z**   1 qui] que **B**, *corr. ex* que
Ps   1 per] *supra* **Ps**, supra *supra* **B**   1 centrum spere] *tr.* **KgOBPsVa**   1 spere transit] *tr.*
V   1–2 est maior] *tr.* **MKgBPsVa**   2 circuli] *om.* **BPsVa**   2 a centro] *supra* **MBPs**, *om.* **Kg**
2–3 sunt equales] *tr.* **Kg**   3 cuius] *supra* **Ps**   3 a centro] *om.* **FiZ**   4 ergo] *om.* **V**, igitur
KgBPsVa   4 *GD*] *corr. ex GA* **Ps**   5 circulorum] aliorum **BPsVa**   6 *EZ*] *corr. ex Z* **Fi**, *corr.*
ex CE **Ps**   6 igitur] ergo **OBPsVa**   6 circulus] *om.* **OBPsVa**   7–9 et quod ... *GD*] *marg.* **BPs**
7 *AB*] et *add.* **PRVFiZ**   7 quoniam] quod **Kg**

الكرة نقطة حـ فهى إذاً مركز دائرة جـدـ وليخرج من نقطة حـ إلى سطحى

دائرتى اـبـ هـزـ عمودا حـطـ حـكـ وليلقيا سطحى الدائرتين على نقطتى طـ كـ

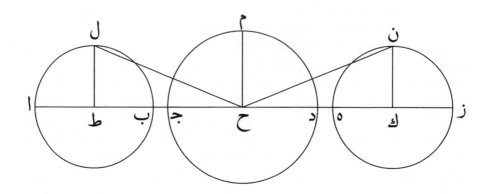

فنقطتا طـ كـ مركزا دائرتى اـبـ هـزـ ولتخرج من نقط طـ كـ حـ إلى الخطوط

المحيطة بدوائر اـبـ جـدـ هـزـ خطوط مستقيمة وهى خطوط طـلـ كـنـ حـمـ 10

وليوصل خطا حـلـ حـنـ فلأن خط حـطـ عمود على سطح دائرة اـبـ فهو

يحدث مع جميع الخطوط المستقيمة التى تخرج من طرفه فى سطح دائرة اـبـ

زوايا قائمة وقد خرج من طرفه خط طـلـ الذى هو فى سطح دائرة اـبـ

فزاوية لـطـحـ قائمة وكذلك أيضاً نبين أن زاوية حـكـنـ أيضاً قائمة وأيضاً

فلأن زاوية لـطـحـ قائمة تكون زاوية لـطـحـ أعظم من زاوية لـحـطـ فخط 15

لـحـ أطول من خط لـطـ وخط لـحـ مساوٍ لخط حـمـ لأن نقطة حـ مركز

الكرة وقد خرج منها إلى سطح الكرة خطا حـلـ حـمـ فخط حـمـ أطول من

7 حـ نقطة ... فهى [marg. A   8 الدائرتين [دارسِ A   8 حـكـ [حـطـ N   9 لـ¹ [الى حـ add. et del.
A   9 طـ² [كـ, supra A   11 حـنـ [حـرـ AN   11 عمود [عمودا A   15 حـطـلـ² [انصا add. N
N محرح [خرج N 17 حـرـ [حـمـ A 16 om. A [خط 16 om. N [لـحـ ... أطول 16 فخط [وحط N 15

8 عمودا [نـا نـا سـ supra A   15 من [سـ اـ سـ supra A   16 أطول [لـبـ اـ سـ supra A

ponam ut centrum spere sit punctum H; ergo ipsum erit centrum circuli GD. A puncto itaque H protraham duas perpendiculares ad duas superficies duorum circulorum $AB\ EZ$, que sint $HT\ HK$, et concurrant duabus superficiebus duorum circulorum super duo puncta $T\ K$. Duo

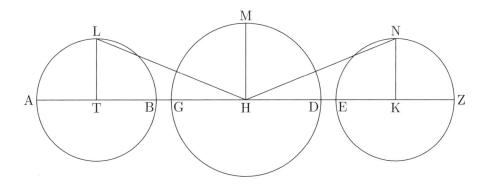

igitur puncta $T\ K$ sunt centra duorum circulorum $AB\ EZ$. Protraham ergo a punctis $T\ K\ H$ ad lineas circulorum circumductas $AB\ GD\ EZ$ lineas rectas, que sint linee $TL\ KN\ HM$, et producam duas lineas HL HN. Et quia linea HT est perpendicularis super superficiem circuli AB, ergo ipsa cum omnibus rectis lineis, que ab eius extremitate egrediuntur, in superficie circuli AB facit rectos angulos. Iam autem ab eius extremitate linea TL protracta est que est in superficie circuli AB: ergo angulus LTH est rectus. Simili quoque modo ostenditur quod angulus HKN est rectus et etiam, quia angulus LTH est rectus, ergo angulus LTH est maior angulo LHT; ergo linea LH est longior linea LT. Sed linea LH est equalis linee HM, quoniam punctum H est centrum spere, a quo ad superficiem spere due linee HL et HM iam sunt protracte.

8 ponam] ponatur **O**, pono **PsVa**   8 ut] quod **Kg**   8–9 circuli *GD*] *tr.* **O**
9 itaque] igitur **MKg**, *om.* **OVa**, *supra* **Ps**   9–10 ad duas superficies] *marg.* **M**   10–
12 que sint ... *EZ*] *marg.* **BPsVa**   10 sint] sit **R**, sunt **Va**   11 duorum circulorum] *illeg.*
BPs   11–12 super ... circulorum] *om.* **Z**   11 *T*] et *add.* **PRVFiZ**   11 *K*] *H* **KgPs**
12 *T*] et *add.* **PRVKgOFiZBPsVa**   13 punctis] puncto **FiZ**   13 *T K H*] *supra, in*
textu add. et del. B **Ps**, *B T K H* **Va**; *H supra* **B**   13 circumductas] *corr. ex* cir-
cumductis **BPs**   14 sint] sunt **BPsVa**   15 *HT*] *seq. illeg.* **Va**   15 circuli *AB*] *tr.*
FiZ   15 *AB*] *om.* **BVa**, *supra* **Ps**   16 ergo] de **Kg**   16 ab] ex **KgOBPsVa**
16 extremitate] terminatione **OBPsVa**   16–17 egrediuntur] progrediuntur **OBPsVa**, protrahuntur **FiZ**   16–18 egrediuntur ... extremitate] *marg.* **R**   17 in] *om.* **Z**   17 circuli] *om.*
OFiZBPsVa   17 rectos angulos] *tr.* **Fi**   18 extremitate] extremicione **Kg**, ext'minatione **BPs**,
ext'matione **Va**   18 circuli] *om.* **FiZ**   19 angulus[1]] *corr. ex* a lineis **BPs**, *hic et ter infra*
19 Simili quoque] Similique **BPsVa**   20 etiam] quod angulus (*tr.* **Fi**) *ATL est rectus add.*
OFiZ   20 quia] quod est quia **O**, quod est **FiZ**, quod **BPsVa**   20 *LTH*] *ATL* B, *ATB* **PsVa**
21 *LTH*] *corr. ex TLH* **R**   21 ergo linea ... *LT*] *add. et del.* **M**, *om.* **PRV**, ergo linea *MLH*
linea *LT* est maior **Fi**   21 est longior] *supra* **OPs**, *om.* **ZB**   21 longior] *om.* **Kg**, maior **Va**
22 linea] *om.* **Va**   22 *HM*] *HDA* **Z**

خط لَطَ وخط حَمَ قد خرج من مركز دائرة جَدَ إلى الخط المحيط بها وخط

طَلَ قد خرج من مركز دائرة اَبَ إلى الخط المحيط بها فدائرة جَدَ أعظم من

دائرة اَبَ وكذلك أيضاً نبين أنها أعظم من دائرة هَزَ فدائرة جَدَ أعظم الدوائر 20

التي في الكرة ،

وأقول أيضاً إن دائرتي اَبَ هَزَ متساويتان وذلك أنه لما كان بعدهما من

المركز متساوياً صار خط حَطَ مساوياً لخط حَكَ ولما كانت أيضاً نقطة حَ مركز

الكرة صار خط حَلَ مساوياً لخط حَنَ فالمربع الكائن من خط حَلَ مساوٍ

للمربع الكائن من خط حَنَ ولكن المربعين الكائنين من خطي لَطَ طَحَ 25

مساويان للمربع الكائن من خط حَلَ والمربعان الكائنان من خطي نَكَ حَكَ

مساويان للمربع الكائن من خط حَنَ فالمربعان الكائنان من خطي لَطَ طَحَ

مساويان للمربعين الكائنين من خطي حَكَ كَنَ والمربع الكائن من خط طَحَ

مساوٍ للمربع الكائن من خط حَكَ فيبقى المربع الكائن من خط طَلَ مساوياً

للمربع الكائن من خط كَنَ فخط طَلَ مساوٍ لخط كَنَ وخط طَلَ قد خرج من 30

مركز دائرة اَبَ إلى الخط المحيط بها وخط كَنَ قد خرج من مركز دائرة هَزَ

إلى الخط المحيط بها فالخط الذي خرج من مركز دائرة اَبَ إلى الخط المحيط

بها مساوٍ للخط الذي خرج من مركز دائرة هَزَ إلى الخط المحيط بها فدائرة

اَبَ مساوية لدائرة هَزَ ،

---

20 دائرة [¹ supra A   20 هَزَ] ومن سائر الدوائر الموازية لها add. N   20 أعظم [² من add.

A   22 متساويتان [متساويان A   24 حَنَ] حَزَ A   24 فالمربع [والمربع N   26 والمربعان [فالمربعان A

26–27 خطي ... نَكَ حَكَ] marg. A   28 كَنَ] نَكَ N   30 مساوٍ N كر ... للمربع [om. N   32 خرج [مخرج

N   33 خرج [مخرج N   33 بها] فدارة add. et del. A

---

26 من [¹ سَ اَ مرَ supra A

Ergo linea *HM* longior est linea *LT*. Linea vero *HM* protracta est a
centro circuli *GD* ad ipsius circumferenciam et linea *TL* producta est   25
a centro circuli *AB* ad ipsius circumferentiam: ergo circulus *GD* maior
est circulo *AB*. Similiter quoque ostenditur quod ipse est maior circulo
EZ. Ergo circulus *GD* est maior circulis qui sunt in spera.

Et dico eciam quod duo circuli *AB EZ* sunt equales. Quod ideo
est quoniam linea *HT* fit equalis linee *HK*, quoniam longitudo circulo-   30
rum a centro est equalis. Et eciam, quoniam punctum *H* est centrum
spere, fit linea *HL* equalis linee *HN*. Quadratum igitur factum ex linea
HL est equale quadrato facto ex linea *HN*. Duo vero quadrata ex duabus lineis *LT TH* facta sunt equalia quadrato ex linea *HL* facto et duo
quadrata facta ex duabus lineis *NK KH* sunt equalia quadrato facto ex   35
linea *HN*. Ergo duo quadrata que fiunt ex duabus lineis *LT TH* duobus
quadratis ex duabus lineis *HK KN* factis equalia existunt. Quadratum autem factum ex linea *TH* equatur quadrato facto ex linea *HK*:
remanet ergo quadratum factum ex linea *TL* equale quadrato facto ex
linea *KN*. Linea igitur *TL* equatur linee *KN*. Sed linea *TL* iam fuit   40
protracta a centro circuli *AB* ad lineam continentem ipsum et linea
KN iam fuit producta a centro circuli *EZ* ad ipsius circumferentiam.
Ergo linea protracta a centro circuli *AB* ad eius circumferentiam est
equalis linee que protrahitur a centro circuli *EZ* ad lineam continentem
ipsum. Ergo circulus *AB* est equalis circulo *EZ*.   45

---

24 longior est] *tr.* **BPsVa**   24 protracta est] *tr.* **BPsVa**   25 ipsius] illius **KgOBPsVa**   25–
26 et . . . circumferentiam] *marg.* **R**   25 producta] protracta **VOFiZ**   25 producta est] *tr.*
BPsVa   26–27 maior est] *tr.* **OBPsVa**   27 est[1]] *om.* **R**   27–28 circulo *AB* . . . maior] *om.*
Va   27 Similiter quoque] Similiterque **BPs**   28 est maior] *tr.* **Kg**   28 circulis] *corr. ex*
circulus **Z**; omnibus *add. supra* **BPs**, *add. in textu* **Va**   29 duo] du **V**   29 Quod] qnd'
B, *in corr.* **Ps**   30 quoniam] quod **OBPsVa** (quod *repet.* **Va**)   30 fit] sit **Fi**   30–
31 linee *HK* . . . equalis] *marg.* **R**   30 quoniam] *illeg.* **B**   31 a centro] *om.* **BVa**, a centro sup
marg. **Ps**   31 quoniam] quia **PRVMOFiZ**   32 igitur] ergo **MKgBPs**   32 factum ex linea] linee
OBPsVa, *hic et saepius*   33–34 duabus] duobus **Fi**   34 *TH*] *HT* **R**   34 facta] *om.*
OBPsVa   34–35 ex linea *HL* . . . quadrato] *om.* **B**   34 facto] *supra* **Ps**, *ante* ex linea *HL* **Va**
34–35 et duo quadrata facta] *om.* **Fi**   35 duabus] duobus **Fi**   36 que . . . lineis] linearum
BPsVa   36 fiunt] sunt **R**   36–37 *LT TH* . . . lineis] *om.* **Z**   37 ex duabus lineis] linearum
OBPsVa   37 *HK*] *BK* **Ps**   37 factis] *om.* **OBPsVa**   37 equalia existunt] sunt equalia **OB**
PsVa   38 autem] ergo **O**, vero **BPsVa**   38 factum] *om.* **OBPsVa**   38 equatur] sequatur
R, est equale **FiZ**   39 remanet] remanetur **Fi**   39 *TL*] *TH* **Fi**   40 *KN*[1]] *KM* **OFi**
40 igitur] ergo **OPsVa**   40 equatur] est equalis **FiZ**   40 fuit] fuerat **OBPsVa**   41 *AB*] ad
AD lineam continentem ipsum et linea *KN* iam fuerat protracta a centro circuli *add. et*
del. **BPs**   41–43 ad lineam . . . *AB*] *repet.* **Kg**   41 ipsum] *corr. ex* ipsam **V**, circumferentiam suam **OBPsVa**, ipsius circumferentiam **FiZ**   42 *KN*] *seq. illeg.* **BPs**   42 iam] *supra*
BPs   42 producta] protracta **KgOFiZBPsVa**   42 ipsius] eius **FiZ**   43 protracta] producta
R   43 *AB*] *supra* **BPs**   43 eius] ipsius **MKgOBPsVa**   44–45 continentem ipsum] *tr.* **FiZ**
45 ipsum] ipsius circumferentiam **MKgOBPsVa**   45 *EZ*] *corr. ex ET* **Ps**

وأيضاً فليكن بعد دائرة اب من مركز الكرة أكثر من بعد دائرة هز ٣٥

منه ، فأقول إن دائرة اب أصغر من دائرة هز ونعمل الأشياء التى عملناها

بأعيانها فلأن بعد دائرة اب من مركز الكرة أكثر من بعد دائرة هز منه يكون

خط حط أطول من خط حك ولأن خط حل مساوٍ لخط حن يكون المربع

الكائن من خط حل مساوياً للمربع الكائن من خط حن ولكن المربعين

الكائنين من خطى حط طل مساويان للمربع الكائن من خط حل والمربعان ٤٠

الكائنان من خطى حك كن مساويان للمربع الكائن من خط حن فمربعا

لط طح مساويان لمربعى حك كن والمربع الكائن من خط طح أعظم من

المربع الكائن من خط حك فيبقى المربع الكائن من خط لط أقل من المربع

الكائن من خط نك فخط لط أصغر من خط كن وخط طل قد خرج من

مركز دائرة اب إلى الخط المحيط بها وخط كن قد خرج من مركز دائرة هز ٤٥

إلى الخط المحيط بها فدائرة اب أصغر من دائرة هز فقد تبين أن ما كان من

الدوائر التى فى الكرة ماراً بالمركز فهو أعظمها وأما الدوائر الباقية فما كان

بعده منها من المركز بعداً متساوياً فهى متساوية وما كان بعده من المركز أكثر

فهو أصغر ، وذلك ما أردنا أن نبين .

ز   إذا كانت دائرة فى كرة ووُصل ما بين مركز الكرة وبين مركز الدائرة بخط

فإن الخط الذى يصل بينهما يكون عموداً على سطح الدائرة .

اكبر [أكثر ٣٥   بعدا add. N   علمنا [عملناها ٣٦   NH دائرة [اب ... دائرة ٣٧   marg. A [اب ... أكبر

حل [حن ٣٨ N   فلان [ولأن A   مساوياً ... خط حن [marg. A (sic)   حك [حك ٤١   ولأن A   ٤١-

٤٢ كن ... فمربعا [om. NH   للمربع [المربعى A   أقصر [أصغر N   الخط [om. N ٤٦   كن ... فمربعا [om. N   كن ٤٢

ومركز [وبين مركز A   ما [om. AN   N ١ وصل [corr. ex وصل N   ووُصل A ١   بعدها [بعده ٤٨

مر أ س [supra A   مر أ س [الكائن ٤١ supra A   مر أ س [الكائن ٤٠

Sit etiam longitudo circuli *AB* a centro spere maior longitudine circuli *EZ* ab eo. Dico igitur quod circulus *AB* est minor circulo *EZ*. Faciam itaque illud idem quod feci. Et quia spacium circuli *AB* a centro spere maius est spacio circuli *EZ* ab eo, ergo linea *HT* est longior linea *HK*. Et quia linea *HL* est equalis linee *HN*, ergo quadratum factum ex linea *HL* est equale quadrato facto ex linea *HN*. Duo autem quadrata facta ex duabus lineis *HT TL* sunt equalia quadrato facto ex linea *HL*, et duo quadrata facta ex duabus lineis *HK KN* equantur quadrato facto ex linea *HN*. Sed quadratum factum ex linea *TH* est maius quadrato facto ex linea *HK*: remanet ergo quadratum factum ex linea *LT* minus quadrato facto ex linea *NK*. Ergo linea *LT* est minor linea *KN*. Linea autem *TL* iam fuit producta a centro circuli *AB* ad lineam continentem ipsum et linea *KN* iam fuit producta a centro circuli *EZ* ad lineam continentem ipsum. Ergo circulus *AB* est minor circulo *EZ*. Iam igitur manifestum est quod circulorum in spera existencium qui per centrum transit eis est maior; et reliqui circuli quorum elongacio a centro est equalis sunt equales, et cuius circulorum a centro elongatio maior est, ipse est minor. Et illud est quod demonstrare voluimus.

7 Cum in spera fuerit circulus et inter centrum spere et centrum circuli linea producta spacium coniungetur, linea que inter ea coniungit erit perpendicularis super superficiem circuli.

46 etiam] et **V**   *47* circuli *EZ*] *tr.* **FiZ**   *47* eo] *in corr.* **RB**; *AB* **Kg**; centro *supra*
O   *47* igitur] ergo **BPsVa**   *48* illud] id **OBPsVa**   *49* spacio] spera **Fi**   *49* circuli] *om.*
OBPsVa   *49* eo] eadem **B**   *49* est longior] *tr.* **MKgOBPsVa**   *50* linea] *om.* **OBPsVa**
50 factum] *om.* **O**   *51 HN*] dico *add. et del.* **P**; d *add. et del.* **Fi**   *52* facta] *om.*
OBPsVa   *52 HT*] *HE* **Fi**   *52* sunt ... linea *HL*] *repet.* **Fi**   *52* facto ex linea] *om.* **OBPsVa**
52 HL] *H* et *L* **Fi**   *53–54* et duo ... *HN*] *om.* **OFiPsVa**   *53–55* et duo ... *HK*] *marg.* **B**
53 duabus lineis] *om.* **B**   *53 HK*] et *add.* **B**   *53* equantur] *corr. ex* dicitur **R**, sunt equalia **B**   *53–54* facto] *om.* **B**   *54* linea[1]] *om.* **B**   *54 HN*] Remanet ergo quadratum factum
ex linea *LT add.* **V**   *54–55* Sed ... *HK*] *marg.* **Ps**   *54* factum] *illeg.* **B**   *54* est maius] *tr.*
OBVa   *55* facto] *om.* **OFiZ**   *55* linea[1]] *om.* **BVa**   *55–56* remanet ... *NK*] *marg.*
R   *55* ergo] *om.* **R**, igitur **FiZ**   *55* factum] *om.* **FiZ**   *56* minor] *corr. ex* maior **R**
57 producta] protracta **FiZ**   *57 AB*] *in corr.* **R**, *corr. ex AD* **Ps**   *57–58* ad ... *EZ*] *marg.*
R   *57–58* lineam continentem ipsum] lineam continentem ipsam *scr. et del.* **Ps**; ipsius circumferentiam **OBVa**, **Ps** (*supra*)   *58 KN*] *HN* **Z**   *58* iam] *om.* **BPsVa**   *58–*
59 lineam continentem ipsum] ipsius circumferentiam **PRVMKgFiZBPs**, **Va** (*om.* circumferenti-
am)   *59* circulo] *om.* **Va**   *59* igitur] enim **Va**   *60* existencium] *om.* **OBPsVa**   *61* eis] *supra*
Z   *61* eis est] *om.* **OBPsVa**   *61* est maior] *tr.* **MKg**   *61* a centro est] *om.* **BVa**, *supra* **Ps**   *61–*
62 a centro est equalis] *permut.* **O**   *61–62* est equalis ... a centro] *om.* **Fi**   *62* maior est] *tr.*
FiZ   *63* est[1]] autem **Va**   *63* Et illud est] *om.* **FiZ**   *63* Et illud ... voluimus] hoc est **O**, *om.*
BPsVa   *63* voluimus] volumus **Z**   *1* et[2]] *om.* **Kg**, *supra* **BPs**   *2* linea producta] lineam pro-
ducendo **PRVM**   *2* spacium] *supra* **B**, *om.* **PsVa**   *2* ea] *supra* **R**   *2* coniungit] coniunge **Fi**

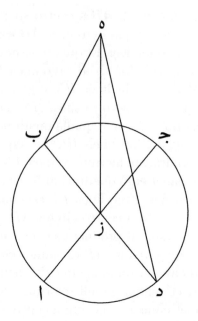

فلتكن الدائرة التى فى الكرة دائرة اٮجد وليكن مركز الكرة نقطة ه

ومركز الدائرة نقطة ز وليوصل خط هز ، فأقول إن خط هز عمود على دائرة

اٮجد فليخرج من مركز الدائرة خطا ازد بزد وليوصل خطا هب هد ٥

فلأن خط زب مساوٍ لخط زد وخط زه مشترك يكون خطا بز زه

مساويين لخطى دز زه كل واحد منهما لنظيره وقاعدة به مساوية لقاعدة

ده وذلك أن نقطة ه مركز الكرة ونقطتى ٮ د على بسيط الكرة فتكون

زاوية بزه مساوية لزاوية دزه وإذا قام خط مستقيم على خط مستقيم

وصيّر الزاويتين اللتين عن جنبتيه مساوية إحداهما للأخرى فكل واحدة من ١٠

الزاويتين المتساويتين زاوية قائمة والخط القائم يقال له العمود على الخط

4 عمود [عمودا N 5 فليخرج [ولٮحرح A 5 من [supra A بزد [وهما قطرا الدائرة add. A

om. A 7 كل [وكل N وصيّر [نصر N 11 زاوية [N

9 دزه [ح أ س add. supra A

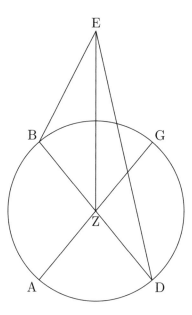

Sit igitur circulus qui est in spera circulus *ABGD*, et sit centrum spere punctum *E* et centrum circuli punctum *Z*; et producam   5
lineam *EZ*. Dico igitur quod linea *EZ* est perpendicularis super circulum *ABGD*. A centro itaque circuli protraham duas lineas *AZG BZD*,
et producam duas lineas *EB ED*. Et quia linea *ZB* est equalis linee
ZD, ergo linea *ZE* communi erunt due linee *BZ ZE* equales duabus
lineis *DZ ZE*, queque videlicet earum sue relative equalis. Sed et basis   10
BE equatur basi *DE*. Quod ideo est quoniam punctum *E* est centrum
spere et duo puncta *D B* sunt super superficiem spere. Ergo angulus
BZE est equalis angulo *DZE*. Sed cum linea recta super rectam erigitur
lineam et fit unus duorum angulorum qui sunt ab utraque parte alteri
equalis, tunc unusquisque duorum angulorum equalium est rectus et   15
linea erecta dicitur perpendicularis super lineam super quam est erec-

4 circulus²] *om.* **OBPsVa**   *4–5* sit centrum] *tr.* **O**   *4–5* centrum] *supra* **B**, *om.* **PsVa**
5 *E*] *corr. ex Z* **Ps**   5 centrum circuli] *tr.* **Va**   6 lineam] *in add.* **Z**   6 igitur] *om.* **BPsVa**
6 est] erit **Kg**   *6–7* circulum] superfi. circuli **Z**   7 A centro] per centrum **B**   7 itaque] igitur
M, *om.* **Va**   7 duas] *repet.* **Fi**   7 *AZG*] *AGZ* **V**   8 duas lineas] *tr.* **PR**   8 linea] *om.*
OBPsVa   8 *ZB*] *corr. ex ZD* **BPs**   8 linee] *corr. ex* linea **Ps**   9 *ZE*] *corr. ex Z* **Ps**;
verb. illeg. add. **B**   9 communi] *om.* **Kg**   9 erunt] *marg.* **R**; *corr. ex* centro, *marg.* **Ps**
9 due] *om.* **OBPs**   10 queque] quia **Fi**, quisque **Z**   10 queque videlicet] utraque scilicet
MKgOBPsVa   10 equalis] *corr. ex* equales **Kg**   11 *BE*] *HE* **Fi**   11 Quod] Et **Va**   11 ideo]
supra **M**   11 quoniam] quod **Kg**, quia **OBPsVa**   13 est equalis] equatur **OBPsVa**   *13–
14* super rectam erigitur lineam] *permut.* **Kg,OB**   14 fit] sit **OB**   14 duorum angulorum] *tr.*
Va   *14–15* qui . . . angulorum] *marg.* **R**   14 ab utraque] ex utraque **Z**, extraque **Fi**
15 equalis] *corr. ex* equales **BPs**   15 unusquisque] uterque **MKgOBPsVa**   16 erecta] recta
OBPsVa   16 dicitur] ducitur **Va**, *corr. ex* ducitur **Ps**   16 perpendicularis] *post* erecta² **BPsVa**
16 super lineam] *om.* **Fi**   16 lineam] *om.* **Z**, rectam *add.* **B**   16 quam] quod **Fi**

الذى قام عليه فكل واحدة من زاويتى بزه دزه قائمة فخط هز عمود على

خط بد وكذلك أيضاً نبين أنه عمود على خط اج أيضاً فلأن خط هز

المستقيم قد قام على الفصل المشترك لخطى اج بد اللذين يقطع أحدهما

الآخر يكون أيضاً قائماً على السطح الذى يمر بخطى اج بد والسطح الذى   15

يمر بخطى اج بد هو دائرة ابجد فخط هز عمود على سطح دائرة

ابجد ، وذلك ما أردنا أن نبين .

ح   إذا كانت دائرة فى كرة وأخرج من مركز الكرة عمود عليها وأنفذ إلى كلتى

الناحيتين فإنه يقع على قطبى الدائرة .

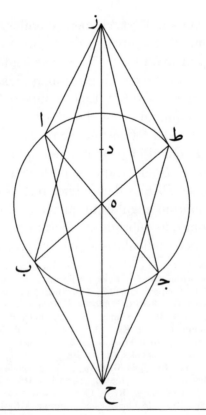

ta; unusquisque igitur duorum angulorum *BZE DZE* est rectus. Ergo
linea *EZ* est perpendicularis super lineam *BD*. Similiter quoque ostenditur quod ipsa est perpendicularis super lineam *AG*; et etiam quia
linea *EZ* recta iam est erecta super communem differentiam duarum   20
linearum *AG BD*, quarum una alteram secat, erit ergo etiam erecta
super superficiem que transit per duas lineas *AG BD*. Sed superficies
que transit per duas lineas *AG BD* est circulus *ABGD*. Ergo linea
EZ est perpendicularis super superficiem *ABGD*. Et illud est quod
demonstrare voluimus.   25

8   Cum in spera fuerit circulus et a centro spere perpendicularis protrahetur super ipsum et pertransiet ad duas partes, tunc ipsa cadet super
polum circuli.

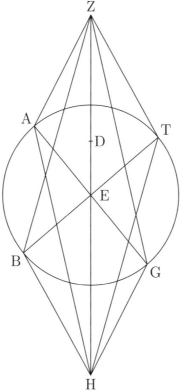

17 unusquisque] uterque   **OB**   17 igitur] ergo   **KgOBPs**   17 *DZE*] *om.*   **VFi**
18 est perpendicularis] *tr.* **MKg**   18 quoque] quod **Fi**   19 est] *supra* **M**, *om.* **Fi**
19 et] *supra* **Ps**; quod, *supra* et **Kg**   19 etiam] *om.* **OBPsVa**; ostenditur scilicet *add.*
Z   20 recta iam] *om.* **BPsVa**   20 recta iam est erecta] erecta **O**, recta **Fi**   20 est] *om.*
PsVa   21–22 quarum una ... *BD*] *om.* **OFiZ**; quarum una ... *ABGD (l. 23) om.* **BPsVa**
22–23 Sed ... *ABGD*] *marg.* **O**   24 *ABGD*] *ABDG* **Z**   24 Et illud est] *om.* **FiZ**, et hoc
est **OBPsVa**   25 demonstrare voluimus] volumus probare **OBPsVa**, demonstrare volumus
Z   *1* spere] *supra* **BPs**   *2* pertransiet] pertranseat **R**   *2* partes] spere *add. supra* **Ps**
2 cadet] ex utraque parte *add. supra* **Ps**, *in textu* **Va**   *3* polum] vel los [*sc.* polos] *supra* **O**

فلتكن الدائرة التى فى الكرة دائرة اٮج وليكن مركز الكرة نقطة د

وليخرج من نقطة د إلى سطح دائرة اٮج عمود ده وليلق بسيط الدائرة

على نقطة ه فنقطة ه مركز دائرة اٮج وليخرج خط ده إلى كلتى الناحيتين

٥ وليلق بسيط الكرة على نقطتى ز ح ، فأقول إن نقطتى ز ح قطبا دائرة

اٮج فليخط خطا اه ٮج به ط ولتوصل خطوط از زج اح حج بز زط

بح حط فلأن خط زه المستقيم عمود على دائرة اٮج ويحدث مع جميع

الخطوط المستقيمة التى تخرج من طرفه فى سطح دائرة اٮج زوايا قائمة

تكون كل واحدة من زوايا زها زهج زه ٮ زهط قائمة وأيضاً فلأن خط اه

١٠ مساوٍ لخط هج وخط هز مشترك وهو على زوايا قائمة تكون قاعدة از

مساوية لقاعدة زج وكذلك أيضاً نبين أن الخطوط التى تخرج من نقطة ز إلى

قوس اٮج مساوٍ بعضها لبعض فنقطة ز قطب لدائرة اٮج وكذلك أيضاً

نبين أن نقطة ح قطب لدائرة اٮج فنقطتا ز ح قطبا دائرة اٮج وقد

تبين أنه إذا كانت دائرة فى كرة وأخرج من مركز الكرة عمود عليها وأنفذ إلى

١٥ كلتى الناحيتين فإنه يقع على قطبى الدائرة ·

بعضا [لبعض 13 النح. *add.* الباقية [الخطوط 12 A. *repet.* الخطوط] 12 ٮ و محرح [وليخرج 4 A. *om.* التى [3

النح N 14-15 *om.* [فنقطتا ... اٮج

12 زج [س د ا *supra* A

Sit itaque circulus qui est in spera circulus *ABG*, et sit cen-
trum spere punctum *K*, a quo ad superficiem circuli *ABG* protraham   5
perpendicularem *KE* et occurrat superficiei circuli super punctum *E*.
Punctum igitur *E* est centrum circuli *ABG*. Protraham autem lineam
KE ad duas partes et occurrat superficiei spere super duo puncta *Z* et
H. Dico igitur quod duo puncta *Z H* sunt duo poli circuli *ABG*. Pro-
traham igitur duas lineas *AEG BET* et producam lineas *AZ ZG AH*   10
HG BZ ZT BH HT. Et quia linea *ZE* recta est perpendicularis super
circulum *ABG* et ex ea cum omnibus rectis lineis que ab eius extre-
mitate in superficie circuli *ABG* protrahuntur recti proveniunt anguli,
ergo unusquisque angulorum *ZEA ZEG ZEB ZET* est rectus. Et etiam
quia linea *AE* est equalis linee *EG*, ergo linea *EZ* communi, que ortho-   15
gonaliter insistit, erit basis *AZ* equalis basi *ZG*. Simili quoque modo
monstratur quod linee que protrahuntur a puncto *Z* ad arcum *ABG* ad
invicem sunt equales. Punctum igitur *Z* est polus circuli *ABG*. Simili-
ter etiam monstratur quod punctum *H* est polus circuli *ABG*. Ergo duo
puncta *Z* et *H* sunt duo poli circuli *ABG*. Iam igitur ostensum est quod   20
cum circulus fuerit in spera et a centro spere perpendicularis super
ipsum protracta fuerit et ad utramque partem pertransierit, super duos
polos circuli cadet. Et illud est quod demonstrare voluimus.

4 qui] quod **R**   4 circulus[2]] *om.* **BPsVa**   4 sit] *supra* **R**, *om.* **BPsVa**   5 *K*] *D* **O**,
hic et saepius   5 ad] in **BPsVa**   5 protraham] producam **OBVa**   6 occurrat] *corr.*
ex occurram **Z**   6 superficiei] perpendiculariter *add.* **OBVa**   6 circuli] *om.* **BPsVa**
7 Punctum igitur] *tr.* **Va**   8 partes] spere *add. supra* **Ps**, *in textu* **Va**   8 et] *supra* **Ps**
8 occurrat] hoc currat **Fi**   8 super] *corr. ex* ad **Kg**   9 igitur] ergo **KgB**   9 puncta] punctum
Fi   9 *Z*] et *add.* **PRVMKgOBVa**   10 duas] *om.* **B**, *supra* **Ps**   10 lineas[1]] *om.* **Va**
10 *AEG*] *EAG* **Z**, *AGE* **B**, *corr. ex AGE* **Ps**   10 lineas[2]] *om.* **BPsVa**   10 *AZ ZG*] *supra*
in corr. **Ps**   11 *ZT*] *supra* **Ps**   11 *BH*] *TBH* **Ps**   11 *ZE*] *corr. ex ZAE* **V**;
est *add.* **Va**, *add. et del.* **Ps**   11 recta] *corr. ex* erecta **Ps**   12 ex ea cum] *supra,*
in textu AB AET **Ps**   12 ea cum] earum **O**   12–13 extremitate] extremitatibus **M**
13 *ABG* protrahuntur] *tr.* **OFiZBPsVa**   14 angulorum] rectangulorum **Va**   15 *EG*] *corr.*
ex AG **V**   15 *EZ*] facta *add.* **OFiZ**   15 communi, que] qm̄ aū **BPs**, Qᵒmām **Va**
15 que] quoniam *add.* **OFiZ**   16 insistit] existit **BPsVa**   16 erit] extra, *supra* erit **Ps**
16 Simili quoque] Similique **BPsVa**   17 monstratur] demonstratur **BPsVa**   17 a puncto] *om.*
BPsVa   17 arcum] lineam **BPsVa**, *supra* circulum **Ps**   17 *ABG*] *AB* **Fi**   18–
19 Similiter … circuli *ABG*] *marg.* **BPs**   19 etiam] quoque **R**   20 puncta] et *add.* **Fi**
20 est] *om.* **BPsVa**   21 perpendicularis] perpendiculariter **Z**   22 ipsum] *in corr.* **BPs**
22 protracta] erecta **BPsVa** (*in corr., supra* **BPs**)   22 pertransierit] *corr. ex* per[...] fue-
rit **R**, transierit **FiZ**   23 polos circuli] *tr.* **PRVFi**   23 Et illud est] *om.* **FiZ**   23 illud] id
Kg, hoc **O**   23 quod … voluimus] *om.* **O**   23 demonstrare voluimus] intendimus **BPsVa**
23 voluimus] volumus **Z**

ط   إذا كانت دائرة فى كرة ووُصل بين أحد قطبيها وبين المركز بخط مستقيم فإن
الخط عمود على الدائرة ٠

برهان هذا الشكل شبيه ببرهان الشكل الذى قبله ٠

ى   إذا كانت دائرة فى كرة وأخرج من أحد قطبيها إليها خط يكون عموداً عليها
فهو يقع على مركز الدائرة ، فإن أخرج إلى الناحية الأخرى فإنه يقع على
القطب الآخر من قطبى الدائرة ٠

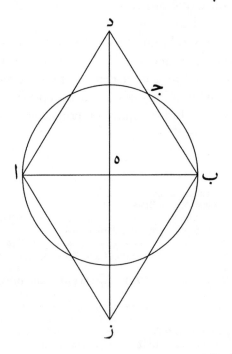

فلتكن الدائرة التى فى الكرة دائرة ا ب ج وليخرج من أحد قطبيها وهو

5   نقطة دَ إليها عمود دهَ وليلق سطح الدائرة على نقطة ه وليخرج خط دهَ

N قطب [¹نقطة   5 A نقطة [¹ يشبه [شبيه 3

لأن الخطوط الخارجة من مركز الدائرة إلى محيطها متساوية ولأن الخطوط الخارجة من القطب أيضاً إلى [قبله 3
محيط الدائرة متساوية add. A; نבّן אן ארדנא מא ודלך כצורתה וצורתה H .add

9   Cum in spera fuerit circulus et coniungetur inter unum duorum polo-
    rum eius et centrum protrahendo lineam rectam, linea erit perpendi-
    cularis super circulum.

        Huius figure probacio declaratur secundum probacionem figure
    que eam precedit.   5

10 Cum in spera fuerit circulus et protrahatur ab uno duorum polorum
    eius ad ipsum linea super ipsum existens perpendicularis, tunc ipsa
    cadet super centrum circuli. Quod si protrahatur ad alteram partem,
    tunc ipsa cadet super alterum duorum polorum circuli.

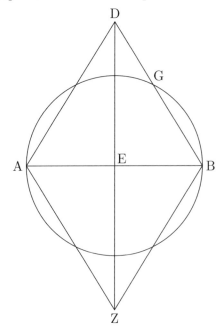

        Sit ergo circulus qui est in spera circulus *ABG*; ab uno polorum   5
    cuius, qui sit polus *D*, protraham ad ipsum perpendicularem *DE* que
    superficiei circuli super punctum *E* concurrat. Et protraham lineam

1 inter unum] cum uno **V**   1 unum] *om.* **FiZ**   2 et] *supra, corr. ex* quia **BPs**
2 protrahendo lineam rectam] protracta linea recta **KgOBPsVa**, protrahendo lineam rectam *su-*
pra **B**, *marg.* **Ps**, *add. in textu* **Va**   2 rectam, linea] rectanguli **Fi**   2 linea] *om.* **OBPsVa**
3 circulum] protrahendo lineam rectam *add.* **O**   4 Huius] cuius **Fi**   4 declaratur] decaratur
R   4 probacionem] *supra, corr. ex* propositionem **B**   5 eam precedit] antecedit **Kg**, antecedit eam **MOBPsVa**, eam procedit **FiZ**   1 Cum] si **PRVMKgOFiZBPsVa**   1 in spera] *supra*
Va   1 duorum] *supra* **M**, *om.* **KgOBPsVa**   1–2 polorum eius] *tr.* **OBPsVa**   2 linea] *corr.*
ex lineam **M**   3 cadet] cadit **OBPsVa**   3 si protrahatur] sic probatur **R**   4 ipsa] *om.*
Fi   5 ergo] igitur **KgOBPsVa**   5 qui] que **R**   5 circulus[2]] *om.* **OBPsVa**   5–
6 polorum cuius] eius polorum **OBPsVa** (*add. et del.* cui[us] **Ps**)   6 protraham] protrahatur
MKgOBPsVa   6 perpendicularem] perpendicularis **MKgOBPsVa**   6 *DE*] *om.* **Fi**   6–
7 que superficiei] *illeg.* **Kg**   7 *E*] est **Kg**   7 concurrat] occurrat **R**

وليلق بسيط الكرة الذى فى الجهة الأخرى على نقطة ز ، فأقول إن نقطة ه

مركز دائرة ا ب ج ونقطة ز القطب الآخر من قطبى دائرة ا ب ج فليخرج من

نقطة ه خطا ا ه ه ب ولتوصل خطوط ا د د ب ا ز ز ب فلأن خط د ه عمود

على دائرة ا ب ج فإنه يحدث مع جميع الخطوط المستقيمة التى تخرج من

طرفه فى سطح دائرة ا ب ج زوايا قائمة وقد خرج من طرفه كل واحد من

خطى ا ه ه ب وهما فى سطح دائرة ا ب ج فتكون كل واحدة من زاويتى

د ه ا د ه ب قائمة فلأن خط ا د مساو لخط د ب يكون المربع الكائن من خط

ا د مساوياً للمربع الكائن من خط د ب ولكن المربع الكائن من خط ا د مساو

للمربعين الكائنين من خطى د ه ه ا والمربع الكائن من خط د ب مساو

للمربعين الكائنين من خطى د ه ه ب فالمربعان الكائنان من خطى ا ه ه د

مساويان للمربعين الكائنين من خطى ب ه ه د وليسقط المربع المشترك وهو

المربع الكائن من خط د ه فيبقى المربع الكائن من خط ا ه مساوياً للمربع

الكائن من خط ه ب فخط ا ه مساو لخط ه ب وكذلك أيضاً نبين أن جميع

الخطوط التى تخرج من نقطة ه إلى خط ا ب ج مساو بعضها لبعض فنقطة ه

مركز دائرة ا ب ج ،

وأقول أيضاً إن نقطة ز هى القطب الآخر من قطبى دائرة ا ب ج فلأن

خط ا ه مساو لخط ه ب وخط ز ه مشترك وهو على زوايا قائمة على هذين

الخطين تكون قاعدة ا ز مساوية لقاعدة ز ب وكذلك أيضاً نبين أن جميع

DE quousque concurrat superficiei spere que est in altera parte super punctum *Z.* Dico igitur quod punctum *E* est centrum circuli *ABG* et punctum *Z* est alter duorum polorum circuli *ABG.* A puncto itaque   10
E protraham duas lineas *EA EB* et producam lineas *AD DB AZ ZB.*
Et quia linea *DE* est perpendicularis super circulum *ABG,* ergo ex ea cum omnibus rectis lineis que ab eius extremitate in superficie circuli *ABG* protrahuntur anguli proveniunt recti. Ab eius autem extremitate iam protracta est unaqueque duarum linearum *AE EB* que sunt in   15
superficie circuli *ABG.* Unusquisque igitur duorum angulorum *DEA DEB* est rectus. Et quia linea *AD* est equalis linee *DB,* ergo quadratum factum ex linea *AD* equatur quadrato facto ex linea *DB.* Quadratum vero ex linea *AD* factum est equale duobus quadratis factis ex duabus lineis *DE EA* et quadratum factum ex linea *DB* est equale duobus   20
quadratis factis ex duabus lineis *DE EB.* Ergo duo quadrata facta ex duabus lineis *AE ED* sunt equalia duobus quadratis factis ex duabus lineis *BE ED.* Minuam autem quadratum factum commune quod est factum ex linea *DE*: remanet ergo quadratum factum ex linea *AE* equale quadrato facto ex linea *EB.* Linea igitur *AE* est equalis linee   25
EB. Similiter quoque monstratur quod omnes linee que a puncto *E* ad lineam *ABG* producuntur ad invicem sunt equales. Punctum igitur *E* circuli *ABG* centrum existit.

Et dico etiam quod punctum *Z* est alter duorum polorum circuli *ABG.* Et quia linea *AE* est equalis linee *EB,* linea *ZE* communi que   30
super has duas lineas orthogonaliter existit, ergo basis *AZ* est equalis

8 concurrat] occurrat **ROFiZBPsVa**   9 *E*] *supra* **Ps**   9 est centrum] in centr̄ **Va**   9–
10 et punctum ... *ABG*] *om.* **Fi**   10 alter] alterum **Kg,** unum **BPsVa**   10 polorum] *ante*
est **Z**   10 A] *supra* **Ps**   10 puncto] pnnab (?) **Va**   11 producam] duas *add.* **OVa,** *add.*
et del. **BPs,** *supra* 4 **B**   11 lineas²] lineam **Z**   11 *AD*] *AB* **Z**   13–14 circuli *ABG*] *tr.*
BPsVa   14 protrahuntur] protrahantur **Va**   14 proveniunt] perveniunt **RFi**   14 recti] *AB*
perpendicularis *add.* **Fi**   14 Ab] Sed ab **OFiVa,** sed *supra* **BPs**   14 eius] *om.* **B,** *su-*
pra **Ps**; perpendicularis *supra* **O**   14 autem] *om.* **KgOBPsVa**   15 unaqueque] utraque
KgOBPsVa   15 *EB*] *EH* **V,** *supra litterae illeg.* **B**   16 Unusquisque] uterque **MKgOB**
PsVa   16 igitur duorum angulorum] *permut.* **Z**   18 factum ex linea] linee **OBPsVa,**
hic et saepius   18 linea¹] *H* **Fi**   18 *AD*] *supra* **BPs**   18 equatur] adequatur
OBPs, equatum **Va**   18–19 equatur quadrato ... *AD*] *om.* **V**   19 est equale] equatur
MKgOBPsVa   19 duobus] duorum **O**   19–20 factis ex duabus lineis] *om.* **OBPsVa**
19 duabus] duobus **Fi**   20 ex] *om.* **Ps**   20 linea] *om.* **OFiZBPsVa**   21 factis ... lineis] *om.*
OFiZBPsVa   21 *DE EB*] *tr.* **MKgOBPsVa**   21–23 Ergo ... lineis *BE ED*] *marg.* **M**
(*BE ED: BE EB*)   21 duo] *corr. ex* dua **Ps**   22 quadratis] quadrata **Va**   22–
23 factis ... lineis] *om.* **OBPsVa**   23 factum] *om.* **VMKgBPsVa**   23 factum commune] *tr.*
OZ   23–24 factum ... remanet] quem factum **Fi**   23–24 quod ... remanet] *om.* **Z**
24 ergo quadratum] *tr.* **FiZ** 24 factum²] *om.* **FiZ** 24 *AE*] *AG* **O**   25–26 Linea ... *EB*] *om.*
BPsVa   26 monstratur] ostenditur **OBPs,** ostendit **Va**   26 omnes linee] *om.* **R,** *vacat* **P**
26 a] *corr. ex* ex **R**   26 *E*] est **R**   27 *ABG*] *AG* **Kg;** *AG, B supra* **B;** *AG* **Ps** (*add. supra*
circuli); circuli *add.* **Va**   27 *E*] *corr. ex Z* **Kg**   28 circuli *ABG* centrum existit] est centrum
circuli *ABG* **MKgOBPsVa** (*ABG: ADG* **Va**)   29–30 Et ... *ABG*] *om.* **KgBPsVa** 30 *AE*] *supra*
R   30 *EB*] *corr. ex AB* **Kg**   31 existit] insistit **MKgOBPsVa**   31 est equalis] equatur
MKgOBPsVa   31 equalis] qualis **P**

الخطوط المستقيمة التى تخرج من نقطة زَ إلى خط ا ب جَ المحيط مساوٍ

بعضها لبعض فنقطة زَ هى القطب الآخر من قطبى دائرة ا ب جَ وقد كان   25

تبين أن نقطة هَ مركز دائرة ا ب جَ فنقطة هَ مركز دائرة ا ب جَ ونقطة زَ هى

القطب الآخر من قطبى دائرة ا ب جَ ، وذلك ما أردنا أن نبين .

يَا   إذا كانت دائرة فى كرة فإن الخط المستقيم الذى يمر بقطبيها هو عمود عليها

وهو يمر بمركزها وبمركز الكرة .

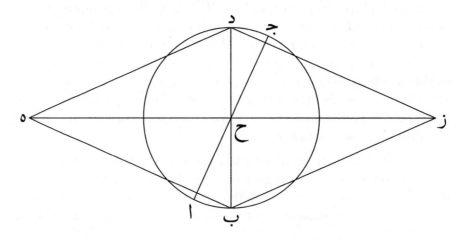

فلتكن الدائرة التى فى الكرة دائرة ا ب جَ دَ وليكن قطباها نقطتى هَ زَ

وليوصل الخط المستقيم الذى يمر بقطبيها وهو هَ زَ ، فأقول إن خط هَ زَ عمود

على دائرة ا ب جَ دَ وهو يمر بمركزها وبمركز الكرة فليمر من سطح دائرة   5

ا ب جَ دَ بنقطة حَ وليخرج من نقطة حَ خطا ا حَ جَ بَ حَ دَ ولتوصل خطوط

ا حَ [ا ب ج د N   3 ا حَ [ا ب ج د N   *supra* ادأ   4 ا حَ [وهو هَ زَ] *om.* N   6 ا ب ج د[ا حَ N

A *supra* على: NH بنقطة[قطة   6

ا حَ ج حَ وليكن ا حَ على استقامة حَ جَ وليخرج أيضاً من نقطة حَ خطا حَ بَ حَ دَ وليكن حَ بَ [ا حَ جَ بَ حَ دَ   6

A على استقامة حَ دَ

basi *ZB*. Simili quoque modo monstratur quod omnes recte linee que
a puncto *Z* ad lineam *ABG* circumductam protrahuntur ad invicem
sunt equales. Punctum igitur *Z* duorum polorum circuli *ABG* alter
existit. Iam autem fuit ostensum quod punctum *E* est centrum circuli   35
ABG: ergo punctum *E* est centrum circuli *ABG* et punctum *Z* est alter
duorum polorum circuli *ABG*. Et illud est quod demonstrare voluimus.

11   Cum in spera fuerit circulus, linea recta que per duos polos eius transit
super eum est perpendicularis et ipsa transit per centrum eius et per
centrum spere.

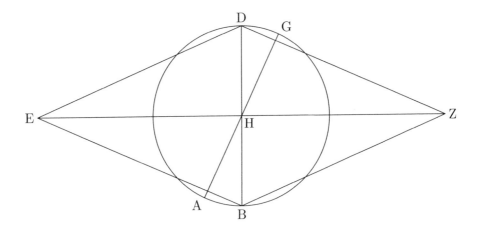

Sit itaque circulus qui est in spera circulus *ABGD* et sint duo
poli eius duo puncta *E* et *Z*; producam autem lineam rectam que   5
transit per polos eius, que sit linea *EZ*. Dico igitur quod linea *EZ*
est perpendicularis super circulum *ABGD* et quod ipsa transit per
centrum eius et per centrum spere. Transeat igitur per punctum *H*
superficiei circuli *ABGD*. Protraham autem a puncto *H* duas lineas

32 Simili quoque modo] Similiter quoque **MKgOFiBPsVa**   *32* modo monstratur] *tr.* **PV**
· *32* monstratur] *marg.* **B**, *supra* **Ps**   *32* recte] *om.* **BPsVa**   *33* *ABG*] *supra* **BPs**   *34–*
35 duorum … existit] *permut.* **MKg,OBPsVa** (circuli *ABG*: *om.* **BPsVa**)   *35* fuit ostensum] *tr.*
R; et dico etiam quod punctum *Z* est alter duorum polorum circuli *ABG* add. **Kg**, *add. et del.*
BPs (*om.* etiam)   *35* quod] quia **OBPsVa**   *36* ergo … *ABG*] *om.* **OBPsVa**   *36* est] *in add.*
Fi   *37* circuli *ABG*] *om.* **BPsVa**   *37* Et … voluimus] quia illud est **Kg**   *37* illud] hoc **OB**
PsVa   *37* est] *om.* **Fi**   *37* quod … voluimus] *om.* **O**   *37* demonstrare voluimus] *tr.* **BPsVa**
37 voluimus] volumus **Z**   *1* duos] *om.* **OBPsVa**   *1* eius transit] *tr.* **OBPsVa**   *2* per²] *om.*
BPsVa   *5* et] *om.* **OBPsVa**   *5* producam autem] et producam **MKgBPsVa**   *5* que] qua **Va**
6 transit] transeat **MKg**   *6* Dico … *EZ*] *om.* **Fi**   *6* igitur] ergo **Kg**   *7–9* et … *ABGD*] *marg.*
BPs   *7* quod] *om.* **OFiZBVa** (**Ps** *non videtur*)   *8* eius … centrum] *om.* **Fi**   *8* per¹] *om.* **OB**
(**Ps** *non videtur*)   *9* superficiei] superficie **Va**   *9* a puncto] per punctum **B**

به ٥د بز زد فلأن خط ٥ب مساوٍ لخط ٥د وخط ٥ز مشترك يكون خطا

به ٥ز مساويين لخطى ده ٥ز كل واحد لنظيره وقاعدة بز مساوية

لقاعدة زد فتكون زاوية ب٥ز مساوية لزاوية د٥ز وأيضاً فلأن خط به

مساوٍ لخط ده وخط ٥ح مشترك يكون خطا به ٥ح مساويين لخطى ده ١٠

٥ح كل واحد لنظيره وزاوية ب٥ح مساوية لزاوية د٥ح فتكون قاعدة

بح مساوية لقاعدة دح ويكون مثلث ب٥ح مساوياً لمثلث ٥دح وتكون

سائر الزوايا مساوية لسائر الزوايا التى توترها الأضلاع المتساوية فزاوية دح٥

مساوية لزاوية ب٥ح وإذا قام خط مستقيم على خط مستقيم وصير الزاويتين

اللتين عن الجانبين متساويتين فكل واحدة من الزاويتين المتساويتين قائمة ١٥

فخط ٥ح قائم على دب على زوايا قائمة وكذلك أيضاً نبين أن خط ٥ح أيضاً

عمود قائم على خط اج على زوايا قائمة فهو أيضاً قائم على السطح الذى

يمر بخطى بد اج أعنى دائرة اب جد على زوايا قائمة فخط ٥ح قائم على

دائرة اب جد على زوايا قائمة ،

وأقول أيضاً إنه يمر بمركز الدائرة وبمركز الكرة وذلك أن دائرة اب جد ٢٠

فى كرة وقد أخرج من أحد قطبيها وهو نقطة ٥ إليها عمود ٥ح ويلقى

سطحها على نقطة ح فنقطة ح مركز دائرة اب جد ،

H ادא [أيضاً 17   om. N 17   عمود [om. N 16   على دب [om. N 16   الحاس [الجانبين 15   N رح [زد 7

N مطب [نقطة 21   del. A 20   وبمركز الكرة [A فاقول [وأقول 20   om. N 20   على دائرة اب جد [18–19

21 وهو عمود [عمود 21   add. NH 21 [ويلقى   in corr. A

فهو 17 [עמוד עלי כטי בד אג אגז ואדא כאן כט עמודא עלי כטין פי סטח ואחד פהו עמוד

פכט הח אדא ימר במרכז דאירה [اب جد 22   add. H עלי אלסטח אלדי ימר בהמא פכט הח

אבנד add. H

AHG BHD et producam lineas *BE ED BZ ZD*. Et quia linea *BE*   10
est equalis linee *ED*, ergo linea *EZ* communi erunt due linee *BE EZ*
equales duabus lineis *DE EZ*, queque videlicet sue relative, sed et basis
BZ est equalis basi *ZD*: ergo angulus *BEZ* est equalis angulo *DEZ*. Et
etiam quia linea *BE* est equalis linee *DE*, ergo linea *EH* communi due
linee *BE EH* sunt equales duabus lineis *DE EH*, queque videlicet sue   15
relative. Angulus quoque *BEH* est equalis angulo *DEH*: ergo basis *BH*
est equalis basi *DH*. Et triangulus *BEH* equatur triangulo *EDH*, reliqui
quoque anguli reliquis equantur angulis quibus latera subtenduntur
equalia; ergo angulus *DHE* est equalis angulo *BHE*. Sed cum linea
recta super rectam erigitur lineam et fiunt duo anguli qui sunt ab   20
utraque parte equales, tunc unusquisque duorum angulorum equalium
est rectus: ergo linea *EH* super lineam *BD* orthogonaliter est erecta.
Eodem quoque modo monstratur quod linea *EH* orthogonaliter erecta
est super lineam *AG*. Ipsa igitur orthogonaliter etiam est erecta super
superficiem que transit per duas lineas *BD AG*, scilicet circulus *ABGD*,   25
ergo linea *EHZ* est orthogonaliter erecta super circulum *ABGD*.

Et dico etiam quod ipsa transit per centrum circuli et centrum
spere. Quod ideo est quoniam circulus *ABGD* est in spera et iam protracta est ab uno polorum eius ad ipsum, qui est *E*, perpendicularis,
que est *EH*, et concurrit superficiei eius super punctum *H*; ergo punc-   30
tum *H* est centrum circuli *ABGD*.

---

10 *AHG*] *ABG* **Va**   10 producam] duas *add.* **OBPsVa**   10–11 *ED BZ ZD ... BE*] *om.*
R   10 Et] Set **O**   11 erunt due linee] erit **BPs**, er' **Va**   12 equales] equalis **BPsVa**
12 *DE EZ*] *tr.* **KgOBPsVa**   12 queque videlicet] utraque **MKgOBPsVa**   13 *BZ*] *corr. ex*
EZ **B**, *corr. ex ET* **Ps**   13 *BEZ*] *BZ* **Z**   14 etiam] *om.* **Va**   14 *BE*] *EZ add. et del.*
OPs; equales duabus lineis *EZ add. et del.* **BPs**   15 *EH*] *corr. ex EZ* **BPs**   15 *DE*] *supra*
BPs   15–16 queque ... relative] *om.* **MKgBPsVa**   16 *BEH*] *DEH* **FiZ**   16 *DEH*] *BEH*
FiZ   16 basi] *supra* **Ps**   17 est] *om.* **OPsVa**   17 est equalis] equatur **B**   17 basi] *verb.*
illeg. add. **M**   17 triangulus] *corr. ex* angulus **BPs**   17 equatur] est equalis **OPsVa**
17 triangulo] *corr. ex* angulo **BPs**   17 *EDH*] *DEH* **OFiZBPsVa**   18 quoque] vero **M**, que
Fi, *om.* **Z**   18 reliquis] *om.* **BPsVa**   18 reliquis equantur] *tr.* **M**   18 equantur] coequantur **Va**
18 subtenduntur] *in corr. supra* **B**, *in corr.* **Ps**   19 *DHE*] *DEH* **KgVa**; *DEH, corr. in DHE* **BPs**
19 est equalis] equatur **B**   19 cum] si **OVa**, **BPs** (*supra*)   20 rectam erigitur lineam] *permut.*
Z   20 erigitur] igitur **FiPs**   20 fiunt] sunt **R**, fuerint **VMKg**, fuerit **Fi**   21 unusquisque] uterque
MKgBPsVa   21 duorum angulorum equalium] eorum **M**, illorum **KgOBPsVa**   22 *EH*] est *add.*
KgOBPsVa   22 est²] *om.* **OBPs**   22–23 est² ... orthogonaliter] *om.* **Va**   23 modo] *supra*
M   23–24 orthogonaliter erecta est] *permut.* **RVMKgOBPs,FiZ**   24 igitur] *supra* **BPs**
24 orthogonaliter etiam est] *permut.* **MKg,R,PsVa** (etiam *om.* **Va**)   24 etiam est] *om.* **B**
25 per] *supra* **BPs**   25 duas lineas] *om.* **BPsVa**   25 scilicet circulus] scilicet circulis **Fi**, que
est circulus **PsVa** (*post* est *add. et del.* lit' **Ps**), que scilicet est circulus **M**, qui scilicet circulus **KgB**
(**B** *in corr.*)   26 *EHZ*] *EAZ* **Va**   26 est orthogonaliter] *tr.* **Fi**   26 orthogonaliter] or gonaliter
Ps   28 est¹] *om.* **Va**   28–33 et iam ... in spera] *repet.* **Kg**   29 qui est *E*] que est **OBVa,Ps**
(que: *corr. ex* qui), quedam **FiZ**   30 que est] *om.* **BPsVa**   30 concurrit] occurrit **RKgOBPsVa**
30–31 ergo punctum *H*] *om.* **Fi**

وأقول إنه يمر بمركز الكرة أيضاً وذلك أن دائرة ا‌ب‌ج‌د فى كرة وقد

أخرج من مركزها إلى سطح الدائرة عمود فهو خط ه‌ح‌ز فمركز الكرة على

خط ه‌ح‌ز لخط ه‌ز يمر بمركز الكرة لخط ه‌ح‌ز عمود على دائرة ا‌ب‌ج‌د وهو   25

يمر بمركزها وبمركز الكرة ، وذلك ما أردنا أن نبين .

ي‌ب   الدوائر العظيمة التى تكون فى الكرة فإنها يقطع بعضها بعضاً بنصفين .

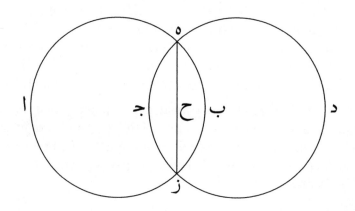

فلتكن دائرتان عظيمتان من الدوائر التى فى الكرة وهما دائرتا ا‌ب‌ج‌د

تقطع إحداهما الأخرى على نقطتى ه‌ ز‌ ، فأقول إن دائرتى ا‌ب‌ج‌د تقطع

كل واحدة منهما الأخرى بنصفين فلنتعلم مركزهما وليكن نقطة ح‌ وهذه

النقطة هى مركز الكرة أيضاً وليوصل خطا ه‌ح ح‌ز ولأن نقطة ه‌ ح‌ ز فى   5

سطح دائرة ا‌ب‌ وهى أيضاً فى سطح دائرة ج‌د تكون نقط ه‌ ح‌ ز فى

سطحى دائرتى ا‌ب‌ ج‌د جميعاً فنقط ه‌ ح‌ ز على الفصل المشترك بينهما

1 فإنها] *om.*   1 تكون] A الداره] الدوائر] N *supra* الكره] A الدائرة] 24   24 على [إلى] A 24

A فلنعلم [فلنتعلم] N 4   على N; *add.* N على الاخر] الأخرى] AN واحد] 4 واحدة [واحدة] A أحدهما [إحداهما] AH 3

A سطحين N, سطح [سطحى] N فلان [فلان] A ولأن [ولأن] 5 A وهذ [وهذ] N وهذه [وهذه] 4   ه N 4 ح‌ [ح‌] 4

Et dico etiam quod ipsa transit per centrum spere. Quod ideo est quoniam circulus *ABGD* est in spera, a cuius centro iam producta est perpendicularis ad superficiem circuli, que est linea *EHZ*. Ergo linea *EZ* transit per centrum spere. Ergo linea *EHZ* est perpendicularis   35 super circulum *ABGD* et ipsa transit per centrum eius et per centrum spere. Et illud est quod demonstrare voluimus.

12 Maiores circuli qui sunt in spera ad invicem se in duo media secant.

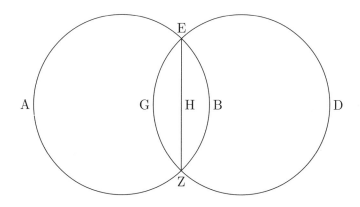

Sint itaque duo circuli maiores <ex> circulis qui sunt in spera circuli *AB GD*, quorum unus alterum super duo puncta *E* et *Z* secet. Dico igitur quod quisque duorum circulorum *AB GD* alterum in duo media secat. Notabo itaque centrum eorum, quod sit punctum *H*. Hoc   5 quoque punctum est centrum spere. Et producam duas lineas *EH HZ*. Et quia puncta *E H Z* sunt in superficie circuli *AB* et etiam sunt in superficie circuli *GD*, ergo puncta *E H Z* simul sunt in superficiebus circulorum *AB GD*. Ergo puncta *E H Z* sunt supra communem differentiam que est inter eos. Communis autem differentia que est inter   10

32 Et] *om.* **Z**, *illeg.* **Ps**   32 etiam] *om.* **BPsVa**   32 etiam quod ipsa] *permut.* **Kg**, etiam *add. et del.* **M**   32 per] centrum circuli et *add.* **Kg**   33 a cuius] ac **Fi**   33 centro] *supra verbum illeg.* **BPs**   33 producta] protracta **Z**   34 ad superficiem] a superficie **V**   34 circuli] *corr. ex* circulum **Ps**   34 *EHZ*] *EZH* **OBPsVa**   34–35 Ergo ... spere] *om.* **O**   35 *EZ*] *H add. supra* **B**, *om.* **Va**   35 *EZ* transit ... linea] *om.* **FiZ**   36 eius et per centrum] *om.* **Fi**   36–37 et per ... spere] *add. corrector* **Ps**   37 Et illud est] *om.* **Fi**   37 Et illud ... voluimus] *om.* **OBPsVa**   37 illud] hoc **Kg**   37 est] *om.* **V**   37 voluimus] volumus **Z**   1 spera] *om.* **Fi**   1 se] *supra* **M**   1 in ... secant] *permut.* **MKgOBPsVa**   2 duo] *in corr.* **O**   2 maiores <ex> circulis] maximi **MKgOBPsVa**   2 qui sunt] *om.* **OBPsVa**   3 *E*] *corr. ex EA* **Z**   3 et] *om.* **BPsVa**   4 quisque] uterque **MKgOBPsVa**, unusquisque **Z**   4 duo] *om.* **Va**   5 Notabo] Notato **OBPsVa**   5 centrum] centra **BPsVa**   5 eorum] illorum **FiZ**   5 Hoc] *H* **BPsVa**   6 punctum] *om.* **BPsVa**   6 centrum] *supra* **O**   6 duas lineas] *tr.* **PV**   7 Et quia ... *Z*] *om.* **Fi**   7 *Z*] *om.* **Z**   7–8 *AB* ... *GD*] *marg.* **B**; *AB ... superficie marg.* **Ps**   7 etiam] *om.* **Z**   7 sunt] *om.* **BPsVa**   8 circuli] *om.* **Ps**   9 *AB GD*] *corr. ex AGD* **Fi**   10 est] *supra* **R**   10 eos] eas **FiZ**   10 autem] etiam **BPsVa**

والفصل المشترك بين كل سطحين هو خط مستقيم فخط ه‍ح‍ز مستقيم ولأن

نقطة ح‍ مركز دائرة ا‍ب‍ يكون خط ه‍ح‍ز قطراً لها فكل واحد من خطى

ه‍ا‍ز ه‍ب‍ز هو قوس نصف دائرة ولأن نقطة ح‍ أيضاً مركز دائرة ج‍د‍ يكون

خط ه‍ح‍ز قطراً لها فكل واحد من خطى ه‍ج‍ز ه‍د‍ز قوس نصف دائرة

فدائرتا ا‍ب‍ ج‍د‍ تقطع كل واحدة منهما الأخرى بنصفين ، وذلك ما أردنا

أن نبين .

10

ما كان من الدوائر التى فى الكرة يقطع بعضها بعضاً بنصفين فهى أعظم   ي‍ج‍

الدوائر التى فيها .

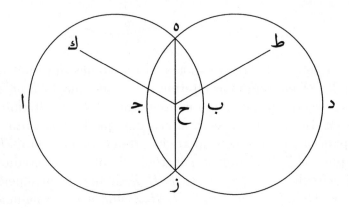

فلتكن فى كرة دائرتا ا‍ب‍ ج‍د‍ تقطع كل واحدة منهما الأخرى بنصفين

على نقطتى ه‍ ز‍ ، فأقول إن دائرتى ا‍ب‍ ج‍د‍ عظيمتان فليوصل الفصل

المشترك لهما وهو خط ه‍ز‍ فخط ه‍ز‍ قطر دائرتى ا‍ب‍ ج‍د‍ وليُقطع خط ه‍ز‍   5

بنصفين على نقطة ح‍ فنقطة ح‍ مركز دائرتى ا‍ب‍ ج‍د‍ ، وأقول إنها مركز

quaslibet duas superficies est linea recta, ergo linea *EHZ* est recta. Et
quia punctum *H* est centrum circuli *AB*, ergo linea *EHZ* est diametrus
eius et unaqueque duarum linearum *EAZ EBZ* est arcus semicirculi.
Et etiam quia punctum *H* est centrum circuli *GD*, ergo linea *EHZ*
est diametrus eius et unaqueque duarum linearum *EGZ EDZ* est ar-　15
cus semicirculi. Unusquisque igitur duorum circulorum *AB GD* secat
alterum in duo media. Et illud est quod demonstrare voluimus.

13　Quicumque circulorum qui sunt in spera ad invicem in duo media se
secant sunt ex circulis maioribus qui sunt in ea.

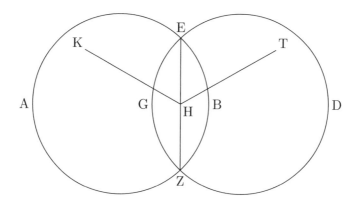

　　Sint itaque in spera duo circuli *AB GD* quorum quisque alterum
in duo media secet super duo puncta *E* et *Z*. Dico igitur quod duo
circuli *AB GD* sunt maiores duo. Describam itaque differentiam eis　5
communem, que sit linea *EZ*; ergo linea *EZ* est diametrus duorum circulorum *AB GD*. Dividam itaque lineam *EZ* in duo media in puncto
H: ergo punctum *H* est centrum duorum circulorum *AB GD*. Et dico

11 duas] *marg.* **R**, *in corr.* **Va**　　11 duas superficies] *tr.* **V**; planas *add.* **FiZB**, *supra* **Ps**　　11 *EHZ*] *corr. ex EZH* **BPs**　　11 Et] *in corr.* **B**, *corr. ex* ideo **Ps**
12–13 diametrus eius] *tr.* **FiZ**　　13 unaqueque] utraque **MKgBPs**　　13 *EAZ EBZ*] *om.*
BPs　　13 est] *om.* **Fi**　　14 Et ... *GD*] *marg. post* semicirculi (*l.* 16) **BPs**　　14–
16 ergo ... semicirculi] *marg.* **BPs**　　15 unaqueque] utraque **MKgOVa**　　15 duarum] *om.*
V　　15 *EGZ EDZ*] *EG ZE* **Fi**, *EAZ EBZ* **BPs**, *om.* **Va**　　15–16 arcus semicirculi] *tr.*
Fi　　16 Unusquisque] uterque **MKgOBPsVa**　　16 *AB GD*] *om.* **BPsVa**　　16 secat] est secans **OPsVa**　　17 duo media] *tr.* **PsVa**, medio **B**　　17 Et illud ... voluimus] *om.* **BPsVa**
17 est] *supra* **P**, *om.* **M**　　17 voluimus] volumus **OZ**　　1 sunt] *om.* **FiZ**　　1 in spera] *om.*
Va　　1 se] *om.* **Fi**　　2 sunt[1]] *om.* **Fi**　　2 circulis maioribus] *tr.* **PRVFiZ**　　2 ea] *om.* **Fi**
3 itaque] *om.* **OBPsVa**　　3 in spera duo circuli] *permut.* **Z**　　3 quisque] uterque **MKg**, *om.* **OB-
PsVa**　　3 alterum] alter **OPsVa** (*corr. ex* alterum **OPs**)　　4 in duo media secet] secet reliquum in
duo media **OBPsVa**　　4 duo] *om.* **BVa**　　4 puncta] punctum **PsVa**　　4 igitur] ergo **MKgBPsVa**
5 maiores duo] magni **OBPs**, mātini (?) **Va**　　5–7 Describam ... circulorum *AB GD*] *om.*
BPsVa　　7 itaque] enim **BPsVa**　　7 lineam] *om.* **V**　　7 media] *supra* **M**, *om.* **KgBPsVa**
7 in[2]] *supra* **Ps**　　8 ergo punctum *H*] *om.* **Fi**

الكرة أيضاً فليقم على نقطة حـ من سطح دائرة جـد خط على زوايا قائمة

وهو خط حـطـ وليقم على هذه النقطة أيضاً من سطح دائرة ا ب خط حـك

على زوايا قائمة فلأن دائرة جـد في كرة وقد أخرج من مركزها على سطح

الدائرة خط على زوايا قائمة وهو خط حـطـ يكون مركز الكرة على خط ١٠

حـطـ وكذلك أيضاً نبين أنه على خط حـك فمركز الكرة على الفصل المشترك

لخطى حـطـ حـك والفصل المشترك لهما هو نقطة حـ فنقطة حـ مركز الكرة

والدوائر التى تمر بمركز الكرة هى عظيمة فدائرتا ا ب جـد عظيمتان ، وذلك

ما أردنا أن نبين .

يد   إذا قطعت دائرة عظيمة فى كرة دائرة أخرى من الدوائر التى فى الكرة على

زوايا قائمة فهى تقطعها بنصفين وتمر بقطبيها .

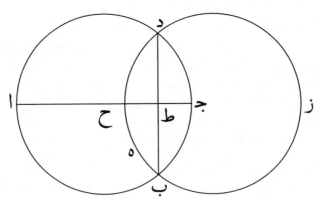

فلتقطع فى الكرة دائرة ا ب جـد العظيمة دائرة أخرى من الدوائر التى

فى الكرة وهى دائرة هـ ب ز د على زوايا قائمة ، فأقول إنها تقطعها بنصفين

سس [فلأن 9   H والدכם N, [وليقم 9   ולקام N, [وليقم 7   خط 7   כשא [כשא H, om. A   8   פלנקים N, [وليقم 7   فليقم

om. N [خط   دلك مں عرص رہ N .add دلك   11   حـطـ [حـطـ[1]   supra N   11   نبين [ايضا A .add   11   خط

فمركز [11 مركز N   العظيمة [3 التى فى الكرة   add. NH

quod etiam est spere centrum. Constituam itaque super punctum *H* superficiei circuli *GD* lineam orthogonaliter, que sit linea *HT*. Super hoc quoque punctum superficiei circuli *AB* constituam lineam *HK* orthogonaliter. Et quia a centro circuli *GD* in spera iam protracta est linea orthogonaliter super superficiem circuli, que est linea *HT*, ergo centrum spere est super lineam *HT*. Similiter quoque ostenditur quod ipsum est super lineam *HK*. Ergo centrum spere est super communem differentiam duarum linearum *HT HK*. Communis autem differentia earum est punctum *H*: punctum igitur *H* est centrum spere. Sed circuli qui transeunt per centrum spere sunt maiores: ergo duo circuli *AB* *GD* sunt maiores. Et illud est quod demonstrare voluimus.

14 Cum maior circulus in spera alium circulorum qui sunt in spera orthogonaliter secat, in duo media secat eum et transit per polos eius.

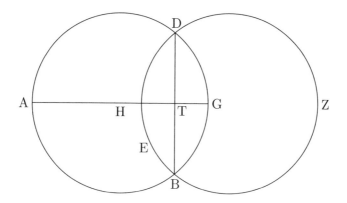

Maior igitur circulus *ABGD* qui est in spera secet alium circulorum qui sunt in spera, qui sit circulus *EBZD*, orthogonaliter. Dico

---

9 etiam] *supra* **Z**   9 etiam est] *tr.* **Kg**   9 etiam ... centrum] *permut.* **BPsVa**
9 spere centrum] *tr.* **KgFiZ**   9 itaque] igitur **KgBPsVa**   9 super] *supra* **BPs**   10 linea] *om.*
OBPsVa   11 hoc quoque] *tr.* **FiZ**   11 quoque] quodcumque **R**   11 punctum] *H add.*
Va   11 circuli] *GD add. et del.* **V**   11–12 orthogonaliter] que sit linea *HK* orthogonaliter *add.* **B**, *add. et del.* **Ps**   12 Et] Que **R**   12 centro] *supra* **BPs**   12 in] *supra* **B**
12 protracta] proposita **OBPsVa**   12 protracta est] *tr.* **MKg**   15 ipsum] *corr. ex* ipsa **B**, ipsa
OPsVa   15 est[1]] sit **OBPsVa**   16 differentiam] differentiarum **O**   16 autem] quoque **BPsVa**
17 earum] eorum **R**   17 H] *supra* **BPs**   17 punctum[2]] *om.* **OBPsVa**   17 Sed] ergo **O**
17–18 circuli qui] *tr.* **FiZ**   18 transeunt ... spere] *permut.* **MKgO,BPsVa**   18 sunt] *corr.*
ex super **Ps**   18–19 ergo ... maiores] *marg.* **R**   18–19 duo ... maiores] etc. **OB**
PsVa   19 Et illud est] *om.* **FiZ**   19 Et ... voluimus] *om.* **BPsVa**   19 illud] hoc **Kg**
19 demonstrare] *om.* **Fi**   19 voluimus] volumus **Z**   1 alium] aliorum **Fi**   1 qui] cum
Fi   1–2 orthogonaliter secat] secat orthogonaliter **OBPsVa**; eum *add.* **V**   2 secat[2]] secet **R**
2 secat eum] *om.* **V**, *tr.* **FiZ**   2 per] duos *add.* **OBPsVa**   3 igitur] *om.* **BPsVa**   3 circulus] sit
add. **OBPsVa**   3 qui est in spera] *om.* **BPsVa**   3 est] sit **MKg**   3 spera] et *add.* **OVa**, *supra*
BPs   4 qui sunt in spera] *om.* **OBPsVa**   4 qui[2] ... EBZD] *marg.* **P**   4 circulus] *om.* **OBPsVa**
4 EBZD] BZD **OBPsVa**

وتمر بقطبيها فليوصل الفصل المشترك لهما وهو خط ب‌د وليجعل مركز دائرة   5

اب‌ج‌د نقطة ح وهى أيضاً مركز الكرة وليخرج من نقطة ح إلى خط ب‌د

عمود ح‌ط ولينفذ فى كلتى الجهتين وليلق بسيط الكرة على نقطتى آ ج فلأن

كل واحد من السطحين قائم على الآخر على زوايا قائمة أعنى سطح دائرة

اب‌ج‌د وسطح دائرة ه‌ب‌ز‌د وقد أقيم على الفصل المشترك لهما وهو خط

ب‌د خط ط‌آ على زوايا قائمة وهو فى أحد السطحين أعنى سطح دائرة   10

اب‌ج‌د يكون خط آج قائماً على سطح دائرة ه‌ب‌ز‌د على زوايا قائمة ولأن

دائرة ه‌ب‌ز‌د فى كرة وقد أخرج من مركز الكرة إليها عمود ح‌ط ولقى سطح

دائرة ه‌ب‌ز‌د على نقطة ط تكون نقطة ط مركز دائرة ه‌ب‌ز‌د فكل واحدة

من قوسى ب‌ه‌د ب‌ز‌د نصف دائرة فدائرة اب‌ج‌د تقطع دائرة ه‌ب‌ز‌د

بنصفين ،   15

فأقول إنها تمر بقطبيها أيضاً وذلك أن دائرة ه‌ب‌ز‌د فى كرة وقد

أخرج من مركز الكرة إليها عمود ح‌ط وأنفذ فى كلتى الجهتين ولقى بسيط

الكرة على نقطتى آ ج وإذا كانت دائرة فى كرة ثم أخرج من مركز الكرة

إليها عمود وأنفذ فى كلتى الجهتين فهو يقع على قطبيها فنقطتا آ ج قطبا

الدائرة فدائرة اب‌ج‌د تمر بقطبى دائرة ه‌ب‌ز‌د وقد كانت قطعتها بنصفين   20

فدائرة اب‌ج‌د تقطع دائرة ه‌ب‌ز‌د بنصفين وتمر بقطبيها ، وذلك ما أردنا

أن نبين .

7 الجهتين] אלנאחיתין H   10 سطح] *marg.* A   10 حط A, خط N, טא H   10 طآ] حطا A   10-
H الجهتين] אלנאחיתין   11 على ¹... سطح] *om.* A   12 ه‌ب‌ز‌د] هدر A   13 فكل] وكل A   19 الجهتين] אלנאחיתין H
20-21 ه‌ب‌ز‌د ... وقد كانت] *om.* A

igitur quod ipse secat eum in duo media et transit per polos eius. Pro-   5
traham ergo differentiam eis communem, que sit linea *BD*, et ponam
ut centrum circuli *ABGD* sit punctum *H*, quod etiam est centrum
spere. Protraham autem a puncto *H* ad lineam *BD* perpendicularem
HT, quam in duas partes transire faciam donec concurrat superficiei
spere super duo puncta *A* et *G*. Et quia unaqueque duarum superfi-   10
cierum super alteram orthogonaliter erigitur, scilicet superficies circuli
ABGD et superficies circuli *EBZD*, et super differentiam communem
eis, que est linea *BD*, iam erecta est linea orthogonaliter que est in
una duarum superficierum, scilicet superficie circuli *ABGD*, ergo linea
AG orthogonaliter erigitur super superficiem circuli *EBZD*. Et quia   15
circulus *EBZD* est in spera et iam protracta est a centro spere ad ipsum perpendicularis *HT* que concurrit superficiei circuli *EBZD* super
punctum *T*, ergo punctum *T* est centrum circuli *EBZD* et unusquisque
arcuum *BED BZD* est arcus semicirculi: circulus igitur *ABGD* secat
circulum *EBZD* in duo media.   20

Dico igitur quod ipse etiam transit per polos eius. Quod ideo est
quoniam ad circulum *EBZD* a centro spere in qua ipse est iam protracta est perpendicularis *HT* et pertransit in duas partes et concurrit
superficiei spere super duo puncta *A* et *G*. Sed cum in spera fuerit
circulus ad quem postea a centro spere protrahitur perpendicularis et   25
pertransit in duas partes, tunc ipsa cadit super duos polos eius. Ergo
duo puncta *A* et *G* sunt duo poli circuli; ergo circulus *ABGD* transit
per duos polos circuli *EBZD*. Sed iam fuit secans eum in duo media:
ergo circulus *ABGD* secat circulum *EBZD* in duo media et transit per
duos polos eius. Et illud est quod demonstrare voluimus.   30

5 igitur] *om.* **OBPsVa**   6 ergo] *supra* **Ps**   6 differentiam eis] *tr.* **V**   6 eis] ei' **Va**
8 autem] ergo **P**, igitur **OBPsVa**   9 quam] quare **Fi**, quas **BPsVa**   9 in] *supra* **B**, *om.* **PsVa**
9 concurrat] *corr. ex* communicant **B**, concurrant **PsVa**   9 concurrat superficiei] concurrant superficies **Kg**   10 *A*] *B, supra A* **Ps**   10 unaqueque] utraque **MKgOBPsVa**   11 alteram] aliam
MKg, *corr. ex* aliquam **O**, *illeg.* **B**   11 orthogonaliter erigitur] *tr.* **OVa**, **B** (orthogonaliter *supra*), **Ps** (orthogonaliter *marg.*)   11 erigitur] eligitur **Fi**   11 scilicet] id est **P**;
set *scr. et del.*, .s. *supra* **Ps**, set **Va**   12 *EBZD*] *EB ZB, supra D* **Ps**   13 est[1]] *om.*
R   13 linea[1]] *om.* **BPsVa**   14 scilicet] in *add.* **FiZVa**, *supra* **Ps**   15 erigitur] *om.* **Va**
16 circulus] *marg.* **R**   16 *EBZD*] *EBZB* **B**   17 circuli] *om.* **BPsVa**   18 *T*] *om.* **Z**
18 unusquisque] uterque **MKgOBPsVa**   19 *BED*] *infra littera illeg.* **V**   19 igitur] ergo
BPsVa   20 in] *supra* **V**   21 igitur] ergo **OBPsVa**; quoniam *add.* **Fi**   21 etiam transit] *tr.*
OBPsVa   22 quoniam] quia **BPsVa**   22 spere] *om.* **Va**   22 ipse] specie **Va**   22–
23 protracta] producta **Va**   23 est] *om.* **Va**   23 pertransit] pertransiit **P**, transit **BPsVa**   23 concurrit] occurrit **RZ**   24 et] *om.* **FiZ**   25 ad quem] *in corr.* **R**
25 postea] postera **Va**   25 postea a centro] *permut.* **R**   25 protrahitur] protrahatur **O**
26 pertransit] protrahitur **OBPsVa**   26 ipsa] *om.* **BPsVa**   26 cadit] cadet **MKgOBPsVa**
27 et] *om.* **BPsVa**   27 circuli] *EBZD add.* **FiZ**   27–28 ergo circulus ... media] *om.*
O   27–29 transit ... *ABGD*] *om.* **BPsVa**   28 duos] *supra* **M**   29 *EBZD*] *corr. ex*
ABZD **B**   29–30 transit ... eius] *permut.* **POBPsVa**   30 Et illud ... voluimus] *om.* **OBPsVa**
30 demonstrare] *om.* **Fi**   30 voluimus] volumus **Z**

يه إذا كانت فى كرة دائرة عظيمة وقطعت دائرة ما غير عظيمة من الدوائر التى فى الكرة بنصفين فإنها تقطعها على زوايا قائمة وتمر بقطبيها ·

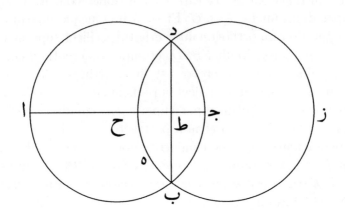

فلتكن الدائرة العظيمة التى فى الكرة دائرة اٮجد ولتقطع دائرة ما من الدوائر التى فى الكرة غير عظيمة وهى دائرة هٮزد بنصفين ، فأقول إنها تقطعها على زوايا قائمة وتمر بقطبيها فليوصل الفصل المشترك لهما وهو خط ٮد فلأن دائرة اٮجد تقطع دائرة هٮزد بنصفين تكون كل واحدة من قوسى ٮهد ٮزد نصف دائرة فليقطع خط ٮد بنصفين على نقطة ط فنقطة ط مركز دائرة هٮزد وليكن مركز دائرة اٮجد نقطة ح وهى مركز الكرة أيضاً وليوصل خط حط ولينفذ فى كلتى الجهتين وليلق بسيط الكرة على نقطتى اٮ جٮ فلأن دائرة هٮزد فى كرة وقد وصل بين مركزها ومركز الكرة بخط حط يكون خط حط عموداً على دائرة هٮزد وجميع السطوح التى تخرج وتمر بخط حط قائمة على دائرة هٮزد على زوايا قائمة وأحد

15   Cum in spera fuerit circulus maior secans aliquem circulorum qui sunt in spera non maiorem in duo media, tunc ipse secat eum orthogonaliter et transit per duos polos eius.

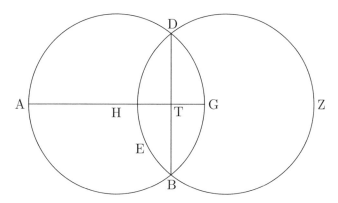

Sit ergo circulus maior qui est in spera circulus *ABGD* qui aliquem circulorum qui sunt in spera non maiorem in duo media secet   5
sitque circulus *EBZD*. Dico igitur quod ipse secat eum orthogonaliter et transit per duos polos ipsius. Producam igitur differentiam eis communem, que sit linea *BD*. Et quia circulus *ABGD* secat circulum *EBZD* in duo media, ergo unusquisque duorum arcuum *BED BZD* est medietas circuli *BZDE*. Dividam itaque lineam *BD* in duo media   10
super punctum *T*. Punctum igitur *T* est centrum circuli *EBZD*. Sit autem punctum *H* centrum circuli *ABGD* et ipsum etiam est centrum spere. Producam itaque lineam *HT*, que in duas pertranseat partes et concurrat superficiei spere super duo puncta *A* et *G*. Et quia circulus *EBZD* est in spera et iam coniunctum est quod est inter centrum eius et   15
centrum spere per lineam *HT*, ergo linea *HT* est perpendicularis super circulum *EBZD*. Et etiam omnes superficies que protrahuntur et transeunt per lineam *HT* orthogonaliter eriguntur super circulum *EBZD*

1 circulus maior] *tr.* **Fi**   2 maiorem] maiorum **OFiZ**, maiores **BPsVa**   2 ipse] *corr. ex* ipsa **BPs**   3 per] super **MKgOBPsVa**   3 duos] *om.* **FiZ**   3 polos eius] polos ipsius **MKgOB**, ipsius polos **PVFiZ**   4 ergo] igitur **OBPsVa**   4 qui … circulus] *om.* **OBPsVa**   4 qui²] secet *add.* **OBPsVa**   5 qui sunt] *om.* **OBPsVa**   5 media secet] *tr.* **PRVFiZ**   5 secet] *om.* **OBPsVa**   6 sitque circulus] et sit **OBPsVa**   6 *EBZD*] *BZD* **BPsVa**   6 igitur] *om.* **BPsVa**   6 ipse] spere **Fi**   7 et transit] intransit **Fi**   7 duos polos ipsius] *permut.* **MKg**, duos polos eius **OBPsVa**, polos eius **FiZ**   7 igitur] ergo **OBPsVa**   7 differentiam eis] *tr.* **VMKg**   7 eis] *corr. ex* eius **Ps**   8 linea] *om.* **OBPsVa**   9–10 in duo … Dividam] *om.* **FiZ**   9 ergo] *om.* **OBPsVa**   9 unusquisque] quisque **PV**, uterque **MKgOBPsVa**   9 *BED*] *EDB* **O**, *corr. ex* EBD **BPs**, *EBD* **Va**   10 *BZDE*] *EBZD* **MKgOBPsVa**   10 itaque] circulum **Kg**   10 media] dividam *add.* **FiZ**   11 Punctum] *om.* **OBPsVa**   11 igitur] ergo **OBPsVa**   11 *T*] *C* **Va**   12 punctum] *om.* **OB PsVa**   13 pertranseat partes] *tr.* **M**; partes transeat **KgOBPsVa**, *tr.* **Z**   14 concurrat] occurrat **OBPsVa**   14 circulus] circulis **Fi**   15 et¹] *supra* **Ps**   15 eius] spere **OBPsVa**   16 spere] eius **OBPsVa**   16 ergo … *HT*] *marg.* **O**   17 etiam] *om.* **V**   17 protrahuntur] protrahantur **Fi**   18 per] *om.* **FiZ**

السطوح التى تمر بخط ح‌ط هو دائرة ا‌ب‌ج‌د فدائرة ا‌ب‌ج‌د قائمة على دائرة ه‌ب‌ز‌د على زوايا قائمة فدائرة ا‌ب‌ج‌د تقطع دائرة ه‌ب‌ز‌د على زوايا قائمة ،

15

وأقول إنها تمر بقطبيها وذلك أنه لما كانت دائرة ه‌ب‌ز‌د فى كرة وقد أخرج من مركز الكرة إليها عمود ح‌ط وأنفذ إلى كلتى الجهتين ولقى بسيط الكرة على نقطتى ا‌ج تكون نقطتا ا‌ج قطبى دائرة ه‌ب‌ز‌د فدائرة ا‌ب‌ج‌د تمر بقطبى دائرة ه‌ب‌ز‌د وقد كانت قطعتها على زوايا قائمة فدائرة ا‌ب‌ج‌د تقطع دائرة ه‌ب‌ز‌د على زوايا قائمة وتمر بقطبيها ، وذلك ما أردنا أن 20 نبين .

يو إذا قطعت دائرة عظيمة فى كرة دائرة ما من الدوائر التى فى الكرة ومرت بقطبيها فهى تقطعها بنصفين وعلى زوايا قائمة .

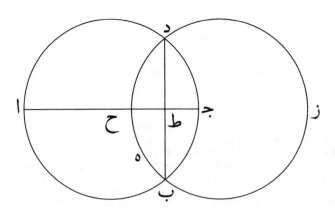

13 هو [هى A 14–13 قائمة ... ا‌ب‌ج‌د فدائرة [om. A 15–14 قائمة ... تقطع ا‌ب‌ج‌د فدائرة [om. H

16 وأقول [فاقول A 19 ه‌ب‌ز‌د [مداره ا‌حد add. et del. N

Una autem superficierum que per lineam *HT* pertranseunt est circulus *ABGD*: ergo circulus *ABGD* orthogonaliter erigitur super circulum *EBZD*. Ergo circulus *ABGD* secat circulum *EBZD* orthogonaliter.   20

Et dico etiam quod ipse transit per duos polos eius. Quod ideo est quoniam idcirco quod circulus *EBZD* est in spera, ad quem a centro spere perpendicularis *HT* iam protracta est, pertransiens in duas partes et concurrens superficiei spere super duo puncta *A G*, sunt duo   25 puncta *A G* duo poli circuli *EBZD*. Ergo circulus *ABGD* transit per duos polos circuli *EBZD*. Sed iam fuit secans ipsum orthogonaliter: ergo circulus *ABGD* orthogonaliter secat circulum *EBZD* et transit per duos polos ipsius. Et illud est quod demonstrare voluimus.

16 Cum circulus maior in spera secat aliquem circulorum qui sunt in spera et transit per duos polos eius, tunc ipse secat eum in duo media et orthogonaliter.

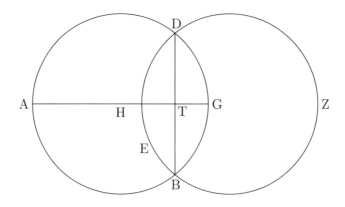

19–21 Una ... *EBZD*] *om.* **BPsVa**   20 ergo circulus *ABGD*] *supra* **V**,   *om.*   **Fi**
21 circulum] *om.* **BPsVa**   21 *EBZD*] *ABZD* **BPsVa**   22 Et] *om.* **MKgOBPsVa**
22 transit] pertransit   **FiZ**   22–23 Quod ... quod] ideo   quoniam   **OBPsVa**
23 quoniam] quia **Va**   23 idcirco] indirecto **Fi**   23 quem] quoniam **Fi**   23 a] in
FiZ   24 *HT* iam protracta est] protracta est scilicet *HT* linea **OBPsVa**   24 est] *om.*
Fi   24 pertransiens] pertransiit **Kg**   25 concurrens] occurrens **OBPsVa**   25 *A*] et *add.*
KgFiZ   25 sunt] *om.* **OFiZBPs**, ergo **Va**   25–26 duo puncta *A G*] *om.* **BPs**, duo puncta
marg. **B**, duo puncta *A* et *G* sunt *marg.* **Ps**   26 *A*] et *add.* **FiZVa**   26 *G*] sunt *add.*
OFiZVa   26 transit] *om.* **MKgOBPsVa**   27 polos circuli] circulos **Kg**   27 circuli] *om.*
OBPsVa   27 *EBZD*] *EBZ* **BPsVa**; transit *add.* **MKgOBPsVa**   27 secans ipsum] *tr.* **KgBPsVa**
28 orthogonaliter] *post EBZD* **FiZ**   29 ipsius] eius **OFiZBPsVa**   29 Et illud est] *om.* **Fi**
29 Et ... voluimus] *om.* **OBPsVa**   29 illud] hoc **MKg**   29 quod demonstrare voluimus] *om.*
Kg   29 voluimus] volumus **Z**   1 secat] secet **OPsVa**   1 circulorum] eorum **OBPsVa**
2 transit] transeat **BPsVa**   2 eius] *supra* **M**   2 secat] *corr. ex* secant **V**

ولتقطع دائرة ا ب جد العظيمة التى فى الكرة دائرة ما من الدوائر التى

فى الكرة وهى دائرة ه ب ز د وتمر بقطبيها ، فأقول إنها تقطعها بنصفين

وعلى زوايا قائمة فليكن قطبا دائرة ه ب ز د نقطتى آ جَ ومن البيّن أن نقطتى 5

آ جَ هما على دائرة ا ب جد وذلك أن دائرة ا ب جد تمر بقطبى دائرة ه ب ز د

وليوصل خط آجَ فدائرة ه ب ز د فى كرة وقد أخرج فى الكرة خط مستقيم

يمر بقطبيها وهو خط آجَ وإذا كانت دائرة فى كرة فإن الخط المستقيم الذى

يمر بقطبيها عمود على الدائرة وهو يمر بمركزها وبمركز الكرة فخط آجَ عمود

على دائرة ه ب ز د فجميع السطوح التى تمر بخط آجَ قائمة على دائرة ه ب ز د 10

على زوايا قائمة وأحد السطوح التى تمر بخط آجَ دائرة ا ب جد فدائرة

ا ب جد تقطع دائرة ه ب ز د على زوايا قائمة فهى تقطعها بنصفين فدائرة

ا ب جد تقطع دائرة ه ب ز د بنصفين وقد كانت قطعتها أيضاً على زوايا قائمة

فدائرة ا ب جد تقطع دائرة ه ب ز د بنصفين وعلى زوايا قائمّة ، وذلك ما

أردنا أن نبين · 15

يز إذا كانت فى كرة دائرة عظيمة فإن الخط الذى يخرج من قطبها إلى الخط

المحيط بها مساوٍ لضلع المربع الذى يرسم فى الدائرة العظيمة ·

فلتكن الدائرة العظيمة التى فى الكرة دائرة ا ب جد ، فأقول إن الخط

المستقيم الذى يخرج من قطبها إلى الخط المحيط بها مساوٍ لضلع المربع الذى

الكرة] N 7 دائرة³ [6 repet. N الكرة] NH: وذلك أن ... هـى سعطـى A 6 تمر بقطبى 6
add. A; فقد add. et del. A مدايره ه ب رد فى كره] .marg A; وإذا كانت ... كرة 8 –12
add. A المستقيم [الخط¹] om. N 1 الخط] على زوايا ... ه ب ز د 13

Circulus itaque *ABGD* maior qui est in spera secet aliquem cir- culorum qui sunt in spera, qui sit circulus *EBZD*, et transeat per duos 5 polos ipsius. Dico igitur quod ipse secat eum in duo media et orthogo- naliter. Sint itaque duo poli circuli *EBZD* duo puncta *A* et *G*. Mani- festum est autem quod duo puncta *A* et *G* sunt supra circulum *ABGD*. Quod ideo est quoniam circulus *ABGD* transit per duos polos circu- li *EBZD*. Protraham vero lineam *AG*. Iam igitur protracta est linea 10 recta transiens per duos polos circuli *EBZD* qui est in spera, que est linea *AG*. Sed cum in spera fuerit circulus, tunc linea recta que transit per duos polos eius est perpendicularis super circulum et ipsa transit per centrum ipsius et centrum spere. Linea igitur *AG* perpendicularis est super circulum *EBZD*: ergo omnes superficies que transeunt per 15 lineam *AG* eriguntur super circulum *EBZD* orthogonaliter. Sed una superficierum que transeunt per lineam *AG* est circulus *ABGD*: ergo circulus *ABGD* secat circulum *EBZD* orthogonaliter; ergo ipse seca eum in duo media. Ergo circulus *ABGD* secat circulum *EBZD* in duo media. Sed iam fuit secans eum etiam orthogonaliter. Ergo circulus 20 *ABGD* secat circulum *EBZD* in duo media et orthogonaliter. Et illud est quod demonstrare voluimus.

17 Cum in spera fuerit maior circulus, linea recta que a polo eius ad ipsius protrahitur circumferentiam erit equalis lateri quadrati quod in maiori circulo signatur.

Sit igitur circulus maior qui est in spera circulus *ABGD*. Dico ergo quod linea recta que ab eius polo ad lineam ipsius circumductam 5 producitur est equalis lateri quadrati quod in maiori circulo signatur.

4 Circulus] Sit **OBPsVa**   4 itaque] igitur **MKgOBPsVa**   4 *ABGD*] *ABG* **FiZ**; circulus *add.* **OBPsVa**   4 maior ... spera] *permut.* **MKg**   4 qui est] *om.* **OBPsVa**   4 spera] et *add.* **OB PsVa**   4 secet aliquem] *tr.* **BPsVa**   4–5 circulorum] eorum **OBPsVa**   5 qui²] et **OBPsVa** 5 qui² ... *EBZD*] *om.* **Fi**   5 circulus] sectus **OBPsVa**   5 duos] dolos **Va**   6 igitur] ergo **M**, *om.* **BPsVa**   6 ipse] *om.* **BPsVa**   7 duo] *om.* **Va**   7 et] *om.* **BPsVa**   8 est] *om.* **PVOFiZBPsVa** 8 autem] *om.* **BPsVa**   8 et] *om.* **BPsVa**   8 *ABGD*] *ADGB* **B**   9 Quod ... quoniam] ideo quia **BPsVa**   9 quoniam] quia **O**   10 Protraham] autem *supra* **BPs**   10 vero] autem **OVa**, *om.* **BPs**   11 recta] *om.* **Kg**   11 recta transiens] *tr.* **R**   11 qui] que **OPsVa** 12 linea recta] *tr.* **OBPsVa**   14 centrum ipsius] *tr.* **P**   14 ipsius] circuli **R**, eius **OBPsVa** 14 et] per *add.* **KgOFiBPsVa**   14–15 perpendicularis est] *tr.* **OBPsVa**   15 ergo] *repet.* **P** 15 omnes] omnis **Fi**   16 eriguntur] erigentur **Fi**   16–17 eriguntur ... lineam *AG*] *marg.* **BPs**   16 orthogonaliter] *post* eriguntur **OFiZBPsVa**   17 que transeunt] transeuntium **OB PsVa**   18 ipse] *om.* **OBPsVa**   19–20 Ergo ... media] *repet.* **P**, *om.* **BPsVa**   19 *ABGD*] *om.* **V**   20 iam] *om.* **BPsVa**   20 etiam] *om.* **OBPsVa**   20–21 Ergo ... orthogonaliter] *om.* **OB PsVa**   21–22 Et illud est] *om.* **FiZ**   21–22 Et ... voluimus] Et ita patet probandum **OB PsVa**   22 est ... voluimus] *om.* **Kg**   22 voluimus] volumus **Z**   2 lateri quadrati] *tr.* **V** 3 signatur] figuratur **KgOBPsVa**, *om.* **Fi**   4 Sit] Signatur sit **Fi**   4 igitur] ergo **FiZ**, *om.* **OB PsVa**   4 qui ... circulus] *om.* **OBPsVa**   4–6 Dico ... signatur] *om.* **BPsVa**   5 ergo] igitur **MKgO**   5 eius] ipsius **M**   6 producitur] producitum **Fi**   6 signatur] figuratur **Kg**, signatum est **OFiZ**

يرسم فى الدائرة العظيمة فليُخرج قطران لدائرة اب‌جد يتقاطعان على زوايا ٥

قائمة وهما خطا اج‌ب‌د فلأن دائرة اب‌جد عظيمة يكون مركزها ومركز

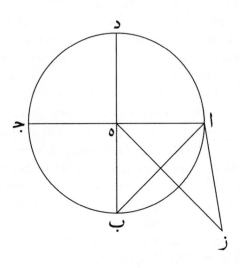

الكرة واحداً بعينه وليكن نقطة ه فنقطة ه مركز الكرة ومركز دائرة اب‌جد

وليقم من نقطة ه من سطح دائرة اب‌جد عمود على الدائرة وهو خط هز

وليلق بسيط الكرة على نقطة ز فنقطة ز قطب دائرة اب‌جد وليوصل خطا

زا اب‌ فخط اب هو ضلع المربع الذى يرسم فى دائرة اب‌جد وخط زا ١٠

يخرج من القطب إلى الخط المحيط بالدائرة ، فأقول إن خط زا مساوٍ لخط

اب وذلك أن خط زه عمود على دائرة اب‌جد فهو يحدث مع جميع الخطوط

المستقيمة التى تخرج من طرفه فى سطح دائرة اب‌جد زوايا قائمة فيكون

خط زه عموداً على كل واحد من خطوط اه هب هج هد ولأن نقطة ه مركز

الكرة يكون خط هب مساوياً لخط هز وخط ها مشترك فخطا هب ها ١٥

Protraham itaque duas diametros circuli $ABGD$ sese orthogonaliter
secantes, que sint due linee $AG\ BD$. Et quia circulus $ABGD$ est maior,
ergo centrum eius et centrum spere est unum et idem sitque illud punc-

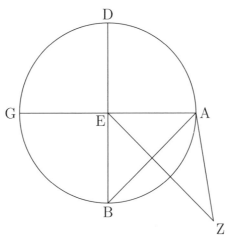

tum E: punctum igitur E est centrum spere et centrum circuli $ABGD$.   10
Super punctum igitur E superficiei circuli $ABGD$ constituam perpendicularem super circulum, que sit linea EZ, et concurrat superficiei spere
super punctum Z. Ergo punctum Z est polus circuli $ABGD$. Producam
autem duas lineas $ZA\ BA$. Linea igitur AB est latus quadrati quod in
circulo $ABGD$ signatur. Sed linea ZA producta est a polo ad lineam   15
circuli circumductam. Dico igitur quod linea ZA est equalis linee AB.
Quod ideo est quoniam linea ZE est perpendicularis super circulum
$ABGD$. Ergo ex ipsa cum omnibus rectis lineis que ab eius extremitate in superficie circuli $ABGD$ protrahuntur proveniunt anguli recti.
Linea igitur ZE perpendicularis existit super unamquamque linearum   20
$AE\ EB\ EG\ ED$. Et quia punctum E est centrum spere, ergo linea
EB est equalis linee EZ. Linea itaque EA communi due linee $EB\ EA$

7 itaque] igitur **Kg**, ergo **OBPsVa**; duas lineas scilicet *add.* **OBPsVa**   7 duas] duos **RZBPsVa**
7 sese] sole **Fi**, se **BPsVa**   7 orthogonaliter] *supra* **B**, *om.* **PsVa**   8 sint] sunt **ZBPsVa**
8 $AG\ BD$] *in corr.* **Ps**, $ABGD$ **Va**   8 circulus] circulis **Fi**, *om.* **BPsVa**   8 maior] circulus
add. **Va**, *add. supra* **Ps**   9 idem] punctum *add.* **OBPsVa**   9 sitque] sit **OBPsVa**   9 illud] *om.*
FiZ   10 punctum igitur . . . $ABGD$] *om.* **FiZ**   10 igitur] *supra* **R**   12 EZ] E, Z *supra*
Ps   12 concurrat] concurrit **PV**, occurrat **OBPsVa**, occurrit **R**   13 Ergo . . . est] cum
Va   13 Ergo punctum Z] *om.* **FiZ**   13–14 Producam autem] Protrahamque **OBPsVa**
14 autem] *om.* **Fi**   14 $ZA\ BA$] $BA\ ZA$ **OBPsVa**   14–15 Linea . . . signatur] ergo latus quadrati
quod in circulo inscribitur est AB linea **OBPsVa**   15 signatur] designatur **M**   15 Sed linea] *om.*
Va   15 producta est] *tr.* **OFiZBPsVa**   15–16 lineam circuli circumductam] circumferentiam
circuli **BPsVa**   16 circumductam] *om.* **O**   16 igitur] ergo **KgOBPsVa**   16 ZA] producta
add. **R**   16 est equalis] equatur **OBPsVa**   17 Quod ideo est quoniam] ideo quia
OBPs, quia **Va**   17 ZE] ZA, *corr. supra* E **Ps**   18–19 Ergo ex . . . $ABGD$] *om.* **Z**
18 ipsa] ipso **PsVa**   19 proveniunt anguli recti] *permut.* **PRV**,**MKgOFiZ**   20 Linea] a
Fi   20 perpendicularis existit] perpendicularis est **MKg**, est perpendicularis **OFiZBPsVa**
20 unamquamque] unaquaque **V**   20 linearum] lineam **OBPsVa**   21 ergo] ideo **OBPsVa**
22 linee[1]] linea **O**   22 communi] km **Va**

مساويان لخطى ه‍آ ه‍ز كل واحد منهما لنظيره وزاوية ب‍ه‍آ القائمة مساوية

لزاوية آه‍ز القائمة فقاعدة ب‍آ مساوية لقاعدة آز وخط زآ هو الذى يخرج

من قطب دائرة آب‍ج‍د إلى الخط المحيط بها وخط آب هو ضلع المربع

الذى يرسم فى دائرة آب‍ج‍د العظمى فالخط الذى يخرج من قطب دائرة

20 آب‍ج‍د إلى الخط المحيط بها مساوٍ لضلع المربع الذى يرسم فى الدائرة

العظيمة ، وذلك ما أردنا أن نبين ·

يح إذا كانت دائرة فى كرة وكان الخط الذى يخرج من قطبها إلى الخط المحيط

بها مساوياً لضلع المربع الذى يرسم فى الدائرة العظيمة فالدائرة أيضاً عظيمة ·

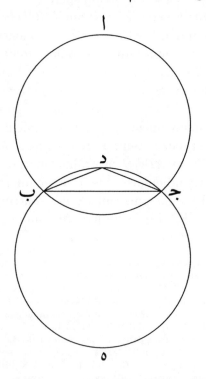

[العظيمة 2 add. A   المستقيم [الخط 1 ¹الخط] om. A   إلى الخط ... آب‍ج‍د 18–19   A وآ [وخط زآ 17
add. A الى تكون فى الكره

sunt equales duabus lineis *EA EZ*, queque scilicet earum sue relative.
Sed angulus *BEA* rectus est equalis angulo recto *AEZ*: ergo basis *BA*
est equalis basi *AZ*. Sed linea *ZA* est ea que protrahitur a polo circuli   25
ABGD ad lineam continentem ipsum et linea *AB* est latus quadrati
quod signatur in circulo maiori *ABGD*: ergo linea que protrahitur a
polo circuli *ABGD* ad lineam continentem ipsum est equalis lateri quadrati quod in maiori signatur circulo. Et illud est quod demonstrare
voluimus.   30

18   Cum in spera fuerit circulus et fuerit linea, que a polo eius ad ipsius
circumferenciam producitur, equalis lateri quadrati quod in maiori circulo signatur, circulus etiam erit maior.

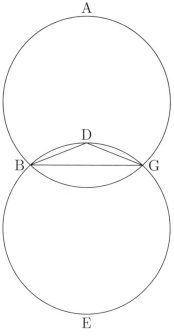

23 queque … relative] *om.* **MKgOBPsVa**   *24* Sed] et *add.* **PR**   *24 BEA*] *BA, E add.supra*
Ps   *24* angulo] alii **OB PsVa**   *24 BA*] ea *add.* **Fi**, etiam *add.* **B**   *25* Sed] linea *EZ* est perpendicularis et *add.* **MKgO FiZBPsVa** (linea *om.* **BPsVa**)   *25* est²] in *add. supra* **Ps**   *25* ea] illa **FiZ**
25 polo] poli **Va**   *26* ipsum] ipsam **V**   *26 AB*] *supra* **BPs**   *27* signatur] figuratur **Kg**, significatur **BPsVa**   *27* ergo] igitur **Fi**   *27–29* ergo … circulo] *om.* **BPsVa**   *27* protrahitur] protrahit
Fi   *28 ABGD*] *marg.* **R**   *29* maiori] minori **R**   *29* signatur] figuratur **Kg**, significatur **BPsVa**
29 Et illud est] *om.* **FiZ**   *29* Et illud … voluimus] et il e q. **M**, et ita patet propositum **OB**
PsVa   *29* illud] id **Kg**   *1* eius] *supra* **BPs**   *1* ad] ab **R**   *2* circumferenciam] circumferentia **R**
2 circumferenciam producitur] *tr.* **PRVMKgFiZ**   *2* producitur] *om.* **Z**   *3* signatur] figuratur
Kg, signatum est **FiZ**, *illeg.* **B**, significatur **PsVa**   *3* erit maior] *tr.* **OBPsVa**

فلتكن فى كرة دائرة اٮج وليكن قطبها علامة د وليكن خط دج

الذى يخرج من قطبها إلى الخط المحيط بها مساوياً لضلع المربع الذى يرسم

في الدائرة العظيمة ، فأقول إن دائرة اٮج أيضاً عظيمة فليخرج سطح يمر

بخط دج وبمركز الكرة يحدث قطعاً يكوّن فى بسيط الكرة دائرة عظيمة

وهى دائرة ٮدجه وليكن الفصل المشترك لها ولدائرة اٮج خط ٮج

وليوصل خط دٮ فخط دٮ مساوٍ لخط دج ولأن خط دٮ مساوٍ لخط دج

وخط جد هو ضلع المربع يكون خط ٮد ضلع المربع أيضاً فكل واحدة من

قوسى ٮد دج ربع دائرة فقوس ٮدج نصف دائرة فخط ٮج قطر دائرة

ده ج فلأن دائرة ده ج العظيمة فى كرة وقد قطعت دائرة ما من الدوائر التى

فى الكرة وهى دائرة اٮج ومرت بقطبيها فهى أيضاً تقطعها بنصفين فدائرتا

اٮج ده ج تقطع كل واحدة منهما الأخرى بنصفين والدوائر التى يقطع

بعضها بعضاً فى الكرة بنصفين نصفين هى عظيمة فدائرة اٮج عظيمة

أيضاً ، وذلك ما أردنا أن نبين ·

3 وليكن قطبها علامة د] *om. AN*   5 العظيمة] *om.* الى تقع فى الكرة   6 يحدث] ومحدث *add. A*   7 لها] عليها *A, marg. corr.*   7 ولدائرة] والدار *A*   8 خط دٮ] خطا دٮ *NH*   8 دج ... ولأن] *repet. A*   8 دج²] ضلع المربع الذى هو *add. N*   9 ند] *N*   9 جد] ند *N*   9 ٮد] ٮج *N*   9 أيضاً] *om. N*   9 واحدة] واحد *A*   10 نصف] قوس *A*   10 فخط] وحط *N*   13 واحدة] واحد *A*   13 والدوائر] الدائره *A,* الدار *corr. ex* *N*   15 أيضاً] *N*   15] *om. A*   15 أن نبين] *om. H*

8-12 ולאן חט דב ... בנצפין] ולאן חט דב וכט אלדי הו צלע אלמרבע אלדי ירסם פי דאירה אבג וכט
בד הו צלע אלמרבע אלמרסום פי דאירה אבג ולאן דאירה הבדג תמאר בקטב דאירה אבג
פהי תקטעהא בנצפין ועלי זאויה קאימה פאקום באג נצף דאירה פכט בג קטר דאירה באג
ולאן כל ואחד מן כטי בד דג צלע אלמרבע אלמרסום פיהא יכון מרבע בג מסאו למרבע בד
H דג פואויה בג קאימה פקום בדג נצף דאירה

Sit ergo in spera circulus *ABG* et sit linea *DG*, que a polo eius ad ipsius circumferenciam protrahitur, equalis lateri quadrati quod signatur in circulo maiori. Dico igitur quod circulus *ABG* etiam est maior. Producam ergo superficiem transeuntem per lineam *DG* et per centrum spere et proveniat sector existens in superficie spere circulus maior, qui sit circulus *BDGE*, sitque differentia communis ei et circulo *ABG* linea *BG*. Protraham autem lineas *DB DG*. Linea ergo *DB* est equalis linee *DG*, et quia linea *DB* est equalis linee *DG*, que est latus quadrati, et linea *BD* est latus quadrati, ergo linea *DG* est latus quadrati. Unusquisque igitur arcuum *BD DG* est quarta circuli. Ergo arcus *BDG* est semicirculus et linea *BG* est diametrus circuli *DEBG*. Et quia circulus *EBG* maior est in spera et iam secuit aliquem circulorum qui sunt in spera, qui est circulus *ABG* et transit per polos eius, ergo ipse etiam secat eum in duo media. Ergo duorum circulorum *ABG EBG* quisque alterum in duo media secat. Circuli autem qui in spera se ad invicem in duo media secant sunt maiores ergo circulus *ABG* etiam existit maior. Et illud est quod demonstrare voluimus.

5

10

15

20

4 ergo] igitur **Z**   4 *DG*] *DE* **Kg**, et *add.* **M**   4 eius] ipsius **OBPsVa**   5 ipsius] eiusdem **OBPs**, eandem **Va**   5 circumferenciam protrahitur] *tr.* **PRVMKgFiZ**   6 signatur] significatur **Ps**, figuratur **Kg** (*post* in circulo)   6 signatur in circulo] *permut.* **M**   6 in circulo] *supra* **Ps**   6 igitur] ergo **R**   6 circulus] etiam *add.* **OBPsVa**   6 etiam est] *tr.* **OZBPsVa**, est m̄ **Fi**   7 Producam] producta **Fi**   7 ergo] vero **Fi**, igitur **PsVa**   8 et] *om.* **Fi**   8 proveniat] proveniet **R**   8 existens] *corr. marg. ex* equalis **Z**   9 qui] quid **Fi**   9 *BDGE*] *BGDE* **KgOB**, *BDGE in corr.* **Ps**, *BDEG* **Va**   9 sitque] sit quoque **Va**   9 communis ei] *tr.* **Va**   10 linea] lineam **Va**   10 *BG*] *supra* **Ps**, *BH* **Fi**   10 lineas *DB DG*] lineam *DB* **Fi**, **B** (*in corr.*), lineam *BD* **Z**   11 et] *om.* **R**   11 est²] esse **Fi**   12 et] ergo quia **Kg**   12 et linea ... quadrati] *marg.* **M**, *om.* **BPsVa**   12–13 ergo linea ... quadrati] *om.* **KgOFi**; est quadratum *add.* **Z**   12 *DG*] *DB* **BPsVa**   13 *BD DG*] *DB GD* **KgOBPsVa**   13 quarta] *in corr.* **Ps**; pars *add.* **B**   14 *DEBG*] *DBEG* **OFiZBPsVa**   15 maior est] *tr.* **KgOBPsVa**   15 est] *om.* **V**   15 iam] *om.* **BPsVa**   15 secuit] secuint **Fi**, secat **BPsVa**   16 qui] que **RVa**   16 qui est] *supra* **Ps**   16 transit] transiit **VMFi**   17 ipse etiam secat] secat ipse **R**   17 eum] *supra* **BPs**   17–18 *ABG EBG*] *ABGE BGD* **Fi**, *om.* **BPsVa**   18 quisque] uterque **NKgOBPsVa**   18 in duo media secat] *permut.* **OBPsVa**   18 autem qui] que **Fi**   18 in spera] *om.* **BPsVa**   18–19 in spera ... invicem] *permut.* **MKg**   19 ad] *om.* **OPsVa**   19 ergo] igitur **Fi**   19 circulus] circulis **Fi**   19 *ABG*] *om.* **Va**   20 etiam] et **Fi**   20 existit] est **OBPsVa**   20 existit maior] *tr.* **MKg**   20 Et illud ... voluimus] etiam id est quod **Kg**   20 est] *om.* **V**   20 demonstrare] *om.* **FiZ**   20 demonstrare voluimus] volumus probare **OBVa**, voluimus probare **Ps**   20 voluimus] volumus **Z**

يط   كيف نجد خطاً مساوياً لقطر دائرة معلومة فى كرة ·

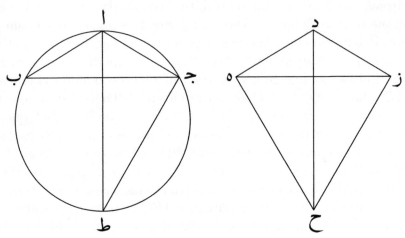

فلتكن الدائرة المعلومة فى كرة دائرة ا ب ج ونريد أن نخط خطاً مساوياً

لقطرها فنتعلم على قوس ا ب ج ثلث نقط كيف ما اتفق وهى نقط آ بّ جّ

ولنعمل من ثلثة خطوط مستقيمة مثلث د ه ز حتى يكون خط د ه مساوياً

للخط الذى يصل بين نقطة آ ونقطة بّ ويكون خط د ز مساوياً للخط

الذى يصل بين نقطة آ وبين نقطة جّ وأيضاً يكون خط ه ز مساوياً للخط

الذى يصل بين نقطة بّ وبين نقطة جّ ولنتوهم أن خطوط ا ج جّ بّ با

موصولة ولنخرج من نقطى ه ز على خطى د ه د ز خطين على زوايا قائمة

وهما خطا ه ح ز ح ولنوصل خط د ح ، فأقول إن خط د ح مساوٍ لقطر

دائرة ا ب ج فلنتوهم قطر الدائرة اط ولنوصل خطى اط جط فلأن خطى

ا ب ب ج مساويان لخطى د ه ه ز كل واحد منهما لنظيره وقاعدة ا ج مساوية

لقاعدة د ز تكون زاوية ا ب ج مساوية لزاوية د ه ز ولكن زواية ا ب ج

3 ما] om. N   3 وهى] A   5-6 نقطة آ … بّ نقطة بين] A. marg   6 نقطة جّ وبين] حّ ونقطه
A   6 وأيضاً يكون] وبكون A   7 وسطه] A   9 د ح¹] د حّ N   10 خطى¹] حط A
10 اط²] د حّ A

19   Qualiter linea equalis diametro circuli dati in spera describatur.

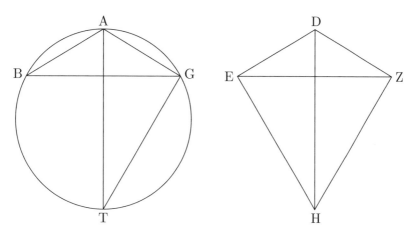

    Sit ergo circulus in spera datus circulus *ABG*. Volo autem descri-
bere lineam diametro ipsius equalem. Super arcum igitur *ABG* notabo
tria puncta quocumque modo contingat, que sint puncta *A B G*. Faciam autem ex tribus lineis rectis triangulum *DEZ* et ponam ut linea   5
DE sit equalis linee que coniungit inter punctum *A* et punctum *B*, et
sit linea *DZ* equalis linee que coniungit inter punctum *A* et punctum
G, et sit etiam linea *EZ* equalis linee que coniungit inter punctum *B*
et punctum *G*. Et imaginabor ut linee *AG GB BA* sint coniuncte. Et
protraham a duobus punctis *E* et *Z* super duas lineas *DE DZ* duas   10
lineas orthogonaliter, que sint linee *EH ZH*, et producam lineam *DH*.
Dico igitur quod linea *DH* est equalis diametro circuli *ABG*. Imaginabor igitur diametrum circuli *AT* et protraham duas lineas *AT GT*
Et quia due linee *AB BG* sunt equales duabus lineis *DE EZ*, queque
videlicet earum sue relative, et basis *AG* est equalis basi *DZ*: ergo   15
angulus *ABG* est equalis angulo *DEZ*. Sed angulus *ABG* est equalis

2 ergo] *om.* **BPsVa**   2 in spera] *supra* **Ps**   2 circulus[2]] *om.* **OBPsVa**   2 *ABG*] *ABTG*
OFiZB, **Ps** (*in corr.*), *ABCD* **Va**   3 ipsius] *om.* **OBPsVa**, ipsi **FiZ**, circuli dati *add. supra*
Ps   3 equalem] equales **Va**   3 Super ... *ABG*] *permut.* **BVa**   3 Super arcum] *marg.* **Ps**
3 igitur] *om.* **Fi**   3–4 notabo tria] *tr.* **PRVFiZ**, noto tria **BPsVa**   4 contingat] contingit **OB**
PsVa   4 sint] sunt **FiZ**   4–5 Faciam] *in corr.* **Ps**; *et add. ante* **B**   5 autem] *om.* **BPsVa**
5 lineis rectis] *tr.* **PRVMKgOFiZ**   5 triangulum] triangulis **Fi**   5 ponam ut linea] *illeg.* **B**
5 ut] *om.* **Va**   5–6 linea *DE*] lineam *D* **Fi**   6 coniungit] contingit **Va**   6 inter punctum] *scr.
et del., deinde scr. iterum* **Ps**   7 *DZ*] *ZDZ* **V**   7–8 que coniungit ... *G*] *om.* **Z**, *AG* **OFi**
BVa, **Ps** (*supra*), *nota illeg. marg.* **Ps**   8–9 et sit ... punctum *G*] *om.* **Ps**   8 sit etiam] *om.*
BVa   8 *EZ*] sit *add.* **BVa**   8–9 que ... punctum *G*] *BG* **BVa**   9 *BA*] *corr. ex GA* **V**
10 a] *infra* **R**   10 et *Z*] *tr.* **PsVa**   10 *Z*] et *add.* **O**   10–11 *DE DZ* duas lineas] *om.* **R**
10 *DZ*] *in corr.* **Ps**   11 sint] sunt **Z**   11 *EH*] *CH* **Va**   11 *DH*] *BH* **Fi**   12 igitur] ergo **Va**
12 linea] *om.* **OBPsVa**   13 igitur] iterum **KgOFiZBPsVa**   13 et] *om.* **Fi**   13 duas lineas] *om.*
BPsVa   14–16 sunt ... angulus *ABG*] *repet.* **Va**   14–15 queque ... relative] *om.* **MKgOFiZB**
PsVa   15–16 basi *DZ* ... equalis[1]] *om.* **Fi**   16 est equalis[1]] equatur **MKgOBPsVa**   16–
17 *DEZ* ... angulo] *marg.* **BPs**   16 *ABG*] *marg.* **R**   16 est equalis[2]] equatur **MKg**

مساوية لزاوية اطج وذلك أنهما فى قطعة واحدة أعنى قطعة اجَ من الدائرة

وزاوية دهز مساوية لزاوية دحز وذلك أن نقط دَ هَ حَ زَ تمر بها دائرة

15 فزاوية اطجَ مساوية لزاوية دحز وزاوية دزح القائمة مساوية لزاوية اجط

القائمة فاجط دحز مثلثان وزاويتا اطج اجط من أحدهما مساويتان

لزاويتى دحز دزح من الآخر كل واحدة لنظيرتها وضلع اجَ من أحدهما

الموتر لإحدى الزوايا المتساوية مساوٍ لضلع دزَ الذى هو نظيره من الآخر

فتكون سائر الأضلاع مساوية لسائر الأضلاع كل ضلع لنظيره فخط اطَ مساوٍ

20 لخط دحَ وخط اطَ قطر دائرة ابجَ فخط دحَ مساوٍ لقطر دائرة ابجَ ،

وذلك ما أردنا أن نعمل ·

كه   كيف نخط خطاً مثل قطر كرة معلومة ·

فلنتوهم الكرة التى نريد أن نخط خطاً مثل قطرها ولنتعلم على بسيط

الكرة نقطتين كيف ما وقعتا وهما نقطتا اَ بَ ونخط على قطب اَ وببعد ابَ

دائرة بجَ دَ فقد يمكننا أن نخط خطاً مثل قطر دائرة دبجَ فليكن خط

5 زح ونعمل من الثلثة الخطوط المستقيمة التى اثنان منها مساويان للخطين

اللذين خرجا من القطب إلى الدائرة والواحد مساوٍ للقطر الذى ذكرنا مثلث

هزح فيكون كل واحد من خطى زهَ هَحَ مساوٍ للخط الذى خرج من قطب

نبين [نعمل N 21 انطح N [ابجَ 2 حَر N 20 corr. ex حَر N [دزَ 18 [ا حَ add. A [دائرة supra A 14 [دهزَ 14
NH 1 كيف [وخط [ونخط N 3 om. N [ما 3 A ولنعلم [ولنتعلم N 2 نجد [نخط A 1 ملسوهم كف [كيف
6 إلى [فى NH 6 ذكرنا [وهو add. marg. A

angulo *ATG*. Quod ideo est quoniam ipsi sunt in portione una, scilicet in portione *AG* circuli. Et angulus *DEZ* est equalis angulo *DHZ*. Quod ideo est quoniam circulus transit per puncta *D E H Z*; ergo angulus *ATG* est equalis angulo *DHZ*. Sed angulus *DZH* rectus est equalis   20 angulo recto *AGT*. Duorum igitur triangulorum *ATG DHZ* duo anguli unius *ATG AGT* sunt equales duobus angulis alterius *DHZ DZH*, quisque videlicet suo relativo. Et latus *AG* unius eorum, quod subtenditur uni duorum angulorum equalium, est equale lateri *DZ*, quod est eius relativum, cum sit alterius. Ergo reliqua latera sunt equalia late-   25 ribus reliquis, unumquodque videlicet suo relativo. Ergo linea *AT* est equalis linee *DH*. Sed linea *AT* est diametrus circuli *ABG*. Ergo linea *DH* est equalis diametro circuli *ABTG*. Et illud est quod demonstrare voluimus.

20   Quomodo linea diametro spere date equalis reperiatur.

     Speram itaque cuius diametro lineam equalem describere volo imaginabor et super spere superficiem duo puncta notabo qualitercumque contingat, que sint duo puncta *A* et *B*, et describam polum *A* et secundum spacium *AB* describam circulum *BGD*. Iam autem possibile   5 est ut describam lineam diametro circuli *DBG* equalem, que sit linea *ZH*. Et faciam ex tribus lineis rectis, quarum due sint equales duabus lineis que protrahuntur a polo in spera et una sit equalis diametro quam nominavimus, triangulum *EZH*. Est igitur unaqueque duarum

---

17 Quod ideo est quoniam] quoniam **M**, quod **Kg**   17 quoniam] quia **OBPsVa**, ip͞iu **Fi** 17 ipsi] *om.* **O**, *illeg.* **B**   17 portione] proportione **KgZB**   18 portione] proportione **KgZ PsVa**   18 est equalis] equatur **OBPsVa**   18–19 Quod ideo est quoniam] ideo quia **OBPsVa**   19 per] *om.* **R**   19 *E*] *om.* **Fi**   19 *Z*] *om.* **R**; uterque enim semicirculorum circumscribentium *E Z* angulos rectos alteri est equale super eandem diametrum *add.* **Fi**   20 est equalis[1]] equatur **B**   20 *DZH*] *DHZ* **Va**   20 rectus est] *tr.* **MKg**, *tr. et corr.* **Ps**   20 est[2]] *add. supra* **M**   20 est equalis[2]] equatur **B**   21 *AGT*] *in corr.* **Ps**   21 igitur] ergo **BPsVa**   21 *ATG DHZ*] *AGT* duorum igitur triangulorum *TDH* et **Fi**   23 quisque] quis **Fi**   23 quisque . . . relativo] *om.* **MKgOBPsVa** 23 videlicet] scilicet **Z**   23 eorum] earum **Va**   24 est equale] equatur **KgOBPsVa** (*repet.* **Ps**)   24–25 quod . . . relativum] suo relativo **MKgOBPsVa** (suo *repet.* **Va**) 25 cum sit alterius] *marg.* **M**, *om.* **KgOBPsVa**   25–26 lateribus reliquis] *tr.* **RVKgO FiZBPsVa**   26 unumquodque . . . relativo] *om.* **KgOBPsVa**, *illeg.* **M**   26 relativo] corell'o **Fi**, correlato **Z**   26 linea] *om.* **OBPsVa**   26 *AT*] *AZ* **Va**   27 linee] *om.* **OBPsVa** 27 *AT*] *HT* **Va**   28 equalis diametro] diameter **O**   28 equalis . . . *ABTG*] *permut.* **MKgB PsVa**   28 Et illud est] *om.* **KgOFiZ**, *illeg.* **M**   28–29 Et illud . . . voluimus] *om.* **BPsVa** 29 voluimus] volumus **Z**   1 reperiatur] recipiatur **KgOBPsVa**, *illeg.***M**; vel reperiatur *supra* **O** 2 itaque] igitur **KgOBPsVa**, **M** illeg.   2 describere] describunt **Va**   3 super] in **KgOBPsVa** 3 spere] *om.* **R**   3 superficiem] superficie **OBVa**, **Ps** (*corr ex* superficiem); spere *add. marg.* **R**   3 duo] *om.* **KgO**   3 puncta notabo] *tr.* **PRVOFiZ**   3–4 qualitercumque] qualicumque **Fi**   4 sint] sunt **Va**   4 et[2]] *om.* **Va**   4 describam] super *add. supra* **Ps**, *in textu* **Va** 5 *BGD*] *corr. ex BDGD* **M**, *ABGD* **KG**, *BDG* **Z**, *BG* **BPsVa**   5–6 possibile est] *tr.* **Va** 6 *DBG*] *ABG* **Va**   6 linea] *om.* **Z**   7 lineis rectis] *tr.* **PRVMKgOFiZ**   7 sint] sunt **ROFiZBPsVa**   8 polo] ad circulum *B add.* **OFiZ**, ad circulum *BG add. supra* **B**   8 in] intra **V** 9 quam] quod, *corr. supra in* quem **B**   9 nominavimus] nominamus **R**   9 *EZH*] *corr. ex DZH* **V**, *EHZ* **OBPsVa**   9 igitur] ergo **Va**   9 unaqueque] utraque **MKgBPsVa**

آ إلى الخط المحيط بدائرة د ب ج ويكون خط ز ح مساوياً للقطر ولنخرج من

نقطتى ز ح من خطى ه ز ه ح خطين على زوايا قائمة وهما خطا ح ط زط

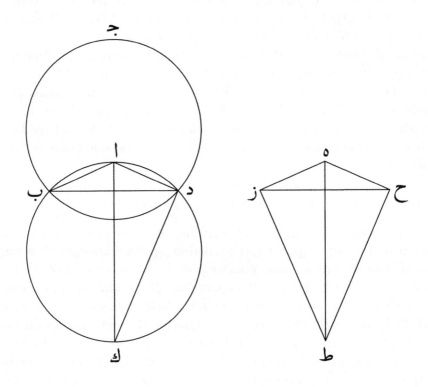

ولنوصل خط ه ط ، فأقول إن خط ه ط مساوٍ لقطر الكرة فلنتوهم قطر   10

الكرة خط آك وليمر بخط آك سطح يحدث قطعاً يكون دائرة عظيمة وهى

دائرة آ ب د ولنوصل خطوط آ ب ب د آ د د ك فلأن خطى آ ب ب د مساويان

لخطى ه ز ز ح كل واحد منهما لنظيره وقاعدة آ د مساوية لقاعدة ه ح تكون

زاوية آ ب د مساوية لزاوية ه ز ح ولكن زاوية آ ب د مساوية لزاوية آك د

وزاوية ه ز ح مساوية لزاوية ه ط ح فزاوية آك د مساوية لزاوية ه ط ح وزاوية   15

8 د ب ج] حـ N 10 خط ه ط] هط A 11 آ اك ١] آل N AN 11 سطح ... وليم [ولعمل خط آل سطحا

A آ اك ل] الاكد N 15 د ل] آد د ك A 12 اكد]H آد: آل. NH,

linearum *ZE EH* equalis linee que a polo *A* ad lineam continentem cir-   10
culum *DBG* protrahitur et linea *ZH* est equalis diametro. Protraham
autem a duobus punctis *Z* et *H* duarum linearum *EZ EH* duas lineas orthogonaliter, que sint linee *ZT HT*, et producam lineam *ET*. Dico

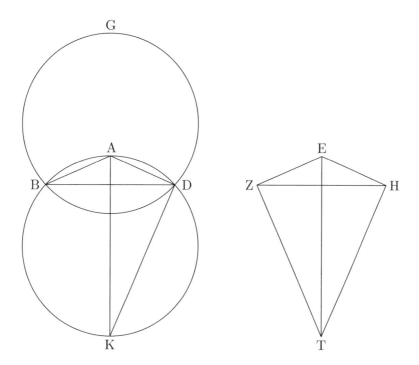

igitur quod linea *ET* est equalis diametro spere. Imaginabor itaque
diametrum spere lineam *AK* et faciam cum linea *AK* superficiem ex   15
qua proveniat sector existens circulus maior sitque circulus *ABD*. Et
producam lineas *AB BD DK*. Et quia due linee *AB BD* sunt equales
duabus lineis *EZ ZH*, queque scilicet earum sue relative, et basis *AD*
est equalis basi *EH*, ergo angulus *ABD* est equalis angulo *EZH*. Sed
angulus *ABD* est equalis angulo *AKD* et angulus *EZH* est equalis   20

10 que] protrahitur *add.* **OBPsVa**   10 *A*] *om.* **KgOBPsVa**   11 *DBG*] *BDG* **M**, *corr.*
ex DBTG **V**   11 et] etiam **OBPsVa**   11 *ZH* est] *om.* **OBPsVa** (*ZH add. supra* **O**)
12 *Z* et *H*] *om.* **BPsVa**   13 *ZT HT*] *HT ZT* **OBPsVa**, *ZT HTZ* **Z**   13 lineam] *su-*
pra **M**, *om.* **BPsVa**   14 igitur] *om.* **BPsVa**   14 linea] *om.* **MKgOBPsVa**   14 est] cum
Va   14 spere] date *add.* **OFiZ**   15 et ... superficiem] *marg.* **Ps**   15 faciam] etiam **Z**,
factam **Va**   15 *AK*] *om.* **Va**   15 ex] a **OBPsVa**   16 proveniat] *corr. ex* provenit **Ps**,
proveniet **KgFi**   16 existens circulus maior] *permut.* **BPsVa**   17 producam] duas *add. et*
del. **MPs**, *add.* **KgOB**   17 *DK*] *BK* **Fi**   17 due] *om.* **BPsVa**   17 due linee] linea **O**
18 lineis] *om.* **BPsVa**   18 queque] que **Fi**   18 queque scilicet ... relative] *om.* **MKgBPsVa**
18 scilicet] *om.* **O**   19 est[1]] cum **Va**   19 basi] *om.* **BPsVa**   19 *ABD*] *corr. ex AD* **O**, *AD* **FiZ**
19 est equalis] equatur **MKgBPsVa**   19–20 *EZH* ... *AKD*] *om.* **Z**   20 est equalis[1]] equatur
MKgOBPsVa   20 *EZH*] *EHZ* **R**, *EZK* **Fi**   20 est equalis[2]] *om.* **BPsVa**

اد‌ك القائمة مساوية لزاوية ه‌ح‌ط القائمة فمثلثا اك‌د ه‌ط‌ح زاويتا اد‌ك د‌ك‌ا

من أحدهما مساويتان لزاويتى ه‌ط‌ح ط‌ح‌ه من الآخر كل واحدة لنظيرتها

وضلع آد من أحدهما وهو الذى يوتر إحدى الزوايا المتساوية مساوٍ لضلع

ه‌ح من الآخر الذى هو نظيره فسائر الأضلاع مساوية لسائر الأضلاع كل

ضلع لنظيره فخط اك‌ مساوٍ لخط ه‌ط وخط اك‌ قطر الكرة فخط ه‌ط مساوٍ لقطر

الكرة المعلومة ، وذلك ما أردنا أن نعمل ·

20

كا   كيف نرسم دائرة عظيمة تمر بنقطتين معلومتين فى بسيط كرة ·

فلتكن النقطتان المعلومتان اللتان فى بسيط الكرة نقطتى آ ب‌ ونريد أن

نرسم دائرة عظيمة تمر بهما فإن كانت هاتان النقطتان على قطر الكرة فقد

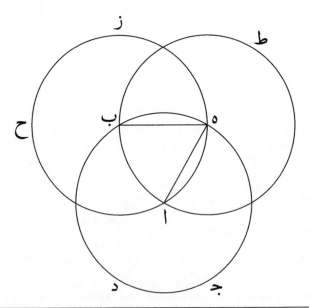

N ومثلا [فمثلثا 16   marg. A [مساوية لزاوية ه‌ح‌ط ... ه‌ط‌ح 16   A قائمة [القائمة[1] 16

om. H [أن نعمل 21   N المعلوم [المعلومة 21   AN, מסמאויה H مساوية [مساوية 19   N وسائر [فسائر 19

om. N [فقد تبين ... قطر الكرة 3-5   A الكرة [كرة 1   N سس [نعمل 21

angulo *ETH*: ergo angulus *AKD* est equalis angulo *ETH*. Sed angulus *ADK* rectus est equalis angulo *EHT* recto. Ergo duorum triangulorum *AKD ETH* unius duo anguli *ADK DKA* sunt equales duobus angulis alterius *ETH THE*, quisque videlicet suo relativo. Sed et latus unius eorum *AD*, quod subtenditur uni duorum angulorum, est equale lateri   25 alterius *EH*, quod est eius relativum. Ergo reliqua latera sunt equalia reliquis lateribus, quodque videlicet suo relativo. Ergo linea *AK* est equalis linee *ET*. Sed linea *AK* est diameter spere; ergo linea *ET* est equalis diametro spere date. Et illud est quod demonstrare voluimus.

21   Qualiter circulus maior in superficie spere per duo puncta data transiens designetur.

Sint itaque duo puncta data in superficie spere puncta *A* et *B*. Volo autem signare circulum maiorem transeuntem per ea. Si ergo hec

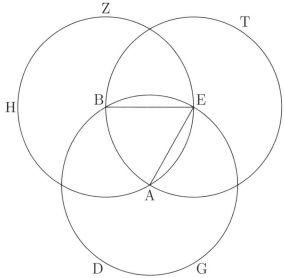

21 *ETH*] *EHT* **O**; *ETH* *corr. ex EHT* **Ps**   21 ergo ... angulo *ETH*] *om.* **Z** 21 ergo ... angulo *EHT*] *marg.* **Ps**, *om.* **Va**   21 est equalis] equatur **MKg**   21– 22 *ETH* ... duorum trian] *marg.* **B**   22 rectus est] est rectus et **MKg**   22 *EHT*] et *HT* **Fi**   22 Ergo] igitur **Fi**   22 duorum] *om.* **M**   22 triangulorum] angulorum **Kg**, *corr. ex* angulorum **Ps**   23 *AKD*] *supra* **Ps**   23 *ETH*] *EHT* **OPsVa**   23 *DKA*] *DKH* **Kg**   23 sunt] sint **Fi**   24 alterius] *om.* **FiZ**   24 *ETH THE*] *EHT ETH* **OBVa**, *ETH EHT* **Kg**, *EHT HTE* **Ps**   24 quisque ... relativo] *om.* **MKgBPsVa**   24 et] etiam **VOFiZ PsVa**, *corr. supra ex* quia **M**, quia **KgB**   25 eorum] scilicet *add.* **PsVa**   25 est] cum **Va** 25 equale] equali **R**   26 *EH*] *TH* **KgVa**   26 quod ... relativum] suo relativo **MKgBPsVa** 26 reliqua] linea **Fi**   27 quodque] utrumque **MKgOBPsVa**   27 videlicet] scilicet **MKgOBPsVa** 27–28 est equalis] equatur **MKgBPsVa**   28 est[1]] cum **Va**   28 spere] scilicet date *supra* **Ps** 29 equalis ... spere] *permut.* **Kg**   29 spere date] *tr.* **OBPsVa**   29 Et ... voluimus] *om.* **OBPsVa**   29 est] *om.* **Kg**   1–2 data transiens] *tr.* **OBPsVa**   2 designetur] assignetur **FiZ**, *corr. ex* designare **Ps**   3 puncta[2]] *om.* **BPsVa**   3 et] *om.* **OZBPsVa**   4 autem] *om.* **BPsVa** 4 signare] significare **OBPsVa**   4 ea] eam **Kg**, illa **OZBPsVa**, ista **Fi**   4 Si] *corr. ex* Set **Ps** 4 ergo] *om.* **BPsVa**

تبين أنه ستُرسم على نقطتى آ بَ دوائر عظيمة غير متناهية فإن لم تكن

5   نقطتا آ بَ على قطر الكرة فلتُرسم على قطب آ وببعد مساوٍ لضلع المربع

الذى يُرسم فى دائرة عظيمة دائرة هجد فدائرة هجد عظيمة وذلك أن الخط

المستقيم الذى يخرج من قطب الدائرة إلى الخط المحيط بها مساوٍ لضلع

المربع الذى يرسم فى دائرة عظيمة وأيضاً فلترسم على قطب بَ وببعد ضلع

المربع الذى يرسم فى دائرة عظيمة دائرة هزح فدائرة هزح عظيمة وذلك أن

10   الخط المستقيم الذى يخرج من قطبها إلى الخط المحيط بها مساوٍ لضلع المربع

الذى يرسم فى دائرة عظيمة وليوصل بين نقطة ه وبين نقطتى آ بَ بخطين

مستقيمين وهما خطا هآ هبَ وكل واحد من خطى اه هبَ مساوٍ لضلع

المربع الذى يرسم فى دائرة عظيمة فخط هآ مساوٍ لخط هبَ فالدائرة التى ترسم

على قطب ه وببعد هبَ تمر بنقطة آ أيضاً من أجل أن خط هآ مساوٍ لخط

15   هبَ ولترسم ولتكن دائرة آبط فدائرة آبط عظيمة وذلك أن الخط المستقيم

الذى يخرج من قطبها إلى الخط المحيط بها مساوٍ لضلع المربع الذى يرسم

5 قطر [قطر H كטבי 7 المستقيم الذى ... إلى الخط [marg. A; من قطب الدائرة: من قطبها A 8–
11 عظيمة ... وأيضاً [om. N 9 فدائرة هزح [marg. A 11 آ بَ [حطا add. supra A 12 وهما [هما N
13 فالدائرة [والدارة A 15 ولترسم [فلرسم N

duo puncta fuerint super extremitatem diametri spere, tunc iam mani- 5
festum est quod super duo puncta *A* et *B* circulos maiores infinitos
signabimus. Quod si duo puncta *A* et *B* non fuerint super extremitatem diametri spere, tunc signabo super polum *A* et cum longitudine
equali lateri quadrati quod in maiori circulo signatur circulum *EGD*.
Circulus igitur *EGD* est maior Quod ideo est quoniam linea recta que 10
protrahitur a polo circuli ad ipsius circumferenciam est equalis lateri quadrati quod in maiori circulo signatur. Et etiam signabo super
polum *B* et cum longitudine lateris quadrati quod in circulo maiori signatur circulum *EZH*. Circulus igitur *EZH* est maior Quod ideo
est quoniam linea recta que a duobus polis eius protrahitur ad ipsi- 15
us circumferentiam est equalis lateri quadrati quod in maiori circulo
signatur. Coniungam autem inter punctum *E* et duo puncta *A B* pro-
ducendo duas lineas rectas, que sint linee *EA EB*; unaqueque igitur
duarum linearum *EA EB* est equalis lateri quadrati quod in circulo
maiori signatur. Linea igitur *EA* est equalis linee *EB*; circulus quoque 20
qui super polum *E* et secundum longitudinem *EB* signatur transit per
punctum *A*, propter hoc quod linea *EA* est equalis linee *EB*. Signetur
ergo et sit circulus *ABT*. Ergo circulus *ABT* est maior. Quod ideo
est quoniam linea recta que protrahitur a polis eius ad ipsius circumferentiam est equalis lateri quadrati quod in circulo maiori signatur. 25
Iam igitur signatus est circulus maior *ABT*, qui per duo puncta *A* et

5 extremitatem] extremitates **OBPsVa**   5 iam] *om.* **OBPsVa**   6 quod] qualiter **B**
6 et] *om.* **Fi**   6–7 circulos ... et *B*] *marg.* **M**   7 signabimus] signavimus **PVMKg**, si-
gnamus **R**, significabimus **BPsVa**   7 fuerint] sunt **Z**   7–8 extremitatem] extremitates
Va   8 signabo] significabo **PsVa**, *illeg.* **B**   8 *A*]] **Z** et super polis *A* describam circulis in spera secundum quantitatem lateris quadrati in maiori circulo *add.* **Fi**   8 et] *om.*
B   8 longitudine] longitudinem **Fi**, *corr. ex* longitudo **Ps**   9 signatur] significatur
BPsVa   9 circulum] significabo speram *supra* **Ps**   9 *EGD*] *EBGD* **OFiZBPs**, *ABGD* **Va**
10 Circulus ... *EGD*] *om.* **Va**   10 igitur] ergo **BPs**   10 *EGD*] *EBGD* **OFiZPs** (*B su-
pra***Ps**)   10 Quod ... quoniam] ideo quia **OBPsVa**   10 ideo est] *tr.* **MKg**   11 circuli] *om.*
KgOVa, *supra* **BPs**   11 ipsius circumferenciam] circumferenciam eius **OBPsVa** (eius *supra*
B)   11 equalis] equale **OBPsVa**   12 signatur] significatur **BPsVa**   12 signabo] significabo
OBPsVa (*corr. ex* longitudo **Ps**)   13 et cum] circulum in **B**   13 cum] in **R**   13–
14 circulo maiori] *tr.* **MKgOBPsVa**   14 signatur] significatur **Ps**   14 *EZH*] *E* et *H* **Fi**;
om., significabo superficiem *EZH add. marg.* **Ps**   14 Circulus igitur *EZH*] *om.* **Va**   14–
15 Quod... quoniam] ideo quia **BPsVa**   15 a duobus polis] polo **Va**   15 duobus polis] polo
FiZ, duobus polo **BPs**   15 protrahitur] producitur **B**   17 signatur] significatur **PsVa**; et
etiam signabo super polum *B add.* **Z**   17 *A*] et *add.* **MKgOBPsVa**   18 lineas rectas] *tr.*
PRVMKgOFiZ   18 sint] sunt **FiZ**   18 unaqueque] utraque **MKgOBPsVa**   19 circulo maiori] *tr.*
BPsVa   20 maiori signatur] *tr.* **KgOFiZ**   20 signatur] significatur **PsVa**   20 est] cum **Va**
20 equalis] qualis **Fi**   20 *EB*] *corr. ex EDB* **Kg**   20 circulus] *corr. ex* circulum **Kg**
21 qui] cum **Fi**; est *add.* **KgOBPs**   21 polum *E*] *tr.* **OBPsVa**   21 et] *om.* **OBPsVa**   21 *EB*] *E*,
supra B **Ps**   21 signatur] significatur **PsVa**   22 *A*] *E* **R**   22 propter hoc quod] quia **OB**
PsVa   22 Signetur] Significetur **OFiZBPs** (*corr. ex* significatur **Ps**)   23 *ABT*[1]] *ABC* **FiZ**
23 *ABT*[2]] *ABC* **FiZ**; qui per duo puncta data (data *om.* **Va**) protrahitur (trahitur **BPsVa**) *add.*
OBPsVa (qui *add. ante* est **PsVa**)   23 est maior] *tr.* **O**, *om.* **Va**   23 maior] maior in spera *marg.*
Ps   23–28 Quod ideo ... voluimus] *om.* **BPsVa**   24 est] *om.* **V**   24 protrahitur] pertrahitur
Fi   24 polis] polo **OFiZ**   25 circulo maiori] *tr.* **MKgOFiZ**

فى دائرة عظيمة فقد رُسمت دائرة اٮٮط العظيمة ومرت بنقطتى آ ٮ

المعلومتين اللتين على بسيط الكرة ، وذلك ما أردنا أن نعمل ·

كب كيف نجد قطب دائرة معلومة فى كرة ·

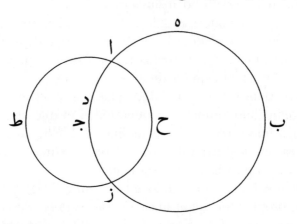

فلتكن الدائرة المعلومة التى فى كرة دائرة اٮٮح ، ونريد أن نجد

قطبها فلنتعلم على الخط المحيط بالدائرة نقطة كيف ما وقعت وهى نقطة آ

ولتُفصل منه قوسان متساويتان وهما قوسا اٮد اٮه ولتُقسم قوس دٮه الباقية

بنصفين على نقطة زٮ فدائرة اٮٮح إما أن تكون عظيمة وإما أن لا تكون

كذلك فلتكن أولاً غير عظيمة ولترسم على نقطتى زٮ آ المعلومتين اللتين على

السطح الكرى دائرة عظيمة وهى دائرة زاٮط فلأن قوس دآ مساوية لقوس

اٮه وقوس دزٮ أيضاً مساوية لقوس زٮه تكون جميع قوس اٮدز مساوية لجميع

قوس اٮهز فدائرة زاٮط تقطع دائرة اٮٮح بنصفين ودائرة ازٮط العظيمة التى

نقطه [وقعت 3 A ولمعلم [فلنتعلم 3 NH الكرة [كرة 2 NH سس [نعمل 18 A الذى [اللتين 18

A supra A الا [ال N; ال N أن ال [ال 5 A بدارة [فدائرة 5 A منها [منه 4 A وهو [وهو 4 A وهى [وهى 3 A .add وهو

om. A أيضاً [8 A مل قوس [مساوية لقوس 7 supra A, om. NH غير [6 NH فلا تكن [فلتكن 6

A مساوا [مساوية 2 8 A مل قوس [مساوية لقوس 8 لقوس اٮه ... وقوس A; (zه sic pro) false scr. N 8

om. A ودائرة ازٮط [9 بنصفين ... اٮد A 9-10 A اٮٮح [اٮٮح 9

B data transit, que sunt super superficiem spere. Et illud est quod demonstrare voluimus.

22 Quomodo in spera polus circuli dati inveniatur.

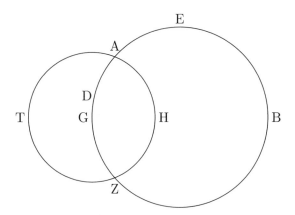

Sit igitur circulus in spera datus circulus *ABG*. Volo autem polos eius invenire. Super lineam igitur continentem circulum notabo punctum, quocumque modo contingat, quod sit punctum *A*, a quo dividam duos arcus equales, qui sint arcus *AD AE*, et dividam reliquum arcum   5
DBE in duo media super punctum *Z*. Circulus igitur *ABG* aut erit maior aut non erit maior. Sit ergo prius non maior. Signabo autem super duo puncta data *Z* et *A*, que sunt super superficiem spere, circulum maiorem, qui sit circulus *AT*. Et quia arcus *DA* est equalis arcui *AE* et arcus quoque *ZD* est equalis arcui *ZE*, ergo totus arcus *ADZ*   10
est equalis toti arcui *AEZ*. Ergo circulus *ZAT* secat circulum *ABG* in duo media. Ergo circulus *AZT* maior qui est in spera secat aliquem

27 data] assignata **OFiZ** 27 super superficiem] in superficie **KgOFiZ** 27 spere] assignate *add.* **OFiZ** 27 illud] hoc **O** 27 est] *supra* **V**, *om.* **Fi** 27–28 quod demonstrare voluimus] *om.* **O** 1 polus] *post* dati **Va** 1 circuli dati] *tr.* **Kg** 2 igitur] ergo **MKgO**, *om.* **FiB PsVa** 2 autem] *om.* **BPsVa** 2 polos] poles **R** 2–3 polos eius invenire] *permut.* **V** 3 lineam igitur] *tr.* **FiZ** 4 modo] *om.* **Va** 5 arcus equales] *tr.* **PRVMKgFiZ** 5 sint] sunt **PsVa** 5 arcus²] altus **Fi** 5 *AD*] arcus *add.* **BPsVa** 6 *DBE*] *DZBE* **FiZ**, *DDE* **Va** 6 media] membra **Va** 6 igitur] ergo **O** 6 aut] *repet.* **Va** 7 maior] circulus *add. marg.* **Ps** 7 non erit maior] minor **BPs**, erit minor **Va** 7 erit maior] *om.* **MKg** 7 ergo] igitur **OFiZ**, *om.* **BPsVa** 7 non] *supra* **BPs** 8 data] *om.* **MKgBPsVa** 8 *Z* et] et *Z* **Fi** 9 *AT*] *DT* **Kg**; *inter A et T add. supra Z* **Ps**, *AZT* **Va** 9 quia] quod **Kg** 10 et] *om.* **MKgOFiZBPsVa** 10 *ADZ*] *EZD* **R** 11 est equalis] equatur **OBPsVa** 11 Ergo] igitur **Z** 11–12 Ergo ... media] *repet.* **R** 12 *AZT*] *ZAT* **FiZ** 12 maior qui est] *permut.* **MKgOBPsVa** 12 aliquem] alium **Fi**

فى الكرة تقطع دائرة ما غير عظيمة من الدوائر التى فى الكرة بنصفين فهى ١٠

تقطعها على زوايا قائمة وتمر بقطبيها فدائرة زاط تقطع دائرة ا ب ج على

زوايا قائمة وتمر بقطبيها فلتُقسم قوس زحا بنصفين على نقطة ح فنقطة ح

قطب دائرة ا ب ج ،

وأيضاً فإنا نجعل دائرة ا ب ج عظيمة وكذلك نبين أن قوس ا د ز

مساوية لقوس ا ه ز ولتقسم قوس ا د ز بنصفين على نقطة ج فكل واحدة من ١٥

قوسى ا ج ج ز ربع الخط المحيط بالدائرة فالدائرة التى تُرسم على قطب ج

وببعد ج ز تمر أيضاً بنقطة ا لأن نقطة ا تقابل نقطة ز فلترسم ولتكن مثل

دائرة ز ا ط فدائرة ز ا ط عظيمة وذلك أن الخط الذى يخرج من قطبها إلى

الخط المحيط بها مساوٍ لضلع المربع الذى يرسم فى دائرة عظيمة ونقطة ج

قطب دائرة ز ا ط فدائرة ا ب ج تقطع دائرة ز ا ط وتمر بقطبيها فدائرة ا ب ج ٢٠

العظيمة فى كرة وهى تقطع دائرة أخرى من الدوائر التى فى الكرة وهى دائرة

ز ا ط وتمر بقطبيها فهى تقطعها بنصفين وعلى زوايا قائمة فدائرة ا ب ج قائمة

على دائرة ز ا ط على زوايا قائمة فدائرة ز ا ط أيضاً قائمة على دائرة ا ب ج

على زوايا قائمة فدائرة ا ط ز العظيمة فى كرة وهى تقطع دائرة أخرى من

الدوائر التى فى الكرة وهى دائرة ا ب ج على زوايا قائمة فهى تقطعها بنصفين ٢٥

وتمر بقطبيها فدائرة ا ط ز تقطع دائرة ا ب ج بنصفين وتمر بقطبيها فلتقسم

---

10 غير عظيمة [بقطبيها ... زاط 11-12 om. N 12 زحا] رجا N 15 واحدة] واحد A
16 بالدائرة المحيط الخط [الخط المحيط بالدائرة A دايره [الدائرة 16 om. N 20 ... تقطع] A والدارة A 20 ... ا ب ج [زاط 2] رط
A 21-22 دائرة زاط] دار رط A وعلى [على 22 A 23 على دائرة ا ب ج] على دائرة A 24 om. A فدائرة] ودائرة
NH 24 وهى] هى A 25 على زوايا قائمة [om. N 25-26 بقطبيها ... تقطعها فهى [om. H

non maiorem circulorum qui sunt in spera in duo media. Ergo ipse
secat eum orthogonaliter et transit per duos polos eius. Ergo circulus
ZAT secat circulum *ABG* orthogonaliter et transit per duos polos   15
eius. Dividam itaque arcum *ZHA* in duo media super punctum *H*:
ergo punctum *H* est polus circuli *ABG*.

Ponam etiam circulum *ABG* maiorem. Et similiter etiam ostendam quod arcus *ADZ* est equalis arcui *AEZ*. Dividam autem arcum
ADZ in duo media super punctum *G*. Unusquisque igitur arcuum *AG*   20
GZ est quarta linee continentis circulum. Circulus autem qui signatur
super polum *G* et secundum longitudinem *GZ* transit etiam per punctum *A*, quoniam punctum *A* opponitur puncto *Z*. Signetur ergo sitque
equalis circulo *ZAT*. Ergo circulus *ZAT* est maior. Quod ideo est quoniam linea que protrahitur a polis eius ad lineam continentem ipsum   25
est equalis lateri quadrati quod in maiori circulo signatur. Punctum
autem *G* est polus circuli *ZAT*. Ergo circulus *ABG* secat circulum
ZAT et transit per duos polos eius. Sed circulus *ABG* maior est in
spera et secat alium ex circulis qui sunt in spera, qui est circulus *ZAT*,
et transit per duos polos eius. Ergo ipse secat eum in duo media et or-   30
thogonaliter. Circulus igitur *ABG* super circulum *ZAT* orthogonaliter
erigitur. Circulus ergo *ZAT* erigitur etiam super circulum *ABG* orthogonaliter, sed circulus *ATZ* est maior in spera et secat alium ex circulis
qui sunt in spera, qui est circulus *ABG*, orthogonaliter. Ergo ipse secat
eum in duo media et transit per duos polos eius. Ergo circulus *ATZ*   35
secat circulum *ABG* in duo media et transit per duos polos eius. Divi-

13 non] *om.* **PRV**   13 maiorem] maiorum **BPsVa**   13 sunt in spera] *permut.***MKgOB**
14 secat] *in corr.* **B**   14 eum] *om.* **Fi**   14 duos] *om.* **Fi**   14 duos polos eius] *permut.*
Kg   14 circulus] angulus **R**   16 Dividam] Dimidiam **OBPs** (*corr. in* dividam **Ps**)
16 itaque] autem **M**, ergo **Kg**, igitur **OBPsVa**, *om.* **Fi**   17 ergo punctum *H*] quod **Z**
17 *ABG*] *lac.* **Fi**   18 etiam] et **Fi**   18 *ABG* maiorem] *tr.* **Kg**   18 Et similiter] *om.*
Kg   18 similiter] simile **Fi**   18 etiam] *om.* **MOBPsVa**   19 est equalis] equatur **OBPsVa**
19 equalis] equales **Fi** 20 igitur] ergo **Kg**, *om.* **Va**   21 continentis circulum] *tr.* **MKgOBPsVa**
(continentes **Va**)   21 autem] igitur **R**   21 signatur] figuratur **OBPsVa**   22 etiam per] in **R**
23 *A*[1]] *H* **Z**   23 *A*[2]] *supra* **Ps**, *H Z*   23 puncto] *om.* **R**   23 Signetur] Significetur **FiZPs**,
significatur **Va**   23 ergo] *marg., in textu add.* BG **O**, *supra* **B**, *illeg. supra* **Ps**   23 sitque] sit
ergo **PRVMFi**, *AT* **Va** 24 equalis circulo] *illeg.* **B**   24 circulus] *om.* **B**   24–25 quoniam] quia
OBPsVa   25 protrahitur] pertrahitur **Fi**   25 polis] polo **FiZ**   25 continentem] *T add.* **Fi**
26 circulo signatur] *tr.***PRVMKg** 26 signatur] significatur **PsVa** 27 *G*] *B* **Va** 27 *ZAT*] *AT*
B, *AZT* **PsVa**   28 eius] *om.* **Va**   28–30 Sed … eius] *om.* **Kg**   28 *ABG*] *AB, G supra*
B   28 est] *supra* **R**   29 alium ex circulis] circulum ex circulis **OPsVa**, alium circulorum
PRVKgFiZ   29 *ZAT*] *AZT* **OBPsVa**   30 ipse] *om.* **MKgOBPsVa**   30 eum] ipsum **MKgOB**
PsVa   30 et[2]] *om.* **OPsVa**, *supra* **B**   31 Circulus … *ABG*] *permut.* **BPs**, abv̄ qui circulus
Va   31 circulum *ZAT*] *AZT* circulum **O**, *AZT* **BPsVa**   32–36 Circulus … eius] *om.* **BPsVa**
32 ergo] igitur **MKg**   32 erigitur etiam] *tr.* **OFiZ**   32–33 orthogonaliter] *ante* erigitur **MKg**
33 est] *om.* **V**   33 ex circulis] circulorum **PRVOFiZ**   34 orthogonaliter] *ante* qui est **OFiZ**
35–36 Ergo … eius] *om.* **OFiZ**   35 circulus] *corr. ex* angulus **M**   36 duos] *om.* **Kg**

قوس زح‌ا بنصفين على نقطة خ‍ فنقطة خ‍ قطب دائرة ا ب ج‍ ، وذلك ما أردنا أن نبين ·

[A] تمت المقالة الأولى من كتاب ثاوذوسوس فى الأكر وهى اثنان وعشرون شكلاً ·

[N] تمت المقالة الأولى من كتاب ثاوذوسيوس فى الكرات إصلاح أبى الحسن ثابت بن قرة الحرانى الصابئ بحمد الله ومنّه وصلوته على محمد وآله ·

[H] תמת אלמקאלה ·

dam itaque arcum *ZHA* in duo media super punctum *H*. Ergo punctum *H* est polus circuli *ABG*. Et illud est quod demonstrare voluimus.

Expleta est pars prima libri theodosii de speris viginti duas continens figuras.   40

---

37 H] huius **Va**   *37–38* Ergo ... est[1]] *H* igitur **Ps**, *G* **Va**   *38* Et illud est] *om.* **Va** *38* illud] id **Kg**, hoc **OBPs** *38* demonstrare] monstrare **O** *38* demonstrare voluimus] volumus declarare **OBPsVa**   *38* voluimus] volumus **Z**

المقالة الثانية من كتاب ثاوذوسيوس فى الأكر

يقال إن الدوائر يماس بعضها بعضاً فى الكرة إذا كان الفصل المشترك
لسطوحها مماساً للدائرتين جميعاً ·

ا   الدوائر المتوازية التى فى الكرة أقطابها واحدة بأعيانها ·

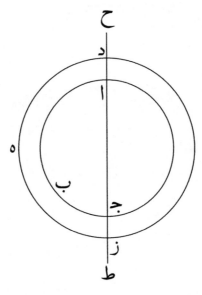

فلتكن فى كرة دائرتا ا ب ج د ه ز المتوازيتان ، فأقول إن قطبى كل   5
واحدة من دائرتى ا ب ج د ه ز هما قطبا الدائرة الأخرى منهما فليكن قطبا
دائرة ا ب ج نقطتى ح ط وليوصل خط ح ط وإذا كانت فى كرة دائرة فإن

الكرات [الأكر   1   A ثاوذوسوس [ثاوذوسيوس   1   praepos. N بسم الله الرحمن الرحيم رب اعن [المقالة   1
N; فاقول ان قطبى [المتوازيتان   5   in corr. A   5   add. N   ترجمه ابى الحسن باس س قره الحرانى [دائرتا   1
داري اح دهر [كل ... منهما   A 5-6   add. et del. A داري اح دهر المواربان واحده باعيانها برهان ذلك انا محد
منهما A 6   وقطى دائرة اح هما قطبا الدائرة الاخرى [منهما   ملكن قطبا ح ط وليوصل كل واحده من داري اح دهر
منهما add. et del. A   7   وليوصل [ولنصل   N   7   ح ط[2   A]خط ح ط مر نمركر دائرة اح ونمركر الكره   add. A
N فادا [واذا   7

Pars secunda libri theodosii de speris incipit

Circuli in spera se ad invicem contingere dicuntur cum communis duarum superficierum differentia duos simul contingit circulos.

1 Circulorum in spera equidistantium uni et idem sunt poli.

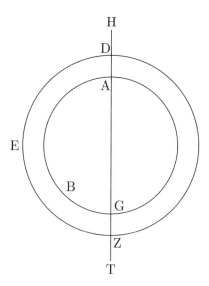

Sint igitur in spera duo circuli *ABG DEZ* equidistantes. Dico   5
igitur quod duo poli cuiusque duorum circulorum *ABG DEZ* sunt duo
poli alterius circuli eorum. Sint itaque duo poli circuli *ABG* duo puncta
H et *T*. Producam autem lineam *HT*. Sed cum in spera fuerit circulus,

---

2 in spera ... contingere] *permut.* **MKg,OZBPsVa**   3 differentia] distantia **Kg**   3 simul] *om.*
BPsVa   3 contingit] coniungit **KgBPs**   3 circulos] *om.* **PsVa**   4 Circulorum] etiam *add.* **Kg**
4 equidistantium] eque distantium **R**, *hic et saepius*   4 idem] eidem **Kg**   5 Sint] sunt **Z**
5 circuli] in spera *add. et del.* **BPs**   6 igitur] *om.* **BPsVa**   6 cuiusque] *supra* **M**, alterius
MOZBPsVa, alterius cuiusque **Kg**   6 duorum circulorum] circulorum, duorum *supra* **M**, angulorum **Kg**, circuli eorum **OZBPsVa**   6 *ABG DEZ*] *om.* **OFiBPsVa**   7 circuli eorum] *om.* **OZBPsVa**
7 itaque] igitur **Fi**   7 duo poli] *om.* **V**   8 spera] ipsa **Kg**

الخط الذى يمر بقطبيها هو عمود عليها وهو يمر بمركزها وبمركز الكرة نخط

ح‌ط عمود على دائرة ا‌ب‌ج وهو يمر بمركزها وبمركز الكرة فلأن خط ح‌ط

عمود قائم على دائرة ا‌ب‌ج ودائرة ا‌ب‌ج موازية لدائرة د‌ه‌ز يكون خط

ح‌ط عموداً على دائرة د‌ه‌ز أيضاً فلأن دائرة د‌ه‌ز فى كرة وقد أخرج من

مركز الكرة إليها عمود ح‌ط وأنفذ إلى كلتى الجهتين ولقى سطح الكرة على

نقطتى ح‌ط تكون نقطتا ح‌ط قطبى دائرة د‌ه‌ز وهما أيضاً قطبا دائرة

ا‌ب‌ج فقطبا كل واحدة من دائرتى ا‌ب‌ج د‌ه‌ز قطبا الدائرة الأخرى منهما ،

وذلك ما أردنا أن نبين ·

ب　الدوائر التى تكون فى كرة على قطبين مشتركين لها هى متوازية ·

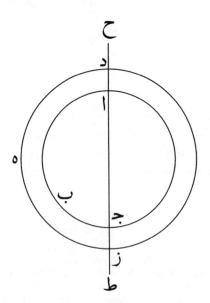

8 لخط ...] om. A　　marg. A [... يمر بقطبيها ... وهو 8　　8-9 المستقيم [الخط add. supra A
A الداره corr. ex [الدوائر 1 add. N　　هما [د‌ه‌ز 14 N ولان [فلأن 11 A أحد [احد 11 A ا‌ب‌ج¹ [ا‌ب‌ج 10

linea que transit per duos polos ipsius est perpendicularis super circulum et transit per centrum eius et centrum spere. Ergo linea *HT* est   10
perpendicularis super circulum *ABG* et transit per centrum eius et per
centrum spere. Et quia linea *HT* perpendiculariter erigitur super circulum *ABG* et circulus *ABG* equidistat circulo *DEZ*, ergo linea *HT* est
perpendicularis etiam super circulum *DEZ*. Et quia circulus *DEZ* est
in spera ad quem a centro spere perpendicularis iam est protracta, que   15
ad utramque pertransiens partem spere superficiei super duo puncta *H*
et *T* concurrit, ergo duo puncta *H* et *T* sunt duo poli circuli *DEZ*. Sed
ipsa etiam sunt duo poli circuli *ABG*. Ergo duo poli cuiusque duorum
circulorum *ABG DEZ* sunt duo poli alterius eorum. Et illud est quod
demonstrare voluimus.   20

2   Circuli qui sunt in spera super duos polos eis communes sunt equidistantes.

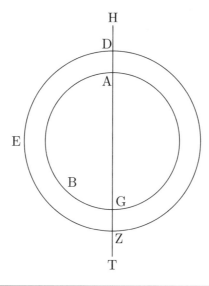

9 ipsius] eius ipsius **M**, eius **OFiZBPs**, cuius **Va**   *10* et[2]] per *add.* **OZBPsVa**   *10 HT* est] *tr.*
Fi   *11* centrum eius] *tr.* **OZBPsVa**   *11* per[2]] *om.* **PRVFi**   *12* linea] *om.* **Fi**
12 perpendiculariter erigitur] est perpendicularis **OZPsVa**; erigitur *add. et del.* **Ps**
12 erigitur] erigantur **Fi**   *13 ABG*] *HBG* **Va**   *13–14* et circulus ... circulum *DEZ*] *add.*
etiam post circulum *ABG* (*lin.* 11) **Fi**   *13 DEZ*] *corr. ex DEG* **P**, *EDZ* **O**, *DZE*
Z   *14* etiam] *om.* **OZ**   *14 DEZ*[1]] et transit per centrum eius et per (per *om.* **O**)
centrum spere *add.* **OZ**   *14* Et] *om.* **Fi**   *14 DEZ*[2]] *DZ* **Fi**, *EZ* **B**; *EZ, supra D* **Ps**
15 spere] *om.* **BPsVa**   *15* perpendicularis iam est] *permut.* **OZBPsVa**   *15* est protracta] *tr.*
MKg   *15* protracta] producta **Va**   *15* que] et quia **Va**   *16* utramque] utrumque **Fi**
17 et[1]] *om.* **BPsVa**   *17* concurrit] occurrit **OZBPsVa**   *18* ipsa etiam] *tr.* **MOZBPsVa**
18 cuiusque] cuiuscumque **V**, uniuscuiusque **BPsVa**   *19 DEZ*] *GEZ* **Z**   *19* illud] hoc
MKgOZBPsVa   *19* est] *om.* **Fi**   *20* demonstrare] d'cl **M**, *om.* **BPsVa**   *20* voluimus] volumus
OZBPsVa   *1* sunt in spera] *permut.* **PRVKgFi**   *1* eis] *in corr.* **BPsVa**

فلتكن فى كرة على قطبى حٰ طٰ دائرتا ابٰجٰ دهٰزٰ ، فأقول إن دائرتى

ابٰجٰ دهٰزٰ متوازيتان فليوصل خط حٰطٰ فلأن دائرة ابٰجٰ فى كرة وقد

أخرج خط يمر بقطبيها وهو خط حٰطٰ يكون خط حٰطٰ عموداً على دائرة

ابٰجٰ وكذلك أيضاً نبين أنه عمود أيضاً على دائرة دهٰزٰ والسطوح التى يقع

عليها خط واحد بعينه فيكون عموداً عليها إذا أخرجت لم تلتق فإذا أخرج

سطحا دائرتى ابٰجٰ دهٰزٰ لم تلتقيا فدائرة ابٰجٰ موازية لدائرة دهٰزٰ ، وذلك

ما أردنا أن نبين ·

جٰ   إذا كانت دائرتان فى كرة تقطعان خطاً محيطاً بدائرة ما عظيمة من الدوائر

التى فيها على نقطة واحدة بعينها وكانت أقطابهما على تلك الدائرة فإن

الدائرتين متماستان ·

فلتقطع فى كرة دائرتا ابٰجٰ دهٰجٰ الخط المحيط بدائرة اجهٰ العظيمة

على نقطة واحدة بعينها وهى نقطة جٰ ولتكن أقطابهما على دائرة اجهٰ ،

فأقول إن دائرتى ابٰجٰ دهٰجٰ متماستان فليكن الفصل المشترك لدائرة اجهٰ

ودائرة ابٰجٰ خط اجٰ والفصل المشترك لدائرة اجهٰ ودائرة جدهٰ خط جهٰ

والفصل المشترك لدائرة ابٰجٰ ودائرة جدهٰ خط زجحٰ ودائرة اجهٰ العظيمة

من الدوائر التى فى الكرة تقطع دائرة أخرى من الدوائر التى فيها وهى دائرة

4 خط [²خط | om. A 5 أيضاً¹ | om. N 5 نبين [سس A 7 تلتقيا [لما A 2 أقطابهما [أقطابها

العظيمة [om. N 4 متماستين corr. ex ستان A; corr. ex منها N متماستان [corr. ex 3

om. [جه ... جده خط 7-8 N احٰ [اجهٰ 7 om. A | ابٰجٰ ... ودائرة 7 5 أقطابهما [اعطابها N

N 8 لدائرة [لسطح A 8 واحه [احهٰ A 8 ودائرة جده [احٰبٰ A ولداره احٰبٰ marg. A

8 زجح [احٰ رٰ A; والفصل المشترك لسطح داره احٰ ولداره حده خط حٰ رٰ add. marg. A 8 ودائرة² [ودارهٰ A

9 دائرة أخرى [اخرى A, داره اخرى عظيمه N 9 وهى [فىٰ A

Sint itaque in spera super duos polos *H T* duo circuli *ABG DEZ*.
Dico igitur quod ipsi sunt equidistantes. Producam ergo lineam *HT*.
Et quia circulus *ABG* est in spera et iam protracta est linea transiens   5
per duos polos eius, que est linea *HT*, ergo linea *HT* est perpendicularis super circulum *ABG*. Similiter quoque ostenditur quod ipsa est
perpendicularis super circulum *DEZ*. Superficies vero, super quas una
et eadem linea cadit super eas existens perpendicularis, cum protrahentur, non concurrent Cum ergo protrahetur superficies circuli *ABG*   10
equidistabit circulo *DEZ*. Et illud est quod demonstrare voluimus.

3   Cum in spera fuerint duo circuli circumductam lineam alicuius maioris
circulorum qui sunt in ea super unum et idem punctum secantes et
poli eorum super illum circulum fuerint, duo circuli sese contingent.

Duo itaque circuli in spera *ABG DEG* lineam circuli *AGE*
circumductam super unum et idem punctum secent, quod sit punctum   5
G, et sint duo poli eorum super circulum *AGE*. Dico igitur quod duo
circuli *ABG DEG* sese contingunt. Sit itaque differentia communis
circuli *AGE* et circuli *ABG* linea *AG*, et differentia communis circuli
AGE et circuli *GDE* linea *GE*, et differentia communis circuli *ABG*
et circuli *GDE* sit linea *ZGH*. Circulus autem *AGE* maior ex circulis   10
qui sunt in spera secat alium ex circulis qui sunt in ea, qui est circulus

3 Sint] Sint ... *DEZ marg.* **P**   *3 H*] et *add.* **PRVMKgFi**   *4* igitur] *om.* **BPsVa**
4 ergo] igitur **OZBPs**   *4* lineam *HT*] *tr.* **Fi**   *5* iam] *om.* **OZBPsVa**   *9* eadem] *marg.*
BPs   *9* linea cadit] *tr.* **OZBPsVa**; cadit *add. et del.* **BPs**   *9* cum] et **Va**   *9–*
10 protrahentur] protrahuntur **KgOFiZBPs**, protrahantur **Va**   *10* concurrent] occurrent **R**,
concurrunt **KgOFiZBPsVa**   *10* protrahetur] protrahitur **OBPsVa**   *11* illud] id **Kg**, hoc
OZBPsVa   *11* quod demonstrare voluimus] *om.* **O**, proposuimus **Z**   *11* demonstrare] *om.*
BPsVa   *11* voluimus] volumus **BPsVa**   *1* fuerint duo] *tr.* **PVM**   *1* lineam] *om.* **Fi**
1 maioris] maiorum **PFiB**, magnorum **OZPsVa**   *3* circulum] *om.* **RFi** 3 contingent] contigent
V, contingunt **OFiZBPsVa**, contingant **Kg**   *4* spera] scilicet *add.* **MKgOZBPsVa** *4 DEG*] *supra*
BPs   *4* circuli] *om.* **OZBPsVa**   *4 AGE*] *AEG* **R**   *5* secent] secant **OBPsVa**   *6* poli] polorum
Fi, poli *corr. ex* polorum **Ps**   *6 AGE*] *corr. ex AEGE* **P**, *om.* **Va**   *7 ABG*] *AEG* **Z**   *7* Sit] Fit
B   *7–10* Sit itaque ... linea *ZGH*] *permut.***M,Kg,O,BPsVa**   *10* maior] ea *add.* **Fi**   *10* ex] eis
Z   *11* ex circulis] circulorum **PRVFiZ**   *11* ea] spera **FiZ**

ا ب ج وتمر بقطبيها فهى تقطعها بنصفين وعلى زوايا قائمة فخط ا ج قطر 10

دائرة ا ب ج وكذلك أيضاً نبين أن خط ج ه قطر دائرة ج د ه فلأن دائرة ا ج ه

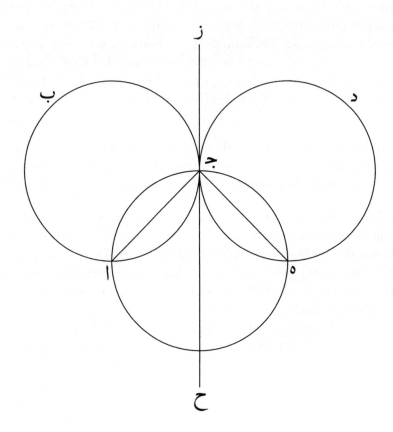

قائمة على كل واحدة من دائرتى ا ب ج ج د ه على زوايا قائمة

تكون كل واحدة من دائرتى ا ب ج ج د ه أيضاً قائمة على دائرة ا ج ه

على زوايا قائمة فيكون الفصل المشترك لهما أيضاً عموداً على دائرة ا ج ه

وذلك أنه إذا قام سطحان على سطح واحد على زوايا قائمة فإن الفصل 15

على زوايا قائمة] om. A   14 قائمة ... تكون [marg. A (om. أيضاً)   13 ا ج ه [A على [وعلى 10

ا ج ه [in corr. A   14 ه ج ه [N   14 فيكون [ويكون A

ABG, et transit per duos polos eius: ergo ipse secat eum in duo media et orthogonaliter. Ergo linea *AG* est diametrus circuli *ABG*. Simili quoque modo monstratur quod linea *GE* est diametrus circuli *GDE*.

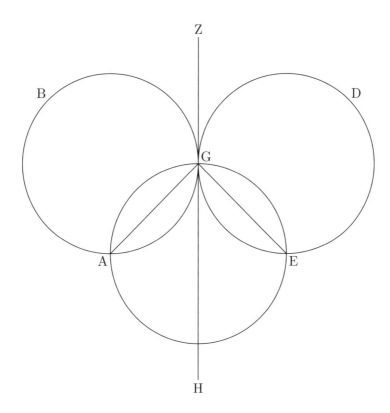

Et quia circulus *AGE* erigitur super unumquemque duorum circulorum *ABG GDE* orthogonaliter, ergo unusquisque duorum circulorum *ABG GDE* erigitur etiam super circulum *AGE* orthogonaliter. Differentia quoque eorum communis est etiam perpendicularis super circulum *AGE*. Quod ideo est quoniam cum due superficies super unam orthogonaliter eriguntur superficiem, tunc etiam differentia communis 20

15

12 transit] pertransit **P**   12 ergo] *repet.* **Va**   12–13 in duo ... orthogonaliter] *permut.* **MKgOBPsVa** (et *om.* **OBPsVa**)   13–14 Simili ... monstratur] similiter ostenditur **OBPsVa**   15 quia] quod **OB**, qui **Va**   15 unumquemque] utrumque **MKgOB PsVa**   15 duorum] *om.* **BPsVa**   16 *ABG*] *AGB* **VMFi**   16 orthogonaliter] *post* erigitur **MKgOBPsVa**   16 unusquisque] uterque **MKgOBPsVa**   17 *GDE*] *DGE* **M**, *GED* **OBPs**   17 etiam] *om.* **BPsVa**   17–20 circulum ... eriguntur superficiem] *marg.* **Ps**   17–21 circulum ... eandem superficiem] *marg.* **B**   17 *AGE*] *supra* **Kg**   18 quoque] *om.* **V**   18 est etiam] *tr.* **FiZ**   18 etiam] *om.* **OBPsVa**   19 est] *om.* **KgVa**   19 super unam] *marg.* **R**   19–20 super unam orthogonaliter] *permut.* **MKg**   20–21 tunc ... superficiem] *et in marg. et in textu* **B** (*om.* tunc)   20 etiam] et **BPs**; si *add.* **PRVMKgOBPsVa**   20–21 differentia communis earum] *permut.* **PRVMFiZ,Va**

المشترك لهما أيضاً يكون عموداً على ذلك السطح بعينه فهو أيضاً عمود على

جميع الخطوط المستقيمة التى تخرج من طرفه فى سطح دائرة اجه على زوايا

قائمة وقد خرج من طرفه كل واحد من خطى اج جه اللذين هما فى سطح

دائرة اجه فخط زح عمود على كل واحد من خطى اج جه فلأنه قد أخرج

20   من طرف قطر دائرة ابج خط زح على زوايا قائمة يكون خط زح مماساً

لدائرة ابج على نقطة جـ وكذلك نبين أن خط زح يماس دائرة جده أيضاً

على نقطة جـ والدوائر التى يقال إن بعضها يماس بعضاً فى الكرة هى التى

يكون الفصل المشترك لسطوحها مماساً لها جميعاً وخط زح مماس للدائرتين

جميعاً على نقطة جـ فدائرتا ابج جده أيضاً تماس إحداهما الأخرى ،

25   وذلك ما أردنا أن نبين .

د   إذا تماست دائرتان فى كرة فإن الدائرة العظيمة التى تمر بأقطابهما تمر أيضاً

بموضع تماسهما .

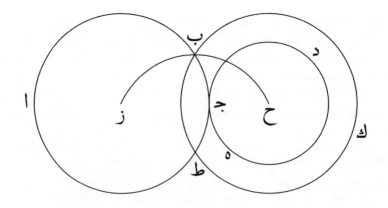

ماست A تماست [1 لداربس A لدائرتين [للدائرتين 22 إن [A ره [A om. A 23 om. N 20 زح [² زح ¹] om. N 20 زح 20

earum est perpendicularis super illam eandem superficiem, tunc ipsa
etiam est perpendicularis super omnes rectas lineas que ab eius extremitate in superficie circuli *AGE* orthogonaliter eriguntur. Sed iam
protracta est ab eius extremitate queque duarum linearum *AG GE*
que sunt in superficie circuli *AGE*: ergo linea *ZH* est perpendicularis   25
super unamquamque duarum linearum *AG GE*. Et quia iam protracta
est ab extremitate diametri circuli *ABG* linea *ZH* orthogonaliter, ergo
linea *ZH* contingit circulum *ABG* super punctum *G*. Similiter quoque
ostenditur quod linea *ZH* contingit etiam circulum *GDE* super punctum *G*. Circuli vero qui se ad invicem in spera contingere dicuntur   30
sunt quorum superficierum differentia communis eos simul contingit.
Sed linea *ZH* simul contingit duos circulos super punctum *G*. Ergo
duorum circulorum *ABG GDE* unus alterum contingit. Et illud est
quod demonstrare voluimus.

4   Cum duo circuli in spera se contingunt, maior circulus qui per polos
    eorum transit transit etiam per locum contactus eorum.

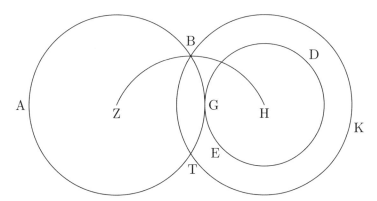

21 illam eandem] *tr.*   **Z**   21 ipsa] ipse   **P**   21–22 ipsa etiam] *tr.*   **Fi**
22 omnes rectas lineas] *permut.*   **MKgOBPsVa**   22 eius] *supra*   **Ps**   23 *AGE*] *ABG*
FiZ   23 eriguntur] *corr. ex* erigitur **Ps**, erigitur **PRVKgOFiZ** (*supra illeg.* **O**), erigig̃t̃ **B**, *illeg.*
M   24 est] *om.* **Va**   24 eius] *om.* **PRVMKgBPsVa**   24 queque] cuiusque **PRV**, cuiusque
supra **M**, *om.* **KgBPsVa**, quam **Fi**   25–26 que … *AG GE*] *om.* **Va**   25 sunt] super *add.*
V   25 *AGE*] *om.* **B**, *supra* **Ps**; linea *ZH* *add.* **PVKgPs**, *marg.* **R**, *supra* **B**, *illeg.* **M**, *add.*
et del. **O**   25 ergo linea *ZH*] *marg.* **B**, *supra* **Ps**   25–26 ergo … *AG GE*] *marg.* **O**, *repet.*
Kg   26 unamquamque] utramque **MKgOBPs**   26 linearum] *supra* **R**   26 *GE*] que sunt
in superficie circuli *AGE* *add.* **KgOFiZ**   26 quia] *repet.* **V**   26 protracta] producta **Va**
28 *ZH*] *ZA* **Va**   28 contingit] *corr. ex* coniungit **Ps**   29 contingit] *corr. ex* coniungit
Ps   29 etiam] *om.* **OFiZBPsVa**   29 circulum] *om.* **FiZ**   30 in spera] *ante* se **OFiZ**, *om.*
BPsVa   30 contingere] *corr. ex* coniungere **Ps**   31 sunt] *supra* **Kg**   31 simul] similis **Fi**
31 simul contingit] *tr.* **OBPsVa**   32 simul] eos *add.* **PsVa**   32 duos circulos] *om.* **BPsVa**
33 *ABG*] et *add.* **O**   33 contingit … voluimus] et cetera **BPsVa**   33 Et … voluimus] Et
hoc est **O**   34 voluimus] volumus **FiZ**   *1* se] sese **MKgOBPsVa**   *2* transit[2]] *om.* **Kg**, *post*
eorum **OBPsVa**   *2* transit etiam] *tr.* **Z**   *2* transit[2] … locum] *permut.* **Fi**   *2* locum] locos
KgOBPsVa; vel .cum *supra* **O**

فلتماس فى كرة دائرتا ا‍ب‍ج‍ جده إحداهما الأخرى على نقطة ج‍

ولتكن نقطة ز قطباً لدائرة ا‍ب‍ج‍ ونقطة ح قطباً لدائرة جده ، فأقول إن

الدائرة العظمى التى تمر بقطبى ز ح تمر أيضاً بنقطة ج ، لا يمكن غير   5

ذلك فإن أمكن فلا تمر بها ولتكن مثل دائرة ز‍ب‍ح‍ ولترسم على قطب ح

وببعد ح‍ب‍ دائرة ب‍ك‍ط‍ فدائرة جده موازية لدائرة ب‍ك‍ط‍ وذلك أنهما

جميعاً على أقطاب بأعيانها فلأن دائرتى ا‍ب‍ج‍ ب‍ك‍ط‍ فى كرة وهما تقطعان

خطاً محيطاً بدائرة ما عظيمة وهو خط ز‍ب‍ح‍ على نقطة ب وأقطابهما على

تلك الدائرة تكون دائرتا ا‍ب‍ج‍ ب‍ك‍ط‍ متماستين وقد تقاطعتا أيضاً وذلك   10

محال فليس يمكن ألا تمر الدائرة العظمى التى تمر بقطبى ز ح بنقطة ج

فالدائرة العظمى التى تمر بأقطاب دائرتى ا‍ب‍ج‍ جده تمر أيضاً بموضع

تماسهما ، وذلك ما أردنا أن نبين .

٥   إذا تماست دائرتان فى كرة فإن الدائرة العظمى التى تمر بقطبى إحدى

الدائرتين وبموضع التماس تمر أيضاً بقطبى الدائرة الأخرى .

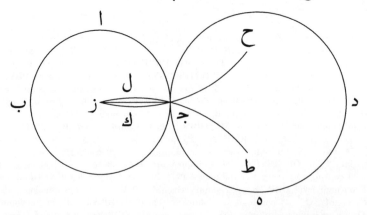

4 نقطة ... ا‍ب‍ج‍ [1 marg. A] تمر 6 تمرز [N وهما 8 تقطعان [N وهى 8 نقطة [NH أيضاً om. 10

NH تماسها A, مماستهما [تماسهما 13 A العظمه [العظمى 12 A

Duorum itaque circulorum *ABG GDE* in spera unus alterum super punctum *G* contingat sitque punctum *Z* polus circuli *ABG* et punctum *H* polus circuli *GDE*. Dico igitur quod circulus maior qui transit per duos polos *Z* et *H* transit etiam per punctum *G*. Non est possibile aliter esse. Quod si possibile fuerit, non transeat per ipsum sitque equalis circulo *ZBH*. Signabo autem super polum *H* et secundum spacium *HB* circulum *BKT*. Circulus igitur *GDE* equidistat circulo *BKT*. Quod ideo est quoniam ipsi simul sunt super eosdem polos Et quia duo circuli *ABG BKT* sunt in spera et secant circumductam lineam alicuius maioris circuli, que est linea *ZH*, super punctum *B* et eorum poli sunt super illum circulum, ergo duo circuli *ABG BKT* sese contingunt Sed iam etiam fuerunt sese secantes, quod est impossibile. Non est ergo possibile quin circulus maior qui transit per duos polos *Z* et *H* transeat per punctum *G*. Circulus ergo maior qui transit per polos duorum circulorum *ABG GDE* transit etiam per locum contactus eorum. Et illud est quod demonstrare voluimus.

5 Cum in spera duo circuli se contingunt, maior circulus qui transit per duos polos unius eorum et per locum contactus transit etiam per polos alterius circuli.

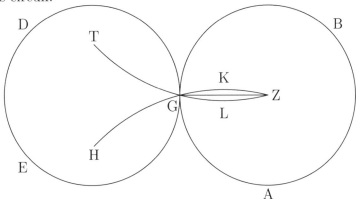

3 Duorum] Quorum **Kg** *3* Duorum itaque] *tr.* **OBPsVa** (duorum *supra* **Ps**) *4* contingat] contingit **BPsVa** *5* punctum *H*] *tr.* **BPsVa** *5* igitur] *om.* **BPsVa** *6* et] *om.* **BPsVa** *6* Non] Nec **M**; enim *add.* **OFiZVa**, *illeg. supra* **Ps** *7* Quod] Sed **Kg** *7* Quod si] si vero **BPs**, si **Va**; non *add.* **Va** *7* possibile] hoc *add.* **PsVa** *7* transeat] transiens **B**; circulus ille *add. marg.* **Ps** *8* *ZBH*] *BZH* **KgB** *8* polum] punctum *Z* *8–9* secundum] supra **OBPsVa** *9* igitur] ergo **OZBPsVa** *9* equidistat] qui distat **Va** *10* quoniam] quia **OBPsVa** *10* ipsi] ipsa **Va** *10* ipsi simul] *om.* **Fi** *10* simul] *om.* **Z** *10* sunt] ē **Kg** *10* eosdem] eodem **M** *11* circuli] *om.* **Kg** *11* *ABG BKT*] *BKT ABG* **OBPsVa** *11* et] *om.* **R** *11* secant] *marg.* **R** *12* linea] *om.* **Kg** *12* punctum] idem *supra* **Ps** *13* illum] *om.* **Kg**, alium **O** *13* *ABG*] *BAG* **OBPsVa** *13* sese] se **MKgOBPsVa** *14* Sed iam] Et si **B**, Et iam **PsVa** *14* etiam] *supra* **M**, *om.* **KgBPsVa** *14* est] cum **Va** *15* Non ... possibile] est **Fi** *15* est ergo] *tr.* **MOZ** *15* ergo] *om.* **B**, *supra* **Ps** *16* et] *om.* **BPsVa** *16* ergo] igitur **ROBPsVa** *17* polos duorum] *tr.* **PsVa** *17* locum] polos **Z** *18* eorum] *om.* **Kg** *18* Et ... voluimus] *om.* **O** *18* est] *om.* **Fi** *18* demonstrare] monstrare **Fi** *18* demonstrare voluimus] volumus demonstrare **BPsVa** *18* voluimus] volumus **Z** *1* spera] sint *add.* **Va** *1* circuli] et *add.* **Va** *1* se] sese **OFiZBPsVa** *1* contingunt] contingant **FiZ** *2* locum] polum **O**

فلتماس فى كرة دائرتا ا ب ج ج ده إحداهما الأخرى على نقطة جـ
وليكن قطب دائرة ا ب جـ نقطة ز وقطب دائرة ج ده نقطة حـ ، فأقول إن
الدائرة العظمى التى تمر بنقطتى ز جـ تمر أيضاً بنقطة حـ فإن لم يكن ذلك
كذلك وأمكن غيره فلتخرج ولتكن مثل دائرة ز ج ط وتخرج دائرة أخرى
عظيمة تمر بقطبى ز حـ فهى تمر بنقطة جـ أيضاً ولأن كل واحدة من دائرتى
ز ج حـ ز ج ط عظيمة صارت كل واحدة منهما تقسم الأخرى بنصفين فكل
واحدة من قوسى ز ك جـ ز ل جـ نصف دائرة ز جـ قطر الكرة لأنه قطر
دائرتى ز ج حـ ز ج ط العظماوين ولكنه أيضاً خرج من قطب دائرة ا ب جـ
وذلك غير ممكن فالدائرة العظمى التى تمر بنقطتى ز جـ تمر بنقطة حـ
أيضاً ، وذلك ما أردنا أن نبين .

و　إذا ماست دائرة عظيمة فى كرة دائرة أخرى من الدوائر التى فى الكرة فإنها
تماس دائرة أخرى مساوية لتلك الدائرة موازية لها .

فلتماس فى كرة دائرة ا ب جـ العظمى دائرة أخرى من الدوائر التى فى
الكرة وهى دائرة ج د على نقطة جـ ، فأقول إن دائرة ا ب جـ تماس دائرة
أخرى مساوية موازية لدائرة ج د فلتتعلم قطب دائرة ج د وليكن نقطة هـ
ولنرسم دائرة عظيمة تمر بنقطتى جـ هـ وهى دائرة ج ه د ب ز حـ ولنفصل منها
قوس ب ز ونجعلها مساوية لقوس ج هـ ولنرسم على قطب ز وببعد ز ب

3 فلتماس ... الأخرى [marg. A 8 بنصفين [פכט זג קטר אלדאירה אלעטימה (fin. ה incert.) marg.
H 10 العظماوين [corr. ex العظمى N, العطمس A 10 أب جـ [الى محطها add. A 12 أيضاً [repet. A
2 لتلك [لدلك, marg. A 2 موازية [موازية A 5 ومواربه A وموارة [ز 7 A] illeg. N

Duorum itaque circulorum *ABG GDE* in spera unus alterum contingat super punctum *G*, et sit polus circuli *ABG* punctum *Z*, et polus   5
circuli *GDE* sit punctum *H*. Dico igitur quod circulus maior qui transit
per duo puncta *Z* et *G* transit etiam per punctum *H*. Quod si hoc non
ita fuerit et fuerit possibile aliter esse, tunc notabo illum sitque equalis
circulo *ZGT*. Producam autem alterum circulum maiorem transeuntem per duos polos *Z* et *H*. Ergo ipse transit etiam per punctum *G*.   10
Et quia unusquisque duorum circulorum *ZGH ZGT* est maior, ergo
quisque eorum secat alterum in duo media; ergo unusquisque duorum
arcuum *ZKG ZLG* est semicirculus Ergo linea *ZG* est diametrus spere
quoniam ipsa est diametrus duorum circulorum maiorum *ZGH ZGT*.
Ipsa autem protrahitur etiam a polo circuli *ABG* quod est impossibile.   15
Circulus igitur maior qui transit per duo puncta *Z G* transit etiam per
punctum *H*. Et illud est quod demonstrare voluimus.

6 Cum circulus maior in spera alium ex circulis qui sunt in spera contingit, ipse contingit circulum alium equalem illi circulo qui ei equidistat.

In spera itaque circulus maior *ABG* contingat alium ex circulis
qui sunt in spera, qui sit circulus *GD*, super punctum *G*. Dico igitur
quod circulus *ABG* contingit circulum alium equalem et equidistantem   5
circulo *GD*. Signabo igitur polum circuli *GD*, qui sit punctum *E*,
et signabo circulum maiorem transeuntem per duo puncta *G E*,
qui sit circulus *GEDBZH*; ex quo abscindam arcum *BZ*, quem ponam
equalem arcui *GE*. Et signabo super polum *Z* et cum longitudine *ZB*

5–6 polus[2] ... punctum *H*] *permut.* **MKgOBPsVa**   6–7 Dico ... punctum *H*] *om.*   **R**
7 punctum] puncta **Z**   7 Quod] Et **FiZ**   7–8 Quod ... fuerit[1]] Quod si ita non fuerit
KgOBPsVa   8 esse] *om.* **B**, *supra* **Ps**   8 tunc] ideo **Z**   9 *ZGT*] *GTZ* **O**, *GZT* **B**,
ZET **Va**   9 autem] igitur **OBPsVa**   9 circulum] *om.* **BPsVa**   9 circulum maiorem] *tr.*
O   10 et] *om.* **KgBPsVa**   10 transit etiam] *tr.* **OFiZBPsVa**   11 unusquisque] uterque
MKgOBPsVa   12 quisque] uterque **MKgOBPsVa**   12 secat alterum] alterum secat et
Va   12 unusquisque] uterque **MKgOBPsVa**   13 arcuum] *marg.* **R**   13 *ZKG*] *ZHG*
R, *ABG* **Fi**   14 quoniam] quia **BPsVa**   14 duorum] *om.* **OBPsVa**   14 *ZGH*] *ZLGH*
V   14 *ZGH ZGT*] *om.* **OBPsVa**   14 *ZGT*] *GT* **Fi**   15 autem] *om.* **OBPsVa**
15 protrahitur etiam] *tr.* **OBPsVa**   15 polo] *Z supra* **Ps**   16 *Z*] et *add.* **Va**, et *add.*
supra **Ps**   16 per] *om.* **V**   17 illud] istud **OVa**   17 est] *om.* **VFi**   17 demonstrare] *om.*
OBPsVa   17 voluimus] volumus **OZBPsVa**   1 maior in spera] *permut.* **KgOBPsVa**
1 ex circulis] circulorum **PRVFiZ**   1 in spera[2]] *om.* **B**, in ea **PsVa**   1–2 contingit] *corr.*
ex contingunt **Kg**   2 circulum alium] *tr.* **KgOBPsVa**   2 alium] *om.* **V**   2 equidistat] in
spera *add.* **KgOBPsVa**   3 In spera] *om.* **KgBPsVa**   3 itaque circulus] *tr.* **Kg**   3 maior] *om.*
BPsVa   3 *ABG*] *repet.* **Va**   3 contingat] continguat **V**   3 ex circulis] circulorum
PRVFiZ   4 qui sit circulus] scilicet circulum **MKgOBPsVa**   5 contingit] contingat **M**
5 circulum alium] *tr.* **RFiZ**; et *add.* **PsVa**   5 et equidistantem] *supra* **Ps**   6 *GD*[1]] *om.*
Fi   6 Signabo] Significabo **PsVa**   6 qui] quod **OBPsVa**   7 signabo] significabo **PsVa**
7 *G*] et *add.* **PMKgOFiBPsVa**   8 *GEDBZH*] *corr. ex GDEDBRZH* **V**, *GDBZH* **Fi**   8 ex] a
OBPsVa   8 abscindam] abscidam **VMZ**   8 quem] *in corr.* **BPs**   8 ponam] ponam[...] **Fi**
9 signabo] significabo **Va**   9 *Z*] *ZI* **Kg**

دائرة بَ‍حَ فلأن دائرتى ابَ‍جَ جَدَ تماس إحداهما الأخرى وهما فى كرة

وقد رسمت فى الكرة دائرة تمر بقطب دائرة جَدَ وهو نقطة هَ وبموضع

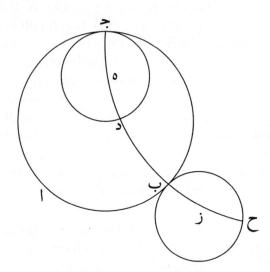

الماسة وهى دائرة جَهَدَبَ‍زَحَ صارت دائرة جَهَدَبَ‍زَحَ تمر بقطبى دائرة

ابَ‍جَ أيضاً ولأن دائرتى ابَ‍جَ بَ‍حَ فى كرة تقطعان خطاً محيطاً بدائرة

أخرى عظيمة على نقطة واحدة بعينها وهى نقطة بَ وقطباهما على الدائرة

تكون كل واحدة من دائرتى ابَ‍جَ بَ‍حَ مماسة للأخرى ولأن قوس جَهَ

مساوية لقوس بَ‍زَ وقوس هَبَ مشتركة تكون كل قوس جَهَبَ مساوية

لكل قوس هَزَ وقوس جَهَبَ نصف دائرة فقوس هَزَ نصف دائرة فنقطة هَ

مقابلة لنقطة زَ ونقطة هَ قطب دائرة جَدَ فنقطة زَ القطب الآخر من دائرة

جَدَ وأيضاً فلأن قوس هَزَ نصف دائرة ونقطة زَ قطب دائرة بَ‍حَ تكون

نقطة هَ أيضاً قطب دائرة بَ‍حَ فدائرتا جَدَ بَ‍حَ على أقطاب بأعيانها

---

10 الماسة] من الدوائر العظيمة .N add. H من اللدائرة אלעטימה .add. N   11 تقطعان] مطع N

12 عظيمة] .om. A   16–17 زَ لنقطة ... زَ] .om. N ¹ دائرة

---

11 أيضاً] من الشكل المعدم .marg. N

circulum *BH*. Et quia duorum circulorum *ABG GD* unus alterum con- 10
tingit in spera et in spera iam signatus fuit circulus maior transiens per
polum circuli *GD*, qui est punctum *E*, et per locum contactus duorum

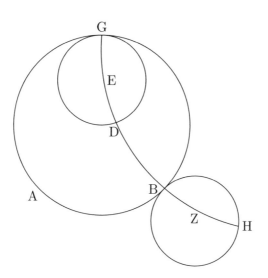

circulorum, qui est circulus *GEDBZH*, ergo circulus *GEDBZH* transit
etiam per polos circuli *ABG*. Et quia duo circuli *ABG BH* in spera
secant lineam alium circulum continentem super unum et idem punc- 15
tum, quod est punctum *B*, et eorum poli sunt super illum circulum,
ergo unusquisque duorum circulorum *ABG BH* contingit alterum. Et
quia arcus *GE* est equalis arcui *BZ*, ergo arcu *EB* communi erit totus
arcus *GEB* equalis toti arcui *EZ*. Sed arcus *GEB* est semicirculus:
ergo arcus *EZ* est semicirculus. Ergo punctum *E* opponitur puncto *Z*. 20
Sed punctum *E* est polus circuli *GD*, ergo punctum *Z* est alter polus
circuli *GD*. Et etiam quia arcus *EZ* est semicirculus et punctum *Z* est
polus circuli *BH*, ergo punctum *E* etiam est polus circuli *BH*. Ergo
duo circuli *GD BH* sunt super eosdem polos. Circuli vero qui sunt super

10 *ABG*] *ABGD* **R**   10 *GD*] *DG* **KgOBPsVa**   11 signatus] signandus **Fi**
11 signatus fuit] *tr.* **OBPsVa**   11 fuit] est **FiZ**   12 circuli *GD*] *tr.* **BPsVa** (*GD corr. ex GB*
B, *GB* **PsVa**)   12 qui] quod **R**   12 locum] punctum **PRVMFiZ**   13 qui est circulus] *marg.* **B**;
maior [...] *add. supra* **Ps**   13 ergo circulus *GEDBZH*] *marg.* **PPs**   13 *GEDBZH*²] *DGEDBZH*
Z   14 etiam] et **Z**   15 alium circulum] *tr.* **FiZ**   16 et] *supra* **Ps**   17 unusquisque] uterque
MKgOBPsVa   18 arcus] *marg.* **R**, *om.* **BPs**, alterum **Va**   18–19 *BZ* ... toti arcui] *om.*
Va   18 ergo] igitur **Fi**   18 arcu] arcus **V**, arcui **ZB**   18 *EB*] *BE* **MKgOBPs**, *BZ* **Z**
20 ergo ... semicirculus] *om.* **Va**   20 opponitur] apponitur **Kg**   22 circuli *GD*] eiusdem
circuli **BPsVa**   22–23 *GD* ... circuli²] *om.* **Fi**   22 arcus] *om.* **BPsVa**   23 *E*] *HE* **Kg**
23 etiam] *supra* **M**   23 etiam est polus] *permut.* **KgOZBPsVa**   23 est] *om.* **M**   23 *BH*²] *TH*
Fi, *H* **Z**   24 *GD*] *DG* **OFiZBPsVa**   24 eosdem] eodem **Fi**   24–25 Circuli ... polos] *marg.*
BPs

والدوائر التى تكون على أقطاب بأعيانها هى متوازية فدائرة جد موازية

لدائرة بح ولأن خط جه مساوٍ لخط بز تكون دائرة جد أيضاً مساوية   20

لدائرة بح وقد كانت موازية لها فدائرة ابج تماس دائرة أخرى مساوية

لدائرة جد وموازية لها ، وذلك ما أردنا أن نبين .

ز   إذا كانت فى كرة دائرتان متساويتان متوازيتان فإن الدائرة العظمى التى تماس

إحداهما تماس الأخرى أيضاً .

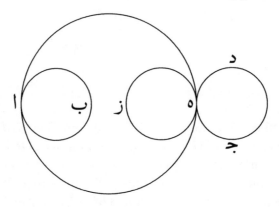

فلتكن فى كرة دائرتان متساويتان متوازيتان وهما اب جد ، فأقول إن

الدائرة العظمى التى تماس دائرة اب تماس أيضاً دائرة جد فإن أمكن ألا

يكون كذلك فلتماس دائرة اه العظمى دائرة اب على نقطة آ ولا تماس دائرة   5

جد فلأن دائرة اه العظمى التى فى الكرة تماس دائرة ما من الدوائر التى فى

الكرة وهى دائرة اب فهى تماس أيضاً دائرة أخرى مساوية لدائرة اب

وموازية لها فلتماس دائرة هز فدائرة اب مساوية موازية لدائرة هز وقد كانت

20 خط جه مساوٍ لخط ر] قوس حه مساوه لهوس A   20 جه] ه N   1 متوازيتان] om. N; illeg. in
marg. N   4 ألا] اب A   5 اه] آه A   5 [اب] N اب 5 ولا] A لا   6 ما] N اما 8 موازيه] وموازيه A
8–9 دائرة ... وقد] هز (وموازية : موازية) A marg. A

eosdem polos sunt equidistantes. Ergo circulus *GD* equidistat circulo　25
BH. Et quia linea *GE* est equalis linee *BZ*, ergo circulus *GD* est etiam
equalis circulo *BH*. Iam vero fuit equidistans ei: circulus igitur *ABG*
contingit alterum circulum equalem circulo *GD* et equidistantem ei.
Et illud est quod demonstrare voluimus.

7　Cum in spera fuerint duo circuli equales et equidistantes, circulus maior
qui contingit unum eorum continget etiam alterum.

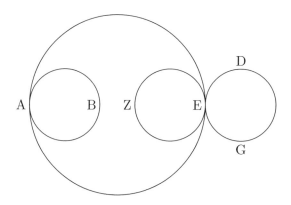

　　　　In spera igitur sint duo circuli equales et equidistantes *AB GD*.
Dico igitur quod circulus maior qui contingit circulum *AB* contingit
etiam circulum *GD*. Quod si est possibile ut non sit ita, contingat　5
maior circulus *AE* circulum *AB* super punctum *A* et non contingat
circulum *GD*. Et quia circulus maior *AE* qui est in spera contingit
unum ex circulis qui sunt in spera, qui est circulus *AB*, ergo ipse etiam
contingit circulum alium equalem circulo *AB* et equidistantem ei; contingat ergo circulum *EZ*. Ergo circulus *AB* est equalis et equidistans　10
circulo *EZ*. Sed circulus *AB* iam fuit equalis circulo *GD*: in spera ergo

25 circulo] *om.* **FiZ**　　26 *GE*] *DE* **R**　　26 est etiam] *tr.* **MKgOFiBPsVa**　　26 etiam] *om.*
RZ　　27 fuit equidistans ei] *permut.* **PVMKgBPsVa,OZ**　　27 igitur] ergo **MKgOB**
PsVa　　28 alterum circulum] *tr.* **OBPsVa**　　28 equidistantem] equidistante **Fi**
29 Et … voluimus] *om.* **OBPsVa**　　29 illud] hoc **FiZ**　　29 demonstrare voluimus] *tr.* **V**
2 contingit] contigit **V**　　2 eorum] *om.* **FiZ**　　2 continget] contingit **FiZ**　　2 etiam] et
VOBPsVa　　2 alterum] reliquum **OBPsVa**　　3 In … sint] Sint in spera **OBPsVa** (Sint: Sunt **Va**)
3 sint] sunt **KgZ**　　3 *AB GD*] *ante* equales **PRVMKgFiZ**　　3 *GD*] *ED* **B**　　4 igitur] ergo
VOFiBPs (**Ps** *supra*)　　4 maior] *supra* **Ps**　　4 contingit[1]] contingit **OBPsVa**　　4 circulum] unum
circulum scilicet **OBPsVa**, unum eorum scilicet circulum **FiZ**　　4 contingit[2]] continget **O**
5 etiam] *om.* **FiZ**; alterum scilicet *add.* **OFiZB**　　5 est possibile] *tr.* **MKgOBPsVa**　　5 ut] *supra*
M　　6 maior] *om.* **Va**　　8 qui est circulus] scilicet circulum **MOKgBPsVa**　　8 ipse etiam] *tr.* **O**
8 etiam] *om.* **V**　　9 circulum] *om.* **BPsVa**　　11 fuit] fuerat **OBPsVa**　　11 equalis] equidistans
supra **M**

دائرة اب مساوية موازية لدائرة جد فتكون فى كرة واحدة ثلث دوائر

متساوية متوازية وذلك غير ممكن فليس يمكن ألا تماس الدائرة العظمى 10

التى تماس دائرة اب دائرة جد فهى مماسة لها ، وذلك ما أردنا أن نبين .

ح   إذا كانت فى كرة دائرة عظيمة مائلة على دائرة أخرى من الدوائر التى فى

الكرة فهى تماس دائرتين مساوية إحداهما للأخرى موازيتين للدائرة الأخرى

التى تقدم ذكرها .

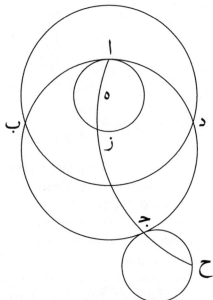

فلتكن دائرة ابج العظمى التى فى الكرة مائلة على دائرة من الدوائر

التى فى الكرة وهى دائرة بد أعنى أن لا تكون مارة بقطبى دائرة بد ، 5

فأقول إن دائرة ابج تماس دائرتين مساوية إحداهما للأخرى موازيتين

لدائرة بد فلأن دائرة ابج مائلة على دائرة بد لا يكون قطب دائرة

و [موازيتين AN 2 الاخرى [للأخرى A 2 ان لا [ألا A] *om.* N 10 [فتكون N *supra*, د [آ—ح [جد 9
N عن [على A 7 *add. supra* A

sunt tres circuli equales et equidistantes, quod est impossibile. Non est igitur possibile quin circulus maior qui contingit circulum *AB* contingat etiam circulum *GD*. Ergo ipse contingit eum. Et illud est quod demonstrare voluimus.   15

8 Cum in spera fuerit circulus maior super alium ex circulis qui sunt in spera inclinatus, ipse contingit duos circulos quorum unus est alteri equalis equidistantes alteri circulo quem prediximus.

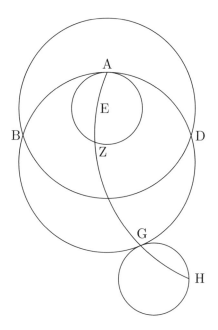

Sit itaque circulus maior *ABG* qui est in spera inclinatus super unum ex circulis qui sunt in spera, qui sit circulus *BD*, scilicet non sit   5 transiens per duos polos circuli *BD*. Dico igitur quod circulus *ABG* contingit duos circulos quorum unus est alteri equalis equidistantes circulo *BD*. Et quia circulus *ABG* inclinatur super circulum *BD*, ergo polus circuli *BD* non est super circulum *ABG*. Signabo ergo polum cir-

12 est[1]] cum **Va**   *13* igitur] ergo **MOBPsVa**   *13* circulus maior] *tr.* **Va**   *13* maior] *supra*
Ps   *13* qui] *om.* **Kg**   *13* qui ... circulum *AB*] *AB* contingens **BPs** (circulum *add. supra*
AB **Ps**, circulum *AB* contingens **Va**)   *14* etiam] *om.* **FiZ**   *14–15* Et ... voluimus] *om.* **O**
14 illud] hoc **FiZBPsVa**   *14* est] *om.* **V**   *14–15* est ... voluimus] *om.* **Kg**   *14* quod] est
add. **M**   *15* demonstrare voluimus] volumus demonstrare **BPsVa**   *1* alium] alterum **BPsVa**
1 ex circulis] circulorum **PRVOFiZ**   *2* contingit] continget **OFiZ**   *3* prediximus] diximus
BPsVa   *4* est] *supra* **M**, *om.* **Kg**   *5* ex circulis] circulorum **PRVFi**, circulum **Z**   *5* qui] cum
Fi   *5* scilicet] sed **OFiZBPsVa**   *6* *BD*] *ABD* **R**, *DB* **BPsVa**   *6* quod circulus] *om.* **BPsVa**
7 est alteri] *tr.* **MKgOBPsVa**   *7* equalis] *in corr.* **R**   *8–9* ergo ... *ABG*] *repet.* **Va** (*prima
vice BD pro ABG*)   *9* Signabo] Significabo **Va**

ب‍د على دائرة ا‍ب‍ج‍ فلنتعلم قطب دائرة ب‍د ولتكن نقطة ه‍ ولنرسم دائرة

عظيمة تمر بنقطة ه‍ وبقطبى دائرة ا‍ب‍ج‍ وهى دائرة ا‍ه‍ح‍ ولنرسم على قطب

ه‍ وببعد ه‍ا‍ دائرة ا‍ز‍ فدائرة ا‍ز‍ موازية لدائرة ب‍د وذلك أنهما على أقطاب   10

بأعيانها ولأن دائرتى ا‍ب‍ج‍ ا‍ز‍ اللتين فى الكرة تقطعان خطاً محيطاً بدائرة

عظيمة من الدوائر التى فى الكرة وهى دائرة ا‍ه‍ز‍ح‍ على نقطة واحدة بعينها

وهى نقطة ا‍ وأقطابهما عليه تكون الدائرتان متماستين فدائرة ا‍ب‍ج‍ تماس

دائرة ا‍ز‍ ولأن دائرة ا‍ب‍ج‍ العظمى فى كرة وتماس دائرة ما من الدوائر التى

فى الكرة فهى أيضاً تماس دائرة أخرى مساوية موازية لدائرة ا‍ز‍ فلتماس دائرة   15

ج‍ح‍ فلأن دائرة ا‍ز‍ مساوية موازية لدائرة ج‍ح‍ ودائرة ا‍ز‍ موازية لدائرة ب‍د

تكون دائرة ج‍ح‍ موازية لدائرة ب‍د فدائرة ا‍ب‍ج‍ تماس دائرتين مساوية

إحداهما للأخرى موازيتين لدائرة ب‍د ، وذلك ما أردنا أن نبين .

ط   إذا كانت فى كرة دائرتان تقطع إحداهما الأخرى ورسمت دائرة عظيمة تمر

بأقطابهما فإنها تقسم القطع التى فصلت من الدوائر بنصفين نصفين .

فلتقطع فى كرة دائرتا ز‍ا‍ه‍ب‍ ز‍ج‍ه‍د‍ إحداهما الأخرى على نقطتى ه‍ ز‍

ولترسم دائرة عظيمة تمر بأقطابهما وهى دائرة ا‍ج‍ب‍د‍ ، فأقول إن دائرة

ا‍ج‍ب‍د‍ تقسم القطع التى فصلت من الدوائر بنصفين نصفين أعنى أن قوس   5

ز‍ا‍ تكون مساوية لقوس ا‍ه‍ وتكون قوس ز‍ب‍ مساوية لقوس ب‍ه‍ وتكون

8 دائرة [2 om. A   8 ه‍] om. N   10 ه‍ا‍] supra A   11 بأعيانها [باعيانها N   11 ولأن [فلان A

N اقطابها [بأقطابهما   2 وموارة [موازية A   16 وموارة [موازية 1]   15 om. A   15 أيضاً [أيضا

culi *BD* sitque punctum *E*. Et signabo circulum maiorem transeuntem   10
per punctum *E* et per duos polos circuli *ABG*, qui sit circulus *AEH*. Et
describam super polum *E* secundum longitudinem *EA* circulum *AZ*.
Ergo circulus *AZ* equidistat circulo *BD*. Quod ideo est quoniam ipsi
sunt super eosdem polos. Et quia duo circuli *ABG AZ* qui sunt in
spera secant lineam continentem circulum maiorem ex eis qui sunt in   15
spera, qui est circulus *AEZH*, super unum et idem punctum, quod est
punctum *A*, super quem sunt eorum poli, ergo duo circuli sese contingunt. Ergo circulus *ABG* contingit circulum *AZ*. Et quia circulus *ABG*
maior est in spera et contingit aliquem ex circulis qui sunt in spera,
ergo ipse etiam contingit circulum alium equalem et equidistantem cir-   20
culo *AZ*. Contingat ergo circulum *GH*. Et quia circulus *AZ* est equalis
et equidistans circulo *GH* et circulus *AZ* equidistat circulo *BD*, ergo
circulus *GH* est equidistans circulo *BD*. Ergo circulus *ABG* contingit
duos circulos, quorum unus alteri est equalis, equidistantes circulo *BD*.
Et illud est quod demonstrare voluimus.   25

9 Cum in spera fuerint duo circuli quorum unus alterum secet et
signabitur circulus maior transiens per polos eorum, ipse dividet porciones circulorum quas secat in duo media et duo media.

Duorum itaque circulorum *AEBZ ZGED* in spera unus alterum
super duo puncta *E* et *Z* secet. Signabo autem circulum maiorem   5
transeuntem per polos eorum, qui sit circulus *AGBD*. Dico igitur quod
circulus *AGBD* dividit porciones quas secuit ex circulis in duo media
et duo media, scilicet quod arcus *ZA* est equalis arcui *AE* et arcus
ZB est equalis arcui *BE* et arcus *ZG* est equalis arcui *GE* et etiam

11 *E*] *B* **Va**   12 longitudinem] linee *add.* **FiZ**   12 *EA*] *corr. ex* eam **B**   12 *AZ*] *AEZ*
BPsVa 13 Quod] Et **OBPsVa**   13 est] *supra* **R**   13 ipsi] circuli illi **BPsVa**   17 punctum] *om.*
OBPsVa   17 *A*] *om.* **KgB**, *supra* **Ps**   17 sese] se **MKgBPsVa**   18 quia] quod **Kg**
19 ex circulis] circulorum **PRVFiZ**, eorum **OBPsVa**   20 ergo] sic **FiZ**   20 etiam] *om.*
Fi   20 etiam contingit] *tr.* **O**   20 et equidistantem] *supra* **Ps**   21 *AZ*] *supra* **Ps**
21 ergo] etiam **OBPsVa**   23 est equidistans] equidistat **MBPsVa**   24 alteri est] *tr.*
FiZ   24 est equalis, equidistantes] equatur et equidistant **KgOBPsVa** (equidistantes **Kg**)
24 equalis] equali **Fi**   24 *BD*] *BV* **Fi**   25 Et … voluimus] Et hoc est **OBPsVa**
25 est … voluimus] *om.* **Kg**   25 quod] quia **Fi**   2 signabitur] significabitur **MOBPsVa**
2 circulus maior] *tr.* **Kg**   2 ipse] *supra* **O**, *om.* **FiZ**   2 dividet] dividit **M**   3 quas] qua
V   3 secat] separat **OFiZ**; divide *supra* **Z**; ex circulis *add.* **MFiZ**   3 et duo media] *om.*
R   4 *AEBZ*] *ABEZ* **Fi**   4 *ZGED*] *ZEGD* **O**, et *GD* **Va**   4 alterum] *om.* **Kg**
5 *E* … secet] secet super *E* et *Z* **OBPsVa**   5 secet] secat **R**   5 Signabo] Significabo **Va**
6 *AGBD*] **OFiZB**   6 igitur] ergo **B**   7 *AGBD*] *ABGD* **OFiZ**   7 ex circulis] *om.*
KgBPsVa 8 *AE*] *BE* **Fi**, *DZ* **Z**, *illeg.* **B** 8–9 *AE … ZB*] *marg.* **P** 8–9 *AE … arcui*[1]] *om.*
Fi   9 est equalis[1]] equatur **OBPsVa**   9 est equalis arcui[1]] *om.* **P**   9 est[2]] *om.* **Kg**
9 est equalis[2]] equatur **OBPsVa**

قوس زج مساوية لقوس جه وتكون قوس زد مساوية لقوس ده فليكن
الفصل المشترك لدائرتى اجبد زاه ب خط اب ولدائرتى اجبد زجه د

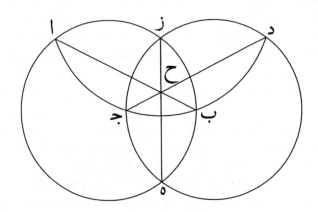

خط جد وليوصل خطا زح حه فلأن نقط ز ح ه فى سطح دائرة زاه ب
وهى أيضاً فى سطح دائرة زجه د تكون نقط ز ح ه على الفصل المشترك

10

لسطحى الدائرتين الأوليين والفصل المشترك لجميع السطوح هو خط مستقيم
فخط زح متصل بخط حه على استقامة ودائرة اجبد عظيمة فى كرة وهى
تقطع دائرة أخرى من الدوائر التى فى الكرة وهى دائرة زاه ب وتمر بقطبيها
فهى تقطعها بنصفين على زوايا قائمة فخط اب قطر دائرة زاه ب وكذلك أيضاً
نبين أن خط جد قطر دائرة زده ج ولأن دائرة اجبد قائمة على كل واحدة

15

من دائرتى زاه ب زده ج على زوايا قائمة تكون أيضاً كل واحدة من دائرتى
زاه ب زده ج قائمة على دائرة اجبد على زوايا قائمة وإذا كانت دائرتان
تقطع إحداهما الأخرى قائمتين على سطح ما على زوايا قائمة فإن الفصل

زاه ب ... اجبد [*om.* 7 وتكون [*add. supra* N؛ الصا [*om.* A 7 قوس [2 رح مساويه [*add. et del.* N 8 اجبد د [*om.*
N 9 ه [ح *om.* N 11 لسطحى [لسطح N 12 عظيمة [*om.* N زاه ب [اهر N؛ ز *supra* A 9 حا [AN وعلى
N وعلى [على A 14 وهو دار [وهى دائرة 13

arcus *ZD* est equalis arcui *DE*. Sit ergo communis differentia duorum   10
circulorum *AGBD AEBZ* linea *AB* et communis differentia duorum
circulorum *AGBD ZGED* linea *GD*. Et producam duas lineas *ZH HE*.

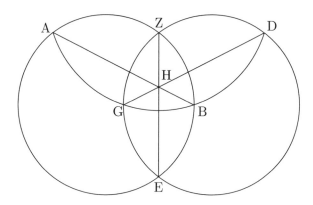

Et quia puncta *Z H E* sunt in superficie circuli *AEBZ* et ipsa etiam
sunt in superficie circuli *ZGED*, ergo puncta *Z H E* sunt super communem differentiam superficierum duorum circulorum primorum. Com-   15
munis autem differentia omnium superficierum est linea recta. Ergo
linea *ZH* est coniuncta linee *HE* secundum rectitudinem. Circulus
vero *AGBD* est maior in spera et secat alium ex circulis qui sunt
in spera, qui est circulus *ZAEB*, et transit per duos polos eius, ergo
ipse secat eum in duo media et orthogonaliter. Ergo linea *AB* est dia-   20
metrus circuli *ZAEB*. Simili quoque modo ostenditur quod linea *GD*
est diametrus circuli *ZDEG*. Et quia circulus *AGBD* erigitur super
unumquemque duorum circulorum *ZAEB ZDEG* orthogonaliter, ergo
etiam unusquisque duorum circulorum *ZAEB ZDEG* orthogonaliter
erigitur super circulum *AGBD*. Sed cum duo circuli quorum unus al-   25
terum secat eriguntur super aliquam superficiem orthogonaliter, tunc

10 est equalis] *scr. et del., deinde scr. iterum* **Ps**   10 arcui] *om.* **OBPsVa**
10 communis differentia] *tr.* **M**   11–12 *AEBZ … AGBD*] *om.* **FiZ**   12 *GD*] et communis differentia duorum circulorum *AGBD AEBZ* linea *AB add.* **FiZ**   12 lineas] lineam
Z, *om.* **Va**   13 *H*] *supra* **Fi**; *L add.* **V**   13 in] *om.* **Fi**   14 sunt[1]] *om.* **Fi**
14 circuli] *om.* **FiZ**   14 *ZGED*] *ZEGD, ZGED supra* **B**   14 *Z H E*] *marg.* **B**, *supra*
Ps   15 duorum] duo **Fi**   16 linea recta] *tr.* **Kg**   17 *ZH*] *ZHE* **FiZ**; *E add. et del.* **O**
17 est coniuncta] *repet.* **Va**   17 linee] *om.* **BPsVa**   18 *AGBD*] *ABGD* **Z**   18 maior] *om.*
PRVFiZ   18 maior in spera] *permut.* **O**   18 alium] circulum *add. supra* **Ps**, *in textu*
Va   18 ex circulis] circulorum **PRVFiZ**, ex eis **O**, ex his **BPsVa**   19 *ZAEB*] *ZAE* **Z**
20 secat] *supra* **BPs**   20 in … orthogonaliter] *permut.* **MKgOBPsVa**   20 et] *supra*
R   21 ostenditur] monstratur **MBPsVa**   22 *ZDEG*] *in corr.* **R**   22 *AGBD*] *ABGD* **B**
23 unumquemque] utrumque **MKgOBPsVa**   23 duorum] *om.* **R**   23 orthogonaliter] *ante*
erigitur **MKgOBPsVa**   23–24 ergo … *ZDEG*] *om.* **Va**   24 etiam] *om.* **KgFi**, et **Z**
24 unusquisque] uterque **MOB**, utrumque **Kg**, **Ps** (*supra vel uterque*)   25 *AGBD*] *in corr.* **B**
26 super] *om.* **V**   26 aliquam] quam **Kg**   26 orthogonaliter] *ante* eriguntur **MKgOBPsVa**

المشترك لهما أيضاً قائم على ذلك السطح بعينه على زوايا قائمة فالفصل

20 المشترك لدائرتى زاه ب زده ج عمود على سطح اجب د والفصل المشترك

لهما هو خط زح ه فخط زح ه عمود على دائرة اجب د فهو يحدث مع جميع

الخطوط المستقيمة التى تخرج من نقطة منه فى سطح دائرة اجب د زوايا

قائمة فكل واحد من خطى اب جد اللذين هما فى سطح دائرة اجب د قد

أخرج من نقطة ح من خط زح ه فخط زح ه عمود على كل واحد من خطى

25 اب جد وكل واحد من خطى اب جد عمود على خط زح ه فلأنه قد خرج

فى دائرة زاه ب خط يمر بالمركز وهو خط اب وقطع خطاً آخر لا يمر

بالمركز وهو خط زح ه على زوايا قائمة فهو يقطعه بنصفين ويكون خط زح

مساوياً لخط ح ه وخط ح ا مشترك لهما وهو قائم عليهما على زوايا قائمة

فقوس ز ا مساوية لقوس ا ه وكذلك أيضاً نبين أن قوس ز ب مساوية لقوس

30 ب ه وأن قوس ز ج مساوية لقوس ج ه وأن قوس ز د أيضاً مساوية لقوس

د ه فدائرة اجب د تقسم القطع التى فصلت من الدائرتين بنصفين نصفين ،

وذلك ما أردنا أن نبين .

ى   إذا كانت فى كرة دوائر متوازية ورسمت دوائر عظيمة تمر بأقطابها فإن

القسى من الدوائر المتوازية التى فيما بين الدوائر العظيمة متشابهة والقسى

من الدوائر العظيمة التى فيما بين الدوائر المتوازية متساوية .

19 قائم] مامه A   21 قائم] illeg. A ¹[زح ه   23 زح ه] marg. A   24 اللذين] الكرين A   24 ح] om. NH   24 فخط زح ه] illeg.

A 25 ... جد] marg. A   25 وكل واحد ... عمود على] N   27 زح ه] رح N   28 مساو] مساوياً N

N 28 وهو قائم] om. N   29 آ ه] illeg. marg. A   29–30 جه ... وكذلك] om. A; repet. قائم] A

30 زد] ره N   31 فصلت] illeg. A   2–3 متشابهة ... العظيمة] marg. A

differentia eis communis erigitur etiam super illam eandem superficiem
orthogonaliter. Ergo communis differentia duorum circulorum *ZAEB*
ZDEG est perpendicularis super superficiem *AGBD*. Communis autem
eorum differentia est linea *ZHE*: ergo linea *ZHE* est perpendicularis   30
super circulum *AGBD*. Ergo ex ipsa cum omnibus rectis lineis que protrahuntur a puncto eius in superficie circuli *AGBD* anguli proveniunt
recti. Sed unaqueque duarum linearum *AB GD* que sunt in superficie circuli *AGBD* iam protracta est a puncto linee *ZHE*: ergo linea
ZHE est perpendicularis super unamquamque duarum linearum *AB*   35
GD. Ergo unaqueque duarum linearum *AB GD* est perpendicularis
super lineam *ZHE*. Et quia in circulo *ZAEB* iam protracta est linea
transiens per centrum, que est linea *AB*, et secat lineam aliam non
transeuntem per centrum, que est linea *ZHE*, orthogonaliter, ergo ipsa
secat eam in duo media; ergo linea *ZH* est equalis linee *HE*. Sed linea   40
HA est eis communis et est super eas orthogonaliter: ergo arcus *ZA* est
equalis arcui *AE*. Similiter quoque ostenditur quod etiam arcus *ZB* est
equalis arcui *BE* et quod arcus *ZG* est equalis arcui *GE* et quod arcus
ZD etiam est equalis arcui *DE*, ergo circulus *AGBD* secat porciones
circulorum, quas secat, in duo media et duo media. Et illud est quod   45
demonstrare voluimus.

10   Cum in spera circuli equidistantes fuerint et signati fuerint circuli maiores per polos eorum transeuntes, arcus circulorum equidistantium qui
erunt inter circulos maiores erunt similes et arcus circulorum maiorum
qui erunt inter circulos equidistantes erunt equales.

27 erigitur] igitur **Fi**   *27* etiam] *om.* **MKgOBPsVa**   *27* eandem] *om.* **KgOBPsVa**
27 superficiem] superficies (*supra*) **M**; differentiam **PRV**   *28 ZAEB*] *in corr.* **B**, *ZABE*
Kg   *29 AGBD*] *in corr.* **B**, *DBGA* **O**   *30* eorum] earum **Z**   *30 ZHE*] *corr. ex ZHEE*
P   *31 AGBD*] *in corr.* **B**, *DGBA* **O**   *32* superficie] superficiem **Z**   *32 AGBD*] *in corr.*
B, *DGBA* **O**   *32* anguli] angulis **Fi**, *illeg.* **B**   *33* unaqueque] utraque **MKgOBPs**, utra **Va**,
unamque **Fi**   *33* duarum] *om.* **BPsVa**   *33* duarum linearum] *tr.* **O**   *34 AGBD*] *ABGD*
Kg, *DGBA* **O**, *in corr.***B**   *34* linea] *om.* **Va**   *35* unamquamque] utramque **MKgOBPsVa**
36 Ergo ... *AB GD*] *marg.* **M**, *supra* **Ps**, *om.* **Va**   *36* unaqueque] utraque **KgOBPsVa**
36 AB ... perpendicularis] *marg.* **B**   *37* lineam] *om.* **OBPsVa**   *37 ZAEB*] *ZAB* **OBPs** (*E*
supra **Ps**)   *37* iam] tamquam **Va**   *38* centrum] eius *add. supra* **Ps**   *38* est] *om.* **V**   *38–*
39 AB ... linea] *post* duo media (*lin. 40*) **R**   *38* lineam] *om.* **BPsVa**   *39* ipsa] ipse **BPsVa**
41 HA] *corr. ex HEA* **V**   *41* eis] *om.* **KgBPsVa**   *41* eis communis] *tr.* **O**   *41* eas] *in corr.* **P**
41–42 est equalis] *tr.* **Va**   *42–43* est equalis] *tr.* **Va**   *43* et[1] ... *GE*] *marg.* **R**   *43* arcus[1]] *om.*
OBPsVa   *43* quod[2]] *om.* **BPsVa**   *44* etiam] *om.* **MKgOBPsVa**   *44* arcui] *GD* et quod arcus
ZD est equalis arcui *add.* **Fi**   *44 AGBD*] *ABGD* **MKgOFiZB**   *45* secat] dividit **PRVMKgO**
FiZ   *45* et duo media] *om.* **R**   *45–46* Et ... voluimus] *om.* **OBPsVa**   *45* illud] hoc **MFiZ**
46 voluimus] *om.* **MKg**, volumus **Z**   *1* fuerint] circuli maiores *add.* **Fi**   *4* qui erunt] *om.* **BPsVa**
4 erunt[1]] *om.* **Fi**

فلتكن فى كرة دائرتان متوازيتان وهما دائرتا اب‌ج‌د ه‌ز‌ح‌ط ولتكن

5   نقطة ك‌ قطباً لهما ولترسم دائرتان عظيمتان تمر بأقطابهما وهما دائرتا اه‌ح‌ج

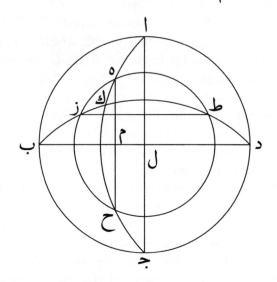

ب‌ز‌ط‌د ، فأقول إن القسى من الدوائر المتوازية التى بين الدوائر العظيمة

متشابهة أعنى أن قوس ب‌ج‌ شبيهة بقوس ز‌ح وتكون قوس ج‌د شبيهة

بقوس ح‌ط وقوس د‌ا شبيهة بقوس ط‌ه وقوس ا‌ب شبيهة بقوس ه‌ز ،

وأقول أيضاً إن القسى من الدوائر العظيمة التى هى فيما بين الدوائر المتوازية

10   متساوية أعنى أن قسى ز‌ب ح‌ج ط‌د ه‌ا الأربع متساوية فليكن الفصل

المشترك لدائرة اب‌ج‌د ولدائرة اه‌ح‌ج خط اج‌ والفصل المشترك لدائرة

ب‌ز‌ط‌د ولدائرة اب‌ج‌د خط ب‌د والفصل المشترك لدائرة ه‌ز‌ح‌ط ولدائرة

ز‌ك‌ط‌ه خط ز‌ط والفصل المشترك لدائرة ه‌ز‌ح‌ط ولدائرة ه‌ك‌ح خط ح‌ه

4 اب‌ج‌د] ا‌ح‌م N   5 احه‌ج] احد N   5 ك‌] ك‌ AN, ك‌ supra A   8 وقوس¹] قوس وبكون N
4 الأربع] الأربعة AN, hic et   9 التى هى] هى om. N, هى om. A   10 وقوس²] قوس وبكون N   8
saepius   11 اب‌ج‌د] اج N   11 اج‌] اب‌ج‌د N   12 marg. A   12 اب‌ج‌د] اهح N   ه‌ز‌ح‌ط] ح add. A
13 ز‌ك‌ط‌ه] ز‌ك‌ط‌د marg. A

Sint itaque in spera duo circuli equidistantes, qui sint duo circuli   5
ABGD EZHT, et sit punctum *K* polus eorum. Describam autem
duos circulos maiores transeuntes per polos eorum, qui sunt duo cir-

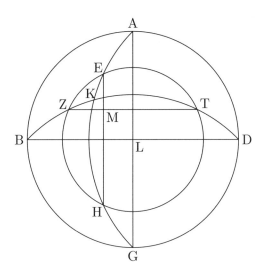

culi *AEHG BZTD*. Dico igitur quod arcus circulorum equidistantium
qui sunt inter circulos maiores sunt similes, scilicet quod arcus *BG* est
similis arcui *ZH* et arcus *GD* similis existit arcui *HT* et arcus *DA* est   10
similis arcui *TE* et arcus *AB* est similis arcui *EZ*. Et dico etiam quod
arcus circulorum maiorum qui sunt inter circulos equidistantes sunt
equales, scilicet quod arcus *ZB HG TD EA* quattuor sunt equales.
Sit itaque differentia communis circuli *ABGD* et circuli *AEHG* linea
AG et differentia communis circuli *BZTD* et circuli *ABGD* sit linea   15
BD et differentia communis circuli *EZHT* et circuli *ZKT* sit linea *ZT*
et communis differentia circuli *EZHT* et circuli *EKH* sit linea *HE*. Cir-

5 Sint] Sunt **Va**   5 in spera] *om.* **R**   5 equidistantes] *in corr.* **B**, *supra* **Ps**, equidem (*hic et saepius*) **Fi**   5 qui ... circuli] *om.* **OPsVa**   5–6 qui ... *EZHT*] *om.* **B**
5 duo circuli] *om.* **MKg**   6 *ABGD EZHT*] *supra* **Ps**   6 *EZHT*] *EHZT* **Kg**   6 polus] polos
Fi   6 polus eorum] *tr.* **PRM**   6 autem] itaque **MKgOBPsVa**   7 duo circuli] *om.*
MKgOB PsVa   8 *BZTD*] *BZCD* **Va**   8 igitur] ergo **OBPsVa**   10 *ZH*] *LZH* **V**
10 existit] existet **Fi**   11 arcui[1]] *om.* **Va**   11 Et] *supra* **Ps**   11 etiam] *om.*
BPsVa   13 scilicet] *in corr.* **R**   13 *TD*] *CD* **Va**   14–16 *ABGD* ... circuli[1]] *marg.*
Ps   14–16 et ... circuli[2]] *marg.* **B**   14–15 linea ... *ABGD*] *om.* **Kg**   15 *AG*] *ALG*
Fi   15 *BZTD*] *ABGD* **OBVa**   15 *ABGD*] *BZTD* **OBVa**   15 sit] *om.* **FiZBVa**
16 differentia] circuli *add.* **Va**   16 circuli[1]] *om.* **PRVKgOBVa**   16 *ZKT*] *BZTD* **FiZ**
16 sit] *om.* **BPsVa**   16 linea *ZT*] *supra* **Ps**   16 *ZT*] *DT* **B**   17 *EZHT*] *BT* **Va**   17–
18 Circulus] Circulis **Fi**

ودائرة اه‍ح‍ج العظمى من الدوائر التى فى الكرة تقطع دائرة ما من الدوائر

التى تكون فى الكرة وهى دائرة اب‍ج‍د وتمر بقطبيها فهى تقطعها بنصفين   15

وعلى زوايا قائمة فخط اج‍ قطر لدائرة اب‍ج‍د وكذلك أيضاً نبين أن خط ب‍د

قطر لدائرة اب‍ج‍د فنقطة ل‍ مركز دائرة اب‍ج‍د وأيضاً فإن دائرة اه‍ح‍ج

العظمى من الدوائر التى فى الكرة تقطع دائرة ما من الدوائر التى تكون فى

الكرة وهى دائرة ه‍ز‍ح‍ط وتمر بقطبيها فهى تقطعها بنصفين وعلى زوايا قائمة

فخط ه‍ح‍ قطر دائرة ه‍ز‍ح‍ط وكذلك أيضاً نبين أن خط ز‍ط أيضاً قطر دائرة   20

ه‍ز‍ح‍ط فنقطة م‍ مركز دائرة ه‍ز‍ح‍ط فلأن سطحى دائرتى اب‍ج‍د ه‍ز‍ح‍ط

المتوازيتين يقطعهما سطح دائرة ب‍ز‍ط‍د يكون الفصلان المشتركان لهما أيضاً

متوازيـين فخط ب‍د موازٍ لخط ط‍ز وكذلك أيضاً نبين أن خط اج‍ أيضاً موازٍ

لخط ه‍ح‍ وخطا ب‍ل ل‍ج‍ اللذان يماس أحدهما الآخر موازيان لخطى ز‍م م‍ح‍

اللذين يماس أحدهما الآخر وليست الخطوط فى سطح واحد فهى تحيط   25

بزاويتين متساويتين فتكون زاوية ز‍م‍ح‍ مساوية لزاوية ب‍ل‍ج‍ وهما على

المركزين وزاوية ز‍م‍ح‍ قاعدتها قوس ز‍ح‍ وزاوية ب‍ل‍ج‍ قاعدتها قوس ب‍ج‍

فقوس ب‍ج‍ شبيهة بقوس ز‍ح‍ وكذلك أيضاً نبين أن قوس ج‍د أيضاً شبيهة

بقوس ح‍ط وقوس اد‍ شبيهة بقوس ه‍ط وقوس اب‍ أيضاً شبيهة بقوس ه‍ز

فالقسى من الدوائر المتوازية التى فيما بين الدوائر العظيمة متشابهة ،   30

16 أيضاً [om. A   17 د [ل N   17 فإن [فلان A   18 تكون [om. A   19 وعلى [على A   21 سطحى [in
corr. A   22 يكون [مكون A   25 اللذين [الكرس A   25 تحيط [محيط A   26 لزاوية [om. N
om. A   30 التى [A   قوس 2 [بقوس N   29 ح‍ط ... 2 بقوس [om. N   29 ب‍ج‍ [لح N   27

culus autem $AEHG$ maior, qui est unus circulorum qui sunt in spera,
secat aliquem circulorum qui sunt in spera, qui est circulus $ABGD$, et
transit per polos eius: ergo secat eum in duo media et orthogonaliter. 20
Linea igitur AG est diametrus circuli $ABGD$. Similiter quoque ostenditur quod linea BD est diametrus circuli $ABGD$. Punctum igitur L est
centrum circuli $ABGD$. Et etiam quia circulus $AEHG$ maior, qui est
unus circulorum qui sunt in spera, secat aliquem circulorum qui sunt
in spera, qui est circulus $EZHT$, et transit per polos eius, ergo ipse 25
secat eum in duo media et orthogonaliter. Ergo linea EH est diametrus circuli $EZHT$. Simili quoque modo monstratur quod linea ZT est
etiam diametrus circuli $EZHT$. Punctum igitur M est centrum circuli
$EZHT$. Et quia duas superficies duorum circulorum equidistantium
$ABGD$ $EZHT$ superficies circuli $BZTD$ secat, communes ergo earum 30
differentie sunt etiam equidistantes Ergo linea BD equidistat linee TZ.
Et similiter etiam monstratur quod etiam linea AG equidistat linee EH
et due linee BL LG, quarum una alteram contingit, equidistant duabus lineis ZM MH, quarum una contingit alteram. Linee autem non
sunt in superficie una: ergo angulos continent equales. Angulus igitur 35
ZMH est equalis angulo BLG; ipsi autem sunt super duo centra. Et
basis anguli ZMH est arcus ZH et basis anguli BLG est arcus BG:
arcus igitur BG est similis arcui ZH. Simili quoque modo monstratur
quod arcus GD est similis arcui HT et quod arcus DA est similis arcui
ET et etiam arcus AB est similis arcui EZ: arcus igitur circulorum 40
equidistantium qui sunt inter circulos maiores sunt similes.

18 *AEHG*] *AEHS* **Fi** 18 maior] *om.* **MKgOBPsVa** 18 unus] cuius **Fi**
18 circulorum] maiorum *add.* **PVMKgOFiZBPsVa** 19 secat] secant **Va** 19 *ABGD*] *A*
Kg 21 Linea] recta *add. supra* **Ps** 21 igitur] ergo **FiZ** 21–22 Similiter … *ABGD*] *om.*
Va 22 linea] recta *add. supra* **Ps** 22 *ABGD*] *ADGD* **Fi** 23 circulus] circulis **Fi**
24 circulorum[1]] maiorum *add. supra* **Ps** 24 aliquem] aliquos **Fi** 25 qui est circulus] scilicet
circulum **KgOBPsVa** 25 ipse] *om.* **FiZ** 26 linea] recta *add. supra* **Ps** 26 *EH*] *corr.*
ex DH **B** 27 quoque] *om.* **Va** 27 *ZT*] *ZEC* **Fi**; *corr. ex ZHT* **Kg** 28 etiam] *om.*
KgBPsVa, *supra* **M** 28 diametrus] super diametrum **PRKg** 29 *EZHT*] *EHT* **Fi** 29–
30 duas … secat] superficies circuli *BZTD* secat duas superficies duorum circulorum
equidistantium *ABGD EZHT* **MBPsVa** 30 secat] *post* Et quia **KgO** 30 ergo] igitur
MKgBPs 30 earum] eorum **Va** 30–31 earum differentie] *tr.* **BPsVa** 31 differentie] etiam
add. **V** 31 etiam] *om.* **PRVMKgOFiZ** 32 Et] *om.* **OBPsVa** 32 Et … linee *EH*] *om.*
FiZ 32 etiam[1]] *om.* **PRFi** 32 etiam[2]] *om.* **RMKgOBPsVa** 33 due linee] *tr.* **PVMKgFiZ**
33 *LG*] *supra* **Ps** 33–34 equidistant … alteram] *marg.* **BPs** 34 contingit alteram] *tr.*
MKg 34 Linee] Linea **Fi** 35 superficie una] *tr.* **PRKgFiVa** 35 continent] contingens **Fi**
35 Angulus] Angulos **Va** 36 *BLG*] *BLD* **Z** 36 autem] *MZL add. supra* **Ps**, *in textu* **Va**
36 super] *om.* **PsVa** 38 igitur] ergo **Va** 38 est] *om.* **Kg** 38 Simili quoque modo] Similiter
OBPsVa 39 et] *om.* **Fi** 40 *ET*] *corr. ex DZ* **B** 40 igitur] ergo **R**

وأقول أيضاً إن القسى من الدوائر العظيمة التى فيما بين الدوائر

المتوازية متساوية وذلك أنه لما كانت نقطة كـ قطب دائرة اٮﺟد صارت

قسى كـا كـب كـﺟ كـد الأربع مساوٍ بعضها لبعض وأيضاً فإن نقطة كـ لما

كانت قطباً لدائرة هزحﻂ صارت قسى كـه كـز كـﺢ كـﻂ الأربع مساوٍ

بعضها لبعض فقسى ﻩا زب ﺣﺟ ﻃد الأربع الباقية يساوى بعضها بعضاً   35

فالقسى من الدوائر العظيمة التى فيما بين الدوائر المتوازية مساوٍ بعضها

لبعض ، وذلك ما أردنا أن نبين ·

يا   إذا عملت على أقطار دوائر متساوية قطع دوائر متساوية قائمة عليها على

زوايا قائمة ثم فصلت منها قسى متساوية مما يلى أطراف الأقطار وكانت تلك

القسى أقل من نصف القطع ثم أخرج من النقط التى تحدث فى موضع

الفصل إلى الخطوط المحيطة بالدوائر الأولى خطوط مستقيمة متساوية فإنها

تفصل من الدوائر الأولى قسياً متساوية مما يلى أطراف الأقطار التى ذكرنا ·   5

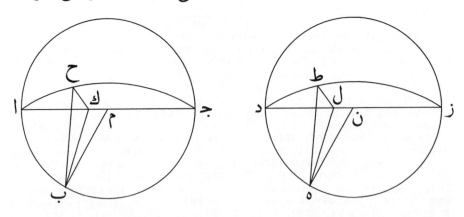

مساوه [مساوٍ 34 A لطا [كـﻂ 34 A لﺢ [كـﺢ 34 N لﻂ [كـﻂ 34 A لﺐ [كـﺐ 33 A لﻚ [كـ 33 A om. لﺐ [كـﺐ 33 العظيمة 31

A مسى [قسياً 5 A مساوه [متساوية 4 A علمٮ [عملت 1 N متساوية A,

Et dico etiam quod arcus circulorum maiorum qui sunt inter circulos equidistantes sunt equales. Quod ideo est quoniam propter hoc quod punctum *K* est polus circuli *ABGD*, fiunt quattuor arcus *KA KB KG KD* ad invicem equales. Et etiam propter hoc quod punctum *K* est polus circuli *EZHT*, fiunt quattuor arcus *KE KZ KH KT* ad invicem equales. Reliqui igitur quattuor arcus *EA ZB HG TD* ad invicem equantur. Ergo arcus circulorum maiorum qui sunt inter circulos equidistantes ad invicem sunt equales. Et illud est quod demonstrare voluimus.

45

50

11 Cum super diametros circulorum equalium equales circulorum porciones signantur super eas orthogonaliter erecte, ex quibus postea arcus equales ab ea parte qua extremitates diametrorum sequuntur separantur, et fuerint arcus illi minores medietatibus porcionum, deinde a punctis que proveniunt in loco communis differentie ad lineas primos circulos continentes linee recte equales protrahuntur, ipse separant ex primis circulis arcus equales ab ea parte qua sequuntur diametrorum extremitates quas nominavimus.

5

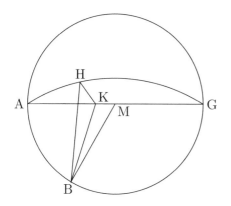 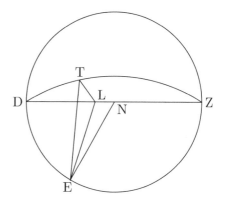

42 sunt] in spera *add.* **O**; in spera qui sunt *add.* **BPsVa**   44 quod] *supra* **R**   44 fiunt] fuerit **R**   44 *KA*] *KHA* **Ps**   45 *KB*] *KL* **Fi**   45 propter hoc] *om.* **V**   45 quod] rectus *add.* **Fi**   46 est] *om.* **Va**   46 fiunt] fuerit **R**   46 *KZ*] *KG* **FiZ**   46 *KH KT*] *supra* **B**   46 *KT*] *HKT* **Fi**; et *add.* **Z**   47 invicem] sunt *add.* **Z**   47 *TD*] *T* **Kg**   48 sunt] *add.* in spera **BPsVa**   49 ad ... equales] equantur **BPsVa**   49 sunt equales] *tr.* **O**   49–50 Et ... voluimus] *om.* **BPsVa**   49 illud] hoc **O**   49–50 quod ... voluimus] *om.* **O**, *illeg.* **Kg**   1 equalium equales circulorum] *om.* **Fi**   1 circulorum[2]] equalium *add. supra* **O**   2 signantur] significantur **Va**   2 eas] diametros **Z**   2 ex] *om.* **Fi**   3 ab] *om.* **F**   3 qua] que **Va**   3 diametrorum] *in corr.* **Ps**   3 sequuntur] sequentur **Fi**   3–4 separantur] seperantur **Fi**   4 fuerint] fuerit **Va**   4 arcus illi] *tr.* **FiZ**   4 minores medietatibus porcionum] *permut.* **MKgOBPsVa**   5 in loco communis differentie] *om.* **PRM**   6 protrahuntur] protrahantur **OFi**; et *add.* **OFiZ**   7 diametrorum] diametrum **BPs**   8 nominavimus] nominamus **BPsVa**; et econverso *add.* **Kg**, *add. et del.* **M**

فلتُعمل على قطرين من أقطار دائرتى اب‍ج‍ ده‍ز‍ المتساويتين قطعتان

من دائرة متساويتان قائمتان عليهما على زوايا قائمة وهما قطعتا اح‍ج‍ دط‍ز‍

وتفصل منهما قوسان متساويتان مما يلى أطراف الأقطار أعنى مما يلى نقطتى

آ د‍ ولتكونا قوسى اح‍ دط‍ ولتكونا أقل من نصفى قوسى اح‍ج‍ دط‍ز‍

وليخرج من نقطتى ح‍ ط‍ إلى الخطين المحيطين بدائرتى اب‍ج‍ ده‍ز‍ الأوليين

خطان مستقيمان متساويان وهما خطا ح‍ب‍ ط‍ه‍ ، فأقول إن قوس اب‍

مساوية لقوس ده‍ فليخرج من نقطتى ح‍ ط‍ إلى سطحى دائرتى اب‍ج‍ ده‍ز‍

عمودان فهو بيّن أنهما يقعان على الفصلين المشتركين لهما أعنى على خطى

اج‍ دز‍ وليكن العمودان عمودى ح‍ك‍ ط‍ل‍ وليكن مركزا دائرتى اب‍ج‍ ده‍ز‍

نقطتى م‍ ن‍ ولتوصل خطوط ك‍ب‍ م‍ب‍ ل‍ه‍ ن‍ه‍ ولأن خط ح‍ك‍ عمود على

سطح دائرة اب‍ج‍ فهو عمود على جميع الخطوط التى تماسه وتكون فى

سطح دائرة اب‍ج‍ ويحدث معها زوايا قائمة فزاوية ح‍ك‍ب‍ قائمة وكذلك

أيضاً نبين أن زاوية ط‍ل‍ه‍ أيضاً قائمة ولأن قطعتى اح‍ج‍ دط‍ز‍ متساويتان

وقوسى اح‍ دط‍ اللتين فصلتا أيضاً متساويتان وقد أخرج عمودا ح‍ك‍ ط‍ل‍

يكون خط اك‍ مساوياً لخط دل‍ وخط ح‍ك‍ مساوياً لخط ط‍ل‍ ولأن خط ب‍ح‍

مساوٍ لخط ط‍ه‍ يكون أيضاً المربع الكائن من خط ب‍ح‍ مساوياً للمربع

الكائن من خط ط‍ه‍ والمربعان الكائنان من خطى ح‍ك‍ ك‍ب‍ مساويان للمربع

الكائن من خط ب‍ح‍ والمربعان الكائنان من خطى ط‍ل‍ ل‍ه‍ مساويان للمربع

مس] نبين¹ 18 N   نقعان N يقعان [, supra 13 يعطعان,   ان هما N أنهما [13   علىها N عليهما [7

N أيضاً²] om. A   18   ال اك N الاك [20   وخط ... ط‍ل‍] om. N   20   ط‍ل‍] in corr. A

22 خطى ... ط‍ه‍ خط] marg. A   23–25 مساويان ... ط‍ل‍ ل‍ه‍] marg. A

Super duas igitur diametrorum duorum circulorum *ABG DEZ*
equalium duas faciam duorum circulorum equalium porciones super   10
eas orthogonaliter erectas sintque due porciones *AHG DTZ*. Et separabo ex eis duos arcus equales ab ea parte qua duarum diametrorum
extremitates sequuntur, scilicet ab ea parte qua sequuntur duo puncta
A et *D*, sintque arcus *AH DT*, qui sint minores duabus medietatibus
duorum arcuum *AHG DTZ*. Protraham autem a duobus punctis *H*   15
et *T* ad duas lineas continentes duos circulos *ABG DEZ* primos duas
rectas lineas equales, que sint due linee *HB TE*. Dico igitur quod arcus
AB est equalis arcui *DE*. Protraham itaque a duobus punctis *H T* ad
duas superficies duorum circulorum *ABG DEZ* duas perpendiculares.
Manifestum est igitur quod ipse cadent super duas earum communes   20
differentias, scilicet super duas lineas *AG DZ*; sintque due perpendiculares perpendiculares *HK TL* et sint centra duorum circulorum *ABG*
DEZ M et *N*. Et producam lineas *KB MB LE NE*. Et quia linea *HK*
est perpendicularis super omnes lineas que eam contingunt et sunt in
superficie circuli *ABG* et proveniunt ex ea cum eis anguli recti, ergo an-   25
gulus *HKB* est rectus. Similiter quoque ostenditur quod angulus *TLE*
est rectus. Et quia due porciones *AHG DTZ* sunt equales et arcus *AH*
DT quos separavimus etiam sunt equales et iam protracte sunt due
perpendiculares *HK TL*, ergo linea *AK* est equalis linee *DL*. Et quia
linea *BH* est equalis linee *TE*, ergo quadratum factum ex linea *BH* est   30
etiam equale quadrato facto ex linea *TE*. Duo autem quadrata facta ex
duabus lineis *HK KB* sunt equalia quadrato facto ex linea *BH* et duo
quadrata facta ex duabus lineis *TL LE* sunt equalia quadrato facto ex

9 Super] ergo *add.* **BPsVa**   9 igitur] ergo **Kg**; *om.* **BPsVa**   9 diametrorum] diametros **FiZ**
9 duorum circulorum] *tr.* **V**   9 *ABG*] *AHG* **O**   10 duorum circulorum equalium] *permut.*
FiZ   11 sintque] sunt **Fi**   12 ex eis] *om.* **Kg**, *post* duos arcus **BPsVa**   12 duarum] duorum
OFiBPsVa, *corr. ex* duorum **R**   13 sequuntur] sequuntur[1]] sequuntur **Fi**   13 parte] *om.* **Fi**   13 duo] *om.*
Va   14 arcus] arcutus **Fi**   14 *AH*] et *add.* **R**   14 *DT*] *TD* **R**   14 sint] sunt **Fi**
15 duorum] duarum **Fi**   15 *AHG*] *ABG* **Z**   15 *DTZ*] *DEZ* **Va**   15 autem] sunt
Fi   15 a] *om.* **Va**   17 sint] sunt **FiZ**   17 *HB*] *HT* **V**   17 igitur] *om.* **FiBPsVa**
17 quod] *supra* **R**   18 est equalis] equatur **OBPsVa**   18 arcui] *om.* **Fi**   18 *H*] et *add.*
OFi   20 est igitur] *tr.* **MKgBPsVa**   20 cadent] cadunt **Fi**   22 perpendiculares] *om.* **VMZ**
22 et] etiam **Fi**   23 producam] duas *add.* **Z**   23 *HK*] linea *add.* **Fi**   24 omnes lineas] *tr.*
BPsVa   25 circuli *ABG*] *tr.* **OBPsVa**   25 proveniunt] perveniunt **Fi**   25–26 angulus] angulis
Fi   26 ostenditur] *om.* **R**   26 *TLE*] *TL, E supra* **B**   27 *AHG*] *ABG* **Va**   27 *DTZ*] *DEZ*
Z; *in corr.* **Ps**   27–28 et . . . sunt equales] *om.* **Fi**   28 iam] tamen **Va**   29 linea] *om.*
OBPsVa   29 *AK*] *HK* **FiVa**   29 linee] *om.* **OBPsVa**   29 *DL*] *TL* **FiZ**   29 quia] *om.*
FiZ   31 etiam] *om.* **MOFiZBPsVa**   31 Duo] *antea* ergo *add.* **OPs**, *corr. supra in set* **Ps**,
Sed **Va**   31–32 Duo . . . linea *BH*] *om.* **Fi**   31 autem] *om.* **OBPs**   31 facta] *om.* **BPsVa**
32 *HK KB*] *BK KH* **OBPsVa**   32 et] *antea* set *add.* **Fi**   33–35 *LE . . . TL LE*] *marg.* **Ps**
33–35 sunt . . . *TL LE*] *marg.* **B**

الكائن من خط طه فالمربعان الكائنان من خطى حك كب مساويان

للمربعين الكائنين من خطى طل له والمربع الكائن من خط حك من هذه ٢٥

الخطوط مساوٍ للمربع الكائن من خط طل فيبقى المربع الكائن من خط كب

مساوياً للمربع الكائن من خط له فخط كب مساوٍ لخط له ولأن خط ام

مساوٍ لخط دن وخط اك من أحدهما مساوٍ لخط دل يكون خط كم الباقى

مساوياً لخط لن الباقى وخط بم مساوياً لخط هن فخطا كم من مب مساويان

لخطى لن نه كل واحد منهما لنظيره وقاعدة كب مساوية لقاعدة له ٣٠

فتكون زاوية كمب مساوية لزاوية لنه وتكون قوس اب مساوية لقوس

ده ،

وكذلك إذا عمل على أقطار دوائر متساوية قطع من الدوائر متساوية

قائمة عليها على زوايا قائمة ثم فصلت منها قسى متساوية مما يلى أطراف

الأقطار أقل من أنصافها وفصلت من الدوائر الأولى قسى متساوية فى جهة ٣٥

واحدة بعينها مما يلى تلك الأطراف من الأقطار ووصلت خطوط مستقيمة

فيما بين النقط الحادثة فى مواضع الانفصال فإن تلك الخطوط تكون

متساوية ،

فلتعمل على دائرتى ابج دهز المتساويتين على قطرى اج دز من

أقطارهما قطعتان متساويتان من الدوائر قائمتان عليها على زوايا قائمة وهما ٤٠

---

linea *TE*: ergo duo quadrata facta ex duabus lineis *HK KB* sunt equalia duobus quadratis factis ex duabus lineis *TL LE*. Sed quadratum   35
factum ex linea *HK*, que est una harum linearum, est equale quadrato
facto ex linea *TL*: remanet ergo quadratum factum ex linea *KB* equale
quadrato facto ex linea *LE*. Ergo linea *KB* est equalis linee *LE*. Et
quia linea *AM* est equalis linee *DN* et linea *AK*, que est unius earum,
est equalis linee *DL*, que est alterius, ergo reliqua linea *KM* est equalis   40
relique linee *LN*. Sed linea *BM* est equalis linee *EN*: ergo due linee
KM MB sunt equales duabus lineis *LN NE*, queque scilicet earum sue
relative. Sed et basis *KB* est equalis basi *LE*: ergo angulus *KMB* est
equalis angulo *LNE* et arcus *AB* est equalis arcui *DE*.

Et similiter cum super diametros circulorum equalium porciones   45
circulorum equalium fiunt, que super eas orthogonaliter sint erecte,
postea separantur ex eis arcus equales ab ea parte qua diametrorum
sequuntur extremitates minores earum medietatibus, et separantur ex
primis circulis arcus equales in una et eadem parte ab ea qua diametrorum extremitates sequuntur et producuntur linee recte inter puncta   50
provenientia in locis ubi se secant linee, erunt equales.

Super duas itaque diametrorum *AG DZ* duorum circulorum
ABG DEZ duas equales faciam duorum circulorum porciones super

34 *TE*] *aliquas lineas corrupte repet.* **FiZ**   35 ex] quod **Va**   35 Sed] *supra* **Ps**
36–38 que ... linea *KB*] *om.* **Va**   36 est una] *tr.* **Fi**   36 harum linearum] *tr.* **OBPs**
37 ergo] igitur **B**   37 factum] *om.* **B**   38 facto] *om.* **Fi**   39 linea[1]] lineam **Fi**
39 est equalis] equatur **B**   39 linee] *om.* **Fi**   39 *DN*] *AN* **Z**   39 unius] una **OFiB**
PsVa   40 est equalis] equatur **B**   41 est equalis] equatur **B**   42 *KM*] *NM* **Fi**   42–
43 queque ... relative] *om.* **MBPsVa**   42 earum] *om.* **R**   43 est equalis] equatur **OBPsVa**
43–44 est equalis] equatur **OBPsVa**   44 est] *om.* **M**   44 est equalis] equatur **OBPsVa**
45 cum] est **Va**   45 circulorum equalium] *tr.* **POFiBPsVa**   46 circulorum equalium] *tr.* **R**
46 fiunt] fuerint **MVa**   46 que] *supra* **BPs**   46 sint] sunt **OFi**   46 sint erecte] *tr.* **OBPs**
Va   46 erecte] recte **Fi**   47 parte] *supra* **BPs**   47 diametrorum] *om.* **Va**   48 earum] eorum
V   48 medietatibus] medie **Fi**   48 separantur] separentur **RV**   49 primis] *in corr.* **Ps**, semis
Va   49 circulis] equalibus *add.* **BPsVa**   49 arcus equales] *supra* **Ps**, *om.* **Va**   49 qua] que **Fi**
50 extremitates sequuntur] *tr.* **POFiBPsVa**   51 ubi se secant] *om.* **V**   51 secant] secantur **Z**
51 erunt] ille *add. supra* **Ps**   52 itaque] vero **Fi**   52 diametrorum] diametros **MOFiZBPsVa**
53–54 duas ... *AHG DTZ*] *om.* **V**   53 equales] lineas *add.* **Z**

قطعتا اح ج دط ز ولتُفصل منهما مما يلى أطراف الأقطار وهما نقطتا آ د

قوسان متساويتان وهما قوسا اح دط ولتكونا أقل من نصفى قطعتى اح ج

دط ز الأوليين ولتفصل من الدائرتين قوسان متساويتان وهما قوسا اب ده

فى جهة واحدة مما يلى أطراف الأقطار وليوصل خطا ح ب ط ه ، فأقول

إن خط ح ب مثل خط ط ه فلتُخرج من نقطتى ح ط إلى سطحى دائرتى   45

اب ج ده ز عمودان فهما يقعان على خطى اج دز اللذين هما فصلان

مشتركان للسطوح وليكونا خطى ح ك ط ل فليكن مركزا الدائرتين نقطتى م

ن ولتوصل خطوط ك ب ب م ل ه ه ن فلأن قوس اب مساوية لقوس ده

تكون زاوية ام ب أيضاً مساوية لزاوية دن ه ولأن قطعتى اح ج دط ز من

الدائرتين متساويتان وقوسى اح دط اللتين فصلتا متساويتان وقد أخرج   50

عمودا ح ك ط ل يكون خط اك مساوياً لخط دل ويكون خط ح ك مساوياً

لخط ط ل فلأن خط ام مساوٍ لخط دن وخط اك مساوٍ لخط دل يبقى خط ك م

مساوياً لخط ل ن وخط ب م مساوٍ لخط ه ن فطا ك م م ب مساويان لخطى ل ن

ن ه كل واحد منهما لنظيره وزاوية ك م ب مساوية لزاوية ل ن ه فقاعدة ك ب

مساوية لقاعدة ل ه وخط ح ك عمود على سطح دائرة اب ج فهو يحدث مع   55

جميع الخطوط التى تماسه وتكون فى سطح دائرة اب ج زوايا قائمة وخط

ك ب مماس له فزاوية ح ك ب قائمة وكذلك أيضاً نبين أن زاوية ط ل ه أيضاً

ولـيكونا امل [دطز] add. et del.   42–43 وهما قوسا ... اح دط [اح دط] om. A   43 دطز] N دهر [دطز 41

N الآ [اك   51 الآ [اك   N هر [ه ن] 48   47 ر [م   N 47 ولكن [فليكن   N اطول من [مثل 45

A دن [دن 2 corr. ex   52 دل] N دن supra, ح [دن   52 خط [2 om. A   51 مساوياً [1 repet. A

53 لن [1] N لن [2] N 54 لز [2] N هر [ه ن] 53 لن [2] N منهما [54   om. A   57 له [N

---

54 لنه [من اقليدس من ح من كو من   marg. N

eas orthogonaliter erectas, que sint due porciones *AHG DTZ*, ex quibus separabo ab ea parte qua diametrorum sequuntur extremitates, que   55
sint duo puncta *A* et *D*, duos arcus equales, qui sint arcus *AH DT*,
et sint minores duabus medietatibus duarum porcionum *AHG DTZ*
primarum. Et separabo ex duobus circulis duos arcus equales, qui sint
arcus *AB DE*, in parte una ab ea parte qua diametrorum sequuntur
extremitates. Et producam duas lineas *HB TE*. Dico igitur quod linea   60
HB est equalis linee *TE*. Protraham autem a duobus punctis *H T* ad
superficies duorum circulorum *ABG DEZ* duas perpendiculares. Ipse
igitur cadent super duas lineas *AG DZ*, que sunt due communes differentie superficierum sintque due linee *HK TL*, et sint centra duorum
circulorum duo puncta *M N*. Et producam lineas *KB BM LE EN*.   65
Et quia arcus *AB* est equalis arcui *DE*, ergo angulus *AMB* est etiam
equalis angulo *DNE*. Et quia due porciones *AHG DTZ* sunt equalium
circulorum et duo arcus *AH DT* quos separavimus sunt equales et iam
protracte sunt due perpendiculares *HK TL*, ergo linea *AK* est equalis
linee *DL* et linea *HK* est equalis linee *TL*. Et quia linea *AM* est equa-   70
lis linee *DN* et linea *AK* est equalis linee *DL*, ergo remanet linea *KM*
equalis linee *LN*. Sed linea *BM* est equalis linee *EN*: due igitur linee
KM MB sunt equales duabus lineis *LN NE*, queque scilicet earum sue
relative. Sed et angulus *KMB* est equalis angulo *LNE*: ergo basis *KB*
est equalis basi *LE*. Sed linea *HK* est perpendicularis super superficiem   75
circuli *ABG*: ex ipsa ergo cum omnibus lineis, que ipsam contingunt
et sunt in superficie circuli *ABG*, anguli proveniunt recti. Linea autem
KB contingit eam: ergo angulus *HKB* est rectus. Simili quoque modo

54 sint] sunt **FiZ**   54 *AHG*] *ABG* **Kg**, *BHG* **Va**   54 *DTZ*] *DTH* **Fi**; *DTZ in corr.* **Ps**
55 ab] ex **Va**   55 qua] quia **Fi**   55–60 que ... extremitates] *repet.* **Fi**   56 sint[1]] sunt **MO**
FiPsVa   56 sint[2]] sunt **OBPsVa**   56 *AH DT*] *supra* **Ps**   56 *DT*] in parte una ab ea qua diametrorum sequuntur extremitates *add.* **OBVa** (= *lin. 59–60*)   57–60 et ... extremitates] *marg.*
Ps   57 sint] sunt **OVa**   57 duabus] duobus **Va**   57 duarum ... separabo] *marg.* **B**
57 duarum ... *DTZ*] duo per comunia *HA TDZ* **Va**   58 primarum] *om.* **OFiZBPsVa**
58 separabo] semper *ABC* **Va**   58 sint] sit **Va**, sunt **Z**   59–60 in ... extremitates] *et in*
marg. et in textu **Ps**   59 una ab ea parte] *marg.* **M**   59 parte[2]] *om.* **B**, *om. in textu* **Ps**
59 qua] que **Ps** (*in textu*)   60 *TE*] *in corr.* **Ps**, *DT* **Kg**   61 *H*] *B* **Fi**   62 duas] *om.*
Fi   63 igitur] ergo **BPsVa**   63 cadent] cadunt **KgPsVa**   63–64 differentie] duarum **Kg**
64 sintque] que sunt **Fi**   64 sint] duo *add.* **BVa**   64 centra] *om.* **Fi**   65 duo ... *N*] *M*
et *N* duo puncta **Fi**   66 est equalis] equatur **B**   66 etiam] *om.* **RKgOFi**   67 angulo] *om.*
V   67 *AHG*] *om.* **Fi**   67 *DTZ*] *TTZ* **Fi**, *DEZ* **OBVa**   67 equalium] duorum *add.* **Fi**
68 et[1]] *supra* **Ps**   68 *AH*] *om.* **Fi**   68 quos] quod **Va**   68 separavimus] separamus
RFi, spere raui' **Va**   68 sunt] *om.* **Fi**   69 est equalis] equatur **OBPsVa**   70 *DL*] *corr.*
ex DB **Z**   71 *DN*] *DM* **R**, *DA* **Kg**   72 *BM*] *HM* **Kg**   72 est] equatur **BPsVa**
73–74 queque ... relative] *om.* **MKgBPsVa**   74 relative] relativi **Fi**   74 et] *om.* **OFiB**
PsVa   74 angulus] angulis **Fi**   74 est equalis] equatur **KgPsVa**   75 est equalis] equatur
OBPsVa   75 basi] *om.* **Va**   75 est perpendicularis] *tr.* **KgOBPsVa**   75 super] *om.* **Fi**
75 superficiem] superficie **Fi**   76 ergo] igitur **OBPsVa**   76 omnibus] rectis *add.* **OBPsVa**, *add.*
et del. **M**   76 lineis] linee **Fi**   76 que] in spera *add.* **R**   77 proveniunt] *marg.* **B**, *supra* **Ps**, *om.*
Va   77 proveniunt recti] *tr.* **OB**   78 angulus] circulus **Va**   78 Simili quoque modo] Similiter
quoque **MOBPsVa**

قائمة فلأن خط ح‌ك مساوٍ لخط ط‌ل وخط ك‌ب مساوٍ لخط ل‌ه فخطا ح‌ك

ك‌ب مساويان لخطى ط‌ل ل‌ه كل واحد منهما لنظيره وهى تحيط بزوايا قائمة

تكون قاعدة ح‌ب مساوية لقاعدة ط‌ه ، وذلك ما أردنا أن نبين ·

60

يب   كيف ترسم على كرة دائرة عظيمة من دوائرها تماس دائرة معلومة وتكون

مماستها لها على نقطة منها معلومة ·

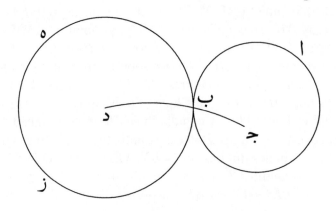

فلتكن فى كرة دائرة معلومة أصغر من الدائرة العظمى وهى دائرة ا‌ب

ولتكن النقطة المعلومة التى على الخط المحيط بها نقطة ب ونريد أن نرسم

دائرة عظيمة تماس دائرة ا‌ب المعلومة وتمر بنقطة ب فلتكن نقطة ج قطباً   5

لدائرة ا‌ب ولنرسم دائرة عظيمة تمر بنقطتى ج ب وهى دائرة ج‌ب‌د ولنفصل

منها قوساً مساوية للقوس التى يوترها ضلع المربع الذى يرسم فى الدائرة

العظمى وهى قوس ب‌د فيبيّن أن قوس ج‌ب ليس يكون ربع الدائرة لأن

الخط الذى يخرج من قطب دائرة ا‌ب إلى الخط المحيط بها ليس هو بمساوٍ

N قوس [قوساً N 7 مهما [منها 7 مماً A] منها 2 N نكون خطا [خطا N 58] add. et del. N 58] مساوٍ [1 مساوٍ 58

N صلح [2 قوس A 8 فتر [فبيّن 8   N, om. H

monstratur quod angulus *TLE* est rectus. Et quia linea *HK* est equalis linee *TL* et linea *KB* est equalis linee *LE*, ergo due linee *HK KB* sunt 80 equales duabus lineis *TL LE*, queque scilicet earum sue relative. Sed ipse continent angulos rectos: ergo basis *HB* est equalis basi *TE*. Et illud est quod demonstrare voluimus.

12 Qualiter super speram maior ex circulis eius signetur contingens circulum datum et sit eorum contactus super unum punctum ei datum.

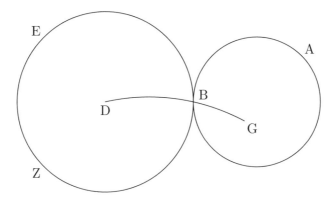

Sit ergo in spera circulus datus minor circulo maiore, qui sit circulus *AB*, et sit punctum datum quod est super lineam continentem ipsum punctum *B*. Volo autem describere circulum maiorem contin- 5 gentem circulum datum *AB* et transeuntem per punctum *B*. Sit itaque punctum *G* polus circuli *AB*. Signabo autem circulum maiorem transeuntem per duo puncta *G B*, qui sit circulus *GBD*, ex quo separabo arcum equalem arcui cui subtenditur latus quadrati quod signatur in circulo maiore, qui sit arcus *BD*. Manifestum est igitur quod latus *GB* 10 non est quarta circuli, quoniam linea que protrahitur a polo circuli *AB* ad lineam continentem ipsum non est equalis lateri quadrati quod

79 monstratur] demonstratur **Va**   79 *HK*] *corr. ex HHK* **V**   79 est equalis] equatur **OBPsVa**
80 est equalis] equatur **OBPsVa**   80 *KB*²] *LEB* **Fi**   81 queque ... relative] *om.* **MKgOB**
PsVa   82–83 Et ... voluimus] *illeg.* **M**   83 illud] id **Kg**, hoc **O**, istud **BPsVa**   83 est] *om.* **Va**
83 quod ... voluimus] *om.* **O**   83 demonstrare] *om.* **BPsVa**   83 voluimus] volumus **BPsVa**
1 signetur] significetur **OBPsVa**   2 sit] sint **R**   2 unum] *supra* **P**   2 unum punctum] *tr.*
MKgOBPsVa   2 ei] e **Fi**   3 ergo] igitur **Kg**   3 datus] *post* maiore **MKgBPs** (**BPs** *in*
corr., dato **Va**)   3 qui] scilicet minor *add. supra* **Ps**   5–6 contingentem] continentem **VFi**
7 Signabo] Significabo **Va**, *hic et saepius*   7–8 circulum ... ex] *om.* **Va**   8 *G B*] *B G* **MKgB**,
Ps (*supra*)   8 *GBD*] *in corr.* **Ps**   9 arcui] *BD add.* **MKg**   9 cui] *corr. marg. ex* qui **P**
9 signatur] significatur **PsVa**   10 circulo maiore] *tr.* **MKgOBPsVa**   10 qui ... *BD*] *om.* **MKg**,
marg. **B**, *supra* **Ps**   10 sit] est **BPsVa**   10 est igitur] *tr.* **Fi**, est ergo **Kg**, ergo est **OBPsVa**
10 latus] arcus **KgOBPsVa**   11 a] qua **Fi**   12 ad] *ADB* **Fi**   12 ipsum] ipsam **V**

لضلع المربع الذى يرسم فى الدائرة العظمى وذلك أن دائرة ا ب حينئذٍ تكون ١٠

عظيمة ولم يكن كذلك فقوس ب ج ليس هى ربع الدائرة لكنها أقل من

الربع فلنرسم على قطب د وببعد د ب دائرة ه ب ز فدائرة ه ب ز عظيمة

وذلك أن الخط الذى يخرج من قطبها إلى الخط المحيط بها مساوٍ لضلع المربع

الذى يرسم فى الدائرة العظمى ودائرتا ا ب ه ب ز فى كرة وهما تقطعان خطاً

محيطاً بدائرة أخرى عظيمة من الدوائر التى فى الكرة وهو خط ج ب د على ١٥

نقطة واحدة وهى نقطة ب وأقطابهما عليها فإحدى الدائرتين تماس الأخرى

فدائرة ا ب تماس دائرة ه ب ز فقد رسمت دائرة عظيمة وهى دائرة ه ب ز

تمر بنقطة ب المعلومة وتماس دائرة ا ب على نقطة ب ، وذلك ما أردنا أن

نبين .

يج   إذا كانت فى كرة دوائر متوازية ثم رسمت فى تلك الكرة دائرتان عظيمتان

تماسان إحدى تلك الدوائر وتقطعان الدوائر الباقية فإن القسى من الدوائر

المتوازية التى فيما بين أنصاف الدائرتين العظيمتين التى لا تلتقى متشابهة

والقسى من الدائرتين العظيمتين التى فى ما بين الدوائر المتوازية متساوية .

فلتكن فى كرة دوائر متوازية وهى دوائر ا ب ج د ه ز ح ط ك ل ولترسم ٥

فى تلك الكرة دائرتان عظيمتان وهما دائرتا ا ه ك ح ج س ب ز ل ط د س تماسان

إحدى تلك الدوائر وهى دائرة ل ك على نقطتى ل ك وتقطعان دائرتى ا ب ج د

١٠ تكون حينئذٍ [in corr. A   ١١ ليس] tr. A   ١٢ فلنرسم] ولنرسم N   ١٣ لضلع [ربع N
١٦ واحدة] واحد A   ١٦ نقطة ² [om. N   ١٦ عليها] علیهما N تماسان [تماس ٢ ,supra بماسان A
٣-٤ العظيمتين ... لا التى] marg. A   ٧ تلك [om. A

signatur in circulo maiore. Quod ideo est quoniam circulus *AB* esset
tunc maior. Sed non est ita: arcus ergo *BG* non est quarta circuli,
immo est minor quarta. Signabo autem super polum *D* cum longi-   15
tudine *DB* circulum *EBZ*. Ergo circulus *EBZ* est maior. Quod ideo
est quoniam linea que protrahitur a polis eius ad lineam continentem
ipsum est equalis lateri quadrati quod signatur in circulo maiore. Duo
vero circuli *AB EBZ* sunt in spera et secant lineam continentem circulum alterum maiorem qui est ex circulis qui sunt in spera, que est   20
linea *GBD*, super punctum unum, quod est punctum *B*, super quem
sunt poli eorum. Unus igitur duorum circulorum alterum contingit.
Circulus ergo *AB* contingit circulum *EBZ*. Iam igitur designatus est
circulus maior, qui est circulus *EBZ*, transiens per punctum datum *B*
contingens circulum *AB*. Et illud est quod demonstrare voluimus.   25

13   Cum in spera fuerint circuli equidistantes et postea in eadem spera duo
circuli maiores signati fuerint unum illorum circulorum contingentes
et reliquos circulos secantes, arcus circulorum equidistantium qui sunt
inter medietates duorum circulorum maiorum que non concurrunt sunt
similes et arcus duorum circulorum maiorum qui sunt inter circulos   5
equidistantes sunt equales.

     Sint itaque in spera circuli equidistantes, qui sint *ABGD EZHT*
KL. In eadem quoque spera duos signabo circulos maiores, qui sint
circuli *EKHGO BZLTDO*, contingentes unum illorum, qui sit circulus *LK*, super duo puncta *L K* et secantes duos reliquos circulos *ABGD*   10

---

13 in circulo] *om.* **Va**   13 circulo maiore] *tr.* **OFiZ**   13–14 esset tunc] *tr.* **FiZ**
14 ita] *supra* **B**, *om.* **PsVa**   15 minor quarte] *tr.* **Fi**   15 quarta] circuli *add.* **MKgOBPsVa**
15 polum] punctum *add.* **Kg**   15 *D*] *supra* **Ps**   15–16 cum longitudine] et secundum longitudinem **Fi**   16 Ergo circulus] *tr.* **KgOBPsVa**   17 que] *supra* **Ps**   18 signatur] significatur **PsVa**
18 circulo maiore] minori circulo **Fi**   19–20 circulum alterum] *tr.* **M**   20 qui] que
R   20 qui ... in spera] *om.* **OBPsVa**   21 *GBD*] *in corr.* **BPs**, *BCD* **Kg**, *BDG*
Va   21 quem] que **KgFi**   22 igitur] ergo **KgOFiBPsVa**   22 circulorum] aliorum
PsVa   22 alterum] alium **KgOVa**   23 ergo] igitur **MKgBPsVa**   23 *AB*] *ABG* **Va**
23 circulum] *om.* **Kg**   23–25 igitur ... voluimus] patet probandum **OB**   23 designatus] de
supra **R**   23–25 designatus ... voluimus] patet probandum **PsVa**   24 *B*] et *add.* **KgFiZ**
25 contingens] contingit **FiZ**   1 fuerint] *om.* **Fi**   1 fuerint circuli] *tr.* **PRVZ**   1 circuli] *om.*
R, *in corr.* **V**   1 postea] *post* eadem **PRVFiZ**   1 spera²] *om.* **MKgBPsVa**   1 duo] *supra* **O**, *om.*
BPsVa   2 circuli maiores] *tr.* **PRVFiZ**   2 signati] significati **BPs**   2 illorum circulorum] *tr.*
R   3 circulos] circulorum **KgOBPsVa**   3 secantes] *in corr.* **R**   3 circulorum] illorum
Fi   3 equidistantium] medietates *add.* **Fi**   3 qui] que **Fi**   4 medietates] *om.* **Fi**
4 maiorum] *om.* **BPsVa**   4 que] qui **RKgBPs**   4 concurrunt] occurrunt **RVMZ** (*hic et*
saepius)   5 similes] Et arcus duorum circulorum maiorum qui non occurrunt sunt similes *add.*
marg. **R**   5 qui]] non occurrunt sunt similes et arcus duorum circulorum maiorum qui *add.* **R**
7 equidistantes] equidem **Fi**, *hic et saepius*   7 sint] sunt **Z**; circuli *add.* **Fi**   8 duos] *om.* **FiZ**
8 sint] sunt **Fi**   9 contingentes] *in corr.* **BPs**, continens **Z**   9 illorum] eorum **KgOBPsVa**
10 reliquos circulos] *tr.* **MKgOBPs**   10 circulos] *om.* **Va**

 هزحط الباقيتين ، فأقول إن القسى من الدوائر المتوازية التى هى فيما بين

أنصاف الدوائر العظيمة التى لا تلتقى متشابهة وإن القسى من الدائرتين

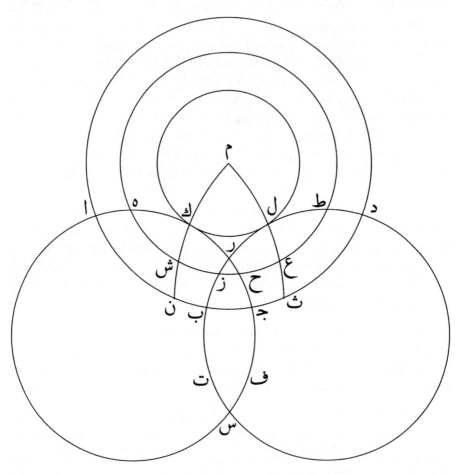

العظيمتين التى هى فيما بين الدوائر المتوازية متساوية ويمكننا أن نعلم 10

القسى التى فيما بين أنصاف الدوائر التى لا تلتقى بما أصف وهو أنه لما

كانت الدوائر العظيمة التى فى كرة يقطع بعضها بعضاً بنصفين صارت قوس

ركاس نصف دائرة فقوس كاس أقل من نصف دائرة فلنضع أن قوس

EZHT. Dico igitur quod arcus circulorum equidistantium qui sunt inter medietates maiorum circulorum que non concurrunt sunt similes et quod arcus circulorum maiorum qui sunt inter circulos equidistantes

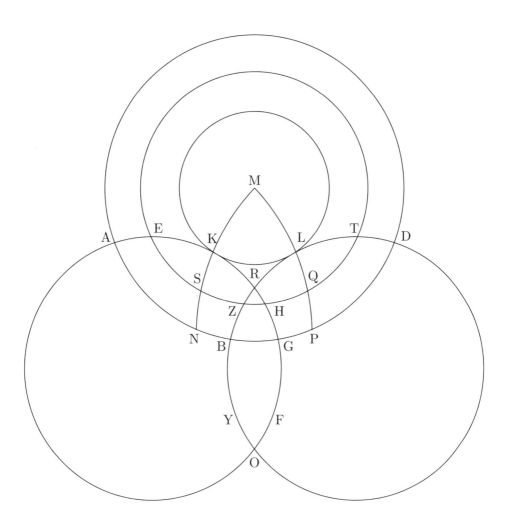

sunt equales. Possibile vero est ut sciamus qui sint arcus qui sunt inter medietates circulorum non concurrentes secundum hoc quod dicam: 15
quod est propter hoc namque quod circuli maiores qui sunt in spera se ad invicem in duo media secant, fit arcus *RKAO* semicirculus. Arcus

11 igitur] ergo **OBPsVa**   12 inter] intus **Fi**   12 que] qui **Va**   14 vero] non **Va**
14 qui] quid **Va**   14 sint] sunt **FiZ**   14 qui²] *supra* **Ps**   16 namque quod] Nam quia **FiZ**
16 circuli maiores] *tr.* **OPsVa**   17 in] *om.* **Fi**   17 secant] *ante* in **Kg**   17–18 Arcus igitur] *tr.*
Va

كاسف مثل نصف دائرة ولأن قوس ربس أيضاً نصف دائرة تكون قوس

15 لربس أكثر من نصف دائرة فلنضع أن قوس لربت نصف دائرة

فنصف الدائرة التى تخرج من نقطة ك إلى ناحية آ وهى قوس كاسف ليس

تلقى نصف الدائرة التى تخرج من نقطة ل إلى ناحية ب وهى قوس

لربت وكذلك أيضاً قوس كرف التى هى نصف دائرة ليس تلقى نصف

الدائرة التى تخرج من نقطة ل إلى ناحية ك وهى قوس لطدست فالقسى

20 من الدوائر المتوازية التى هى فيما بين أنصاف الدوائر العظيمة التى لا تلتقى

هى قسى كل هزآب حط جد ، فأقول إن قسى كل هزآب يشبه بعضها

بعضاً ، وأقول أيضاً إن القسى من الدوائر العظيمة التى هى فيما بين الدوائر

المتوازية متساوية أعنى أن أربعاً منها وهى آه زب حج طد مساوٍ بعضها

لبعض وأن أربعاً منها وهى كه كح زل لط مساوٍ بعضها لبعض ،

25 فلنتعلم قطباً للدوائر المتوازية وليكن نقطة م ولنرسم دائرتين عظيمتين

تمران بنقطة م وبكل واحدة من نقطتى ك ل وهما دائرتا مكشن ملعث

فلأن دائرتى اهكحج كل فى كرة تماس إحداهما الأخرى على نقطة ك وقد

رسمت دائرة عظيمة تمر بقطب دائرة واحدة منهما وهى دائرة كل وعلى

موضع الماسة من الأخرى وهى دائرة مكشن صارت دائرة مكشن تمر

30 أيضاً بقطبى دائرة اهكحج وتكون قائمة عليها على زوايا قائمة وكذلك نبين

الذى [التى N 16   om. N 16   نصف [¹ 15   رلس N 15   ربس [فلان A 14   [ولأن A, om. H 14   ملا [مثل 14
A   كرف [18   كرو [كرف A; illeg., marg. N   الدى [التى A 17   ستحرح [تخرج N 17   -18
يشبه بعضها بعضا [جد 21   A وهو [وهى A   الدى [التى in corr. A 19   om. N 19   ليس ... لطدست [19
add. marg. A 22   فاقول [وأقول A 22   أيضاً [om. N 23   أعنى [واعى N 25   فلنتعلم قطباً للدوائر [
N ولدلك [وكذلك 30   اهكح [اهكحج N 30   اهمحح A 30   اهكحج [اهكحج 27   مكشن [in corr. A,   مكس N, hic et saepius   وكل :H [وبكل 26   وكل AN 26   مكشن [A فلعلم قطب الدوائر

igitur KAO est minor semicirculo. Ponam autem ut arcus $KAOF$ sit
sicut medietas circuli. Et quia etiam arcus RYO est semicirculus, ergo
arcus $LRYO$ est maior semicirculo. Ponam autem ut arcus $LRZBY$ 20
sit medietas circuli. Medietas igitur circuli que producitur a puncto K
ad partem A, que est arcus $KAOF$, non concurrit medietati circuli que
producitur a puncto L ad partem B, qui est arcus $LRZY$. Similiter
quoque arcus KRF, qui est medietas circuli, non concurrit arcui LDY,
qui est medietas circuli. Ergo arcus circulorum equidistantium qui sunt   25
inter medietates circulorum maiorum que non concurrunt sunt arcus
$KL\ EZ\ AB\ HT\ GD$. Dico igitur quod arcus $KL\ EZ\ AB$ ad invicem
sunt similes. Et dico etiam quod arcus circulorum maiorum qui sunt
inter circulos equidistantes sunt equales, scilicet quod quattuor eorum,
qui sunt $AE\ ZB\ HG\ TD$, sunt ad invicem equales et quattuor eorum,   30
qui sunt $KE\ KH\ ZL\ LT$, ad invicem sunt equales.

Notabo igitur polum circulorum equidistantium, qui sit punctum
M, et signabo duos circulos maiores transeuntes per punctum M et per
unumquodque duorum punctorum K et L, qui sint duo circuli $MKSN$
$MLQP$. Et quia duorum circulorum $AEKHG\ KL$ in spera unus alterum   35
contingit super punctum K et iam signatus est circulus maior transiens
per polum unius eorum, qui est circulus KL, et super locum contactus
alterius, qui est circulus $MKSN$, fit circulus $MKSN$ transiens etiam per
duos polos circuli $AEKHG$ et est super eum orthogonaliter erectus. Et

18 minor] quatuor **Fi**   18 autem] aut **Fi**   18 $KAOF$] $KAEF$ **Fi**   18 sit] *in corr.* **P**, *supra*
BPs   19 etiam] *om.* **PRKg**   19–20 RYO ... arcus¹] *marg.* **Ps**   19–20 est ... arcus¹] *marg.*
B   20 autem] *supra* **Ps**, *om.* **FiZ**   20 $LRZBY$] $LTZBY$ **R**, $LRZB$ **KgOFiZBPsVa**   21 sit] *om.*
Kg   21 Medietas] etas **Fi**   21 igitur] ergo **OFiZ**   21 producitur] protenditur **MKgOBPsVa**
21 a puncto] *marg.* **R**   23 producitur] *illeg.* **Kg**, procedit **OPsVa**, procediitur **B**   23 L] K
Kg   23 B] *supra* **BPs**   23 qui] que **KgOFiZBPsVa**   23 $LRZY$] $LRZB$ **KgOFiZBVa**   23–
27 Similiter ... GD] *marg.* **BPs**   24 KRF] RKF **RVMFiZ**   24–25 non ... circuli] *om.*
V   24 LDY] LDB **FiZBPsVa**, LOY **R**, *illeg.* **Kg**   26 circulorum] terminorum **Va**
26 maiorum] *om.* **PRVMKgBPsVa**   26 que] qui **R**   27 Dico] *om.* **B**   27 Dico ... AB] *om.*
O   27 igitur] *supra* **Ps**, itaque **B**   27 quod] *om.* **B**   27 AB] $HT\ GD$ *add.* **FiZB**
PsVa   29 sunt] *om.* **Fi**   29 equales] similes **Kg**   30 sunt ad invicem] *permut.* **FiZ**
30 eorum] *om.* **Z**   31 $ZL\ LT$] $ZT\ BT$ **Va**   32 Notabo] Vocabo **PsVa**   32 sit] sunt **Fi**
34 unumquodque] utrumque **MOBPsVa**   34 sint] sunt **OFiBPsVa**   34 $MKSN$] $MKLN$ **Va**
35 unus] *supra* **B**   35–36 alterum contingit] *tr.* **OBPsVa**   36 iam] ideo **R**   37 polum] *in corr.*
BPs, polos **FiZ**   37 est circulus] *tr.* **Fi**   37 contactus] eorum *add.* **V**   38 fit circulus] *illeg.* **Kg**
38 circulus ²] etiam *add.* **M**   38 $MKSN$²] *in corr.* **BPs**   38 etiam] *supra* **M**, *om.* **KgOBPsVa**
39 $AEKHG$] $AEKH$ **PVMFiZ**, *corr. ex* $AEZH$ **R**   39–41 $AEKHG$ ... circuli] *om.* **Va**

أن دائرة م ل ع ث أيضاً تمر بقطبى دائرة ب ز ل ط د س وتكون قائمة عليها
على زوايا قائمة فقد عمل فى دوائر متساوية أعنى دوائر ا ه ك ح ج س
ب ز ل ط د س على الأقطار التى تخرج من نقطتى ك ل قطعتان متساويتان
من دوائر وهما قطعتا ل م م ك وهما قائمتان عليها على زوايا قائمة والقطع
التى تتصل بهذه لتمام نصفى دائرتين وقد فصل منهما قوسان متساويتان   35
وهما قوسا ك م م ل وهى أصغر من نصفى القطعتين المعمولتين والخط الذى
يصل بين نقطة م وبين نقطة ا مساوٍ للخط الذى يصل بين نقطة م ونقطة
د وذلك أنهما جميعاً يخرجان من قطب دائرة ا ب د إلى الخط المحيط بها فهى
تفصل قسياً متساوية فقوس ا ك مساوية لقوس ل د ومن قبل ذلك أيضاً
تكون قوس ه ك مساوية لقوس ل ط ولأن دائرتى ا ب ج د ا ه ك ح ج س فى   40
كرة وإحداهما تقطع الأخرى وقد رسمت دائرة عظيمة تمر بأقطابهما وهى
دائرة م ك ش ن صارت دائرة م ك ش ن تقسم القطع التى فصلت بنصفين
نصفين فقوس ا ه ك مساوية لقوس ك ح ج وقوس ا ن مساوية لقوس ن ج
وكذلك أيضاً نبين أن قوس ب ل مساوية لقوس ل د وأن قوس ب ث مساوية
لقوس ث د ولأن قوس ا ه ك مساوية لقوس ل ط د وقوس ا ه ك ج ضعف   45
قوس ا ه ك وقوس د ط ل ب ضعف قوس ل ط د تكون قوس ا ك ح ج
مساوية لقوس د ط ل ب والدائرتان متساويتان وذلك أنهما عظيمتان فالخط

31 دائرة²] دار A   33 ب ز ل ط د س]   34 وهما¹] N وهى   34 قائمة ... وهما²] posuit
post دائرتين (l. 35) A   35 دائرتين] الدارس N   37 وبين نقطة] N ونقطه   38 ا ب د] N
ا ه ك ح ج س] اهكحس N   41 وإحداهما] N واحده بهما   41 رسمت] N رسم   42 تقسم] قطع N
43 ن ج ... مساوية لقوس ك ح ج] marg. A   45 لقوس ... ث د ولأن] marg. A   45 ولأن] فلان N
47 والدائرتان متساويتان] والدواير مساوية A, والدارة مساوية supra N   47 عظيمتان] عظيمة AH, supra N

similiter ostenditur quod circulus $MLQP$ etiam transit per duos polos   40
circuli $BZLTDO$ et est erectus super eum orthogonaliter. Iam ergo in
circulis equalibus, scilicet circulis $AEKHO\ BZLTDO$, super diametros
que a duobus punctis K et L protrahuntur due circulorum porciones
equales sunt constitute, que sunt porciones $KM\ ML$, et sunt super eas
orthogonaliter erecte. A porcionibus autem que istis adiunguntur ad   45
medietates circulorum complendas iam separantur arcus equales, qui
sunt arcus $KM\ ML$, qui sunt minores medietatibus porcionum datarum. Linea quoque que coniungit inter punctum M et punctum A
est equalis linee que coniungit inter punctum M et punctum D. Quod
ideo est quoniam ipse ambe producuntur a polo circuli AN ad lineam   50
continentem ipsum. Ergo ipse secant arcus equales. Ergo arcus AK est
equalis arcui LD, et propter hoc etiam erit arcus EK equalis arcui LT.
Et quia duorum circulorum $ABGD\ EKGO$ in spera unus alterum secat
et iam signatus est circulus maior transiens per polos eorum, qui est
circulus $MKSN$, fit circulus $MKSN$ secans porciones quas separavit in   55
duo media et duo media. Ergo arcus AEK est equalis arcui KHG et
arcus KN est equalis arcui NG. Et similiter etiam monstratur quod
arcus BL est equalis arcui LD et quod arcus BP est equalis arcui PD.
Et quia arcus AEK est equalis arcui LTD et arcus $AEKG$ est duplus
arcus AEK et arcus $DTLB$ est duplus arcus LTD, ergo arcus $AKHG$ 60
est equalis arcui $DTLB$. Sed circuli sunt equales, quod ideo est quoniam

---

40 quod circulus] *om.* **Z**   40 etiam] et **V**   41 *BZLTDO*] super diametros *add.* **O**   41 est] *om.*
FiVa   42 *AEKHO*] *AEKO* **R**, *AEHKO* **VVa**   42 *BZLTDO*] *LTDO* **Kg**   43 due] *in corr.*
B   43–44 circulorum porciones equales] *permut.* **OBPsVa**   44 *KM*] *HM* **Fi**   44 *ML*] *LML*
Z   44 et] *om.* **FiZ**   44 sunt[2]] etiam *add.* **Fi**   45 autem] *ML* et *KM add. supra* **Ps**
46 separantur] separant **R**   47 arcus ... sunt] *om.* **BPsVa**   47 minores] *verbum illeg.*
add. **Kg**   47–48 medietatibus porcionum datarum] *permut.* **MKgBPsVa**   48 Linea] Linearum
Fi   48 coniungit] adiungit **O**, adiungitur **BPsVa**   49 est equalis] equatur **OBPsVa**
49 coniungit] contingit **Va**   51 continentem] *illeg.* **Kg**   51–52 est equalis] equatur **OBPsVa**
52 etiam erit] *tr.* **Kg**   54 est[1]] *om.* **PsVa**   54 qui] que **Va**   55 fit circulus *MKSN*] *marg.*
P   55 secans] *supra* **R**   55 separavit] separat **O**, separatur **B**   56 et duo media] *om.* **R**
56 *KHG*] *KRHG* **PRVMKgOPs**, *KTHG* **Va**   57 *KN*] *AN* **VMFiZVa**   57 est equalis] equatur
OBPsVa   57–58 *NG* ... arcui] *om.* **Va**   58 est equalis[1]] equatur **OBPs**   58 quod] quia **O**,
quidem **Va**   58 est equalis[2]] equatur **OBPsVa**   59 est equalis] equatur **OBPsVa**   59 arcui] *om.*
B, *supra* **Ps**   59 *AEKG*] *AKG* **Fi**   60 arcus[1]] arcui **OFiZBPsVa**   60 *DTLB*] *illeg.* **Kg**, *DAB*
Va   60–92 duplus ... arcus *KH* est] *om.* **Fi**   60 arcus[3]] arcui **OBPsVa**   60 *LTD*] *ATD* **R**
60 arcus[4]] *om.* **BPsVa**   60 *AKHG*] *ARHG* **PsVa**

الذى يصل بين نقطة اّ ونقطة جّ مساوٍ للخط الذى يصل بين نقطة دّ

ونقطة بّ فقوس ان ب ج مساوية لقوس ب ث د وذلك أن الخطوط

50 المستقيمة التى توترها متساوية وهى من دائرة واحدة بعينها وقوس ان نصف

قوس ان ب ج وقوس ب ث نصف قوس ب د فقوس ان مساوية لقوس

ب ث ونزيد قوس ب ن المشتركة فكل قوس ان ب مساوية لكل قوس

ن ب ث وهى من دائرة واحدة بعينها فقوس ان ب تشبه قوس ن ب ث

وقوس ن ب ث تشبه قوس كل وذلك أنه إذا كانت فى كرة دوائر متوازية

55 ورسمت دوائر عظيمة تمر بأقطابها فإن القسى من الدوائر المتوازية التى هى

فيما بين الدوائر العظيمة متشابهة والقوسان من الدوائر المتوازية اللتان فيما

بين م ن م ث واللتان هما من الدوائر العظيمة التى تمر بأقطابهما هما قوسا

كل ن ث وقوس ان ب أيضاً شبيهة بقوس كل فمن قبل ذلك أيضاً تكون

قوس كل شبيهة بقوس ه ز فقوس ه ز أيضاً شبيهة بقوس اب لأن سبيلهما

60 واحدة وذلك أنهما فيما بين دائرتى اه ج دزب فقوس ه ز شبيهة بقوس كل

وكذلك أيضاً نبين أن قوس جّ ث د شبيهة بقوس ح ع ط وأن هذه القوس

شبيهة بقوس ه ز وذلك أن قوس ح ع ط أيضاً شبيهة بقوس كل فالقسى من

---

48 يصل [N   49 ان ب ج] الصا add. A   51 سد [ب د N, בנתד H   51 فقوس [وموس N

53 ن ب ث ²[N   53 وهى ... ²ن ب ث] om. A   53 المشتركه] ر add. et del. N; نبت [¹ن ب ث N

54 ن ب ث] سب N   55 هى [هو A   56 واللتان [اللس N   57 in corr. N   58 ن ث] نحب N

58 وقوس [فموس A   59 شبيهة ¹] هر انضا سبه A   59 add. et del. A   59 بقوس ²[بحط N   59 اب [ار N

59-60 دزب ... سبيلهما [سبيلهما NH: سلها marg. A   60 لأن سبيلهما [... دزب marg. A   60 دزب] corr. ex دزه

A 60 كل ... فقوس] مساهه الله هر كل هر آب فمسى A قوس [61 A حط] add. supra A   61 جّ ث د [

N حط حد   ط

ipsi sunt magni. Ergo linea que coniungit inter punctum A et punctum G est equalis linee que coniungit inter punctum D et punctum B. Ergo etiam arcus ABG est equalis arcui BPD. Et hoc ideo quoniam linee recte que subtenduntur eis sunt equales et sunt unius et eiusdem cir- 65 culi. Sed arcus AN est medietas arcus ABG et arcus BP est medietas arcus BPD: ergo arcus AN est equalis arcui BP. Addam autem arcum BN communem: ergo totus arcus ANB est equalis toti arcui NBP. Sed ipsi sunt unius et eiusdem arcus: ergo arcus ANB est similis arcui NBP. Sed arcus NP est similis arcui KL. Quod ideo est quoniam cum 70 in spera fuerint circuli equidistantes et signati fuerint circuli maiores transeuntes per polos eorum, tunc arcus circulorum equidistantium qui sunt inter circulos maiores erunt similes. Et duo arcus circulorum equi- distantium qui sunt inter MN et MP, qui sunt circulorum maiorum qui transeunt per polos eorum, sunt arcus $KL\ NGP$. Et etiam arcus ANB 75 est similis arcui KL: propter hoc igitur etiam arcus KL est similis arcui EZ. Ergo arcus EZ etiam est similis arcui AB. quoniam eorum modus est unus. Quod ideo est quoniam ipsi sunt inter duos circulos AEG DZB; ergo arcus EZ est similis arcui KL. Et similiter etiam ostendi- tur quod arcus $HT\ GPD$ sunt similes arcui HQT et quod isti arcus 80

---

62 ipsi sunt magni] *permut.* **V** 62 Ergo linea] *illeg.* **Kg** 62 linea] *om.* **V** 62 coniungit] contingit **Va** 64 est equalis] equatur **OBPsVa** 64 arcui *BPD*] *illeg.* **Kg** 64 hoc] est *add.* **BPsVa** 64 ideo] est *add.* **O** 65 unius et eiusdem] *permut.* **OBPsVa** 66 arcus[1]] *om.* **BPsVa** 66–67 medietas[1] ... *BP*] equalis **R** 66–67 *ABG* ... arcus[1]] *om.* **Va** 66 arcus *BP*] *illeg.* **Kg** 67 est equalis] equatur **OBPsVa** 67 arcui] *infra* **B**, *supra* **Ps**, *om.* **Va** 68 *BN*] *in corr.* **Ps** 68 est equalis] equatur **OBPsVa** 68 toti] *illeg.* **Kg** 68 *NBP*] *NPB* **P** 70 *NBP*] *NPB* **P** 71 signati] significati **BPs** 72 equidistantium qui] *illeg.* **Kg** 72–73 qui sunt] *om.* **V** 73 circulos] *om.* **OBPsVa** 73 similes] *om.* **Kg** 74 *MN*] *N, supra* *M* **Ps** 74 sunt[2]] arcus [...] *supra* **Ps** 74 circulorum maiorum] *tr.* **Va** 74 maiorum] *om.* **O** 75 polos eorum] *illeg.* **Kg** 75 Et] Ergo **FiBPsVa** 75 arcus *ANB*] *tr.* **OBPsVa** 76 igitur] ergo **VO**, *om.* **BPsVa** 76 igitur etiam] *tr.* **Kg** 76 etiam] et **OFiZ** 77 Ergo ... *AB*] *primum false, deinde recte* **Va** 77 etiam] *om.* **ROBPsVa**, *post* est **Z** 77 est similis] *illeg.* **Kg** 77 *AB*] *corr. ex ZB* **R** 77–79 quoniam ... arcui *KL*] *marg.* **R** 77 eorum] *om.* **Va** 77 eorum modus] *tr.* **KgOBPs** 78 quoniam] quod **V** 78 ipsi] *om.* **OBPsVa** 79 est similis] *tr.* **Va** 79 arcui] *om.* **Va** 79 *KL*. Et] *illeg.* **Kg** 79– 80 etiam ostenditur] *tr.* **RVM** 80 similes] equales **Va** 80 arcui *HQT*] *om.* **PRVa** 80 quod[2]] quia **KgB**

الدوائر المتوازية التى فيما بين أنصاف الدوائر العظيمة التى لا تلتقى

متشابهة ،

65   وأقول أيضاً إن القسى من الدوائر العظيمة التى هى فيما بين الدوائر

المتوازية متساوية وذلك أن القسى الأربع أعنى قسى اهك كح ج بزل

لط د مساوٍ بعضها لبعض وأربع منها وهى هك كح زل لط مساوٍ بعضها

لبعض وذلك أن دائرة كن العظمى تقسم قطعتى هك ه ش ح اللتين فصلتا

بنصفين نصفين وكذلك تقسم أيضاً قطعتى زل ط زع ط فقوس هك مساوية

لقوس كح وقد كان تبين أن قوس هك مساوية لقوس لط فقوس كح   70

مساوية لقوس طل وقوس طل مساوية لقوس لز فقوس لز أيضاً مساوية

لقوس كح فقسى هك كح زل لط الأربع متساوية وقسى اه بز جح

دط الأربع الباقية مساوٍ بعضها لبعض فالقسى من الدوائر المتوازية التى هى

فيما بين أنصاف الدوائر العظيمة التى لا تلتقى متشابهة والقسى من الدوائر

العظيمة التى هى فيما بين الدوائر المتوازية متساوية ، وذلك ما أردنا أن   75

نبين .

يد   إذا كانت فى كرة دائرة معلومة أصغر من الدائرة العظمى وكانت على سطح

الكرة نقطة معلومة فيما بين الدائرة التى ذكرنا وبين الدائرة التى تساويها

وتوازيها وأردنا أن نرسم دائرة عظيمة تمر بالنقطة المعلومة وتماس الدائرة التى

ليست بعظيمة .

72 –om. A   71 أيضاً [marg. A   71 لقوس طل [N   69 تقسم [سقم N   68 وكذلك [وذلك أن
73 الأربع ... متساوية [marg. A   73 حط [دط N   73 مساوٍ [supra A   74 فيما [om. A
75 هى [supra A, om. N

sunt similes arcui *EZ*. Quod ideo est quoniam arcus *HQT* est etiam similis arcui *KL*. Ergo arcus circulorum equidistantium qui sunt inter medietates circulorum maiorum que non concurrunt sunt similes.

Et dico etiam quod arcus circulorum maiorum qui sunt inter circulos equidistantes sunt equales. Quod ideo est quoniam quattuor arcus, scilicet arcus *AEK KHG LTD BZL*, ad invicem sunt equales et quattuor eorum qui sunt *EK KH ZL LT* ad invicem sunt equales. Et hoc ideo quoniam circulus maior *KN* dividit duas porciones *EKH ESH*, quas separavit, in duo media et duo media. Et similiter etiam dividit porciones duas *ZLT ZQT*: ergo arcus *EK* est equalis arcui *KH*. Sed iam fuit ostensum quod arcus *EK* est equalis arcui *LT*, ergo arcus *KH* est equalis arcui *LT*. Sed arcus *TL* est equalis arcui *LZ* et arcus *LZ* est etiam equalis arcui *KH*: ergo quattuor arcus *EK KH ZL LT* sunt equales et quattuor arcus *AE BZ GH DT* reliqui sunt ad invicem equales. Ergo arcus circulorum equidistantium qui sunt inter medietates circulorum maiorum que non concurrunt sunt similes et arcus circulorum maiorum qui sunt inter circulos equidistantes sunt equales. Et illud est quod demonstrare voluimus.

14 Cum in spera fuerit circulus datus existens minor circulo maiore et fuerit super superficiem spere punctum datum inter circulum quem prediximus et circulum qui est ei equalis et equidistans et voluerimus signare circulum maiorem transeuntem per punctum datum et contingentem circulum qui non est maior.

---

81 sunt] *supra* **O**   81 arcui] *supra* **R**   81 est[1]] *supra* **R**   81 *HQT*] *illeg.* **Kg**
81 etiam] *supra* **R**   81–82 etiam similis] *tr.* **MKgOBPsVa**   83 circulorum maiorum] *tr.*
MKgOFiZBPsVa   83 que] qui **PROVa**   84 circulorum] *illeg.* **Kg**   85 Quod] quoniam **O**   85–
86 quattuor arcus] *tr.* **Z**   86 *AEK*] *in corr.* **Ps**   86 *KHG*] *in corr.* **BPs**, *KAG* **Z**, *om.* **Va**
87 et ... equales] *om.* **ZVa**   87 *KH*] *KZ* **B**, *KR* **Ps**   88 ideo] est *add.* **R**   88 *KN*] *KO*
Kg   89 quas] scilicet *add. supra* **Ps**, quam **Va**   89 et duo media] *om.* **R**   89 etiam] *om.*
OBPsVa; circulus *MP add.* **MKgOFiZBPsVa**   90 dividit] in duo media et duo media *add.*
supra **Ps**   90 est equalis] equatur **BPsVa**   91 fuit ostensum] ostenditur **OBPsVa** (*in corr.*
B)   91 est equalis] equatur **OBPsVa**   91–92 ergo ... *LT*] *om.* **R**   92 est equalis[1]] equatur
OBPsVa   92 arcui] *supra* **BPs**   92 est equalis[2]] equatur **OBPsVa**   92 *LZ*] *AZL* **B**
92 et] *om.* **Fi**   93 est etiam] *tr.* **OBPsVa**   93 est etiam equalis] *permut.* **Kg**   93 etiam] *om.*
R   93 etiam equalis] *tr.* **MFi**   93 *KH*] *KB* **Fi**   93 quattuor arcus] *tr.* **OBPsVa**   94 *LT*] *EL*
Z   94 quattuor] *illeg.* **Kg**   94 arcus] *om.* **V**   94–95 sunt ad invicem] *permut.* **OBPsVa**
95 inter] duas *add.* **KgOBPsVa**, *add. et del.* **M**   96 et] *supra* **B**   98 illud] hoc **O**, alliud **Fi**, istud
BPsVa   98 quod ... voluimus] *om.* **O**   98 demonstrare] *om.* **BPs**   98 voluimus] volumus **ZB**
PsVa   1 minor circulo maiore] *permut.* **MKgOBPsVa**   2 super] *om.* **Z**   3 circulum] circulus
R, *corr. ex* circulus **BPs**   3 est ei] *tr.* **Z**   3 ei equalis] *tr.* **OBPsVa**   3 voluerimus] volumus
Va   4 signare] significare **PsVa**, *hic et saepius*   4 maiorem] *supra* **O**, *om.* **BPsVa**   4–
5 contingentem] continentem **Va**   5 est] *om.* **Z**

فإنا نجعل الدائرة المعلومة التى فى الكرة التى هى أصغر من الدائرة 5
العظمى دائرة اب والنقطة المعلومة التى على سطح الكرة التى فيما بين دائرة

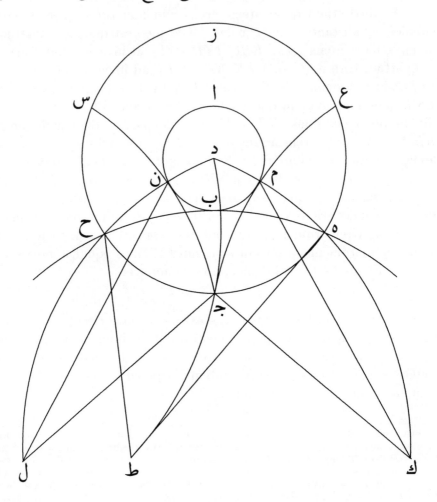

اب والدائرة التى تساويها وتوازيها نقطة ج ونريد أن نرسم دائرة عظيمة تمر
بنقطة ج وتماس دائرة اب فلنتعلم قطب دائرة اب وليكن نقطة د ولنرسم
على نقطة د وعلى بعد دج دائرة جه زح ولنرسم دائرة عظيمة تمر بنقطتى

Ponam ut circulus datus qui est in spera qui est minor circulo maiore sit circulus *AB* et sit punctum datum quod est super superficiem spere quod est inter circulum *AB* et circulum qui est ei equalis et

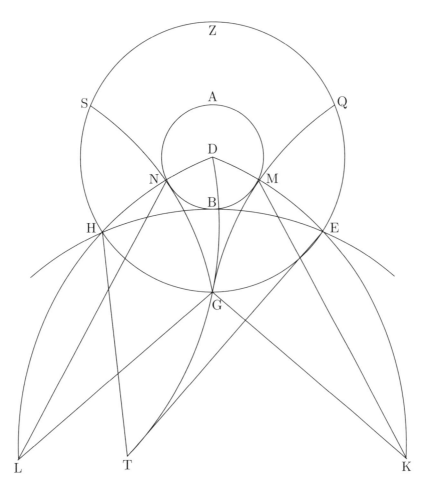

equidistans punctum *G*. Volo autem signare circulum maiorem transeuntem per punctum *G* et contingentem circulum *AB*. Signabo ergo   10
polum circuli *AB*, qui sit punctum *D*, et describam super punctum *D*
secundum spatium *DG* circulum *GEZ*. Et signabo circulum maiorem
transeuntem per duo puncta *D* et *G*, qui sit circulus *DGT*, ex quo sepa-

6 est in spera] *permut.* **KgOB**   6 qui est [2]] *om.* **O**, *scr. et del.* **B**   7 quod] *supra*
B   8 spere] *om.* **VFiZ**   8 circulum[1]] datum *add.* **O**   8 est ei] *tr.* **Z**   9–
10 Volo … punctum *G*] *om.* **Va**   10 contingentem] continentem **Fi**, *corr. ex* continentem **Ps**   10 Signabo] Significabo **Va**, *hic et saepius*   10 ergo] autem **OBPsVa**
11 et … punctum *D*] *supra* **Ps**, *om.* **Va**   11–12 describam … secundum] *marg.* **B**
11 super punctum *D*] *om.* **FiZ**   11 punctum[2]] polum **OBPs**   13 *DGT*] *DGZ* **Va**
13–14 separabo] *in corr.* **Ps**

د ج وهى دائرة د ب ج ط ولنفصل منها قوساً مساوية للقوس التى يوترها 10

ضلع المربع الذى يرسم فى الدائرة العظمى وهى قوس ب ط ولنرسم على

قطب ط وببعد ط ب دائرة ه ب ح فدائرة ه ب ح عظيمة وذلك أن الخط

الذى يخرج من قطبها إلى الخط المحيط بها مساوٍ لضلع المربع الذى يرسم فى

الدائرة العظمى وبيّن أنها تماس دائرة ا ب وذلك أنها تقطع الخط المحيط

بالدائرة العظمى وهو خط د ب ج ط على نقطة واحدة وهى نقطة ب 15

وأقطابها على ذلك الخط المحيط بهذه الدائرة ونرسم دائرتين عظيمتين تمران

بنقطة د وبنقطتى ه ح وهما دائرتا د م ه ك د ن ح ل ولنفصل كل واحدة من

قوسى ه ك ح ل مساوية لقوس ج ط فلأن دائرتى ه ب ح ز ه ج ح فى كرة

وإحداهما تقطع الأخرى وقد رسمت دائرة عظيمة تمر بقطبيها وهى دائرة

د ب ج ط تقسم القطع التى فصلت بنصفين نصفين فقوس ه ج مساوية 20

لقوس ج ح وقوس ه ب مساوية لقوس ب ح فلأن قسى ه د د ج د ح الثلث

يساوى بعضها بعضاً وذلك أنها تخرج من قطب د الذى هو قطب للدائرتين

جميعاً وقسى د م د ب د ن أيضاً يساوى بعضها بعضاً تكون قسى م ه ب ج

ن ح الباقية مساوٍ بعضها لبعض وقسى ه ك ج ط ح ل مساوٍ بعضها لبعض

فقسى م ك ب ط ن ل مساوٍ بعضها لبعض وقوس ب ط مساوية للقوس التى 25

للقوس [لقوس N 10   ولنفصل [ولنوصل N 10   د ب ج ط [حط *corr. ex* A 10   د [ر N 10

بالدائرة [بالدار A   انها يعطعان [أنها تقطع A 15   أنها [اه N 14   أنها [1] N 14   الذى يرسم فى [الذى يرم فى 11 *repet.* N

هل [ه ك N 18   واحدة [واحد N *marg.* A 17   د ن ح ل [دنحل N 17   د م ه ك [دمهك A 17   ويعطى [وبنقطتى 17

om. N   قطب [2] N 18   لانها [أنها A 22   التى [الدى A *marg.*   التى [الدى 20   وهمحح [ز ه ج ح N 18

لبعض [بعضاً N 23   د ن [دن N 23   د ر [د د A, *text. illeg., marg.* د ب [د ب N 23   للدائرتين [الدائرين N 22

A التى [التى الذى, *marg.* N 25   مل [مل N 25   هل [ه ك N 25   ه ك [ه ك 24 *repet.* N   تكون [... لبعض [1] N 23-24

rabo arcum equalem arcui cui subtenditur latus quadrati quod signatur
in circulo maiore, qui sit arcus *BT*; et signabo super polum *T* secun-   15
dum spacium *TB* circulum *EBH*. Ergo circulus *EBH* est maior. Quod
ideo est quoniam linea que protrahitur a polis eius ad lineam continentem ipsum est equalis lateri quadrati quod signatur in circulo maiore.
Et ostendam quod ipse contingit circulum *AB*. Quod ideo est quoniam
ipse secat lineam continentem circulum maiorem, que est linea *DBGT*,   20
super punctum unum, quod est punctum *B*, et eius poli sunt super
illam lineam hunc circulum continentem. Signabo autem duos circulos
maiores transeuntes per punctum *D* et duo puncta *E* et *H*, qui sunt
duo circuli *DMEK DNHL*; et separabo unumquemque duorum arcuum
EK HL equalem arcui *GT*. Et quia duorum circulorum *EBH EGH*   25
unus alterum secat et iam signatus fuit circulus maior transiens per
polos eorum, qui est circulus *DBGT*, ergo circulus *DBGT* dividit
porciones, quas separavit, in duo media et duo media. Ergo arcus *EG*
est equalis arcui *GH* et arcus *EB* est equalis arcui *BH*. Et quia tres
arcus *DE DG DH* ad invicem sunt equales, quod ideo est quoniam ipsi   30
protrahuntur a polo *D*, qui est polus duorum circulorum, ad lineam
continentem circulum *EGHZ*, et arcus *DM DN DB* etiam ad invicem
sunt equales, ergo reliqui arcus *ME BG HN* ad invicem sunt equales.
Sed arcus *EK GT HL* sunt ad invicem equales: ergo arcus *MK BT*
NL ad invicem equantur. Sed arcus *BT* est equalis arcui cui latus quad-   35

---

14 cui subtenditur] *supra* **Ps**   15 sit] fit **Va**   16 maior] *om.* **Va**   17 polis] polo **FiZ**
17 lineam] *om.* **Fi**   17–18 continentem] continente **Fi**   19 quoniam] *om.* **Va**   20 ipse] ipsi
Z   20 secat] secant **Fi**   20 continentem] contingentem **Fi**   21 punctum unum] *tr.*
FiZ   21 eius] *om.* **FiZ**   21 poli] eorum *add.* **FiZ**   22 illam lineam] *tr.* **OBPsVa**
22 hunc circulum continentem] *om.* **FiZ**   22 Signabo] Significabo **FiZ**   23 qui] que
Va   24 separabo] *in corr.* **Ps**   24 unumquemque] utrumque **MKgBPsVa**, unumquo-
dque **Fi**   24 duorum arcuum] *tr.* **V**   26 signatus] significatus **BPs**   26 fuit] fuerit
RMOBPs, est **FiZ**   27 circulus²] *om.* **OFiZB**, *supra* **Ps**   28 separavit] separat
FiZ   29 GH ... arcui²] *om.* **Va**   29 et ... BH] *false scriptum, corr. in marg.*
R   30 DE DG] *tr.* **BPsVa**   30–31 ipsi ... qui est] *false scriptum, deinde recte* **Fi**
31 D] AD **Fi**   31 qui ... circulorum] *marg.* **M**, *om.* **KgOBPsVa**   31 polus] *om.*
PRM   31 duorum] omnium trium **FiZ**   31–32 ad ... EGHZ] *om.* **VFiZ**   31–
32 lineam continentem circulum] circumferentiam   circuli **PR**,   circumferentiam **Kg**
32 EGHZ] EGH **Kg**, GEZ **OPsVa** (**Ps** *supra*), *illeg.* **B**; circumferenciam *add.* **Va**   32 et] *om.* **B**
32 DM DN DB] *permut.* **OBPsVa**   32 etiam ad invicem] *om.* **B**   33 equales¹] *om., supra
verb. illeg.* **Ps**   34 invicem] sunt *add.* **R**   34–35 ergo ... equantur] *primum false, deinde recte*
R

يوترها ضلع المربع الذى يرسم فى الدائرة العظمى فكل واحدة من قوسى م ك

ن ل مساوية للقوس التى يوترها ضلع المربع الذى يرسم فى الدائرة العظمى

ولأن دائرة د ب ج ط عظيمة فى كرة وهى تقطع دائرة من الدوائر التى فى

الكرة وتمر بقطبيها وهى دائرة ز ه ج ح فهى تقطعها بنصفين وعلى زوايا قائمة

30   فدائرة د ب ج ط قائمة على دائرة ز ه ج ح على زوايا قائمة وكذلك أيضاً نبين

أن دائرة د ن ح ل أيضاً قائمة على دائرة ز ه ج ح على زوايا قائمة ودائرة

د م ه ك قائمة على دائرة ز ه ج ح على زوايا قائمة ولنوصل خطوط ل ن ل ج

ط ه فقد عمل على قطرين من أقطار دائرة ز ه ج ح يخرجان من نقطتى ج ح

قطعتان من دائرتين متساويتين قائمتان عليها وعلى زوايا قائمة وهما قطعتا

35   ج ط ح ل وتمامهما وكل واحدة من ج ط ح ل أقل من نصف دائرة وقوس

ه ج مساوية لقوس ج ح فخط ط ه مساو لخط ل ج وضلع المربع الذى يرسم فى

الدائرة العظمى مساو لخط ط ه فخط ل ج أيضاً مساو لضلع المربع الذى يرسم

فى الدائرة العظمى وخط ل ن هو ضلع المربع الذى يرسم فى الدائرة العظمى

فخط ل ج مساو لخط ل ن فالدائرة التى ترسم على قطب ل وببعد ل ج تمر

40   أيضاً بنقطة ن فلتمر ولتكن مثل دائرة ج ن س وهذه الدائرة من الدوائر

العظيمة وذلك أن الخط الذى يخرج من قطبها إلى الخط المحيط بها مساو

---

27 ن ل] بل N   28 عظيمة] om. N   28 فى كرة] om. A   29 وعلى] على A   30 وكذلك] ولدلك N

31 ودائرة] ودار A   32 د م ه ك] رمهل N   32 قائمة [1] om. A   32 دائرة] supra A   32 ولنوصل] فلنوصل N

32 ل ن] لر N   33 ز ه ج ح] add. AN اللدس   34 A داره N; in corr. A;   34 دائرتين] على

A   34 وعلى] على N   35 وكل] فكل A   35 دائرة] ذلك A   36 ج ح] مح N

N   35 وتمامهما] in corr. A, om. H   36–37 لخط ط ه ... العظمى] مساو لخط ط add. A man. rec.

  38 لخط ط ه] marg. A   38 وخط ل ن هو] add. A   39 ل ن] لر N; انضا add. A   40 ج ن س] حس N

A   38 ل ن] لر N   39 ل ن] لر N

---

35 وتمامهما] وكانت add. A القطعة تامة ... دائرة ز ه ج ح محيط يلقى الطرف الآخر حتى وج ط ح تمم لو موهوم يعمل لم الذى

rati quod signatur in maiore circulo subtenditur: ergo unusquisque
duorum arcuum *MK NL* est equalis arcui cui latus quadrati quod in
circulo maiore signatum est subtenditur. Et quia circulus *DBGT* est in
spera et secat unum circulorum qui sunt in spera et transit per polos
eius, qui est circulus *ZEGH*, ergo ipse secat eum in duo media et 40
orthogonaliter. Ergo circulus *DBGT* erectus est orthogonaliter super
circulum *ZEGH*. Et similiter etiam demonstratur quod circulus *DNHL*
orthogonaliter erigitur super circulum *ZEHG* et circulus *DMEK* super
circulum *ZEGH* existit orthogonaliter. Producam ergo lineas *LN LG*
TE. Iam ergo fecimus super duas diametrorum circuli *EGH*, que a duo- 45
bus punctis *G H* protrahuntur, duas circuli porciones equales super eas
orthogonaliter erectas, que sunt due porciones *GT HL*, et unaqueque
duarum porcionum *GT HL* est minor medietate circuli. Et arcus *EG*
est equalis arcui *GH*: ergo linea *TE* est equalis linee *LG* Sed latus quadrati quod in circulo maiore signatur est equale lateri *TE*: ergo etiam 50
linea *LG* est equalis lateri quadrati quod signatur in circulo .maiore
Linea autem *LN* est latus quadrati quod signatur in circulo maiore:
ergo linea *LG* est equalis linee *LN*. Ergo circulus qui signatur super
polum *L* secundum spacium *LG* transit etiam per punctum *N*. Signetur ergo et sit equalis circulo *GNS*. Circulus ergo hic est ex circulis 55
maioribus. Quod ideo est quoniam linea que protrahitur a polis eius ad lineam continentem ipsum est equalis lateri quadrati quod signa-

36 maiore circulo] *tr.* **Fi**   36 subtenditur] *post* cui **OFiZ**   36 unusquisque] uterque **MKgB**
PsVa, unius quisque **Fi**   37 *NL*] *HL* **Va**   38 circulo maiore] *tr.* **OFiBPsVa**, minori circulo **Z**
38 signatum est] signatur **OB**, significatur **PsVa**   40 qui] quod **Ps**   40 *ZEGH*] *in corr.* **V**,
EGH **OBPsVa**   40 et] duo media *add.* **Va**   41 erectus est] *tr.* **OBPsVa**   41 orthogonaliter] est
add. **Ps**   42 *ZEGH*] *ZEHG* **P**, *hic et saepius*   42 *DNHL*] *DNHR* **Va**   43 *ZEHG*] *ZEGH*
RKgOBVa   43 et circulus] *repet.* **Va**   44 existit orthogonaliter] *tr.* **MKgOFiZBPsVa**
45 duas] duos **Va**   45 circuli] *om.* **FiZ**   46 *G*] et *add.* **FiZ**   47 unaqueque] utraque **MKgB**
PsVa   49 *TE*] *in corr.* **Ps**; *E, T supra* **B**   50 circulo maiore] *tr.* **OPsVa**   50 *TE*] *T* **Fi**   50–
51 etiam linea] *tr.* **PsVa**   51 linea] *om.* **B**   51 maiore] est maior *add.* **Fi**   52 Linea] *illeg.* **B**
52 Linea ... maiore] *om.* **FiPs**   52 est] *om.* **BVa**   52–53 in ... linee] *marg.* **B**   54 *L*] et *add.*
B   54–55 Signetur] Significetur **OFiZBPsVa**   54–55 Signetur (*sc.* Significetur) ergo] *tr.* **PsVa**
55 circulo] *supra* **Z**; id est sit ille **Fi**   55 *GNS*] *GNL* **Kg**   55 circulis] *om.* **Va**   56 est] *om.*
Va   56 quoniam] quia **BPsVa**   57 continentem] contingentem **Fi**   57 equalis] equale **BPsVa**

لضلع المربع الذى يرسم فى الدائرة العظمى ولأن دائرتى اب جنس فى كرة

وهما تقطعان خطاً محيطاً بدائرة ما عظيمة على نقطة واحدة وهى نقطة ن

وقطباهما على الدائرة تكون الدائرتان متماستين فدائرة جنس تماس دائرة

45 اب وكذلك أيضاً نبين أن الدائرة التى ترسم على قطب كـ وبعد كـ جـ تمر

أيضاً بنقطة مـ وذلك أنا إن وصلنا خطى جـكـ طـحـ يكون أحدهما مساوياً

للآخر وخط طـحـ خط ضلع مربع وذلك أنه يخرج من قطب دائرة ه بـحـ العظمى

إلى الخط المحيط بها وخط جـكـ أيضاً ضلع مربع وكذلك أيضاً خط كـمـ فخط

كـمـ مساوٍ لخط كـجـ فالدائرة التى ترسم على قطب كـ وببعد كـمـ تمر أيضاً

50 بنقطة جـ وهى جـمـعـ ومن البيّن أنها تماس دائرة اب فإن قال قائل إن

القوس التى تُفصل أعنى قوس بـجـ مساوية للقوس التى يوترها ضلع المربع

الذى يرسم فى الدائرة العظمى بيّنّا له ذلك على هذه الجهة وذلك أنه إذا كانت

كل واحدة من قوسى ده دح مساوية لقوس دجـ وكل واحدة من قوسى دم

دن مساوية لقوس دب تصير قوس بـجـ الباقية مساوية لكل واحدة من

55 قوسى نـح مـه وقوس بـجـ يوترها ضلع المربع فكل واحدة من قوسى نـح

مـه أيضاً يوترهما ضلع مربع فلأن قوس نـح يوترها ضلع المربع وكذلك قوس

جـح أيضاً يوترها ضلع مربع يكون خط نـح مساوياً لخط جـح فالدائرة التى

42 اب] آر N 43 ما] om. A 44 الدائرتان] داربان A 44 illeg. N 45 وكذلك] ولدلك N 46-

المربع [مربع A 48 فخط] خط A 48 وخط] حرح N 47 المربع] NH 47 يخرج] حرح N 47 طـح ... يكون] om. N 47

لضلع :H 48 أيضاً [²كـمـ] om. N 48 كـمـ] انضا add. N 49 كـمـ] كـج supra A 51 ضلع ... للقوس] :H AN

ذلك] om. A 52 إذا] لا NH 53 وكل] فكل N 55 المربع] ... وقوس [:H ضلع بـجـ قوس ووتر

H قـسـى¹ [قوس 56 om. AN ... المربع :H 56 فلأن ... المربع :H om. AN 56 المربع (N) 56 المربع] (AN) مربع

56-57 وكذلك ... مربع om. A 56 قوس²] קוסי H مربع 57 [مربع] אלמרבע H

tur in circulo maiore. Et quia circuli *AB GNS* sunt in spera et secant lineam continentem aliquem circulum maiorem super punctum unum, quod est punctum *N*, quorum poli sunt super circulum, ergo duo circuli   60
sunt se contingentes. Ergo circulus *GNS* contingit circulum *AB*. Et similiter etiam monstratur quod circulus qui signatur secundum polum *K* et secundum longitudinem *KG* transit etiam per punctum *M*. Quod ideo est quoniam si coniunxerimus puncta protrahendo duas lineas *GK TH*, erit una earum equalis alteri. Sed linea *TH* est latus quadrati: 65
quod ideo est quoniam ipsa protrahitur a polo circuli *EBH* maioris ad lineam continentem ipsum. Et etiam linea *GK* est latus quadrati, et similiter etiam linea *KM*: ergo linea *KM* est equalis linee *KG*. Ergo circulus, qui signatur super polum *K* et secundum longitudinem *KM*, transit etiam per punctum *G*, qui est *GMQ*. Et manifestum est   70
quod ipse contingit circulum *AB*. Quod si quis dixerit quod arcus qui separatur, scilicet arcus *BG*, est equalis lateri quadrati quod signatur in circulo maiore, ostendam ei illud secundum hunc modum, qui est: quoniam unusquisque duorum arcuum *DE DH* est equalis arcui *DG* et unusquisque duorum arcuum *DM DN* est equalis arcui *DB*, fit reli-   75
quus arcus *BG* equalis unicuique duorum arcuum *NH ME*. Sed corda arcus *BG* est latus quadrati: ergo unicuique duorum arcuum *NH ME* etiam subtenditur latus quadrati. Et quia arcui *NH* subtenditur latus quadrati et similiter etiam arcui *GH* subtenditur latus quadrati, ergo

---

59 continentem] contingentem **Fi**   59 unum] *L* **R**   60 punctum] *in corr.* **B**   61 se] *om.* **Z**, sese **OBPsVa**   62 polum *K*] *tr.* **OBPs**   63 *K*] et secundum polum *K repet.* **Fi**   63 *KG*] *BG* **Fi**   65 linea *TH*] una earum **OBPsVa** (*corrupte* **Va**); linea *TH add. supra* **Ps**   66 ipsa] *om.* **OBPsVa**; linea *TH add. supra* **Ps**   67 continentem] contingentem **Fi**   68 linea[1]] lineam **Fi**   68 ergo linea *KM*] *marg.* **B**, *supra* **Ps**   68 *KG*] *KH* **Fi**   70 qui] quod **RV**   70 est[1]] circulus *add. supra* **Ps**   70 *GMQ*] *GM* **OFiZBPsVa**   71 ipse] *in corr.* **Ps**   71 dixerit] dixit **Fi**; scilicet quod ponatur *add.* **P**   72 separatur] *in corr.* **Ps**   73 ostendam] ostenditur **FiZ**   73 illud] *om.* **R**   74 unusquisque] uterque **MKgOBPsVa**   74–75 *DE* ... arcuum] *om.* **Va**   75 unusquisque] uterque **MKgOBPs**   75 *DB*] et uterque duorum arcuum *add.* **Va**   75 fit] sit **RVMKgB**   76 equalis] duorum *add.* **Kg**   76 unicuique] utrique **MKgOBPs**, *om.* **Va**   76 unicuique duorum arcuum] *permut.* **KgBPs**   76 arcuum] *BH add. et del., marg.:* In alio *NH* **R**   76 *NH*] *BH* **PRV**, In alio *NH add. marg.* **PR** (*hic et ter iterum*)   77 *BG*] *NH* **V**   77 ergo] igitur **PsVa**   77 unicuique] utrique **MKgOBPsVa**   78 etiam] et **Va**   78–80 Et ... linee *GH*] *om.* **B**   78–79 Et ... quadrati[2]] *om.* **PsVa**   78–59 quia ... et] *om.* **Fi**   79 ergo] et quia **O**, et etiam **Fi**, et enim **Z**

ترسم على قطب ح̄ وببعد ح̄جّ تمر أيضاً بنقطة ن̄ وكذلك أيضاً نبين أن

الدائرة التى ترسم على قطب ة̄ وببعد ة̄جّ تمر بنقطة م̄ أيضاً فقد تبين أن

60   الذى أردنا أن نعمل يكون على ضربين ، وذلك ما أردنا أن نبين ٠

يه   الدوائر العظيمة التى تفصل فى كرة من دوائر ما متوازية قسياً متشابهة فيما

بينهما فهى إما أن تمر بأقطاب الدوائر المتوازية وإما أن تماس دائرة واحدة

بعينها من الدوائر المتوازية ٠

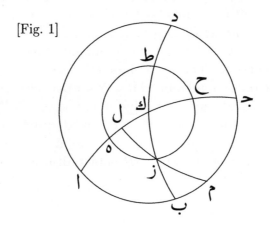

[Fig. 1]

فلتكن فى كرة دائرتان عظيمتان وهما دائرتا ا ح̄جّ ب̄ طّ د̄ ولتفصل من

5   دائرتى ا ب̄جّ د̄ ة̄ زح̄طّ المتوازيتين قسياً متشابهة فيما بينهما ولتكن قوس ا ب̄

شبيهة بقوس ة̄ ز̄ ولتكن قوس ب̄جّ شبيهة بقوس ز̄ح̄ ولتكن قوس جّ د̄

شبيهة بقوس ح̄طّ وقوس ا د̄ شبيهة بقوس ة̄طّ ، فأقول إن دائرتى ا ح̄جّ

فهى اما على قطب [فهى ... بعينها   2–3   N بينها [بينهما   2   om. AH   ما [1   A   مكذلك :NH] وكذلك   58

(الداره corr. ex [الدوائر (الدوائر) الدوائر المتوازية وإما فهى تمر باقطاب الدوائر المواربه وتماس مماسه لداره واحده بعنها

N   دارس مواربس وهما دارتا احد هرحط ولكن [كرة   4   A او [وإما أن   2   om. A   [إما أن   2   add. N

N سها [بينهما   5

---

linea *NH* est equalis linee *GH*. Ergo circulus qui signatur super polum   80
H et cum longitudine *GH* transit etiam per punctum *N*. Simili quoque
modo demonstratur quod circulus qui signatur super polum *E* et cum
spacio *EG* transit etiam per punctum *M*. Iam igitur manifestum est
quod illud quod facere voluimus secundum duas existit figuras. Et illud
est quod demonstrare voluimus.   85

15 Circuli maiores, qui in spera separant ex circulis equidistantibus arcus similes inter se, transeunt per polos circulorum equidistantium aut
contingunt unum et eundem ex circulis equidistantibus.

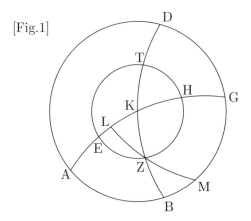

[Fig.1]

Sint itaque in spera duo circuli maiores, qui sint circuli *AHG*
BTD, et separent ex circulis *ABGD EZHT* equidistantibus arcus simi-   5
les inter se: sitque arcus *AB* similis arcui *EZ* et arcus *BG* sit similis
arcui *ZH* et sit arcus *GD* similis arcui *HT* et arcus *AD* similis arcui *ET*. Dico igitur quod duo circuli *AHG DTB* aut transeunt per polos

80 est equalis] *repet.*   **Fi**   80–82 Ergo ... quod] *marg.*   **BPs**   80 qui] *om.*
Va   80 signatur] significat **Va**   81 et] ideo *add.* **Va**   81 cum] secundum **B**
81 longitudine] longitudinem **FiB**   81 longitudine *GH*] *om.* **Va**   82 modo] *om.* **O**
82 demonstratur] monstratur **M**   82 quod] ergo *add.* **Va**   82 circulus] ergo circulus **Ps**
83 *M*] *in corr.* **Kg**   83 Iam] lineam **Fi**   84 facere] demonstrare **FiZ**   84 voluimus] volumus
OBPsVa   84–85 Et ... voluimus] *om.* **OBPsVa**   1 maiores] *om.* **R**   1 separant] secant
BPsVa, *supra* separant **B**, *supra* vel separant **Ps**; vel separant *add.* **Va**; per contactum vel secando
add. **V**, *marg.* **M**   2 transeunt] transeuntes **Z**   2 equidistantium] *om.* **R**   3 contingunt] unum
earum *add. et del.* **Va**   4 in spera] *scr. et del.* **B**, *om.* **PsVa**   4 sint] sunt **ZVa**   4 *AHG*] *ABG*
V, *HAG* **Kg**   5 *BTD*] *DTB* **B**, *corr. ex BZD* **Ps**   5 separent] separantur **PsVa**; *hic repet.*
prop. 13, l. 60-92 (est duplus ... arcus *KH*) **Fi**   5 equidistantibus] equidem **Fi**, *hic et saepius*
6 sitque] sit itaque **OBPsVa**   6 arcus[2]] *om.* **Va**   6 sit] *om.* **KgOBPsVa**   7 sit arcus] *tr.* **V**
7–8 arcui[3]] *HT* et arcus similis arcui *add.* **V**   8 igitur] ergo **MBPsVa**   8 *AHG*] *ABG* **Z**
8 *DTB*] *DTH* **Va**

ب ط د إما أن تمرا بأقطاب الدوائر المتوازية وإما أن تماسا دائرة واحدة

بعينها من الدوائر المتوازية وذلك أن دائرة ا ح ج إما أن تمر بأقطاب الدوائر

المتوازية وإما أن لا تمر ،

10

فلتمر أولاً بأقطاب الدوائر المتوازية كما فى الصورة الأولى ، فأقول

إن دائرة ب ط د أيضاً تمر بأقطاب الدوائر المتوازية أعنى أن نقطة ك تكون

قطباً لدائرتى ا ب ج د ه ز ح ط المتوازيتين فإن لم يكن ذلك كذلك وأمكن غيره

فلتكن نقطة ل قطباً لهاتين الدائرتين المتوازيتين ولترسم دائرة عظيمة تمر

بنقطتى ل ز وهى دائرة ل ز م فقوس ا ب شبيهة بقوس ه ز وقوس ه ز   15

شبيهة بقوس ا ب فقوس ا ب شبيهة بقوس م ا وهى من دائرة واحدة بعينها

وذلك غير ممكن فليست نقطة ل بقطب للدائرتين المتوازيتين وكذلك أيضاً

نبين أنه لا يمكن أن يكون قطبها نقطة أخرى غير نقطة ك فنقطة ك

قطب للدائرتين المتوازيتين فدائرتا ا ح ج ب ط د تمران بأقطاب الدائرتين

20   المتوازيتين ،

[Fig. 2]

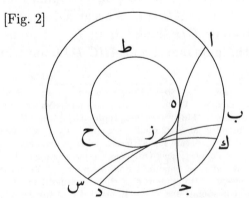

A لهما بن [لهاتين   14   كر [ك in corr. A   12   marg. A [وذلك أن ... المتوازية 9–10   N تمر [تمرا 8

A الدائرتى [للدائرتين   19   A الدائرين [للدائرتين   17   N ملس [فليست 17

circulorum equidistantium aut contingunt unum et eundem circulorum equidistantium. Quod ideo est quoniam circulus *AHG* aut transit per polos circulorum equidistantium aut non transit per eos.

Transeat igitur primum per polos circulorum equidistantium sicut in prima apparet figura. Dico igitur quod etiam circulus *BTD* transit per polos circulorum equidistantium, scilicet quod punctum *K* est polus duorum circulorum equidistantium *ABGD EZHT*. Quod si illud non ita fuerit et fuerit possibile aliter esse, sit punctum *L* polus horum circulorum. Signabo igitur circulum maiorem transeuntem per duos polos *L Z*, qui sit circulus *LZM*. Ergo arcus *ABM* est similis arcui *EZ*. Sed arcus *EZ* est similis arcui *AB*: ergo arcus *AB* est similis arcui *MA*; ipsi vero sunt unius et eiusdem circuli, quod est impossibile. Punctum ergo *L* non est polus circulorum equidistantium. Simili quoque modo monstratur quod non est possibile ut sit polus eorum aliud punctum preter punctum *K*: ergo punctum *K* est polus duorum circulorum equidistantium. Ergo duo circuli *AHG BTD* transeunt per polos duorum circulorum equidistantium.

[Fig. 2]

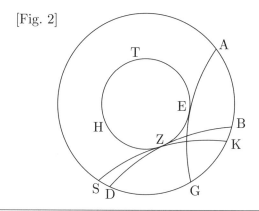

9–11 contingunt ... aut] *marg.* *sup.* **M**   *10* Quod ideo] *om.* **Fi**   *10–*
11 Quod ... equidistantium] *om.* **V**   *10* *AHG*] *in corr.* **O**   *12* igitur] ergo **BPsVa**
12 polos] polum **PRFiZ**, circulum **V**   *12* circulorum equidistantium] *tr.* **KgOBPsVa**
13–14 sicut ... equidistantium] *marg.* **Ps**   *13* igitur] ergo **B**   *13* etiam] *om.* **OB**
13 etiam circulus] cuicumque **Va**   *13* *BTD*] *in corr.* **Z**, *BCD* **Va**   *14* circulorum] *om.*
Va   *14* circulorum equidistantium] *tr.* **OB**   *14* scilicet] secundum **PsVa**; per punctum
K add. **FiZ**; sed *add.* **B**   *14* quod punctum *K*] *repet.* **O**   *15* equidistantium] *om.* **FiZ**
15 *ABGD*] *ABG* **Fi**   *16* non ita] *tr.* **MFiBPsVa**   *16* ita] *om.* **Kg**   *16* et fuerit possibile] si
possit **OBPsVa**   *16* esse] *om.* **Va**   *16* sit] fit **FiPsVa**   *16* punctum] punctus **OBPsVa**
16 polus] polos **Fi**   *17* horum] *om.* **FiZ**   *17* circulorum] equidistantium/equidem *add.* **ZFi**
17 Signabo] Significabo **Va**   *17* igitur] ergo **ZBPsVa**   *17* maiorem] *om.* **FiZ**   *18* polos] vel
per duo puncta *add.* **VOBPsVa**, **M** (*supra*)   *18* qui ... *LZM*] *primum false, deinde recte scr.*
Fi   *18* *ABM*] *AMB* **VB**   *19* est similis[2]] *tr.* **FiZ**   *20* arcui] *supra* **BPs**   *20* vero] eius **Fi**
21 *L*] *om.* **FiZ**   *21–24* Simili ... equidistantium] *om.* **M**   *21–22* Simili quoque] Similique
BPsVa   *22* possibile ut sit] *om.* **OBPsVa**   *23* punctum[2]] *om.* **Fi**   *23* polus] *om.* **R**
23 duorum] *om.* **OBPsVa**, *post* equidistantium **Z**   *24–25* Ergo ... equidistantium] *om.* **PsVa**
24 *AHG*] *HAG* **Kg**, *ABG* **Fi**   *24* *BTD*] *BED* **R**   *25* duorum] *om.* **OFiZ**

وأيضاً فلا تمر دائرة اه‍ج بأقطاب الدائرتين المتوازيتين فهى إما أن

تماس دائرة ه‍زح‍ط وإما أن تكون مائلة عليها فلتماسها أولاً على نقطة ه‍ كما

فى الصورة الثانية ، فأقول إن دائرة زب‍ أيضاً تماسها فإن أمكن فلا تماسها

ولترسم على نقطة ز‍ دائرة عظيمة تماس دائرة ه‍زح وهى دائرة زك‍ وليكن

نصف دائرة زك‍ لا يلقى نصف دائرة ه‍ا‍ فقوس اك‍ شبيهة بقوس ه‍ز وقوس

ه‍ز شبيهة بقوس اب‍ فقوس اك‍ شبيهة بقوس اب‍ وهى من دائرة واحدة

بعينها وذلك غير ممكن فليس يمكن أن لا تماس دائرة زب‍ أيضاً دائرة ه‍زح

فهى تماسها ،

[Fig. 3]

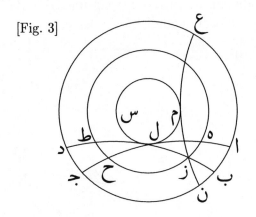

فلتكن دائرة اح‍ج‍ مائلة على الدوائر المتوازية كما فى الصورة الثالثة

فهى تماس دائرتين مساوية إحداهما للأخرى موازيتين لدائرتى اب‍ج‍د

ه‍زح‍ط ، فأقول إن دائرة ب‍زط‍د مماسة لهما فإن أمكن فلتماس دائرة

اه‍ح‍ج‍ إحدى الدائرتين المتوازيتين اللتين ذكرنا وهى دائرة م‍ل‍س على نقطة

21 أيضاً] A درب‍ [زب‍ 23 A ملمسها [فلتماسها 22 N هرح‍ [ه‍زح‍ط 22 N ممرن [تمر 21 om. N
24 زك‍ [ركس A, زل‍ [زل‍ H N, ال‍ [ال‍ H 25 الك‍ [الك‍ NH 26 ال‍ [ال‍ N رل‍ [رل‍ 27 يمكن [يمكن NH om. N
27 أن لا [الا N 27 زب‍ [درب‍ A 27 ه‍زح‍ [هر N فلتكن [فلتكن 29 A وليكن [وليكن A 30 متساوه [مساوية in
corr. A 31 لهما [illeg. A 32 م‍ل‍س [مكس A

Sit etiam circulus *AEG* non transiens per polos duorum circulorum equidistantium: ergo ipse aut contingit circulum *EZHT* aut est inclinatus super eum. Contingat ergo ipsum prius super punctum *E* sicut in figura secunda continetur. Dico igitur quod circulus *ZB* contingit eum. Quod si possibile fuerit, non contingat ipsum. Signabo autem   30
super punctum *Z* circulum contingentem circulum *EZH*, qui sit circulus *ZL*. Sitque circuli *ZL* medietas non concurrens medietati circuli *EA*: ergo arcus *AL* est similis arcui *EZ*. Sed arcus *EZ* est similis arcui *AB*: ergo arcus *AL* est similis arcui *AB*. Ipsi vero unius et eiusdem circuli sunt, quod est impossibile. Non est igitur possibile ut etiam circulus   35
ZB non contingat circulum *EZ*: ergo ipse contingit eum.

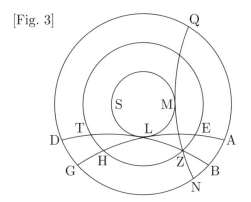

[Fig. 3]

Sit autem circulus *AHG* inclinatus super circulos equidistantes, sicut in tercia apparet figura: ergo ipse contingit duos circulos, quorum unus est alteri equalis, equidistantes duobus circulis *ABGD EZHT*. Dico igitur quod circulus *BZTD* contingit eos. Quod si fuerit possibile,   40
sit circulus *AEHG* contingens unum duorum circulorum equidistantium quos prediximus, qui sit circulus *MLS*, super punctum *L* et

26 etiam] itaque **Va**   26 *AEG*] *in corr.* **Ps**   26 duorum circulorum] *tr.* **O**   26–27 duorum circulorum equidistantium] eorum **FiZ**   28 eum] ipsum **BPsVa**   28 *E*] *supra* **BPs**   29 figura secunda] *tr.* **BPsVa**   29 igitur] ergo **BPsVa**   29 *ZB*] *DZB* **O**   30 si possibile fuerit] non possibile est **Va**   30 possibile fuerit] *tr.* **Fi**   30 contingat] contingit **BPs**, *corr. in* contingat **Ps**   30 ipsum] eum **Fi**   30 Signabo] Significabo **BPs**   30 autem] aut **Fi**   31 punctum] *om.* **BPsVa**   31 *Z*] *ZH* **KG**   31 circulum[2]] *marg.* **R**   32 Sitque] Sit quod **Fi**, Sit itaque **OBPs**, Sit ita ut **Va**   32 circuli[1]] circulus **R**   32 *ZL*] *L, Z supra* **OB**   32 medietas] ut *add.* **OBVa**, *scr. et del.* **Ps**   32 concurrens] concurrat **OZBPsVa**, concurunt **Fi**   33 arcui[1]] *om.* **OBPsVa**   35 sunt] *post* vero **BPsVa**   38 in tercia] *infra* **Va**   38 ipse] *om.* **FiZ**   39 est alteri] *tr.* **Fi**   41 circulus] *om.* **Fi**   42 sit] ergo **Fi**, fit **Va**   42 *MLS*] *LMS* **KgOBPsVa**   42 *L*] *om.* **Va**

ل فلا تماسها دائرة ب ز ط د إن أمكن ذلك ولترسم دائرة عظيمة تمر بنقطة ز

التى هى فيما بين دائرة ل م س وبين الدائرة المساوية الموازية لها وتماس دائرة

35   ل م س فلتماسها على نقطة م وهى دائرة ن ز م ع وقوس ا ب ن شبيهة بقوس

ه ز وقوس ه ز شبيهة بقوس ا ب فتكون قوس ا ب ن أيضاً شبيهة بقوس ا ب

وهى من دائرة واحدة بعينها وذلك غير ممكن فليس يمكن أن لا تماسها دائرة

ز ط د أيضاً فهى تماسها فدائرتا ا ه ح ج ب ز ط د تماسان دائرة واحدة بعينها

من الدوائر المتوازية التى تقع فى الكرة ، وذلك ما أردنا أن نبين .

يو   الدوائر المتوازية التى فى الكرة التى تفصل من دائرة عظيمة قسياً متساوية

مما يلى الدائرة العظمى من الدوائر المتوازية هى متساوية والدوائر التى

تفصل قسياً أعظم فهى أصغر .

فلتكن فى كرة دائرتان متوازيتان وهما دائرتا ا ب ج د ولتفصلا من

5   دائرة ا ب ج د العظمى قسياً متساوية أولاً وهما قوسا ب ز ز د مما يلى

الدائرة العظمى من الدوائر المتوازية وهى دائرة ه ز ، فأقول إن دائرة ا ب

مساوية لدائرة ج د فليكن الفصل المشترك لدائرة ا ب ودائرة ا ب ج د خط

ا ب والفصل المشترك لدائرة ه ز ودائرة ا ب ج د خط ه ز والفصل المشترك

لدائرة ج د ودائرة ا ب ج د خط ج د فلأن سطحى ه ط ز ج ك د المتوازيين قد

فلتماسها [فلماس 35 المساوية الموازه [المساوية الموازية 34 فلترسم [ولترسم 33 A لَك [لَ 33 corr. in A

ولتفصلا [ولعصلان 4 س س N supra ,يعمل [نبين 39 N ربط [زطد 38 N الا [الا أن 37 N

N 5 وهما قوسا ب ز ز د [om. N 8-9 خط ه ز ... ا ب ج د [om. A 9 سطحى [in corr. A

A جكد marg., in corr. 9 حكل [جك د 9

هذ <ا> والفصل المشترك لداره [ج د 9 add. A مرب الداره العظيمه يعطى الدوار المواربه او لم يمر [أصغر 3

جب وداره ا نحد خط جد marg. A

non contingat ipsum circulus *BZTD*, si illud fuerit possibile. Circulum
itaque maiorem signabo transeuntem per punctum *Z*, quod est inter
circulum *MLS* et circulum equalem et equidistantem ei et contingen-   45
tem circulum *LMS* – contingat ergo ipsum super punctum *M* –, qui
sit circulus *NZMQ*. Ergo arcus *ABN* est similis arcui *EZ*. Sed arcus
EZ est similis arcui *AB* ergo arcus *ABN* etiam est similis arcui *AB*.
Ipsi vero sunt unius et eiusdem circuli, quod est impossibile. Non est
ergo possibile ut non contingat ipsum etiam circulus *BZTD*: ergo ipse   50
contingit ipsum. Ergo duo circuli *AEHG BZTD* contingunt unum et
eundem ex circulis equidistantibus qui cadunt in spera. Et illud est
quod demonstrare voluimus.

16 Circuli equidistantes, qui sunt in spera et separant ex circulo maiori
arcus equales ab ea parte qua sequuntur circulum maiorem qui est
ex circulis equidistantibus, sunt equales et circuli qui separant arcus
maiores sunt minores.

Sint itaque in spera duo circuli equidistantes, qui sint duo circuli   5
AB GD, et separent ex circulo maiore *ABGD* primo arcus equales ab
ea parte qua existit circulus maior qui est ex circulis equidistantibus,
qui est circulus *EZ*. Dico igitur quod circulus *AB* est equalis circulo
GD. Sit igitur differentia communis circulo *AB* et circulo *ABGD* linea
AB et differentia communis circulo *EZ* et circulo *ABGD* linea *EZ* et   10
differentia communis circulo *GD* et circulo *ABGD* sit linea *GD*. Et quia
duas equidistantes superficies *ETZ GKD* iam aliqua secat superficies,

---

43–44 Circulum ... signabo] *permut.* **FiZ**   44 itaque] igitur **OBPsVa**   45 *MLS*] *ML*
Fi, *LMS* **KgOBVa**   45–46 *MLS* et ... circulum] *om.* **R**   45 circulum²] ei *add.* **FiPsVa**
45 et²] *supra* **M**, *om.* **Fi**   45 equidistantem ei] *tr.* **OBPsVa**   45 et³] *per add.* **PsVa**
48 arcus] *om.* **Va**   48 *ABN*] *HBN* **Va**   48 etiam] *om.* **OFiZ**   48 etiam est] *tr.* **KgBPsVa**
50 ipsum] eum **FiZ**   50 etiam] *om.* **FiZVa**   50 *BZTD*] *illeg.* **Kg**   51 ipsum] eum
VMKgO FiZBPs   51 *BZTD*] *om.* **Z**, *HZTD* **Va**   51 contingunt] contingent **OPsVa**
(*corr. ex* contingit **Ps**)   51 unum] *infra* **B**, *om.* **PsVa**   52 cadunt] sunt **Kg**   52–
53 Et ... voluimus] *om.* **BPsVa** 53 voluimus] volumus **OFi**   1 separant] *in corr.* **Ps**, sperant
Fi (*hic et saepius*)   3 equidistantibus] equidem **Fi**, *hic et saepius*   5 qui ... circuli] *om.* **Va**
5 duo circuli] *om.* **MKgOBPs**   6 arcus] *supra* **BPs**   7 parte] *om.* **R**   7 circulus] circulis
Fi   7 est] *supra* **MVa**, *post* equidistantibus **O**   8 igitur] ergo **V**   9 Sit igitur] Sitque **OBPs**
Va   9 igitur] ergo **MFiZ**   9 differentia communis] *tr.* **MKgOBPsVa**   9 *AB*] *A, B supra* **O**
9 *ABGD*] *ABDG* **Z**, *hic et saepius*   9 linea] recta *add. supra* **Ps**, *in textu* **Va**   10 et¹] est
Z   10 differentia communis] *tr.* **OBPsVa**   10–11 *EZ* ... circulo¹] *om.* **Fi**   10 *ABGD*] sit
add. **OVa**, *marg.* **BPs**   10 linea] recta *add. supra* **Ps**   10–11 linea ... *ABGD*] *marg.*
BPs   10 et³] est **Z**   11 differentia communis] *tr.* **OBPsVa**   11 *GD* et circulo] *repet.*
R   11 linea] recta *add.* **Va**   12 equidistantes] *om.* **PRV**   12 superficies¹] *om.* **OBPsVa**
12 *ETZ*] et etiam **Fi**   12 aliqua] equalia **R**, alia **O**, aliā **Z**   12 aliqua secat] *tr.* **MKgOBPsVa**

قُطعا بسطح ما وهو سطح دائرة اٮحد يكون الفصلان المشتركان لهما　10

متوازيين فخط هز موازٍ لخط حد وكذلك أيضاً نبين أن خط اٮ أيضاً موازٍ لخط

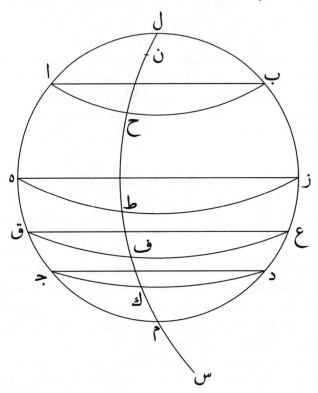

حد ولخط هز ولأنه قد أخرج فى دائرة اٮحد خطان متوازيان وهما خطا

هز حد تكون قوس دز مساوية لقوس هح وذلك أنّا إن وصلنا نقطة ه

بنقطة د تكون الزوايا المتبادلة متساوية والزوايا المتساوية فى الدوائر

المتساوية تكون على قسى متساوية فتكون قوس هح مساوية لقوس زد　15

وكذلك نبين أن قوس ٮز أيضاً مساوية لقوس اه وقوس ٮز مساوية

لقوس زد فقوس اه مساوية لقوس هح فقوسا اه ٮز جميعاً مساويتان

المساداﯨ [المتبادلة N 14 تكوس [تكون N 13 فلاٮه [ولأنه 12 om. A 2] أيضاً [repet. A 11 وهو] 10

marg. , مساوه [مساويتان 17 om. A 17 جميعاً] add. A ورصٮ 16 ٮز 2] A المتبادلتان marg.

A مساوٮاس

que est superficies circuli $ABGD$, ergo due differentie earum communes sunt equidistantes. Ergo linea EZ equidistat linee GD. Et similiter

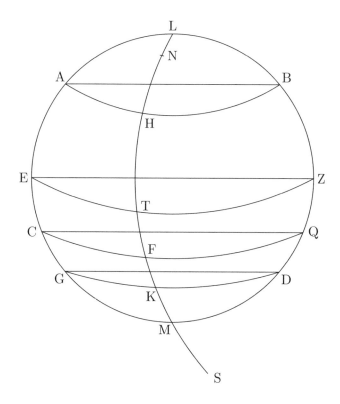

monstratur quod linea AB etiam equidistat linee GD et linee EZ. 15
Et quia a circulo $ABGD$ iam protracte sunt due linee equidistantes,
que sunt linee $EZ\ GD$, ergo arcus DZ est equlis arcui EG. Quod ideo
est quoniam cum coniunxerimus punctum E puncto D erunt anguli
coalterni equales Sed anguli equales qui sunt in circulis equalibus sunt
super arcus equales: ergo arcus EG est equalis arcui ZD. Simili quoque 20
modo monstratur quod arcus BZ est equalis arcui AE. Sed arcus BZ
est equalis arcui ZD et arcus AE est equalis arcui EG: ergo duo arcus

13 que est superficies] *om.* **Fi** 13 *ABGD*] *ABDG* **O** 13 due] *om.* **FiZ** 15 *AB* etiam] *ABZ*
R 15 et] etiam *add.* **Fi** 16 due linee equidistantes] *permut.* **KgBPsVa** 17 linee] *om.* **OBPsVa**
17 ergo] linea *add. omnes MSS* 17 est] *om.* **Va** 17 est equalis] equatur **BPs** 17 arcui] linee
arcus **OFiZ** 18 cum] *supra* **Ps**, si **B** (*supra*) 18 coniunxerimus] coniunximus **VFiZ**
18 *E*] autem *add.* **Kg** 20 Simili quoque] Similique **BPs**, Simili **Va** 21 monstratur] ostenditur
VFiZ 21 *BZ*] *illeg.* **Kg** 22 et arcus *AE*] *supra* **M**

لقوسى هج زد جميعاً فلأن جميع قوس هال‍بز مساوية لجميع قوس

هجم‍دز وذلك أن دائرتى هط‍ز اب‍جد عظيمتان وقوسا اه بز مجموعتان

منهما مساويتان لقوسى هج زد مجموعتين منهما تكون قوس ال‍ب الباقية 20

مساوية لقوس جم‍د الباقية وهى من دائرة واحدة بعينها فخط اب مساوٍ لخط

جد ودائرة اب‍جد إما أن تقطع دائرتى اح‍ب جك‍د وتمر بأقطابهما وإما أن

تقطعهما ولا تمر بأقطابهما ،

فلتقطعهما أولاً وتمر بقطبيهما فهى إذاً تقطعهما بنصفين فخط اب قطر

دائرة اح‍ب وخط جد قطر دائرة جك‍د وخط اب مساوٍ لخط جد فدائرة 25

اح‍ب مساوية لدائرة جك‍د ،

وأيضاً فلتقطع دائرة اب‍جد دائرتى اح‍ب جك‍د ولا تمر بقطبيهما

ولنتعلم قطب الدائرتين المتوازيتين ولتكن نقطة ن ولنرسم دائرة عظيمة تمر

بنقطة ن وبأحد قطبى دائرة اب‍جد وهى دائرة لن‍ط‍م‍س ولنفصل قوس

م‍س مساوية لقوس لن فلأن قوس لن مساوية لقوس م‍س وقوس ن‍ك‍م 30

مشترك تكون كل قوس لك‍م مساوية لكل قوس ن‍ك‍م‍س وقوس لك‍م

نصف دائرة فقوس ن‍ك‍م‍س أيضاً نصف دائرة فنقطة ن مقابلة لنقطة س

ونقطة ن قطب الدوائر المتوازية فنقطة س أيضاً هى القطب الآخر من قطبى

الدوائر المتوازية ولأن دائرتى اب‍جد جك‍د فى كرة وإحداهما تقطع

18 قوس [¹ موسى A　　18 حمعا [جميع A　　18 om. N [فلأن ... 18–19 هجم‍دز [هج N
18 هال‍بز [corr. ex ر A 19 م [supra A, هح رد [هجم‍دز 19 A هح رد [أن N 20 مساويتان [مساويان N ‍لان
om. [اح‍ب ... دائرة 25 A منهما [om. A 20 ال‍ب [in corr. A 21 الباقية ... مساوية [marg. A 25 دائرة ... اح‍ب [
A 27 وأيضاً فلتقطع [ولمعطمهما انضا A 28 ولنتعلم [ملعلم A 28 الدائرتين [للدارس N 29 لن‍ط‍م‍س [
in corr. A 30 لن [لر N 30 م‍س [supra A 31 لكل قوس [لموس A 33 الدوائر [لر ال طس A
N 34 وإحداهما [ملان A 34 ولأن [احداهما N

AE BZ simul sunt equales duobus arcubus *EG ZD* simul. Et quia totus arcus *EALZ* est equalis toti arcui *EGMZ* – quod ideo est quoniam duo circuli *ETZ ABGD* sunt maiores et duo arcus eorum coniuncti *AE BZ* 25 sunt equales duobus arcubus eorum *EG ZD* coniunctis –, ergo reliquus arcus *ALB* est equalis reliquo arcui *GMD*. Ipsi autem sunt unius et eiusdem circuli: ergo linea *AB* est equalis linee *GD*. Circulus autem *ABGD* aut secat duos circulos *AHB GKD* et transit per polos eorum aut secat eos et non transit per polos eorum.   30

Secet ergo eos primum et transeat per polos eorum. Ipse igitur secat eos in duo media: ergo linea *AB* est diametrus circuli *AHB* et linea *GD* est diametrus circuli *GKD*. Sed linea *AB* est equalis linee *GD*: ergo circulus *AHB* est equalis circulo *GKD*.

Sit etiam circulus *ABGD* secans duos circulos *AHB GKD* et non 35 transeat per polos eorum. Signabo igitur polum duorum circulorum equidistantium, quod sit punctum *N*, et signabo circulum maiorem transeuntem per punctum *N* et per unum duorum polorum circuli *ABGD*, qui sit circulus *LTKMS*, et separabo arcum *MS* equalem arcui *LN*. Et quia arcus *LN* est equalis arcui *MS*, ergo arcu *NKM* communi 40 totus arcus *LKM* est equalis toti arcui *NKMS*. Sed arcus *LKM* est semicirculus: ergo arcus *NKMS* est etiam circuli medietas. Ergo punctum *N* est oppositum puncto *S*. Punctum vero *N* est polus circulorum equidistantium: ergo etiam punctum *S* est alter duorum polorum circulorum equidistantium. Et quia duorum circulorum *ABGD GKD* in 45

---

23 simul[1]] similiter **Fi**   23 simul[2]] similis **Fi**, similiter **Va**   24 *EALZ*] *EABZ* **Va**   24 toti] totali **OBPsVa**, *hic et iterum*   24 quoniam] quia **OBPsVa**   25 *ETZ*] *in corr.* **Ps**, *EGZ* **FiZ**   26 eorum] *om.* **Va**   26 coniunctis] coniunctus **Fi**   27 autem] etiam **OBPsVa**   27–28 et eiusdem] *om.* **OBPsVa**   28 linea] recta *add. supra* **Ps**   28 *AB*] que est *AB* **Va**   28 linee] recte que est *add. supra* **Ps**   28 autem] *om.* **R**   30–31 aut ... eorum] *marg.* **Ps**   31 ergo] *om.* **R**   31 eos primum] *tr.* **V**   31 primum] prius **Va**   31 igitur] sibi **Fi**, ergo **OBVa**   32 *AHB*] *AHD* **FiZ**   33 *GKD*] *GDK* **BPs**   34 circulo] *om.* **Fi**   35 *ABGD*] *ABGH, supra D* **Ps**   36 Signabo] Significabo **Ps**, *hic et saepius*   36 igitur] ergo **OBPsVa**   37 quod] qui **M**   37–38 et ... punctum *N*] *om.* **Va**   38 unum] utrumque *marg.*, *pro* unum **Ps**, *in textu* **Va**   38 polorum] *corr. ex* circulorum **Ps**   39 *ABGD*] *ABDG* **Va**   39 qui sit circulus] *om.* **FiZ**   39 *LTKMS*] *LTMKS* **KgOBVa**, *in corr.* **Ps**   39 arcui] *om.* **Fi**   40 Et] *supra* **Ps**   40 arcu] arcus **V**, arcui **Z**   40 *NKM*] *in corr.* **R**, *NK* **B**   41 *LKM*[1]] *NKM* **V**, *LKON* **Va**   41 *NKMS*] *NMS* **Va**   42 est] *supra* **Ps**   42 etiam] *om.* **MKgOBPs Va**   42 circuli medietas] semicirculus **MKgOBPsVa**   42 medietas] medietates **Fi**   43 Punctum] Punctus **OBPs**   44 etiam] *om.* **V**, et **Z**

35 الأخرى وقد رسمت على أقطابهما دائرة عظيمة وهى دائرة لطكس صارت

دائرة لطكس تقسم القطع التى فصلت من الدوائر بنصفين نصفين فقوس

جـم مساوية لقوس مـد فقوس جـمـد ضعف قوس مـد وكذلك نبين أن قوس

الـب أيضاً ضعف قوس الٓ وقوس جـمـد مساوية لقوس الـب فقوس مـد

مساوية لقوس الٓ ولأنه قد عمل على قطر دائرة ابـجـد الذى خرج من

40 نقطة لٓ إلى نقطة مٓ قطعتان من دائرة قائمتان عليها على زوايا قائمة

متساويتان وهما قطعتا لـن مـس مع القطعة التى تتصل بهذه لتمام نصف

الدائرة ثم فصلت منهما قوسان متساويتان وهما قوسا لـن مـس وهما أقل من

نصفهما وفصل من الدائرة الأولى قوسان متساويتان وهما قوسا الٓ دم يكون

الخط المستقيم الذى يصل بين نقطة نٓ ونقطة آ مساوياً للخط الذى يصل

45 بين نقطة سٓ ونقطة دٓ والخط الذى يصل بين نقطة نٓ ونقطة آ يخرج من

قطب دائرة احـب إلى الخط المحيط بها والخط الذى يصل بين نقطة سٓ

وبين نقطة دٓ يخرج من قطب دائرة جـكـد إلى الخط المحيط بها فالخط الذى

يخرج من قطب دائرة احـب إلى الخط المحيط بها مساوٍ للخط الذى يخرج

من قطب دائرة جـكـد إلى الخط المحيط بها والدوائر التى تكون الخطوط

50 المستقيمة الخارجة من أقطابها إلى الخطوط المحيطة بها متساوية هى أيضاً

---

35 الدى [التى N   35 وقد] فقد N   A

36 القطع] الصلع add. supra N   36 لطكس] in corr. A

41 لـن مـس] marg. A, in textu H, لطم مـس N; فانصان من داره من add. et del. A   41 القطعة] اللفظه N

42 منهما [منها N   42 لـن] in corr. A   45 سٓ] illeg. N   45 ونقطة ... (pro والخط) marg. A

46 احـب] in corr. A   46 بها] in corr. A   46–48 والخط ... بها] om. N

spera unus alterum secat et iam signatus est super polos eorum circulus maior, qui est circulus *LTKS*, fit circulus *LTKS* secans porciones
quas ex circulis separavit in duo media et duo media. Ergo arcus
GM est equalis arcui *MD*: ergo arcus *GMD* est duplus arcus *MD*. Et
similiter ostenditur quod etiam arcus *ALB* est duplus arcus *AL*. Sed   50
arcus *GMD* est equalis arcui *ALB*: ergo arcus *MD* est equalis arcui
AL. Et quia super diametrum circuli *ABGD*, que protrahitur a puncto
L ad punctum *M*, iam facte sunt due porciones circuli equales super
ipsam orthogonaliter erecte, que sunt due porciones *LTM NMS* cum
porcione que eis adiungitur ad complendum semicirculum, et postea   55
separati sunt ex eis duo arcus equales qui sunt arcus *LN MS* qui sunt
minores semicirculo, et separant ex circulo primo duos arcus equales
qui sunt duo arcus *AL DM*, ergo linea recta que coniungit inter punctum *N* et punctum *A* est equalis linee que coniungit inter punctum *S*
et punctum *D*. Sed linea que coniungit inter punctum *N* et punctum   60
A protrahitur a polo circuli *AHB* ad lineam continentem ipsum, et
linea que coniungit inter punctum *S* et punctum *D* producitur a polo
circuli *GKD* ad lineam continentem ipsum. Ergo linea que protrahitur
a polo circuli *AHB* ad lineam continentem ipsum est equalis linee que
protrahitur a polo circuli *GKD* ad lineam continentem ipsum. Circuli   65
vero, a quorum polis linee recte ad lineas continentes ipsos protracte

46 iam] *in corr.* **O**   46 signatus est] *tr.* **OBPsVa** (signatus: significatus **OPsVa**)   46 est] *post*
polos eorum **Fi**   47 fit] sit **VMKg**   47 fit circulus *LTKS*] *marg.* **P**   48 quas] *om.*
R   48 ex circulis separavit] *permut.* **FiZ**   48 et duo media] *om.* **Fi**   48 arcus] *om.*
BPsVa   48 arcus *GM*] *tr.* **O**   49 ergo arcus *GMD*] *om.* **Va**   49–51 est² ... *GMD*] *marg.*
BPs   49 arcus²] arcui **OFi**, ad **Va**   50 quod etiam] *tr.* **R**   50 etiam] *om.* **MKgOB**
PsVa   50 arcus²] ad **Va**   51 *ALB* ... arcui] *om.* **Va**   52 protrahitur] protrahuntur
B, *corr. ex* protrahuntur **Ps**   53 iam facte sunt] manifeste erunt **Fi**   53 super] *om.* **V**
54 *LTM*] *in corr.* **Va**   54 cum] *om.* **Kg**, et **Va**   55 eis adiungitur] *tr.* **Z**   56 ex] ab
KgOBPsVa   56 arcus] *om.* **Va**   57 semicirculo] semicirculi **PsVa**   57 separant] separantur
PsVa   57 circulo primo] *tr.* **FiZ**   57 duos] duo **Va**; *ABDG add. supra* **Ps**   58 duo arcus] *om.*
MKgOBPsVa   58 recta] tercia **Z**   60 Sed] *hinc usque ad* a polo (*l.* 65) *elementa sen-*
tentiorum confudit **Va**   60 Sed ... punctum *A*] *om.* **V**   60–62 Sed ... punctum *D*] *marg.*
BPs   60 linea] recta *add. supra* **Ps**   61 protrahitur] producitur **VOBPs**   63 *GKD*] *GDK*
M   63–65 Ergo ... ipsum] *om.* **BPs**   64 *AHB*] *ABH* **M**, *ADBH* **Fi**   65 Circuli] circulum **Va**
66 vero] autem **BPs** (**Ps** *supra*), *om.* **Va**   66 ipsos] ip̄us **Fi**

متساوية فدائرة اح ب مساوية لدائرة جكد فلتكن قوس دز أعظم من

قوس زب ، فأقول إن دائرة جكد أصغر من دائرة اح ب وذلك أن قوس

دز أعظم من قوس زب فتفصل من قوس دز قوس مساوية لقوس زب

وهى قوس زع ولترسم دائرة موازية لدائرة هطز تمر بنقطة ع وهى دائرة

ع ف ق فدائرة ع ف ق مساوية لدائرة اح ب وذلك أن قوس زع مساوية

لقوس زب ودائرة ق ف ع أعظم من دائرة جكد وذلك أن دائرة ق ف ع

أقرب إلى مركز الكرة من دائرة جكد فدائرة اح ب أعظم من دائرة جكد

فدائرة جكد أصغر من دائرة اح ب ، وذلك ما أردنا أن نبين ·

يز   الدوائر المتساوية المتوازية التى فى الكرة هى تفصل من دائرة ما عظيمة قسياً

متساوية مما يلى الدائرة العظمى من الدوائر المتوازية والدوائر التى هى أعظم

تفصل قسياً هى أصغر ·

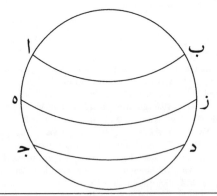

فأقول ... ¹زب] *marg.* A   52–53   جكد] *add.* N   51   وانصا] *add.* A, انصا *add.* N   51 ¹اح ب] انصا 51
دائرة 57] *om.* N   52   ¹أن] N لان 53   ³قوس] وسا AN 54   متوازية [موازية] A 57   إلى] *supra* A 57   ¹دائرة [*om.* N   52
المتساوية [المتوازية] *marg.* N 1   ما] *om.* A 1   ¹ما [*om.* A

لأن هذه الخطوط تفصل قسياً متساوية من نصف الدائرة التى تمر بقطبيها فيكون الخط الذى يصل [متساوية 51
بين النقطتين المشتركتين لمحيط الدائرة التى تمر بالقطبين وكل واحد من محيطى الدائرتين موازيا لقطر الكرة فيكون
العمودان الخارجان من هاتين النقطتين إلى قطر الكرة متساويين وهما الخطان الخارجان من مركز كل واحدة من
الدائرتين إلى محيطها *add.* A

sunt equales, etiam sunt equales. Ergo circulus *AHB* est equalis circulo *GKD*. Sit etiam arcus *DZ* maior arcu *ZB*. Dico igitur quod circulus *GKD* est minor circulo *AHB*. Quod ideo est quoniam arcus *DZ* est maior arcu *ZB*. Et separabo ex arcu *DZ* arcum equalem arcui *ZB*, qui   70 sit arcus *ZQ*, et signabo circulum equidistantem circulo *ETZ* transeuntem per punctum *Q*, qui sit circulus *QFC*. Ergo circulus *QFC* est equalis circulo *AHB*; et hoc ideo quoniam arcus *ZQ* est equalis arcui *ZB*. Sed circulus *CFQ* est maior circulo *GKD*. Quod ideo est quoniam circulus *CFQ* est propinquior centro spere quam circulus *GKD*, ergo   75 circulus *AHB* est maior circulo *GKD*. Ergo circulus *GKD* est minor circulo *AHB*. Et illud est quod demonstrare voluimus.

17   Circuli equales et equidistantes in spera separant ex aliquo circulo maiore arcus equales ab ea parte qua existit circulus maior qui est ex circulis equidistantibus; et circulus qui est maior separat arcum qui est minor.

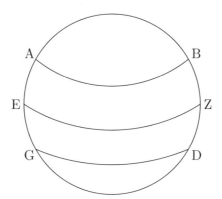

67 equales²] equalis **Fi**   67 circulus] circulis **Fi**   67 circulo] *supra* **BPs**   68–70 Dico ... arcu *ZB*] *om.* **B**   68 igitur] ergo **OPsVa**   69 *GKD*] *BKD* **Va**   69 *AHB*] *AHG* **Z**   69 *DZ*] quia *add.* **KgOFiZ**, *scr. et del.* **MPs**   70 equalem arcui] *marg.* **M**   72 Ergo circulus *QFC*] *marg.* **B**, *supra* **Ps**, *om.* **Va**   73 hoc] *supra* **Ps**   73 est] erit **V**   74 *CFQ*] *QFC* **MKgOBPsVa** (*corr. ex QEF* **Ps**), *corr. ex CFLQ* **V**   74 maior] minor **Fi**   75 *CFQ*] *corr. ex CQF* **Ps**, *CQF* **MKgB**, *CTQ* **Va**   76 *AHB*] *ABH* **B**   76 minor] maior **P**   77 *AHB*] *ABH* **Fi**   77 illud] hoc **O**, istud **BPsVa**   77 quod ... voluimus] *om.* **O**   77 demonstrare voluimus] volumus probare **BPsVa**   2–6 arcus ... maiore] *om.* **R**   2 existit] exstitit **OBPsVa**, *hic et saepius*; vel existit *supra* **O**   3 separat] sperat **Fi**   3 arcum] *in corr.* **Ps**

فلتكن فى كرة دائرتا اب جد المتوازيتان المتساويتان ولتفصلا من دائرة

ما عظيمة وهى دائرة اب جد قوسى زب زد مما يلى الدائرة العظمى من

الدوائر المتوازية ، فأقول إن قوس زب مساوية لقوس زد فذلك أنه إن لم

تكن قوس زب مساوية لقوس زد لم تكن دائرة اب مساوية لدائرة جد

ولكنها مساوية لها فقوس بز مساوية لقوس زد ·

وأيضاً فلتكن دائرة اب أعظم من دائرة جد ، فأقول إن قوس بز

أصغر من قوس زد وذلك أنه إن لم تكن قوس بز أصغر من قوس زد لم

تكن دائرة اب أيضاً بأعظم من دائرة جد وهى أعظم منها فقوس بز

أصغر من قوس زد ، وذلك ما أردنا أن نبين ·

إذا كانت فى كرة دائرة عظيمة وقطعت بعض الدوائر المتوازية التى فى الكرة

ولم تمر بقطبيها فإنها تقسمها بأقسام غير متساوية خلا الدائرة التى هى أعظم

الدوائر المتوازية وأما القطع التى تُفصل فى أحد نصفى الكرة فما كان منها

فيما بين أعظم الدوائر المتوازية وبين القطب الظاهر فكل واحدة منها أعظم

من نصف دائرة وأما القطع الباقية التى فى ذلك النصف من الكرة فكل

واحدة منها أصغر من نصف دائرة والقطع المتبادلة من الدوائر المتوازية

المتساوية مساوٍ بعضها لبعض ·

كان [وذلك أنه إن 10   قوسى A   قوس [corr. ex   9 repet. A   9 إن] N 9   فأقول [معمول   9 أقول [om. A   5 ما]

N واحدة [واحدة   4 المتوازية [om. N   4 مما [فما   3 فا [احدى A   3 أحد [marg. N   3 هى أعظم [N 2

A منهما [منها   6 AN واحد [واحدة   6 تحت [in corr.)   (تحت الأرض add. A   5 الكرة [A منها [منها   4

N مساوه [مساوٍ   7

Sint ergo in spera duo circuli *AB GD* equidistantes et equales et   5
separent ex aliquo circulo maiore, qui sit circulus *ABGD*, duos arcus
ZB ZD ab ea parte qua existit circulus maior qui est ex circulis equidistantibus. Dico igitur quod arcus *ZB* est equalis arcui *ZD*. Quod ideo
est quoniam si non fuerit arcus *ZB* equalis arcui *ZD*, non erit circulus *AB* equalis circulo *GD*. Sed ipse est equalis ei: ergo arcus *ZB* est   10
equalis arcui *ZD*. Sit igitur etiam circulus *AB* maior circulo *GD*.

Dico igitur quod arcus *BZ* est minor arcu *ZD*. Quod si arcus *BZ*
non fuerit minor arcu *ZD*, non erit circulus *AB* maior circulo *GD*. Ipse
autem est maior eo: ergo arcus *BZ* est minor arcu *ZD*. Et illud est
quod demonstrare voluimus.   15

18   Cum in spera fuerit circulus maior et diviserit aliquem circulorum equidistantium qui sunt in spera et non transierit per polos eorum, dividet
eum in sectiones inequales preter circulum qui ex circulis equidistantibus est maior. Portionum autem quas separabit in una medietatum
spere quecumque fuerit inter maiorem circulorum equidistantium et   5
polum manifestum, unaqueque erit maior semicirculo; reliquarum vero
portionum que erunt in illa medietate spere unaqueque erit minor semicirculo, portiones quoque coalterne circulorum equidistantium et equalium ad invicem erunt equales.

5 ergo] *om.* **Fi**   5 duo circuli] *tr.* **FiZ**   6 separent] separant **OBPsVa**   6 maiore] *om.* **FiZ**
6 circulus] circulis **Fi**   6 *ABGD*] *ABGDO* **BPs**, *ABDG* **Va**   7 circulus] *in corr.* **Ps**, circulis
Fi   7 maior] *EZ add.* **Va**   8 igitur] ergo **OBPsVa**   9 arcus] *om.* **Va**   9–10 circulus *AB*] *tr.*
OBPsVa   10 *GD*] *corr. ex ZD* **Z**   10 *ZB*] *AB* **BPsVa**   11 Sit] Sed **Z**   11 igitur] *om.*
VMKgOFiBPsVa   11 *AB*] est *add.* **Z**   12 igitur] ergo **OPsVa**   12–13 Quod ... arcu *ZD*] *om.*
Fi   13 erit] esset **R**   14 est[1]] *om.* **OBPsVa**   14 maior] est *add. supra* **Ps**   14 maior eo] *tr.*
MKgOBPsVa   14 eo] est *add. supra* **B**, *in textu* **Va**   14 arcus] *in corr.* **O**   14 illud] hoc **O**, istud
BPs   15 quod ... voluimus] *om.* **O**   15 voluimus] volumus **KgZBPsVa**   1 circulus maior] *tr.* **V**
1 aliquem] aliquos **M**, vel aliquos *supra* **OPs**   1 circulorum] *om.* **BPsVa**   2 transierit] transit
Fi, transerit **Va**   3 eum] eos **MKgBPsVa**   3 in] duas *add.* **MOBPsVa**   4 autem] *supra* **R**,
om. **Kg**   5–6 et polum] *om.* **Fi**   6 manifestum] maximum **Va**   6 unaqueque erit] *tr.* **MKgOB**
PsVa   7 que erunt] *om.* **FiZ**   7 erunt] erit **OBPsVa**   7 illa] alia **MKg**   7 unaqueque erit] *tr.*
MKgOBPsVa   8 quoque] vero **KgOBPsVa**   8 et] *om.* **FiZ**   8–9 equalium] equales **Kg**

فلتكن فى كرة دائرة عظيمة وهى دائرة اب‌جد تقطع بعض الدوائر

المتوازية التى تكون فى الكرة وهى داوئر اد هز بج ولا تمر بقطبيها ولتكن

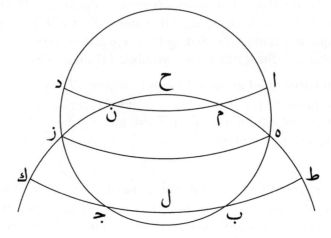

أعظم الدوائر المتوازية دائرة هز ، فأقول إن دائرة اب‌جد تقسم هذه   10

الدوائر بأقسام غير متساوية ما خلا دائرة هز التى هى أعظم الدوائر المتوازية

وإن كل قطعة من القطع التى تفصل فى أحد نصفى الكرة ما كان فيما

بين دائرة هز وبين القطب الظاهر فهو أعظم من نصف دائرة وكل قطعة من

القطع الباقية أصغر من نصف دائرة وإن القطع المتبادلة من الدوائر المتوازية

المتساوية مساوٍ بعضها لبعض فليكن القطب الظاهر من قطبى الدوائر المتوازية   15

نقطة ح ولترسم دائرة عظيمة تمر بنقطتى ه ح وهى دائرة طه‌ح فدائرة

حه‌ط إذا تممت تمر بنقطة ز أيضاً فلتخط ولتكن مثل دائرة حنزك ولتتمم

دائرة بج حتى تنتهى إلى نقطتى ط ك فلأن دائرة طه‌حنزك العظمى فى

A فدائر [فدائرة 16   tr. N [المتوازية المتساوية 14-15   N فان [وإن 14   A قطع [قطعة 13

N لـ et saepius ل pro ك حرل [حنزك 17

(خ) وذلك لأن الدوائر [فلأن 18   A .add   18 لأنها تقطع دائرة هز بنصفين وقوس هز نصف دائرة هز [أيضاً 17

marg. N المظام يقاطع بعضها بعضاً ونقطة ... (.illeg) الفاصلة للدائرة التى هى عليها بنصفين

Sit ergo in spera circulus maior, qui sit circulus *ABGD*, secans ali-   10
quem circulorum equidistantium qui sunt in spera, qui sunt circuli *AD*
EZ BG, neque transeat per polos eorum; et sit circulorum equidistan-

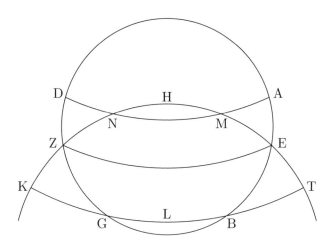

tium maior circulus *EZ*. Dico igitur quod circulus *ABGD* secat hos circulos in partes inequales preter circulum *EZ*, qui est maior circulorum
equidistantium, et quod portionum quas ipse separat in una medieta-   15
tum spere quecumque est inter circulum *EZ* et polum manifestum est
maior semicirculo et unaqueque reliquarum portionum earum est minor semicirculo, et quod portiones coalterne circulorum equidistantium
et equalium ad invicem sunt equales. Sit ergo polus ex polis circulorum equidistantium manifestus punctum *H*. Signabo autem circulum   20
maiorem transeuntem per duo puncta *E* et *H*, qui sit circulus *EHT*,
ergo circulus *EHT*, cum complebitur, transibit per punctum *Z* etiam.
Describatur ergo et sit equalis circulo *ENZK*, et signetur circulus *BG*
et producatur donec perveniat ad duo puncta *T* et *K*. Et quia circulus
maior *TEMZK* secat in spera equidistantes circulos ex circulis qui   25

10 ergo] igitur **Fi**   10 in spera] *om.*   **Fi**   10 maior . . . circulus] *om.*   **Fi**
10 qui sit circulus] *om.* **MKgOBPsVa**   10–11 aliquem] aliquos **MKg**, alios **OPsVa**, *illeg.*
B   11 sunt[2]] sint **MKgO**   12 *BG*] *HG* **B**   12 sit] fit **Ps**   13 circulus[1]] circulis
Fi   13 igitur] *om.* **OBPsVa**   14 circulum] circulos **Fi**   15 quas] quos **B**, *in corr.* **Ps**
15 ipse] *om.* **Va**   15–16 medietatum] medietate **BPs**   16 spere] sue **V**   16 est[1]] *om.* **OPsVa**
16 manifestum] maximum **Va**   17–18 et . . . semicirculo] *marg.* **BPs**   17 earum] eorum **Va**
18 et] vel **Z**   19 Sit] Si **Va**   19 ex] quod **Fi**   20 manifestus] manifestum **MKgOFiZB**
PsVa; est *add.* **PsVa**   21 *EHT*] *ETHT* **V**, *EHZ* **KgOBPsVa**   22 ergo circulus *EHT*] *om.*
FiZBPsVa   22 *EHT*] *EHZ* **KgO**   22 transibit] transit **R**   22 *Z*] *illeg.* **V**, *A* **Kg**, *N*
OBPsVa   22 etiam] *om.* **Kg**, et **PsVa**   23 ergo] *om.* **Va**   23 *ENZK*] *ENZL* **PRVM**
24 perveniat] proveniat **RMKg**, proveniunt **Fi**, conveniat **Z**   24 circulus] circulis **Fi**
25 *TEMZK*] *TEMZL* **PRZ**, *TEMGZL* **V**, in *EZMK* **Fi**

كرة تقطع دوائر متوازية من الدوائر التى تكون فى الكرة وهى دوائر ا م ن د

20 ه ز ط ب ل ج ك وتمر بأقطابها فهى تقطعها بنصفين نصفين وعلى زوايا قائمة

فكل واحدة من قطع من ه ز ط ب ل ج ك نصف دائرة فلأن قطعة من نصف

دائرة تكون قطعة ا م ن د أعظم من نصف دائرة وكذلك أيضاً نبين أن جميع

القطع التى فيما بين دائرة ه ز وبين قطب ح أعظم من نصف دائرة وأيضاً

فلأن قطعة ط ل ج ك نصف دائرة تكون قطعة ب ج أصغر من نصف دائرة

25 وكذلك أيضاً نبين أن جميع القطع التى فيما بين دائرة ه ز وبين القطب الخفى

مما فى هذا النصف من الكرة بعينه أصغر من نصف دائرة ،

وأيضاً فلتكن دائرة ا د مساوية لدائرة ب ج وموازية لها ، فأقول إن

القطع المتبادلة من دائرتى ا د ب ج مساوٍ بعضها لبعض فلأن دائرة ا د

مساوية لدائرة ب ج وموازية لها تكون قوس ا ه مساوية لقوس ه ب وقوس

30 د ز لقوس ز ج فقوسا ا ه د ز إذا جمعتا مساويتان لقوسى ه ب ز ج إذا جمعتا

وقسى ه ا ا د د ز إذا جمعت مساوية لقسى ه ب ب ج ج ز إذا جمعت لأن

كل واحدة من قوسى ه ا د ز ه ب ج ز نصف دائرة وذلك أن دائرتى ا ب ج د ه ز

عظيمتان فقوس ا د الباقية مساوية لقوس ب ج الباقية وقوسا ا د ب ج من

دائرة واحدة بعينها فالخط المستقيم الذى يصل بين نقطة ا ونقطة د مساوٍ

35 للخط المستقيم الذى يصل بين نقطة ب ونقطة ج والخط المستقيم الذى

يصل بين نقطة ا ونقطة د هو الذى يوتر قوس ا م د والخط المستقيم الذى

20 بأقطابها [اعطابهما N 21 واحد [واحدة N 21 مَنْ [مِنْ et saepius ر pro ن N 24 فلأن [لان A
24 ط ل ج ك [طلحلح N 25 الخفى [المحمى A 26 النصف من الكرة [الصفكره N 27 دائرة [N add. supra
A 27 لدائرة [لداه A 28 ا د [1 آد N 29 آب [N 30 دز [2 حَرْ N 30 جمعتا [1 حمعا N
32 ا ب ج د [احٯ N 33 الباقية [2 الباٯ N

sunt in spera, qui sunt circuli *AMND EZ TBGK*, et transit per polos
eorum, ergo ipse secat eos in duo media et duo media et orthogonaliter.
Ergo unaqueque portionum *MN EZ TBGK* est semicirculus. Et quia
portio *MN* est semicirculus, ergo portio *AMND* est maior semicirculo.
Simili quoque modo ostenditur quod omnes portiones que sunt inter   30
circulum *EZ* et polum *H* sunt maiores semicirculo. Et etiam quia portio *TBGK* est semicirculus, ergo portio *GB* est minor semicirculo. Et
similiter etiam monstratur quod omnes portiones que sunt inter circulum *EZ* et polum occultum ab ea parte in hac eadem spere medietate
sunt minores semicirculo.   35

Sit etiam circulus *AD* equalis circulo *BG* et equidistans ei. Dico
igitur quod portiones coalterne circulorum *AD BG* ad invicem sunt
equales. Et quia circulus *AD* est equalis circulo *BG* et equidistans
ei, ergo arcus *AE* est equalis arcui *EB* et arcus *DZ* est equalis arcui
ZG. Ergo duo arcus *AE DZ*, cum coniunguntur, sunt equales duobus   40
arcubus *EB ZG*, cum coniungentur. Et arcus *EA AD DZ*, cum
coniunguntur, sunt equales arcubus *EB BG GZ*, cum coniunguntur,
quoniam unusquisque arcuum *EADZ EBGZ* est semicirculus Quod
ideo est quoniam duo circuli *ABGD EZ* sunt maiores, reliquus ergo
arcus *AD* est equalis reliquo arcui *BG*. Sed duo arcus *AD BG* sunt 45
unius et eiusdem circuli: linea igitur recta que coniungit inter punctum
A et punctum *D* est equalis linee recte que coniungit inter punctum *B*
et punctum *G*. Sed linea recta que coniungit inter punctum *A* et punctum *D* est ea que subtenditur arcui *AMD* et linea recta que coniungit

---

26 *EZ*] *marg.* **M**, *om.* **OVa**   26 *TBGK*] *corr.* *ex* *TBLGK* **PB**, *TBLGK* **KgO**
PsVa   27 et duo media] *om.* **FiZ**   28 Ergo] igitur **Fi**   28 *TBGK*] *corr.* *ex*
TBLGK **P**, *TBLGK* **KgOBPsVa**   28–29 Et . . . *MN*] *illeg.* **Kg**   29 *AMND*] *MND*
KgOBPsVa   29 est maior semicirculo] *permut.* **BPsVa**   29 maior semicirculo] *tr.*
MKgO   30 Simili quoque] Similique **OBPs**, Simili **Va**   30 modo] monstratur vel
add. **R**   30 ostenditur] monstratur **KgOBPsVa**   30 que] qui **B**   30 inter] intus **Fi**
31 etiam] *om.* **KgBPs**, in **Fi**   32 *TBGK*] *corr.* *ex* *TBLGK* **P**, *TBLGK* **KgOBPsVa**   32–
33 minor . . . similiter] *illeg.* **Kg**   32 Et] *om.* **R**   33 etiam] *om.* **BPsVa**   36 *BG*] *G* **Va**
36 ei] *om.* **Fi**, *EL* **Va**   37 igitur] ergo **OBPsVa**   38 circulus] circulis **Fi**   39 ei] eis **B**; ei,
corr. *ex* eis **Ps**, *EL* **Va**   39 *EB*] *corr.* *ex* *AB* **R**   40 *DZ*] simul coniunguntur *add.* *et del.*
P, coniunguntur *add.* *et del.* *marg.* **P**; simul *add.* **KgBPsVa**   40 coniunguntur] simul *add.*
O   40 duobus] *om.* **V**   41 coniungentur] coniunguntur **VOZVa**   43 unusquisque] uterque
MKgBPsVa   43 *EADZ*] *EAD*, *Z* *supra* **Ps**   44 reliquus] reliqua **Fi**, reliquis **Z**   44 ergo] autem
P   45 arcus[1]] *om.* **Fi**   45 reliquo arcui] *tr.* **BPsVa**   45 Sed] hii *add.* **OVa**   45 *AD BG*] *om.*
MKgOBPsVa   46 circuli] qui sunt arcus (arcus *om.* **OBVa**, scilicet arcus *supra* **Ps**) *AD BG*
add. **MKgOBPsVa**   47 *D*] *T* **R**   48–49 Sed . . . *AMD*] *marg.* **Ps**; Sed . . . *AMD* et *marg.* **B**
48 recta] tunc **Z**   49 ea] illa **OBPsVa**   49 *AMD*] *AMND* **OBPsVa**, *ANMD* **FiZ**

يصل بين نقطة بـ ونقطة جـ هو الذى يوتر قوس بـلـجـ والخطوط المستقيمة

المتساوية التى فى الدوائر المتساوية تفصل قسياً متساوية والخط الأطول

يفصل قوساً أعظم والخط الأقصر يفصل قوساً أصغر العظمى منها للعظمى

40 والصغرى للصغرى فالقوس العظمى من دائرة امـد مساوية للقوس العظمى

من دائرة بـلـجـ والقوس الصغرى من دائرة امـد مساوية للقوس الصغرى

من دائرة بـلـجـ وقطعة امـد أعظم من نصف دائرة وقطعة بـلـجـ أصغر

من نصف دائرة فالقطع المتبادلة من الدوائر المتساوية المتوازية مساوٍ بعضها

لبعض ، وذلك ما أردنا أن نبين ·

يط   إذا كانت فى كرة دائرة عظيمة تقطع بعض الدوائر المتوازية التى

تكون فى الكرة ولا تمر بقطبيها فإن القسى التى تنفصل فى أحد نصفى

الكرة ما كان منها أقرب إلى القطب الظاهر فهو أعظم من قوس من تلك

الدائرة شبيهة بما ينفصل مما هو أبعد من ذلك القطب ·

5   فلتكن فى كرة دائرة عظيمة وهى دائرة اهـزب تقطع بعض الدوائر

المتوازية من الدوائر التى تكون فى الكرة ولا تمر بقطبيها ولتكن الدوائر

المتوازية دوائر اب جـد هـز ، فأقول إن القسى التى تنفصل فى أحد نصفى

الكرة ما كان منها أقرب إلى القطب الظاهر فهو أكثر من القوس من تلك

متسانة 38 [متساوية *corr. ex* والمتساوه A   [1المتساوية 38   المستقيم A *corr. ex*   [المستقيمة 37

الاعظم *corr. ex* [الأقصر 39   نور N (*ed.*): [يفصل 39   *om.* AH [والخط ... أصغر 39-38   A

قطعه 1 [وقطعة 42   N آد [امـد 41   امـد A [امـن 40   *om.* N [العظمى ... للصغرى 40-39   N

N من 2 [*om.* N   3 قوس [*in corr.* A   3 من [*om.* N   وذلك ... نبين 44   الداره A [الدوائر 43   A

شبيه بالقوس الى سفصل من الداره الى هى ابعد ;(.H *om*[ينفصل مما N, الشبيه NH [شبيهة ... القطب 4

من ذلك القطب A 6 من [ولتكن 6   N 7 دائرا [دوائر 7   *add.* N [اب 7   *om.* N

inter punctum B et punctum G est ea que subtenditur arcui BLG; et 50
linee recte equales que sunt in circulis equalibus separant arcus equales et longior linea separat arcum maiorem et brevior linea minorem
separat arcum Ergo maior arcus circuli AMD est equalis maiori arcui circuli BLG et minor arcus circuli AD est equalis minori arcui
circuli BLG. Sed portio AMD est maior semicirculo et portio BLG 55
est minor semicirculo: ergo portiones coalterne circulorum equalium et
equidistantium ad invicem sunt equales. Et illud est quod demonstrare
voluimus.

19 Cum in spera fuerit circulus maior secans aliquos circulorum equidistantium ex circulis qui sunt in spera neque transierit per polos eorum,
arcuum qui in una duarum medietatum spere separantur quicumque
fuerit propinquior polo manifesto erit maior arcu illius circuli simili ei
qui separatur ex circulo qui est magis remotus ab illo polo. 5

Sit itaque in spera circulus maior, qui sit circulus $AEZB$, secans
aliquos circulos equidistantes ex circulis qui sunt in spera neque transeat per polos eorum, sintque ex circulis equidistantibus circuli $AB\ GD$
EZ. Dico igitur quod arcuum qui separantur in una medietatum spere
quicumque polo manifesto est propinquior est maior arcu illius circuli 10

---

50 punctum¹] *om.* **P**   50 *B*] *G* **OBPsVa**   50 *G*] *B* **OBPsVa**   50 *BLG*] *BG* **PRVM** (*corr.*
ex BLG **P**)   51 in] *om.* **R**   52 separat] secat **Va**   52–53 brevior ... arcum] *permut.* **OBPsVa**
52 minorem] breviorem **Kg**   52–53 minorem separat arcum] minor est **Fi**   53 separat] *om.*
O   53 separat arcum] *om.* **Z**   53 *AMD*] *AMND* **OFiZ**   53–54 maiori arcui] *tr.*
Va   53–54 arcui circuli] *tr.* **V**   54 *BLG*] *BG* **PRVM**   54–55 et ... *BLG*] *om.* **ZVa**
54 arcus circuli] *tr.* **MKgOBPs**   55 *BLG*¹] *BG* **PRVM** (*corr. ex BLGD* **P**)   55 *BLG*²] *BG*
PRVM (*corr. ex BLG* **P**)   57–58 Et ... voluimus] *om.* **BPsVa**   57 illud] hoc **O**   57–
58 quod ... voluimus] *om.* **O**   58 voluimus] volumus **Z**   1 aliquos circulorum] *tr.* **M**
1 circulorum] *om.* **Kg**   2 circulis] neque *supra* **R**   2 sunt] *om.* **R**   2 transierit] transit
Va   3 duarum medietatum] *tr.* **OBPsVa**, medietate **FiZ**   3 medietatum] medietatibus **Kg**
3 separantur] *om.* **Va**   3 quicumque] arcus *supra* **O**   4 propinquior] propinquiorum **R**
4 illius] eiusdem **Va**, *add. supra* **OPs**   4 simili] *om.* **Va**   5 separatur] separantur **O**   5 ex] a
KgOBPsVa   6 Sit] Sint **Kg**   7 circulos] *om.* **Z**   9 quod] *repet.* **Fi**   9 medietatum] medietate
Z   10 manifesto] maiori **Z**   10 arcu] arcui **Va**   10 arcu illius] *tr.* **R**   10 illius] eiusdem *supra*
Ps

الدائرة الشبيهة بالقوس التى تنفصل من الدائرة التى هى أبعد من القطب

10   الظاهر أعنى أن قوس ا ب أعظم من القوس الشبيهة بقوس جـ د من دائرتها

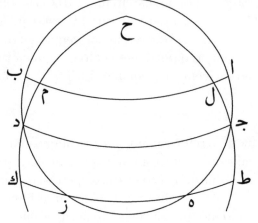

وقوس جـ د أعظم من القوس التى تشبه قوس ه ز من دائرتها فليكن القطب

الظاهر من قطبى الدوائر المتوازية نقطة حـ ولترسم دائرة عظيمة تمر بنقطة

حـ وبنقطة جـ وهى دائرة حـ ل جـ ط ولترسم دائرة عظيمة تمر بنقطة حـ

وبنقطة دـ وهى دائرة حـ م د ك فدائرتا حـ ل جـ ط حـ م د ك تفصلان فيما بينهما

15   قوسين متشابهتين فقوس ل م شبيهة بقوس جـ د فقوس ال م ب أعظم من

القوس من دائرتها الشبيهة بقوس جـ د وكذلك نبين أن قوس جـ د أيضاً أعظم

من القوس من دائرتها الشبيهة بقوس ه ز إذا نحن رسمنا دائرتين عظيمتين

تمران بنقطة حـ وبكل واحدة من نقطتى ه ز وقد يمكن أيضاً أن نبين ذلك

من غير أن نرسم هاتين الدائرتين بأن نقتصر على أن نتمم دائرة ه ز فقط كما

20   عملنا فى الشكل الذى قبل هذا ·

9 الدائرة [² داره A   10 الظاهر] om. N   10 أن] om. A   10 من القوس] 15 A marg.   و جـ د] .add

لم [جـ د 16   A الشبيه [الشبيهة 16   om. A   16 من دائرتها] om. N   15 من] N الملز [الملز] N   15 ال م ب [ال م ب

om. A   من دائرتها [17   A in corr.   16 قوس] N كذلك [وكذلك 16   A .add من داره الملز وهى موس

N دار] [دائرتين] 17   H فاننا [17   A .add إذا [17   من داره حد [ه ز] 17   A الشبيه [الشبيهة 17

simili arcui qui separatur ex circulo qui est magis remotus, scilicet
quod arcus *AB* est maior arcu sui circuli simili arcui *GD* et arcus *GD*

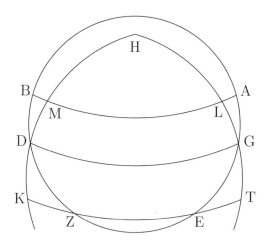

est maior arcu sui circuli simili arcui *EZ*. Sit ergo polus manifestus ex
duobus polis circulorum equidistantium punctum *H*. Signabo autem
circulum maiorem transeuntem per punctum *H* et per punctum *G*,   15
qui sit circulus *HLGT*. Et signabo circulum maiorem transeuntem per
punctum *H* et per punctum *D*, qui sit circulus *HMDK*. Ergo inter duos
circulos *HLGT HMDK* separantur duo similes arcus; ergo arcus *LM*
est similis arcui *GD*; ergo arcus *ALMB* est maior arcu sui circuli simili
arcui *GD*. Et similiter ostenditur quod etiam arcus *GD* erit maior arcu   20
sui circuli simili arcui *EZ*, cum nos signaverimus duos circulos maiores
transeuntes per punctum *H* et per unumquodque duorum punctorum *E*
et *Z*. Est tamen possibile ut hoc demonstretur etiam si hos duos circulos
non signaverimus. Sufficiet enim nobis ut circulum *EZ* compleamus,
quemadmodum fecimus in figura que hanc precedit. Et illud est quod   25
demonstrare voluimus.

11 separatur] *in corr.* **MPs**   12 circuli simili] *tr.* **Fi**   12 arcui] *om.* **Fi**   12–
13 et … *EZ*] *marg.* **R**   13 sui circuli] *tr.* **B**   13–14 ex duobus polis] *om.* **FiZ**
14 equidistantium] equidem **Fi**, *hic et saepius*   14–15 Signabo … punctum *H*] *om.*
Va   15 *H*] et per punctum *D false add.* **R**   15 per²] *om.* **Z**   16 circulus] *HMDK. Ergo inter*
duos arcus *false add.* **R**   16 *HLGT*] *ALGT* **Kg**, *in corr.* **B**   16–17 Et … *HMDK*] *primum
false scriptum, deinde recte* **KgB**, *scr. et del.* **Ps**   16–18 Et … circulos *HLGT*] *om.* **Va**
16 signabo] significabo **O**   16 transeuntem] *om.* **FiZ**   17 per] *om.* **FiZ**   17 circulus] *supra*
P   17–18 Ergo … *HMDK*] *marg.* **BPs**   18 separantur] separentur **Fi**, *in corr.* **Ps**; ergo
add. **BPsVa**   19 circuli] *supra* **R**   19 simili] *om.* **Fi**   20 maior] minor **Fi**   21 nos] *om.*
Va, vero **OB**, vero *scr.et del.* **Ps**   21 signaverimus] signavimus **Fi**, significaverimus **Va**
22 transeuntes] *post H* **Va**   22 unumquodque] utrumque **MKgOBPsVa**   22 duorum] *om.*
Va   23 et] *om.* **KgOBPsVa**   23 Est] Et est **OBPsVa**   23 tamen] hoc *add.* **OBPsVa**
23 demonstretur] demonstremus **Fi**, demonstratur **PsVa**   23 etiam] et **Z**   24 non] *marg.* **B**,
supra **Ps** 24 signaverimus] significaverimus **Va** 24 Sufficiet] Sufficit **OFiBVa** 24 enim] *supra*
BPs   24 circulum *EZ* compleamus] *permut.* **MKgOBPsVa**   25 quemadmodum] *in corr.* **Ps**,
quem **KgB**   25 fecimus] *om.* **FiZ**   25–26 Et… voluimus] *om.* **OBPsVa**   25 illud] hoc **FiZ**

كا

إذا كانت على أكر متساوية دوائر عظيمة مائلة على دوائر أخرى عظيمة فأى

دائرة اتفق أن يكون قطبها أعلى فهى أكثر ميلاً على صاحبتها وأما الدوائر

التى بعد أقطابها من سطوح الدوائر التى هى قائمة عليها بعد متساوٍ فإن

ميلها يكون ميلاً متساوياً .

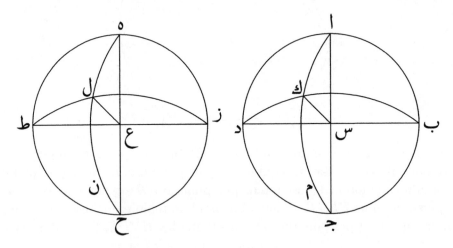

فلتكن فى أكر متساوية دائرتان عظماوان من الدوائر التى فى الكرة 5

وهما دائرتا ب‌ك‌د ز‌ل‌ط مائلتان على دائرتى ا‌ب‌ج‌د ه‌ز‌ح‌ط العظماوين

وليكن قطب دائرة ب‌ك‌د نقطة م‌ وقطب دائرة ز‌ل‌ط نقطة ن‌ وليكن قطب

م‌ أعلى من قطب ن‌ ، فأقول إن ميل دائرة ب‌ك‌د على دائرة ا‌ب‌ج‌د أكثر

من ميل دائرة ز‌ل‌ط على دائرة ه‌ز‌ح‌ط فلترسم دائرة عظيمة تمر بنقطة م‌

وبأحد قطبى دائرة ا‌ب‌ج‌د وهى دائرة ا‌ك‌م‌ج ودائرة أخرى عظيمة تمر 10

بنقطة ن‌ وبأحد قطبى دائرة ه‌ز‌ح‌ط وهى دائرة ه‌ل‌ن‌ح فهى تمر بأقطاب

---

1 أخرى [احر N   2 فهى [فهو A   5 عظماوان [عظماان A   6 على [ان *add. et del.* A

---

2 صاحبتها [المائلة عليها أطول وإذا كان العمودان متساويين كان ‹ال›ميلان متساويين *add.* A

يعنى بقوله إن قطب الدائرة أعلى إذا كان ‹ال›عمود الواقع من قطب الدائرة المائلة على سطح الدائرة

20   Cum in speris equalibus circuli maiores super alios circulos maiores inclinati fuerint, cuiuscumque circuli poli fuerint altiores ipse erit maioris inclinationis super comparem suum. Circuli vero, spacium polorum quorum a superficiebus circulorum super quos sunt erecti fuerit spatium equale, similis erunt declinacionis.   5

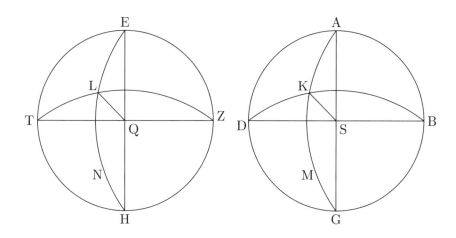

Sint itaque in speris equalibus duo circuli maiores ex circulis qui sunt in spera, qui sunt duo circuli *BKD ZLT*, inclinati super duos circulos maiores *ABGD EZHT*; et sit polus circuli *BKD* punctum *M* et polus circuli *ZLT* punctum *N*; et sit polus *M* altior polo *N*. Dico igitur quod inclinatio circuli *BKD* super circulum *ABGD* est maior declina-   10
tione circuli *ZLT* super circulum *EZHT*. Signabo itaque circulum maiorem transeuntem per punctum *M* et per unum duorum polorum circuli *ABGD*, qui sit circulus *AKMG*, et circulum alium maiorem transeuntem per punctum *N* et per unum duorum polorum circuli *EZHT*, qui sit circulus *ELNH*. Ipsi ergo transeunt per polos circuli *BKD* et   15

1–2 maiores] *om.* **M**   2 fuerint] sunt **Fi**   2 altiores] *et add.* **OFiZ**   2 ipse erit] *tr.* **KgPsVa**
2 erit] *om.* **B**   3 maioris] maiores **Fi**   3 comparem suum] *in corr.* **Ps**, comprehensum **B**
4 quorum] quarum **Va**   4 fuerit] sunt (*in corr.*) **B**, fit (*cum lac.*) **Ps**, fuit **Va**   5 erunt] *in corr.*
B, erit **PsVa** (*defective* **Ps**)   6 itaque] *om.* **V**, enim **Fi**   6 equalibus] *om.* **OBPsVa**   7 sunt] sint
KgZ   7 *ZLT*] *om.* **Fi**   8 et ... *BKD*] *om.* **Fi**   9 *ZLT*] *ZBT* **Va**   10 inclinatio] declinatio
FiZ   10 *ABGD*] *ABG* **BPsVa**   10–11 declinatione] *in corr.* **Ps**   11 *ZLT*] *ZBD* **Va**   13–
14 *ABGD* ... circuli] *om.* **R**   13 *AKMG*] *AKMH* **V**, *AKMDG* **FiZ**   13 alium] *in corr.* **Ps**,
alum **Fi**   14 polorum] *om.* **Fi**   14 *EZHT*] *AEZHT* **R**   14–16 *EZHT* ... et circuli] *om.*
Va   15 circulus] *om.* **Ps**   15 circulus *ELNH*] *tr.* **O**   15 *ELNH*] *in corr.* **B**, *EBZH* **Kg**
15 ergo] igitur **VFiZ**   15 et circuli] *om.* **V**

دائرة ب‍ك‍د ودائرة ز‍ل‍ط وتقطعهما بنصفين نصفين وعلى زوايا قائمة وليكن

الفصل المشترك لدائرة ا‍ب‍ج‍د ودائرة ب‍ك‍د خط ب‍د والفصل المشترك

لدائرة ا‍ب‍ج‍د ودائرة ا‍ك‍م‍ج خط ا‍ج والفصل المشترك لدائرة ب‍ك‍د ودائرة

ا‍ك‍م‍ج خط ك‍س وأيضاً فليكن الفصل المشترك لدائرة ه‍ز‍ح‍ط ودائرة ز‍ل‍ط

15

خط ز‍ع‍ط والفصل المشترك لدائرة ه‍ز‍ح‍ط ودائرة ه‍ل‍ن‍ح خط ه‍ح والفصل

المشترك لدائرة ز‍ل‍ط ودائرة ه‍ل‍ن‍ح خط ل‍ع ولأن دائرة ا‍ك‍م‍ج فى كرة

وهى تقطع بعض الدوائر العظام التى تكون فى الكرة وهما دائرتا ا‍ب‍ج‍د

ب‍ك‍د وهى تمر بأقطابهما فهى تقطعهما بنصفين نصفين وعلى زوايا قائمة

فدائرة ا‍ك‍م‍ج قائمة على كل واحدة من دائرتى ا‍ب‍ج‍د ب‍ك‍د على زوايا

20

قائمة فكل واحدة من دائرتى ا‍ب‍ج‍د ب‍ك‍د قائمة على دائرة ا‍ك‍م‍ج على

زوايا قائمة وإذا كان سطحان يقطع أحدهما الآخر وكانا قائمين على سطح

آخر على زوايا قائمة فإن الفصل المشترك لهما أيضاً قائم على ذلك السطح

بعينه على زوايا قائمة فالفصل المشترك لدائرتى ا‍ب‍ج ب‍ك‍د عمود على دائرة

ا‍ك‍م‍ج والفصل المشترك لهما خط ب‍د فخط ب‍د عمود على دائرة ا‍ك‍م‍ج فهو

25

يحدث مع جميع الخطوط المستقيمة التى تمر بطرفه وتكون فى سطح دائرة

ا‍ك‍م‍ج زوايا قائمة وكل واحد من خطى ك‍س س‍ا يمر بطرفه وهو فى سطح

13 ا‍ب‍ج‍د [ودائرة N 13–14 ا‍ج [حط 14 add. A 14–
13 ا‍ب‍ج‍د [ودائرة ... خط ب‍ك‍د] marg. A
15 سلط [ز‍ل‍ط N 15 N hic et saepius ولدائره A، ودار [²ودائرة 14 ب‍ك‍د ودائرة ا‍ك‍م‍ج] tr. A
A الكره [كرة 16 N لط [ل‍ع 17 N, ۵ H ز‍ل‍ط ودائرة ه‍ل‍ن‍ح 17 tr. A ل‍ع [ل‍ط N 17 رع [ز‍ع‍ط N
N تقطعهما [تقطعها 19 A وهى تمر [وهى ومر 19 corr. ex وهو [وهى 19 A دائرتا [ودائرتا 18
A وان [وإذا 22 marg. A ²قائمة [قائمة N 21 على زوايا ... ب‍ك‍د] om. A 20–21 نصفين [نصفين 19
om. A أيضاً [23 قائم [قائم ... لهما 23–25 دائره على دارى مام بكد ماذن ساطح دارى ا‍ب‍ح‍د مام على دلك السطح
ويقاطعهما المشترك وتقاطعهما اكمح marg. A عمود [عمود A 25 س [ك‍س N 27 N, hic et saepius

circuli *ZLT* et secant eos in duo media et duo media et orthogonaliter. Sitque differentia communis circulo *ABGD* et circulo *BKD* linea *BD*, et differentia communis circulo *ABGD* et circulo *AKMG* linea *AG*, et differentia communis circulo *BKD* et circulo *AKMG* sit linea *KS*, et sit etiam differentia communis circulo *EZHT* et circulo *ZLT* linea *ZQT*, et differentia communis circulo *EZHT* et circulo *ELNH* sit linea *EH*, et differentia communis circulo *ZLT* et circulo *ELNH* sit linea *LQ*. Et quia circulus *AKMG* est in spera et secat aliquos circulos maiores qui sunt in spera, qui sunt duo circuli *ABGD BKD*, et transit per polos eorum, ergo ipse secat eos in duo media et duo media et orthogonaliter; ergo circulus *AKMG* erectus est super unumquemque duorum circulorum *ABGD BKD* orthogonaliter et unusquisque duorum circulorum *ABGD BKD* orthogonaliter est erectus super circulum *AKMG*. Sed cum duarum superficierum una alteram secat et fuerint orthogonaliter erecte super aliam superficiem, tunc etiam communis earum differentia erigetur super illam eandem superficiem orthogonaliter; ergo communis differentia duorum circulorum *ABGD BKD* est perpendicularis super circulum *AKMG*. Communis vero eorum differentia est linea *BD*, ergo linea *BD* est perpendicularis super circulum *AKMG*: ex ipsa igitur cum omnibus lineis rectis, que transeunt per eius extremitatem et sunt in superficie *AKMG*, anguli proveniunt recti. Unaqueque autem duarum linearum *KS SA* transit per eius extremitatem et est in superficie cir-

20

25

30

35

16 et duo media] *om.* **FiZ**   16 et³] *om.* **V**   17 Sitque] Sit itaque **OFiZBPsVa**   17–19 differentia ... *AKMG*] *haec elementa confuderunt* **OBPsVa**   18 *ABGD*] *ABG* **FiZ**   19 *BKD*] *AKD* **Z**   19 et²] *om.* **OBPsVa**   20 etiam] *supra* **P**   20 *EZHT*] *in corr.* **B**   20 *ZLT*] *in corr.* **B**   21 et¹ ... *EH*] *om.* **Va**   21 circulo²] *om.* **V**   23 aliquos] quos **B**, alios **Va**   25 et duo media] *om.* **FiZ**   26 *AKMG*] *corr. ex ADMG* **R**   26 erectus est] *tr.* **KgOBPsVa**   26 unumquemque] utrumque **MKgOBPsVa**   26 duorum] *om.* **KgBPsVa**   27 orthogonaliter] *om.* **PsVa**   27–28 et ... *BKD*] *marg.* **Ps**   27–28 et ... orthogonaliter] *marg.* **B**   27 unusquisque] uterque **MKg**   29 et fuerint] *supra* **O**   29 fuerint] *in corr.* **B**, sunt **PsVa**   30 erecte] *om.* **Kg**   30 aliam] aliquam **OVa**   30 tunc etiam] *tr.* **O**   30 etiam] *om.* **FiZ**   31 erigetur] erigitur **OFiZBPsVa** (*in corr.* **Z**)   31 illam] *in corr.* **M**, *om.* **OBPsVa**   31 superficiem] superficie **Fi**   31 ergo] tunc **Z**   33 *AKMG*] *AKM* **P**   33 vero] *supra* **V**   33 eorum] earum **KgOFiZBPsVa**   33 eorum differentia] *tr.* **BPsVa**   33–34 ergo linea *BD*] *repet.* **Va**   34 linea *BD*] *om.* **FiZ**   34 est] *ante* ergo **Z**   34 ex] Et **R**, vel **V**   34–35 cum omnibus] communibus **Kg**   35 lineis rectis] *tr.* **KgOBPsVa**   35 extremitatem] extremitates **OBPsVa**   36 *AKMG*] *ABMG* **Fi**   36 Unaqueque] Utraque **MKgOBPsVa**   37 eius extremitatem] *tr.* **OFiZ**

دائرة اكمج لخط بٮد قائم على كل واحد من خطى كس سا على زوايا

قائمة ولأن سطحى ابجد بكد يقطع أحدهما الآخر وقد أخرج إلى خط

٣٠ بٮد وهو الفصل المشترك خطا كس سا على زوايا قائمة وخط كس منهما

فى سطح دائرة بكد وخط سا فى سطح دائرة ابجد

فزاوية كسا هى ميل سطح بكد على سطح ابجد وكذلك نبين أن

زاوية لعه أيضاً ميل سطح زلط على سطح هزحط ،

وأقول إن زاوية كسا أصغر من زاوية لعه فلأن نقطة م أعلى من

٣٥ نقطة ن يكون العمود الذى يخرج من نقطة م إلى سطح دائرة ابجد

أطول من العمود الذى يخرج من نقطة ن إلى سطح دائرة هزحط والعمود

الذى يخرج من نقطة م إلى سطح دائرة ابجد يقع على الفصل المشترك

لدائرتى اكمج ابجد أعنى على خط اج لأن سطحى اكمج ابجد

أحدهما قائم على الآخر على زاوية قائمة والعمود الذى يخرج من نقطة ن

٤٠ إلى سطح دائرة هزحط يقع على خط هح فالعمود الذى يخرج من نقطة م

إلى خط اج أطول من العمود الذى يخرج من نقطة ن إلى خط هح ولأن

قطعتى اكمج هلنح من الدوائر متساويتان وقد تعلمت عليهما نقطتا م ن

كيف ما وقعتا والعمود الذى يخرج من نقطة م إلى خط اج أطول من

marg. 29 من إلى] A 32 من ابجد... فزاوية [ ] 29 *om.* N إلى] بكد 29 A سطح [سطحى] سطح 29 N ولأن [29

A (ولأن فان زاويه *pro* فزاوية) 33 أصغر [أعظم N 35 م] م , *supra* N 38 م A ابجد] [¹]

om. A دائرة [40 فالعمود [والعمود A 39 سطح [سطحى] سطح *add. et del.* N 38 مع على الفصل المسرك

A دوار [الدوائر] قطعتى [قطعتى , *supra* 42 ملان N ولأن [ولأن A 41 *corr. ex* ز] ح ه [41 ح

A أعظم [أطول] *marg.* A 43 م إلى... ح ه 43-44 A وبعلمت [وقد تعلمت 42

36-37 ابجد... أطول [ ] *om.* A; فان زاويه كسا هى ميل سطح بكد على سطح ابجد *marg.* A

culi *AKMG*: ergo linea *BD* orthogonaliter est erecta super duas lineas
KS SA. Et quia duarum superficierum *AB KD* una alteram secat et ad
lineam *BD*, que est communis differentia, iam protracte sunt due linee   40
KS SA orthogonaliter et linea *KS*, que est una earum, est in superficie
circuli *BKD*, et linea *SA* est in superficie circuli *ABGD*, ergo angulus
KSA est inclinatio superficiei *BKD* super superficiem *ABGD*. Et similiter ostenditur quod etiam angulus *LQE* est inclinatio superficiei *ZLT*
super superficiem *EZHT*.   45

Et dico quod angulus *KSA* est minor angulo *LQE*. Et quia punctum *M* est altius puncto *N*, ergo perpendicularis que protrahitur a
puncto *M* ad superficiem circuli *ABGD* est longior perpendiculari que
protrahitur a puncto *N* ad superficiem circuli *EZHT*. Perpendicularis
vero que protrahitur a puncto *M* ad superficiem circuli *ABGD* cadit   50
super communem differentiam circulorum *AKMG ABGD*, scilicet
super lineam *AG*, quoniam duarum superficierum *AKMG ABGD* una
super alteram erigitur orthogonaliter et perpendicularis, que producitur a puncto *N* ad superficiem circuli *EZHT*, cadit super lineam *EH*.
Perpendicularis ergo que a puncto *M* ad lineam *AG* protrahitur lon-   55
gior existit perpendiculari que producitur a puncto *N* ad lineam *EH*.
Et quia due portiones *AKMG ELNH* sunt circulorum equalium, super
quas duo puncta *M* et *N* quolibet modo sunt signata, et perpendicularis que protrahitur a puncto *M* ad lineam *AG* est longior perpendiculari

39 AB] *AL* **Kg**, *ABGD* **OFiZPsVa**   *39 KD*] *BKD* **OPsVa**, *KBD* **Fi**, *KDB* **Z**
40 differentia] earum *add. supra* **Ps**, *in textu* **Va**   41 ²] *supra* **Ps**; et **KgOFi**, *illeg.* **R**
om. **B**   42 circuli *BKD*] *tr.* **FiZ**   43 *KSA*] *KS* **Fi**   43 *BKD*] *in corr.* **B**   43 super] *om.* **Fi**
43 Et] est **Fi**   43–44 similiter] etiam *add.* **KgBPsVa**, *add. et del.* **M**   44 ostenditur] ostendit
Fi; etiam *add.* **O**   44 etiam] *supra* **M**, *om.* **KgPsVa**   44 inclinatio] in circulo **Fi**, *hic et*
saepius   45 *EZHT*] *ZHT* **Z**   46 Et] quia **Kg**; Et, *corr. ex* quia **BPs**   46 quod] *om.* **Va**
46 *KSA*] *KFA* **Ps**   47 altius] alterius **VB**, alius **MZ**, alicuius **Fi**   48–50 est ... *ABGD*] *marg.*
M   49 *EZHT*] *in corr.* **B**   51 communem differentiam] *tr.* **BPsVa** (communem *supra*
BPs)   51 *AKMG*] *in corr.* **B**   52 duarum superficierum] *tr.* **BPsVa** (duarum: ductarum
PsVa)   52 *AKMG*] *in corr.* **B**   53 erigitur orthogonaliter] *tr.* **PVKgOBPsVa**   54–
56 ad ... puncto *N*] *marg.* **BPs**   55 Perpendicularis ergo] *tr.* **FiZ**   55 que] *supra* **RVa**
55 protrahitur] *ante* ad **MKg**; protrahuntur **Fi**   56 existit] est **OFiZBPsVa**   56 *EH*] *EK*
Fi, *TH* **Va**   57 *ELNH*] *ELNS* **Kg**, *ELMH* **Fi**   57 sunt] duorum *add.* **PRVFiZ**, *supra* **M**
58 sunt signata] *tr.* **MKgOBPsVa**   59 *AG*] *in corr.* **B**

العمود الذى يخرج من نقطة نَ إلى خط هَحَ تكون قوس مَجَ أعظم من

قوس نحَ وقوس مَكَ مساوية لقوس نلَ وذلك أن كل واحدة منهما   45

مساوية للقوس التى يوترها ضلع المربع الذى يرسم فى الدائرة العظمى وذلك

أن كل واحدة منهما خرجت من قطب من أقطاب دائرتى بَكَد زلَط لجميع

قوس كَمَجَ أعظم من جميع قوس لنحَ فلأن كل قوس اكَمَجَ مساوية

لكل قوس هَلنحَ وقوس كَمَجَ من إحداهما أعظم من قوس لنحَ من

الأخرى تبقى قوس اكَ أصغر من قوس هَلَ وزاوية كَسَا قاعدتها قوس اكَ   50

وزاوية لَعَهَ قاعدتها قوس لَهَ وهاتان الزاويتان على مركزى الدائرتين

فزاوية كَسَا أصغر من زاوية لَعَهَ وزاوية كَسَا هى ميل سطح دائرة

بَكَد على سطح دائرة ابَجَد وزاوية لَعَهَ هى ميل سطح دائرة زلَط

على سطح دائرة هَزحَط فميل دائرة بَكَد على سطح دائرة ابَجَد أكثر

من ميل دائرة زلَط على دائرة هَزحَط فدائرة بَكَد أميل على دائرة   55

ابَجَد من دائرة زلَط على دائرة هَزحَط ،

وأيضاً فليكن بعد أقطاب دائرتى بَكَد زلَط من السطوح التى هى

قائمة عليها متساوياً أعنى أن يكون العمود الذى يخرج من نقطة مَ إلى

سطح دائرة ابَجَد مساوياً للعمود الذى يخرج من نقطة نَ إلى سطح دائرة

A. *add.* اكمح مساوبه لكل قوس هلح وفصل موس [¹قوس 48   A. *add.* الى محيطها [زلَط 47

A. *om.* [³قوس ... لنح 48-49   جميع [48   A. *om.* [كل 48   N. *om.* [لنح 48-49   A. *add.* من احدهما [كَمَج 48

امل [أكثر 54   N. عن [على 54   A. وان راوه [وزاوية 53   A. مركز [مركزى 51   N. لقوس [لكل قوس 49

N. 55 على [²سطح   A. *add.* 55-56 هَزحَط [فدائرة ... هَزحَط 55-56   A. *om.* [

A. *add. marg.* لان كل واحدة منهما مساونة للقوس الى يورها صلع المربع المرسوم فى الدايره العطيمة [نلَ 45

A. حرحت من قطب الدوار العطمى الى محيطها [مساوية ... العظمى 46

que producitur a puncto N ad lineam EH ergo arcus MG est maior   60
arcu NH. Sed arcus MK est equalis arcui NL. Quod ideo est quoniam
unusquisque eorum est equalis arcui cui subtenditur latus quadrati
quod in circulo maiore signatum est. Et hoc ideo quoniam unusquisque eorum protrahitur ab uno polorum duorum circulorum BKD ZLT.
Totus igitur arcus KMG est maior toto arcu LNH. Et quia totus arcus   65
$AKMG$ est equalis toti arcui $ELNH$, sed arcus KMG, qui est unius
eorum, est maior arcu LNH, qui est alterius eorum, remanet arcus AK
minor arcu EL. Anguli vero KSA basis est arcus AK et anguli LQE
basis est arcus LE, et hii duo anguli sunt super centra duorum circulorum: ergo angulus KSA est minor angulo LQE. Angulus vero KSA 70
est inclinatio superficiei circuli BDK super superficiem circuli $ABGD$
et angulus LQE est inclinatio superficiei ZLT super superficiem circuli
$EZHT$: ergo inclinatio circuli BKD super circulum $ABGD$ est maior
inclinatione circuli ZLT super circulum $EZHT$. Circulus igitur BKD
magis est inclinatus super circulum $ABGD$ quam circulus ZLT super   75
circulum $EZHT$.

Sit etiam spacium duorum polorum circulorum BKD ZLT a
superficiebus super quas sunt erecti equale, scilicet sit perpendicularis
que producitur a puncto M ad superficiem circuli $ABGD$ equalis perpendiculari que protrahitur a puncto N ad superficiem circuli $EZHT$.   80

60 producitur] protrahitur **Z**   60 a puncto] *marg.* **M**   60 *EH*] *EG* **V**, *EK* **Fi**,
TH **Va**   61 *NH*] *in corr.* **BPs**, *NG* **Fi**   62 unusquisque] uterque **MKg**, *in corr.*
O   62–63 unusquisque ... quoniam] *om.* **BPsVa**   63 ideo] iam **Kg**; est *add.* **V**   63–
64 unusquisque] unumquodque **MSS**   64 uno] duorum *add.* **M**   64 polorum] polo **BPsVa**
65 igitur] ergo **FiZ**   66 *AKMG*] *ABMG* **Fi**   66 *ELNH*] *in corr.* **B**   66 unius] unus **R**
67 remanet] ergo *add.* **OB**; ergo arcus remanet **Va**   67 arcus] *in corr.* **BPs**   68 *EL*] *ES*
Fi   69 est] *om.* **PsVa**   69 super] duo *add.* **Fi**   69 centra] duo *add.* **Z**, centrum **OPs**
71 *BDK*] *BKD* **MVFiZBPsVa**   71 *ABGD*] *ABG* **P**   72 *ZLT*] *om.* **M**   73 *EZHT*] *EHT*
BPsVa   73 *BKD*] *in corr.* **Kg**   73 maior] minor **KgOFiBPsVa**   74 *ZLT*] *ZHT* **Fi**
74 *EZHT*] *EZH* **R**   75 magis est] *tr.* **OBPs**   75 *ABGD*] *in corr.* **Kg**   77 spacium] *om.*
Fi   77 duorum polorum] *tr.* **PVKgO**   78 equale] *om.* **B**, *add. in marg.* **Ps**, circuli equales **Z**
78 scilicet sit] *tr.* **Fi**, sitque **Z**   78 sit] *om.* **KgBPs**   79 que] qui **Z**   79 a puncto] *in corr.* **B**
79 puncto] recto **PsVa**   79 *ABGD*] *ABG* **Va**

ه‍زح‍ط ، فأقول إن ميل دائرتى ب‍ك‍د ز‍ل‍ط على دائرتى اب‍ج‍د ه‍زح‍ط   60

ميل متشابه أعنى أن زاوية ك‍س‍ا تكون مساوية لزاوية ل‍ع‍ه فإذا عملنا هذه

الأشياء بأعيانها بيّنّا على هذه الجهة أن زاوية ك‍س‍ا هى ميل سطح دائرة

ب‍ك‍د على سطح دائرة اب‍ج‍د وأن زاوية ل‍ع‍ه هى ميل سطح دائرة ز‍ل‍ط

على سطح دائرة ه‍زح‍ط ،

فأقول إن زاوية ك‍س‍ا مساوية لزاوية ل‍ع‍ه ولأن العمودين اللذين   65

يخرجان من نقطتى م‍ ن‍ إلى سطحى دائرتى اب‍ج‍د ه‍زح‍ط متساويان

والعمودان اللذان يخرجان من نقطتى م‍ ن‍ إلى سطحى دائرتى اب‍ج‍د

ه‍زح‍ط يقعان على خطى اج‍ ه‍ح يكون العمودان اللذان يخرجان من

نقطتى م‍ ن‍ إلى خطى اج‍ ه‍ح متساويين ولأن قوسى اك‍م‍ج‍ ه‍ل‍ن‍ح

قطعتان من دائرتين متساويتين وقد تعلمت نقطتا م‍ ن‍ عليهما كيف ما وقعتا   70

والعمود الذى يخرج من نقطة م‍ إلى خط اج‍ مساوٍ للعمود الذى يخرج من

نقطة ن‍ إلى خط ه‍ح تكون قوس م‍ج‍ مساوية لقوس ن‍ح وقوس ك‍م‍ أيضاً

مساوية لقوس ن‍ل‍ وذلك أن كل واحدة منهما مساوية للقوس التى يوترها

ضلع المربع الذى يرسم فى الدائرة العظمى فجميع قوس ك‍م‍ج‍ مساوية لجميع

قوس ل‍ن‍ح وجميع قوس اك‍م‍ج‍ مساوية لجميع قوس ه‍ل‍ن‍ح فقوس اك‍   75

الباقية مساوية لقوس ه‍ل‍ الباقية وزاوية ك‍س‍ا قاعدتها قوس اك‍ وزاوية

ه‍زح‍ط [زل‍ط 60 على دائرتى ن‍ك‍د ر‍ل‍ط .add N 61 متشابه [متساوية AN, מחשמא' H 61 تكون [زاوية ك‍س‍ا .tr
N عن [على 63 .om A 62 بيّنّا [.om A 61 هذه [علما A 61 عملنا [مساوية A 61 مساوية [.om N
ه‍ح...يكون [.om A 65 ولأن [ط‍لان N 67 سطحى [سطح A 68 ه‍ح [.corr ex ه‍ح‍م A 68-69
قوس ل‍ن‍ح...الباقية 74-76 اك‍م‍ج [ل‍ح A 74 ك‍م‍ج [.marg (sequuntur 7 lineae
deletae), in textu: لقوس هل الباقه A 75 فقوس [فاذأ قوس A

Dico igitur quod inclinatio duorum circulorum *BKD ZLT* super duos circulos *ABGD EZHT* est inclinatio similis, scilicet quod angulus
KSA est equalis angulo *LQE*. Cum ergo eadem fecerimus, monstrabi-
tur secundum hunc modum quod angulus *KSA* est inclinatio superficiei
circuli *BKD* a superficie circuli *ABGD* et quod angulus *LQE* est incli-   85
natio superficiei circuli *ZLT* super superficiem circuli *EZHT*.

    Dico igitur quod angulus *KSA* est equalis angulo *LQE*. Et quia
due perpendiculares, que protrahuntur ad superficies duorum circulorum *ABGD EZHT* a duobus punctis *M* et *N*, sunt equales et due
perpendiculares, que a duobus punctis *M* et *N* producuntur ad super-   90
ficies duorum circulorum *ABGD EZHT*, cadunt super duas lineas *AG*
et *EH*, ergo due perpendiculares que producuntur a duobus punctis *M*
et *N* ad duas lineas *AG EH* sunt equales. Et quia duo arcus *AKMG*
ELNH sunt due portiones duorum circulorum equalium, super quas
duo puncta *M* et *N* quolibet modo sunt notata, et perpendicularis que   95
protrahitur a puncto *M* ad lineam *AG* est equalis perpendiculari que
producitur a puncto *N* ad lineam *EH*, ergo arcus *MG* est equalis
arcui *NH*; arcus autem *KM* etiam est equalis arcui *NL*. Quod ideo est
quoniam unusquisque eorum est equalis arcui cui latus quadrati quod
signatur in maiore circulo subtenditur: totus igitur arcus *KMG* est   100
equalis toti arcui *LNH*. Sed totus arcus *AKMG* toti arcui *ELNH* equatur: ergo reliquus arcus *AK* est equalis reliquo arcui *EL*. Anguli autem

83 *KSA*] *KFA* **Ps**   83 *LQE*] *LQC* **Va**   83 ergo] igitur **BPsVa**   83 fecerimus] fecimus
Fi   84 secundum] super **V**   84 *KSA*] *KFA* **Ps**   84–86 *KSA* ... *EZHT*] *marg.*
B   84–86 est ... *EZHT*] *marg.* **Ps**   85 *BKD*] *in corr.* **ZVa**   85 a] super **OBPsVa**
85 superficie] superficiem **OBPsVa**   85 quod] quia **Fi**   85 *LQE*] *LGE* **Va**   87 igitur] ergo
OBPsVa (**Ps** *supra*)   87 quia] quod **PsVa**   88 superficies] *in corr.* **Ps**, superficiem **Va**   89–
93 sunt ... et *N*] *om.* **OBPsVa**   89–93 et ... sunt equales] *om.* **FiZ**   90 a duobus ... *N*] *post*
EZHT **VMKg**   90 producuntur] protrahuntur **M**   91 duorum circulorum] *tr.* **R**
91 duas lineas] *tr.* **V**   92 *EH*] *EG* **V**   93 *AG*] et *add.* **OVa**   94 *ELNH*] *in corr.*
B   94 duorum] *om.* **B**   94 circulorum equalium] *tr.* **PRV**   95 notata] notatata **Fi**   96–
97 *M* ... a puncto] *om.* **Z**   96 perpendiculari] perpendicularis **Fi**   97 producitur] protrahitur
M   97 *EH*] *in corr.* **V**   97 *MG*] *MHG* **BPsVa**   98 *NH* ... arcui] *om.* **Fi**   98 autem] *om.*
BPsVa   98 *KM*] *in corr.* **Ps**   99 unusquisque] uterque **MKgOBPsVa**   100 maiore circulo] *tr.*
MKgOBPsVa   100 subtenditur] subtendatur **Fi**   101 toti] tot **Va**   101 toti arcui] *tr.*
OBPsVa   101–102 Sed ... equatur] *primum false, deinde recte* **KgBPs**   101 *ELNH*] *in corr.*
B   101–102 equatur] *post* toti **R**   102 *EL*] *EB* **FiZ**

لَعَه قاعدتها قوس هَلَ فزاوية كسآ مساوية لزاوية لَعَه وزاوية كسآ

هى ميل سطح دائرة بَكَد على سطح دائرة ابجد وزاوية لَعَه هى ميل

سطح دائرة زلَطَ على سطح دائرة هزحطَ فميل دائرة بَكَد على دائرة

ابجد مساوٍ لميل دائرة زلَطَ على دائرة هزحطَ فميل دائرتى بَكَد زلَطَ

على دائرتى ابجد هزحطَ ميل متشابه وقد علمنا أنه إنما يقال إن ميل

سطح على سطح آخر شبيه بميل سطح آخر على سطح آخر إذا كانت الخطوط

المستقيمة التى تخرج من الفصول المشتركة للسطوح على زوايا قائمة فى كل

واحد من السطوح تحيط بزوايا متساوية ، وذلك ما أردنا أن نبين .

إذا كانت فى كرة دائرة عظيمة تماس دائرة من الدوائر التى تكون فى الكرة

ليست بعظيمة وتقطع دائرة أخرى موازية لتلك من الدوائر التى فيما بين

مركز الكرة وبين الدائرة التى تماسها الدائرة الأولى وكان أيضاً قطب الدائرة

العظمى فيما بين الدائرتين المتوازيتين ورسمت دوائر عظيمة تماس أعظم

الدائرتين المتوازيتين فإن هذه الدوائر تكون مائلة على الدائرة العظمى

وتكون أكثر هذه الدوائر ارتفاعاً الدائرة التى تكون مماستها على وسط

القطعة العظمى من قطعتى تلك الدائرة وأكثرها انخفاضاً الدائرة التى تكون

مماستها على وسط القطعة الصغرى من قطعتى الدائرة وأما سائر الدوائر فما

كان منها بعد موضع مماسته من أحد وسطى القطعتين أيهما كان بعداً

فميل دائرى [هزحطَ 81 *om.* A هزحطَ [80 *add.* A سطح N; عن [على 80 A لدارة [ميل دائرة 80

H: الدائرة [NH 5 السطح]ين [A: السطوح] [الطوح [N 84 سطح مساوه A, مشابه [متشابه *add. et del.* N 81

N الدوائر [6 مماستها [A *in corr.* 6 وسط] *in corr.* A 6 *in corr.* A 9 أحد] N احدى [A, *in corr.*

KSA basis est arcus *AK* et basis anguli *LQE* est arcus *EL*: angulus
ergo *KSA* equatur angulo *LQE*. Sed angulus *KSA* est inclinatio superficiei circuli *BKD* super superficiem circuli *ABGD* et angulus *LQE* est   105
inclinatio superficiei circuli *ZLT* super superficiem circuli *EZHT*: ergo
inclinatio circuli *BKD* super circulum *ABGD* est equalis inclinationi
circuli *ZLT* a circulo *EZHT*. Ergo inclinationes duorum circulorum
BKD ZLT super duos circulos *ABGD EZHT* sunt similes. Iam igitur
novimus quod non dicitur quod inclinatio superficiei super superficiem   110
sit similis inclinationi alterius superficiei super aliam superficiem nisi
cum linee que protrahuntur a communibus differentiis superficierum in
unaquaque duarum superficierum continent angulos equales. Et illud
est quod demonstrare voluimus.

21 Cum in spera fuerit circulus maior contingens aliquem circulorum
qui sunt in spera non maiorem et secuerit alium equidistantem illi ex
circulis, qui sunt inter centrum spere et circulum quem primus circulus
contingit, et fuerit etiam polus circuli maioris inter circulos equidistantes et signati fuerint circuli maiores contingentes maiorem circulorum   5
equidistantium, isti circuli erunt inclinati super circulum maiorem. Et
qui ex hiis circulis maioris erit altitudinis, erit circulus qui continget
super medium maioris duarum portionum illius circuli; et maioris
depressionis erit circulus qui continget super medium minoris duarum
portionum circuli. Reliquorum vero circulorum, quorum spacium loci   10
contactus ab una duarum medietatum duarum portionum, quecumque

---

103 KSA] *KFA* **PsVa**   *103* angulus] anguli **R**   *104 KSA*[1]] *KFA* **Ps**   *105 ABGD*] *AHGD*
R   *107* circuli … inclinationi] *primum bis false, deinde recte* **B**   *108* circuli] *om.* **Va**
108 a circulo] super circulum **Va**   *108 EZHT*] *corr. ex EZKHT* **P**   *109* sunt] *supra*
BPs   *109* igitur] ergo **FiZBPsVa**   *110* inclinatio] inclinationi **V**, inclusio **Va**
111 super aliam superficiem] *om.* **BPsVa**   *112–113* in … superficierum] *marg.* **R**
113 unaquaque] utraque **MKgOBPsVa**, unamque **Fi**   *113* superficierum] ficierum *add.*
Z   *113* continent] continet **Fi**   *113* continent angulos] *tr.* **MKgBPsVa**   *113* illud] hoc
OBPsVa   *114* demonstrare] *om.* **OBPsVa**   *114* voluimus] volumus **ZOBPsVa**   *1–*
2 contingens … secuerit] *om.* **Va**   *1* aliquem] *in corr.* **P**   *2* maiorem] maior est
Fi   *3* inter] intra **Kg**, *hic et saepius*   *3* centrum] centra **Fi**   *3* quem] quod **Fi**
4 continget] contigerit **OBPs**   *4* etiam] *supra* **M**, *om.* **KgOBPsVa**   *4* maioris] maiores **Fi**
5 circulorum] circulum **PsVa**   *5–6* circulorum equidistantium] *tr.* **O**   *7* circulus] *om.* **Fi**
7 continget] contingit **OFiVa**; circulus *add.* **Fi**   *8* illius] *supra* **M**   *8* illius circuli] *om.*
B   *8–11* illius … portionum] *marg.* **Ps**   *8–10* et … circuli] *marg.* **M**   *8–*
11 et … portionum] *marg.* **B**   *9* depressionis] depressioni **Fi**   *9* circulus] *in corr.* **B**
9 continget] contingit **Fi**

متساوياً فميله ميل متشابه والدائرة التى يكون موضع مماستها أبعد من وسط   10

القطعة العظمى تكون اكثر ميلاً من الدائرة التى موضع مماستها أقرب

وأقطاب الدوائر العظيمة أيضاً إنما تكون على دائرة واحدة موازية للدائرتين

اللتين ذكرنا هى أصغر من الدائرة التى تماسها الدائرة الأولى ·

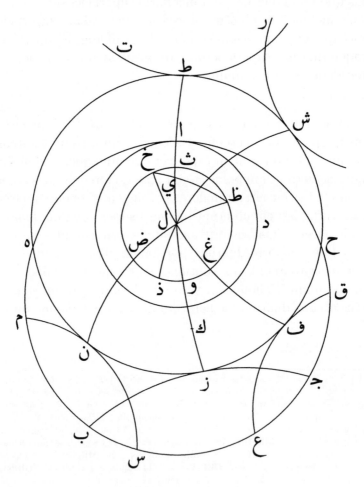

فلتماس فى كرة دائرة اب‌ج العظمى دائرة من الدوائر التى تكون فى

الكرة غير عظيمة وهى دائرة اد على نقطة آ ولتقطع دائرة أخرى موازية   15

om. N إنما] 12   supra A 12   أيضاً] AH والدوائر   والدائرة] 10 marg. A   مساوٍ [متشابه 10

N وسطح [ولتقطع 15   N آك [آد 15   add. infra A 15   و [هى 13

fuerit, erit equale, inclinationes erunt similes. Circulus autem, cuius contactus locus fuerit magis remotus a medio portionis maioris, erit maioris inclinationis quam circulus cuius locus contactus fuerit vicinior. Poli quoque circulorum maiorum erunt super unum duorum cir-   15 culorum equidistantium quos prediximus qui erit minor circulo quem ipse primus contingit.

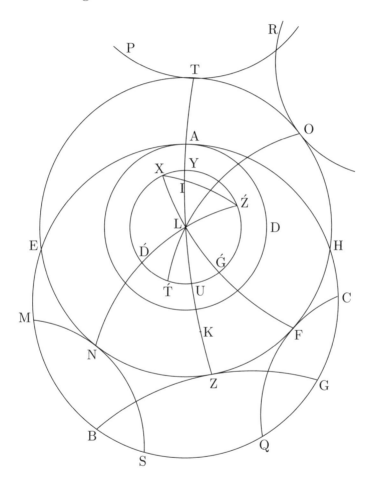

Circulus itaque ABG maior in spera contingat unum circulorum qui sunt in spera non maiorem, qui sit circulus AD, super punctum

12 erit equale] *tr.* **Fi**   12 equale] *in corr.* **Ps**; inclinationis *add.* **V**   12 erunt] *in corr.* **B**   12 cuius] *in corr.* **B**   13 locus] *om.* **Fi**   13 locus fuerit] *tr.* **KgOBPsVa** 13 fuerit] fuit **BPsVa** 13–14 magis … fuerit] *marg.* **M** 13 erit] *repet.* **Fi** 14 locus] *om.* **Fi** 14 locus contactus fuerit] *permut.* **KgOBPsVa**   14 fuerit] fuit **Va**   15 Poli quoque] polique **Fi** 15 erunt super] *supra* **BPs** 16 minor] *in corr.* **BPs** 16 quem] quera **Fi** 17 ipse] ipsi **PRVMKg**   17 primus] *supra* **B** (*add.* circulus), *om.* **PRVMKgPsVa**   17 contingit] *corr. ex* contingunt **Ps**, contingunt **PRVMKg**   18 *ABG*] *supra* **Kg**; *A, BG supra* **BPs**; *post* in spera **Z** 19 *AD*] *ND* **Va**

لهذه الدائرة من الدوائر التى فيما بين مركز الكرة وبين دائرة اد وهى دائرة

ه‍زح‍ط وليكن قطب دائرة اب‍ج فيما بين دائرتى اد ه‍زح‍ط ولترسم دوائر

عظيمة تماس دائرة ه‍زح‍ط التى هى أعظم الدائرتين المتوازيتين وهى دوائر

م‍ن‍س ب‍ز‍ج ع‍ف‍ق ت‍ط رش ولتكن دائرة ب‍ز‍ج مماسة لدائرة ه‍زح‍ط

20   فى موضع النصف من القطعة العظمى من قطعتى دائرة ه‍زح‍ط وهى قطعة

ه‍زح على نقطة ز ولتكن دائرة ت‍ط مماسة لها فى موضع النصف من

القطعة الصغرى منهما وهى قطعة ه‍ط‍ح على نقطة ط‍ وليكن بعد موضع

مماسة دائرتى م‍ن‍س ع‍ف‍ق لها من نقطة أحد النصفين أيهما كان بعداً

متساوياً وليكن ذلك كيف ما وقع ،

25   فأقول إن دوائر م‍ن‍س ب‍ز‍ج ع‍ف‍ق ت‍ط رش تكون مائلة على

دائرة اب‍ج وإن أكثرها ارتفاعاً دائرة ب‍ز‍ج وأكثرها انخفاضاً دائرة ت‍ط

وإن دائرتى م‍ن‍س ع‍ف‍ق يكون ميلهما متشابهاً وإن دائرة رش أميل على

دائرة اب‍ج من دائرة ع‍ف‍ق وإن أقطاب دوائر م‍ن‍س ب‍ز‍ج ع‍ف‍ق رش

ت‍ط تكون على دائرة واحدة موازية لدائرتى اد ه‍زح‍ط هى أصغر من

30   الدائرة التى تماسها دائرة اب‍ج الأولى فلنتعلم قطباً من قطبى دائرتى اد

ه‍زح‍ط المتوازيتين وليكن نقطة ل ولترسم دائرة عظيمة تمر بنقطتى ا‍ل وهى

A. illeg. [¹ا | NH داره 19 [ب‍ز‍ج] A add. et del. 18 [دوائر 19 داره | N داره [تماس 18 | N آاك [اد 17 | N آاك [اد 16

N منها [منها | A هرج 22 [ه‍رج | A هى 21 [ه‍زح | supra وهو [وهى, | A add. N 19 [ب‍ز‍ج] ² | N ح [ح 20 | وهى [وهى | ع add. N 19 [رش

deest [لها من نقطة ... من جميع 113–22 | N illeg. 25–22 [فأقول ... وليكن] repet. marg. A 113–23 [وليكن ... فأقول | illeg. N 22–25 [وهى 22

in N 26 [اب‍ج] H: marg. A 27 [ميلهما | A ملها 28 [ع‍ف‍ق] A marg. 29 [وهى | و

add. infra A 30 [فلنتعلم] H: ملعلم A 31 [بنقطتى] بعطى, marg. A بنقطتى

وبعد موضع مماسة دائرة رش لها من نقطة ز أبعد من موضع مماسة دائرتى م‍ن‍س ع‍ف‍ق لها [متساوياً 24
add. A

A, et secet alium circulum huic circulo equidistantem ex circulis qui   20
sunt inter centrum spere et circulum *AD*, qui sit circulus *EZHT*. Polus
quoque circuli *ABG* sit inter duos circulos *AD EZHT*. Signabo itaque
circulos maiores contingentes circulum *EZHT*, qui est maior circulorum equidistantium, sintque circuli *MNS BZG QFC PT RO*, et sit
circulus *BZG* contingens circulum *EZHT* in loco medii maioris dua-   25
rum portionum circuli *EZHT*, que est portio *EZH*, super punctum *Z*.
Et sit circulus *PT* contingens eum in loco medii portionis que ex eis est
minor, que est portio *ETH*, super punctum *T*. Et sit spacium loci, in
quo duo circuli *MNS QFC* contingunt ipsum, a puncto unius duarum
medietatum, quecumque fuerit, equale sitque illud quocumque modo   30
contingat.

Dico igitur quod circuli *MNS BZG QFC PT RO* sunt inclinati
super circulum *ABG* et quod circulus, qui ex eis est maioris altitudinis,
est circulus *BZG*, et maioris depressionis est circulus *PT*, et quod duo
circuli *MNS QFC* sunt similis inclinationis et quod circulus *RO* est   35
magis inclinatus super circulum *ABG* quam circulus *QFC* et quod
poli circulorum *MNS BZG QFC RO PT* sunt super unum circulum
equidistantem duobus circulis *AD EZHT*, qui est minor circulo quem
primus circulus *ABG* contingit. Signabo itaque unum duorum polorum
duorum circulorum *AD EZHT* equidistantium, qui sit punctum *L*, et   40
signabo circulum maiorem transeuntem per duo puncta *A* et *L*, qui

22 quoque] itaque **OBPsVa**   22 circuli] *om.* **Fi**   22 *ABG*] *ABGD* **Va**   22 circulos] *om.* **Va**
23 *EZHT*] *EZBT* **V**   23 est maior] *tr.* **BPsVa**   24 *BZG*] *GZB* **KgOBPsVa**   24 *RO*] *CO* **R**
25 *BZG*] *in corr.* **Ps**   25 circulum] *om.* **Z**   25–27 maioris … medii] *om.* **Fi**   26 *EZH*] *corr.*
ex EZHT **V**   27 loco medii] medio loco **MKgOBPsVa**   28 est] *om.* **Fi**   28 *ETH*] *in corr.*
Va   28 punctum *T*] *tr.* **O**   28 *T*] *supra* **B**, *marg.* **Ps**, *in corr.* **Va**   29 *QFC*] *QTE* **R**
30 quecumque] quocumque **Fi**   32 igitur] ergo **OBPs**   32 quod] *supra* **R**   32 *QFC*] *QFE*
R   32 *PT*] *PC* **Fi**   33 quod] *om.* **OZ**   33 circulus] *supra* **M**   33 eis] circulis *add.* **KgOB**
PsVa, *add. et del.* **M**   34 depressionis] *om.* **Kg**   34 quod] quia **KgBPs**, vel quod *add. su-*
pra **Ps**   35–36 sunt … *QFC*] *marg.* **M**, *om.* **Z**   35 quod] quia **BPs**, vel quod *add. supra* **Ps**
36 inclinatus] inclinationis **Fi**   36 quod] *om.* **OFiBPsVa**; duo *add.* **OBPsVa**   37 *QFC*] *QFO* **B**
38 equidistantem] *in corr.* **Ps**, equidistantes **Va**   38 *EZHT*] *EZH* **Fi**   39 primus circulus] *tr.*
Kg   40 duorum circulorum] *marg.* **BPs**   40 qui] quod **R**   40 punctum] punctus **OBPsVa**
41 *L*] *B* **Fi**

دائرة آل فلأن دائرتى ا ب ج ا د فى كرة وإحداهما مماسة للأخرى وقد

رسمت دائرة عظيمة تمر بقطب إحداهما وبموضع المماسة وهى دائرة آل

تكون دائرة آل مارة بقطبى دائرة ا ب ج أيضاً فتكون قائمة عليها على زوايا

قائمة فلتكن نقطة كـ قطباً لدائرة ا ب ج فدائرة آل إذا تممت تمر بنقطة كـ   35

أيضاً فلتمر ولتكن مثل دائرة ال كـ ودائرتا ا ب ج ه ز ح ط فى كرة وإحداهما

تقطع الأخرى وقد رسمت دائرة عظيمة تمر بأقطابهما وهى دائرة ال كـ

فدائرة ال كـ تقسم القطع التى فصلت من الدائرتين بنصفين نصفين وموضع

النصف من قطعة ه ز ح نقطة ز وموضع النصف من قطعة ه ط ح نقطة طـ

فدائرة ال كـ إذا تممت تمر أيضاً بنقطتى ز طـ فلتمر ولتكن مثل دائرة   40

طـ ال كـ ز ولأن نقطة كـ قطب دائرة ا ب ج ودائرة ا ب ج من الدوائر العظيمة

يكون الخط الذى يوتر قوس ا كـ ضلع المربع الذى يرسم فى الدائرة العظمى

فقوس ا كـ ز أعظم من القوس التى يوترها ضلع المربع الذى يرسم فى الدائرة

العظمى ولأن دائرة ه ز ح ط أصغر من الدائرة العظمى لأنها فيما بين مركز

الكرة وبين دائرة ا د وقطبها نقطة لَ تكون قوس لَ ز أصغر من القوس التى   45

يوترها ضلع المربع الذى يرسم فى الدائرة العظمى ولأن قوس ال ز أعظم من

القوس التى يوترها ضلع المربع الذى يرسم فى الدائرة العظمى وقوس لَ ز

أصغر من القوس التى يوترها ضلع المربع الذى يرسم فى الدائرة العظمى فإنّا

إذا فصلنا من قوس ال ز عند نقطة ز قوساً مساوية للقوس التى يوترها ضلع

35 كـ ¹ [*supra* A   43 التى [*corr. ex* الذى A   44–47 العظمى ... ولأن [*om.* H   44 العظمى [*om.* H

من عد [عند A   49 الز [أكر, *marg* الز A   47 يرسم [*om.* A H: 49 الز [ولأن ... العظمى [*repet.* A   46–48

A 49 التى [الذى, *supra* التى A   49 التى [*in corr.* A   49 للقوس [A

sit circulus AL. Et quia duorum circulorum ABG AD in spera unus
alterum contingit et iam signatus est circulus maior transiens per
polum unius eorum et per locum contactus eorum, qui est circulus AL,
ergo circulus AL transit etiam per polos circuli ABG et est super eum   45
orthogonaliter erectus. Sit itaque punctum K polus circuli ABG: ergo
cum circulus AL complebitur transibit etiam per punctum K. Compleatur igitur et sit sicut circulus ALK. Duorum autem circulorum
ABG $EZHT$ in spera unus alterum secat, et iam signatus fuit circulus
maior transiens per polos eorum, qui est circulus ALK. Circulus ergo   50
ALK secat portiones quas ex duobus separat circulis in duo media
et duo media; locus vero medietatis portionis EZH est punctum Z et
locus medii portionis ETH est punctum T. Ergo circulus ALK, cum
complebitur, transibit etiam per duo puncta Z et T. Compleatur ergo
et sit sicut circulus $TALKZ$. Et quia punctum K est polus circuli   55
ABG et circulus ABG est ex circulis maioribus, ergo linea que subtenditur arcui AK est latus quadrati quod signatur in circulo maiore.
Arcus igitur AKZ est maior arcu cui subtenditur latus quadrati quod
signatur in circulo maiore. Et quia circulus $EZHT$ est minor circulo
maiore, quoniam ipse est inter centrum spere et circulum AD et polus   60
eius est punctum L, ergo arcus LZ est minor arcu cui subtenditur latus
quadrati quod in maiore circulo signatur. Et quia arcus ALZ est maior
arcu cui subtenditur latus quadrati quod signatur in circulo maiore et
arcus LZ est minor arcu cui subtenditur latus quadrati quod in maiore
circulo signatur, ergo cum nos separaverimus ex arcu ALZ a puncto   65

---

42 unus] unius **B**; unus, *corr. ex* unius **Ps**   43 iam signatus est] *permut.* **FiZ**   44 polum] polos
OFiZ   44 est] *om.* **Va**   45 ergo circulus AL] *om.* **R**   45 et] qui **FiZ**   45 est] ita *add.* **OBPs**
45 eum] est **Fi**   46 polus] polos **Fi**   47 cum] *supra* **M**, *post* AL **OVa**, *om.* **BPs**   47 AL] cum
add. supra **Ps**   48 sicut] *om.* **FiZ**   48 ALK] *in corr.* **Kg**   49 unus] unius **BPsVa**   49 fuit] *in
corr.* **Ps**   49–50 circulus ... ALK] *marg.* **BPs**   50 circulus] circulis **Fi**   51 separat circulis] *tr.*
PsVa   52 portionis] qui est *add.* **Va**   52 et] *om.* **Fi**   53 portionis] qui est *add.* **Va**
53 ALK] AKL **Kg**   54 ergo] circulus *add.* **OBPsVa**   55 sicut] *om.* **OBPsVa**   55 $TALKZ$] *in
corr.* **B**   55 polus] *in corr.* **BPs**   56 et circulus ABG] *primum false, deinde recte* **FiZ**, *om.*
Va   57 signatur] significatur **Ps**, *hic et saepius*   58–59 Arcus ... maiore] *om.* **OZBPsVa**   58–
60 Arcus ... maiore] *om.* **Fi**   58 AKZ] $AKHZ$ **Kg**   58 arcu] *om.* **Kg**   61 L] *om.* **Kg**,
supra **B**   61 arcu cui] arcui **Fi**   62 maiore circulo] *tr.* **OBPsVa**   62 signatur] *post* quod
OBPsVa   62 quia] *supra* **B**, quod **Fi**   62 maior] *in corr.* **B**   63 signatur] *post* maiore **OB**
PsVa; signatur ... maiore *permut.* **Fi**   64–65 in maiore circulo signatur] *permut.* **RMKgOBPsVa**
65 separaverimus] speramus **Fi**, separavimus **B**, separemus **Va**

المربع الذى يرسم فى الدائرة العظمى وقع طرفها الآخر فيما بين نقطتى اٰ لٰ

فلنفصل قوساً مساوية للقوس التى ذكرنا وهى قوس ثٰ زٰ ولنرسم على قطب

لٰ وببعد لٰ ثٰ دائرة ثٰ خٰ ذٰ ظٰ فهى موازية لدائرتى اٰ دٰ هٰ زٰ حٰ طٰ ولتريم دوائر

عظيمة تمر كل واحدة منها بنقطة لٰ وبكل واحدة من نقط نٰ فٰ شٰ وهى

دوائر نٰ لٰ ظٰ فٰ لٰ خٰ ذٰ لٰ شٰ ولأن قوس نٰ لٰ مساوية لقوس لٰ زٰ وذلك أنهما

أخرجتا من قطب دائرة هٰ زٰ حٰ طٰ إلى الخط المحيط بها وقوس لٰ ثٰ مساوية

لقوس لٰ ظٰ وذلك أنهما أخرجتا من قطب دائرة ذٰ ظٰ خٰ إلى الخط المحيط بها

تكون جميع قوس نٰ لٰ ظٰ مساوية لجميع قوس زٰ لٰ ثٰ وقوس زٰ لٰ ثٰ مساوية

للقوس التى يوترها ضلع المربع الذى يرسم فى الدائرة العظمى فقوس نٰ لٰ ظٰ

مساوية للقوس التى يوترها ضلع المربع الذى يرسم فى الدائرة العظمى

وكذلك أيضاً نبين أن كل واحدة من قسى خٰ لٰ فٰ ولٰ طٰ ذٰ لٰ شٰ مساوية

للقوس التى يوترها ضلع المربع الذى يرسم فى الدائرة العظمى ولأن دائرتى

مٰ نٰ سٰ هٰ زٰ حٰ طٰ فى كرة وإحداهما تماس الأخرى وقد رسمت دائرة عظيمة

تمر بقطبى إحداهما وبموضع المماسة وهى دائرة نٰ لٰ ظٰ صارت دائرة نٰ لٰ ظٰ

تمر أيضاً بقطبى دائرة مٰ نٰ سٰ وتكون قائمة عليها على زوايا قائمة ولأن دائرة

مٰ نٰ سٰ عظيمة تكون القوس التى خرجت من قطبها إلى الخط المحيط بها

مساوية للقوس التى يوترها ضلع المربع الذى يرسم فى الدائرة العظمى

50

55

60

65

حرحنا [أخرجتا 55 ندص [نٰ لٰ ظٰ 54 مهما [منها 53 supra A للقوس [51 A وقع [50 A وعت :H

لوس [للقوس 58 illeg. A ذٰ ظٰ خٰ [56 marg. A وقوس لٰ ثٰ ... بها [55-56 H אכרונא A, A

A العظمى ... وكذلك [repet. sine 60-61 A لوس [للقوس A 59 الذى [التى A 58

A العظمى الدائرة in corr. A نٰ لٰ ظٰ [63 in corr. A نلض [1 A 66 للقوس A, om. H 61

طلط [ولٰ طٰ 60

Z arcum equalem arcui cui subenditur latus quadrati quod in maiore
circulo signatur, cadet altera eius extremitas inter duo puncta *A* et *L*.
Separetur ergo arcus equalis illi quem prediximus, qui sit arcus *ZY*.
Signabo autem super polum *L* secundum longitudinem *LY* circulum
YXT́ et qui est equidistans duobus circulis *AD EZHT*, et signabo   70
circulos maiores quorum quisque transeat per punctum *L* et per unum
punctorum *N F O*, qui sint circuli *NLŹ FLX OLT́*. Et quia arcus *NL*
est equalis arcui *LZ* – quod ideo est quoniam ipsi producuntur a polo
circuli *EZHT* ad lineam ipsum continentem – et arcus *LY* est equalis
arcui *LŹ* – et hoc ideo quoniam ipsi producuntur a polo circuli *T́Ź*   75
ad lineam continentem ipsum – ergo totus arcus *NLŹ* est equalis toti
arcui *ZLY*; arcus vero *ZLY* est equalis arcui cui latus quadrati subtenditur quod in circulo maiore signatur. Ergo arcus *NLŹ* est equalis arcui
cui subtenditur latus quadrati quod in maiore signatur circulo. Simili
quoque modo monstratur quod unusquisque arcuum *XLF T́LO* est   80
equalis arcui cui subtenditur latus quadrati quod signatur in circulo
maiore. Et quia duorum circulorum *MNS EZHT* in spera unus alterum contingit et iam signatus est circulus maior transiens per polos
unius eorum et per locum contactus eorum, qui est circulus *NLŹ*, fit
circulus *NLŹ* transiens etiam per polos circuli *MNS* et est super eum   85
orthogonaliter erectus. Et quia circulus *MNS* est maior, ergo arcus qui
producitur a polo eius ad lineam continentem ipsum est equalis arcui
cui subtenditur latus quadrati quod in circulo maiore signatur. Arcus

67 cadet] *supra* **BPs**   67 *L*] *B* **Fi**   68 Separetur] Separentur **V**   68 arcus[1]] *om.* **BPsVa**
68 illi] isti **BPsVa**   68 *ZY*] *in corr.* **RB**   69 Signabo] Significabo **Va**   69 autem] *om.*
V   69 polum] punctum **Z**   69 *L*] *supra* **BPs**; *Z* **R**   69 secundum] *supra* **BPs**
70 et qui est] *om.* **M**   70 et ... equidistans] equidistantem **KgOBPsVa**   70 qui] *om.* **Z**
70 circulis] *om.* **Fi**   71 unum] *in corr.* **P**   72 sint] sunt **BPs**   73 *LZ*] *in corr.* **R**,
LKZ **OVa**; *LZ*, *K supra* **BPs**   73 ipsi] *om.* **OBPsVa**   73 producuntur] producitur **OBPsVa**
74 ipsum] *in corr.* **Ps**, ipsam **Va**   74 *LY*] *in corr.* **B**, *LI* **Fi**   75 ideo] *supra* **Kg**; est
add. **OBPsVa**   75 ipsi] *om.* **OBPsVa**   75 producuntur] producitur **OBPsVa**   75 *T́Ź*] *T́ŹX*
O, *T́Ź*, *supra X* **Ps**, *T́ŹYX* **Va**   76 continentem] contingentem **Z**   76 ipsum] *in corr.*
Ps   76 totus] *om.* **OBPsVa**   76 toti] *om.* **BPs**   77–78 *ZLY[1]* ... arcui] *marg.* **BPs**
77 arcus vero] sed arcus **OBPs**   77 arcus vero *ZLY*] *om.* **Va**   77 equalis] toti *add.* **Z**
77–78 latus quadrati subtenditur] *permut.* **VMKgOBPsVa**   78 circulo maiore] *tr.* **MKgOB**
PsVa   78 signatur] significatur **Va**, *hic et saepius*   79 signatur circulo] *tr.* **KgOBPsVa**
80 quoque] *om.* **BPsVa**   80 *XLF*] *ZLF* **V**   80 *T́LO*] *in corr.* **P**, *T́O* **Kg**; *T́O*, *L supra* **BPs**
82 *EZHT*] *HZ add. supra* **B**, *in corr.* **Ps**, *EHTZ* **Va**   82 in] *om.* **Fi**   82 unus] unius **B**, *in*
corr. **Ps**   82–83 alterum] alteram **Fi**   84 unius] *om.* **O**, *supra* **BPs**   84 fit] sit **VMKgFiZ**
84–85 fit circulus *NLŹ*] *om.* **RVa**   86 est maior] *tr.* **O**   87 lineam] *supra* **BPs**   88 cui] *om.*
M   88 circulo maiore] *tr.* **OBPsVa**

وقوس نلظ مساوية للقوس التى يوترها ضلع المربع الذى يرسم فى دائرة

عظيمة فالخط الذى يخرج من نقطة ن إلى نقطة ظ هو مثل الخط الذى

يخرج من الخط المحيط بدائرة منس إلى قطبها فنقطة ظ قطب دائرة منس

وكذلك نبين أن نقطة ث أيضاً قطب دائرة بزج وأن نقطة خ قطب دائرة　70

عفق وأن نقطة ذ قطب دائرة رش وأن نقطة و قطب دائرة تط فأقطاب

دوائر منس بزج عفق رش تط على دائرة ثخذ الموازية لدائرتى آد

هزحط التى هى أصغر من دائرة آد ،

فأقول إن دوائر منس بزج عفق رش تط مائلة على دائرة

ابج وإن أكثرها ارتفاعاً دائرة بزج وأكثرها انخفاضاً دائرة طت وإن　75

دائرتى منس عفق متشابهتا الميل وإن دائرة رش أكثر ميلاً على دائرة

ابج من ميل دائرة عفق عليها ولأن قوس نز مساوية لقوس فز وهما

من دائرة واحدة بعينها تكون قوس نز شبيهة بقوس فز ولكن قوس نز

شبيهة بقوس وض وقوس زف شبيهة بقوس وغ فقوس وض شبيهة بقوس

وغ وهما من دائرة واحدة بعينها فقوس وض مساوية لقوس وغ ولكن　80

قوس وض مساوية لقوس ثظ وذلك أنها مقابلة لها فيما بين قوسين من

دائرتين عظيمتين تمران بقطبهما وقوس وغ مساوية لقوس ثخ فقوس

ثظ مساوية لقوس ثخ فقد عمل فى دائرة ثخ ذ ظ على قطر ث و

לקום H [للقوس 67　مساوי H [مساوية 67　*om.* A [وقوس ... عظيمة 67-68

69 منس ... قطبها [إلى A [الموازية]H:　خذ A 72 [ثخذ　*marg.* A 72 [دوائر]الدوائر A 71-72

corr. سمة [شيبة]ها *supra* A 78　المواربه A 74 [فأقول]واقول H:　الدوائر A 75 [أكثرها　H: دوائر A 74 [دوائر

ex سمبا A 78 [ولكن]ولكن A 80 [وهما]وهي A 81 *supra* هما A [لها]بقطبيهما بקמביהמא *supra* A 82

H 82 [وغ]*illeg.* A 82 [فقوس]وقوس A 83-84 [قطعة¹ ... فى دائرة]*marg.* A

autem $NL\acute{Z}$ est equalis arcui cui subtenditur latus quadrati quod in
circulo maiore signatur, et linea que protrahitur a puncto N ad punc-   90
tum $\acute{Z}$ transit a linea continente circulum MNS ad polum eius. Ergo
punctum $\acute{Z}$ est polus circuli MNS. Similiter quoque monstratur quod
punctum Y est polus circuli BZG et quod punctum X est polus circuli
QFC et quod punctum $\acute{T}$ est polus circuli RO et quod punctum U est
polus circuli PT. Ergo poli circulorum MNS BZG QFC RO PT sunt   95
super circulum $YX\acute{T}$ equidistantem duobus circulis AD $EZHT$, qui est
minor circulo AD.

Et dico quod circuli MNS BZG QFC RO PT sunt inclinati super
circulum ABG, et quod ille qui est ex eis maioris altitudinis est circulus
BZG et maioris depressionis est circulus TP, et quod duo circuli MNS 100
QFC sunt similis inclinationis et quod circulus RO est maioris inclinationis super circulum ABG quam sit inclinatio circuli QFC super
eum. Et quia arcus NZ est equalis arcui FZ – ipsi sunt unius et eiusdem circuli– ergo arcus NZ est similis arcui FZ. Arcus autem NZ est
similis arcui $U\acute{D}$ et arcus FZ est similis arcui $U\acute{G}$: ergo arcus $U\acute{D}$ est   105
similis arcui $U\acute{G}$. Ipsi autem sunt unius et eiusdem circuli: ergo arcus
$U\acute{D}$ est equalis arcui $U\acute{G}$. Sed arcus $U\acute{D}$ est equalis arcui $Y\acute{Z}$. Quod
ideo est quoniam ipse opponitur ei inter duos arcus duorum circulorum
maiorum qui transeunt per polos eius, et arcus $U\acute{G}$ est equalis arcui
YX: ergo arcus $Y\acute{Z}$ est equalis arcui YX. In circulo igitur $YX\acute{T}\acute{Z}$ super   110

89 NLŹ] *NL,* *Ź* *supra* **Ps**   *89* cui] *om.* **Fi**   *90* circulo maiore] *tr.* **MKgOBPsVa**
91 transit] fuerit **Fi**   *91* continente] continentem **M**, contingente **BPsVa**; ipsum *add.* **FiZ**
92 Similiter quoque] Simili quoque **Fi**, Simili quoque modo **Z**, Similiterque **Ps**   *93* *Y*] *om.*
Kg, *supra* **B**   *93* quod] quia **B**; quia, vel quod *supra* **Ps**   *93* *X*] *Z* **Va**   *93* circuli[2]] *supra*
BPs   *94* quod[2]] quia **B**; quia, vel quod *supra* **Ps**   *94* *U*] *in corr.* **B**   *95* *PT*[1]] *in corr.*
R   *95* *QFC*] *QFT* **Z**, *QFO* **Va**   *95* *RO* *PT*] *tr.* **Fi**   *96* *YXŤ*] *in corr.* **B**, *YŹŤ* **VKgOPs**
96 qui] *in corr.* **Ps**   *98* circuli] circulo **Fi**   *98* *QFC*] *QFT* **Fi**   *98* *RO*] *TO* **Z**   *99* est[1]] *post*
eis **VOBPsVa**   *99* est circulus] *tr.* **Va**   *100* *BZG*] *AZG* **Z**   *100* depressionis] dispo posicionis **Fi**
100 circulus *TP*] *tr.* **Z**   *100* quod] quia **B**; quia, vel quod *supra* **Ps**   *101* quod] quia **B**; quia,
vel quod *supra* **Ps**   *101* *RO*] *TO* **Z**   *102* circulum] *om.* **BPsVa**   *102* circulum *ABG*] *tr.* **O**
102 inclinatio] in circulo **Fi**   *102* *QFC*] *QFT* **Fi**   *103* ipsi] et ipsi **VKgZVa**   *103* et] et *supra*
R   *104* *NZ*[1]] *in corr.* **B**, *corr. marg.* **Ps**   *104* Arcus autem] Sed arcus **P**   *105* *UD́*[1]] *in corr.*
Ps   *105* *UD́*[2]] *in corr.* **Ps**   *106* *UǴ*] *UFǴ* **B**, *in corr.* **Ps**   *106–107* Ipsi ... arcui *UǴ*] *om.*
BPsVa   *110* *YX*[1]] *in corr.* **B**, *YQC* **Kg**, *ZYI* **Z**   *110* est ... *YX*[2]] *om.* **Va**   *110* *YX*[2]] *ZX*
Kg   *110* *YXŤŹ*] *YXŤ* **KgOBPs**

قطعة من دائرة قائمة عليها على زوايا قائمة وهى قطعة و كـز وما يتصل بهذه

القطعة وفصل منها قوس أصغر من نصف جميع القطعة وهى قوس 85

وكـ وفصل من الدائرة الأولى قوسان متساويتان وهما قوسا ثـخ ثـظ

فالخط المستقيم الذى يصل بين نقطة كـ ونقطة خـ مساوٍ للخط المستقيم الذى

يصل بين نقطة كـ ونقطة ظـ فالدائرة التى ترسم على قطب كـ وببعد كـخ

تمر بنقطة ظـ أيضاً فلتمر ولتكن مثل دائرة خـظ فدائرة خـظ موازية لدائرة

ا بـجـ وذلك أنهما على أقطاب بأعيانها وذلك أن نقطة كـ قطب دائرة ا بـجـ 90

ولأن دائرة خـظ موازية لدائرة ا بـجـ يكون العمود الذى يخرج من نقطة خـ

إلى سطح دائرة ا بـجـ مساوياً للعمود الذى يخرج من نقطة ظـ إلى سطح

دائرة ا بـجـ وكذلك أيضاً يكون مساوياً للعمود الذى يخرج من نقطة ىـ إلى

سطح دائرة ا بـجـ والعمود الذى يخرج من نقطة ىـ إلى سطح دائرة ا بـجـ

أطول من العمود الذى يخرج من نقطة ثـ إلى سطح دائرة ا بـجـ والعمود 95

الذى يخرج من نقطة ظـ إلى سطح دائرة ا بـجـ أطول من العمود الذى

يخرج من نقطة ثـ إلى سطح دائرة ا بـجـ وكذلك أيضاً العمود الخارج من

نقطة خـ وذلك أن كل واحد منهما مساوٍ للعمود الذى يخرج من نقطة ىـ

فنقطة ظـ أعلى من نقطة ثـ ونقطة ظـ هى قطب دائرة مـنـس ونقطة ثـ

قطب دائرة بـزجـ فقطب دائرة مـنـس أعلى من قطب دائرة بـزجـ 100

والدوائر التى تكون أقطابها أعلى هى أكثر ميلاً على السطوح التى هى عليها

---

86 وكـ [*in corr.* A　86 الدائرة [*corr. ex* داره A　87 للخط [الخط A　92 للعمود [العمود A　95 ثـ] *corr.*

ex ـ A　95 والعمود [فالعمود A :H　96-97 أطول ... ا بـجـ] *marg.* A　99 ثـ [2] *supra* A

diametrum *YU* iam facta est portio circuli erecta super eam orthogonaliter que est portio *UKZ* et quod cum hac portione iunctum est.
Et separatus est ex ea arcus minor medietate totius portionis, qui est
arcus *UK*, et separavit ex circulo primo duos arcus equales, qui sunt
duo arcus *YX YŹ*. Linea igitur recta que coniungit inter punctum *K*   115
et punctum *X* est equalis linee recte que coniungit inter punctum *K*
et punctum *Ź*. Circulus igitur qui signatur super polum *K* secundum
longitudinem *KX* transit etiam per punctum *Ź*. Transeat ergo et sit
sicut circulus *XŹ*: ergo circulus *XŹ* equidistat circulo *ABG*. Quod ideo
est quoniam sunt super unum et eundem polum, et hoc ideo quoniam   120
punctum *K* est polus circuli *ABG*. Et quia circulus *XŹ* equidistat circulo *ABG*, ergo perpendicularis que protrahitur a puncto *X* ad superficiem circuli *ABG* est equalis perpendiculari que protrahitur a puncto
Ź ad superficiem circuli *ABG*; et propter hoc etiam est equalis perpendiculari que protrahitur a puncto *I* ad superficiem circuli *ABG*. Sed   125
est longior perpendiculari que protrahitur a puncto *Y* ad superficiem
circuli *ABG*; et perpendicularis que producitur a puncto *Ź* ad superficiem circuli *ABG* est longior perpendiculari que protrahitur a puncto
Y ad superficiem circuli *ABG*, et similiter etiam perpendicularis que
producitur a puncto *X*. Quod ideo est quoniam unaqueque earum est   130
equalis perpendiculari que protrahitur a puncto *I*. Punctum igitur *Ź*
altius est puncto *Y*; sed punctum *Ź* est polus circuli *MNS* et punctum
Y est polus circuli *BZG*: ergo polus circuli *MNS* est altior polo circuli
BZG. Circuli vero quorum poli sunt altiores sunt maioris inclinatio-

111 portio] proportio **Z**, *hic et saepius*   112 et quod] *om.* **Fi**   112 cum] *in corr.* **B**
113 separatus est] *in corr.* **RPs**   113 ea] eo **OBPsVa**   113 minor] minorum **V**   114 *UK*] *NB*
R, *LUK* **Va**   114 circulo] *supra* **Va**   115 duo arcus] *om.* **MKgBPsVa**   115 *YX YŹ*] *in corr.*
B   115 recta] *om.* **Fi**   115 que coniungit] *om.* **BPsVa**   116 *X*] *Y* **V**   117 *Ź*] et punctum *Ď add.* **V**   117 igitur] ergo **FiZ**   117 secundum] *in corr.* **Ps**   118 *KX*] *KY* **V**, *om.*
Z   118 sit] *om.* **Va**   119 sicut] *om.* **ROBPsVa**   119 ergo circulus *XZ*] *add. infra* **B**, *su-*
pra **Ps**   119 *XŹ*] *om.* **Va**   120 est] *om.* **B**   120 sunt] *post* polum **OVa**, *add. supra post*
sunt **BPs**   120 et[1]] *idem add.* **R**   120 ideo] est *add.* **OBPsVa**   120 quoniam] quia **M**
121 quia] quoniam **OBPsVa**   121 *XŹ*] *in corr.* **B**, *XY* **V**, *AZ* **Kg**   121 equidistat] equidistant
Fi   123 protrahitur] protrahuntur **Fi**   124 circuli *ABG*] *tr.* **BPsVa**   124 etiam] *om.* **Fi**
124 est] eis *add.* **PRVMPsVa**   124 equalis] etiam *add.* **Fi**   124 equalis perpendiculari] *tr.*
PRMBVa   124–125 perpendiculari] *in corr.* **PsVa**, perpendicularis **PRVMKgB**   125 *I*] *in corr.*
B   125 Sed] ergo **FiZ**   125–127 Sed … *ABG*] *marg.* **P**, *om.* **OBPsVa**   126 est longior] *tr.*
FiZ   126 *Y*] *X* **VKg**, *I* **PROBPsVa**   127 et] ergo **KgOBPsVa**   127 producitur] protrahitur
MKgOBPs   127–128 *Ź* … a puncto] *om.* **Va**   128 circuli] *om.* **BPs**   128 *ABG*] *AB, G su-*
pra **B**   128 protrahitur] producitur **MKgOBPs**   129 etiam] est **OBPsVa**   130 est] *om.* **M**
131 perpendiculari] perpendicularis **Kg**   131 *I*] *in corr.* **O**, *YI* **Z**   131 Punctum] Punctus
Z   131 *Ź*] *L add.* **Kg**   132 altius est] *tr.* **MKgOBPsVa**   133 circuli[1]] *supra* **M**, *om.* **BPsVa**
133 circuli *BZG*] *tr.* **O**

فدائرة م‌ن‌س أكثر ميلاً على دائرة ا‌ب‌ج من دائرة ب‌ز‌ج فدائرة ب‌ز‌ج

أكثر ارتفاعاً من دائرة م‌ن‌س وكذلك نبين أيضاً أن دائرة ب‌ز‌ج أكثر ارتفاعاً

من جميع الدوائر التى تماس دائرة ه‌زح‌ط فدائرة ب‌ز‌ج أكثر ارتفاعاً من

جميع هذه الدوائر ،

وأقول إن دائرة ت‌ط أكثر انخفاضاً من جميعها وذلك أن العمود الذى

يخرج من نقطة و إلى سطح دائرة ا‌ب‌ج أطول من العمود الذى يخرج من

نقطة ذ إلى سطح دائرة ا‌ب‌ج فنقطة و أرفع من نقطة ذ ونقطة و قطب

دائرة ت‌ط ونقطة ذ قطب دائرة ر‌ش فقطب دائرة ت‌ط أرفع من قطب

دائرة ر‌ش والدوائر التى تكون أقطابها أعلى هى أكثر ميلاً على السطوح

التى هى عليها فدائرة ت‌ط أكثر ميلاً على دائرة ا‌ب‌ج من دائرة ر‌ش فدائرة

ت‌ط أخفض من دائرة ر‌ش وكذلك نبين أنها أيضاً أخفض من جميع الدوائر

التى تماس دائرة ه‌زح‌ط فدائرة ت‌ط أخفض من جميع هذه الدوائر ولأن

العمود الذى يخرج من نقطة ظ إلى سطح دائرة ا‌ب‌ج مساوٍ للعمود الذى

يخرج من نقطة خ إلى سطح دائرة ا‌ب‌ج يكون بعد نقطتى ظ خ من

سطح دائرة ا‌ب‌ج بعداً متساوياً ونقطة ظ قطب دائرة م‌ن‌س ونقطة خ

قطب دائرة ع‌ف‌ق فبعد أقطاب دائرتى م‌ن‌س ع‌ف‌ق من سطح دائرة

ا‌ب‌ج بعد متساوٍ والدوائر التى بعد أقطابها من السطوح التى هى مائلة

A الداره [دائرة 107   *om.* H [الدوائر ... جميع 104-105   الجميع [*corr. ex* جميع 104

دار [دائرة 113   A مدار [فدائرة 111   فى [*corr. ex* هى 111   A سطوح [السطوح 110

مساواً [مساوٍ 114   *om.* A [دائرة 114   A عمود [العمود 114   *incip. iterum* N [هذه 113   A *corr. ex*

A يكون [ا‌ب‌ج 116   A (*supra*) خ ث [ظ خ 115   (فان : يكون) A *marg.* [للعمود 114   بعد ... بعد 114-115 A

supra A 118   مائلة [قائمة ANH, مائلة *supra* N   A بعدا متساو [بعد متساوٍ 118

105

110

115

nis super superficies super quas sunt: ergo circulus *MNS* est maioris   135
inclinationis super circulum *ABG* quam circulus *BZG*. Ergo circulus
BZG est maioris altitudinis quam circulus *MNS*. Et similiter ostenditur quod circulus *BZG* est maioris altitudinis quam omnes circuli qui
contingunt circulum *EZHT*. Ergo circulus *BZG* est maioris altitudinis
quam omnes hii circuli.   140

Et dico quod circulus *PT* est maioris depressionis quam omnes
hii. Quod ideo est quoniam perpendicularis que producitur a puncto
U ad superficiem circuli *ABG* est longior perpendiculari que producitur a puncto *T́* ad superficiem circuli *ABG*; ergo punctum *U* est
altius puncto *T́*. Punctum autem *U* est polus circuli *PT* et punctum   145
T́ est polus circuli *RO*. Ergo polus circuli *PT* est altior polo circuli
RO. Circuli vero quorum poli sunt altiores sunt maioris inclinationis
super superficies super quas sunt. Ergo circulus *PT* est maioris inclinationis super circulum *ABG* quam circulus *RO*. Ergo circulus *PT*
est magis depressus circulo *RO*, et similiter etiam ostenditur quod est   150
magis depressus quam omnes circuli qui contingunt circulum *EZHT*.
Ergo circulus *PT* est magis depressus omnibus istis circulis. Et quia
perpendicularis que producitur a puncto *Ź* ad superficiem circuli *ABG*
est equalis perpendiculari que protrahitur a puncto *X* ad superficiem
circuli *ABG*, ergo spacium duorum punctorum *Ź X* a superficie circuli   155
ABG est spacium equale. Punctum autem *Ź* est polus circuli *MNS*
et punctum *X* est polus circuli *QFC*: ergo spacium polorum duorum
circulorum *MNS QFC* a superficie circuli *ABG* est spacium equale.
Circulorum vero quorum spacium polorum a superficiebus super quas
sunt erecti est spacium equale, inclinationes sunt equales. Et etiam   160

136 *BZG*] *BGZ* **OBPsVa**   137 circulus] *om.* **Kg**   137 *MNS*] *in corr.* **O**   141 omnes] omnis
Fi   142 hii] circuli *add.* **OBPsVa**   142 quoniam] quia **BPsVa**   143 *U*] *supra* **BPs**
143 circuli] *supra* **RBPs**   145 altius] altior **Fi**   145–146 Punctum … *T́*] *om.*
R   146 polo] *in corr.* **BPs**   150–151 circulo … depressus] *om.* **KgBVa**   150–
152 circulo … depressus] *om.* **Ps**   150 et] *om.* **O**   151 circulum] *om.* **BPsVa**
151 *EZHT*] *ERHT* **Fi**   152 Ergo … circulis] *om.* **Fi**   152 est magis] *tr.* **PRVOFi**;
magis, *supra* est **B**   152 depressus] quam *add.* **Va**   153 producitur] protrahitur **MKgOPsVa**
153 *ABG*] *in corr.* **Kg**   155 punctorum] *in corr.* **O**, *corr. in* polorum **BPs**, polorum **Va**
155 *Ź*] et *add.* **Kg**   155 a superficie] ad superficies **FiZ**   156–160 Punctum … equale] *om.*
PsVa   156–160 polus … equales] *marg.* **B**   157 punctum] punctus **Z**   157 *X*] *Y* **V**
157 polorum duorum] *tr.* **OFiZ**   160 spacium] *om.* **OFiZB**   160 inclinationes] ergo *add.*
ante **PsVa**; duorum circulorum *MNS QSC add. marg.* **Ps**, *in textu* **Va**   160 equales] ergo
inclinationes circulorum *MNS QFC* sunt *add. marg.*, equales *in textu* **B**

عليها بعد متساوٍ فميلها ميل متشابه وأيضاً فإن العمود الذى يخرج من نقطة

ذ إلى سطح دائرة ا ب ج لما كان أعظم من العمود الذى يخرج من نقطة خ   120

إلى سطح دائرة ا ب ج صارت نقطة ذ أعلى من نقطة خ ونقطة ذ قطب

دائرة رش ونقطة خ قطب دائرة ع ف ق فقطب دائرة رش أعلى من قطب

دائرة ع ف ق والدوائر التى تكون أقطابها أعلى هى أكثر ميلاً على السطوح

التى هى عليها فدائرة رش أكثر ميلاً على دائرة ا ب ج من دائرة ع ف ق

فدوائر م ن س ب ز ج ع ف ق رش ت ط مائلة على دائرة ا ب ج وأكثرها   125

ارتفاعاً دائرة ب ز ج وأخفضها دائرة ت ط ودائرتا م ن س ع ف ق متشابهتا

الميل ودائرة رش أكثر ميلاً على دائرة ا ب ج من دائرة ع ف ق وأيضاً فإن

أقطابها على دائرة واحدة من الدوائر المتوازية هى أصغر من دائرة ا د ،

وذلك ما أردنا أن نبين ·

إذا كانت هذه الأشياء بأعيانها على ما وصفنا وكانت القسى التى تخرج فيما

بين مواضع العقدة أعنى فيما بين مواضع مماسة الدوائر وبين قطعها للدائرة

الأولى متساوية فإن الدوائر العظيمة التى تقدم ذكرها متشابهة الميل ·

ומילהא N, نكون ملها ميلا مساها] A (marg.), فميلها ميل متساوٍ] A فميلها ميل متشابه 119 بعدا] A بعد 119

marg. [لما كان ... ا ب ج 120–121   ذ] A, supra خ, ص N in hac prop.   120–121 ميل מתשאבהה H   120 ذ] N

infra A   121 خ] ونقطه supra A   121–122 رش ... ونقطة] marg. infra A   122 رش] ۱ رس ع N

hic et saepius زش ع] رش ۲ [122   marg. illeg. A فقطب ... ع ف ق 122–123   A .om [خ 122

N تكون] om. A   123 هى] supra A هى] 124 من, supra فهى A, כם H   123 من] om. A   123 مائلة] om.

N مساويا, supra مساها A مساها, مساها] A ودارة 126   متشابهتا] 126 ودائرتا N الى [على 126   125 N

مكاب وكانت] AN إذا ۱ [1 إذا H: وإذا 1 add. supra A المتوازية] التى 128 A الداره corr. ex الداره A 128 الدوائر 128

A لداره [للدائرة ۲ N

---

add. A فميل دائرتا (sic) م ن س ع ف ق منهما على دائرة ا ب ج متساوٍ ... (illeg.) متشابه] 119

quia perpendicularis que producitur a puncto $\acute{T}$ ad superficiem circuli
ABG est maior perpendiculari que protrahitur a puncto X ad superficiem circuli ABG, fit punctum $\acute{T}$ altius puncto X. Sed punctum $\acute{T}$
est polus circuli RO et punctum X est polus circuli QFC. Ergo polus
circuli RO est altior polo circuli QFC. Circuli vero quorum poli sunt   165
altiores sunt maioris inclinationis super superficies super quas existunt:
ergo circulus RO est maioris inclinationis super circulum ABG quam
circulus QFC. Circuli igitur MNS BZG QFC RO PT sunt inclinati ad
circulum ABG, et qui ex eis est maioris altitudinis est circulus BZG
et maioris depressionis est circulus PT. Et duo circuli MNS QFC sunt   170
similis inclinationis, et circulus RO est maioris inclinationis super circulum ABG quam circulus QFC; poli quoque eorum sunt super unum
circulorum equidistantium qui est minor circulo AD. Et illud est quod
demonstrare voluimus.

22 Cum hec eadem fuerint secundum quod prediximus et fuerint arcus qui
producuntur inter loca nodorum, scilicet inter loca contactus circulorum et inter sectiones primi, equales, circuli maiores quos prediximus
erunt similis inclinationis.

---

161 quia] *supra* **M**   *161* quia perpendicularis] *om.* **KgBPsVa**   *161* producitur] *corr. marg.*
ex protrahitur **R**   *163* fit] sit **Kg**   *163* fit ... puncto X] *marg.* **O**   *163* puncto] *om.*
BPsVa   *164–165* QFC ... circuli[1]] *om.* **Fi**   *164–165* Ergo ... QFC] *marg.* **M**, *om.* **KgB**
PsVa   *165* altior polo] polo altius **Fi**, altius polo **Z**   *169* eis] his **M**   *169* BZG] AZG
Z   *171* et ... inclinationis] *om.* **Va**   *172* eorum] eum **Fi**   *173* illud] hoc **MKgOB**
PsVa   *173–174* quod ... voluimus] *om.* **OBPsVa**   *1* prediximus] diximus **KgOBPsVa**
2 nodorum] modorum **Fi**   *4* erunt] *in corr.* **Ps**

فلتكن القوسان اللتان تخرجان من عقدتى ن ف أعنى من موضع

5 التماس إلى موضع تقاطع دائرة ا ب ج ودائرتى م ن س ع ف ق وهما قوسا

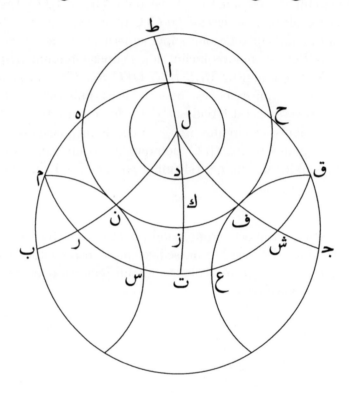

من ف ق متساويتين ، فأقول إن ميل دائرتى م ن س ع ف ق على دائرة

ا ب ج ميل متشابه فلنتعلم قطب دائرتى ا د ه ز ح ط المتوازيتين ولتكن نقطة

ل ولنرسم دائرة عظيمة تمر بنقطتى ا ل وهى دائرة ط ا ل ز ت فيبيّن أنها ستمر

بنقطة ك التى هى قطب دائرة ا ب ج ولنرسم دائرتين عظيمتين تمر كل

10 واحدة منهما بنقطة ل وبواحدة من نقطتى ن ف وهما دائرتا ل ن ب ل ف ج

اللتان [التى A   4   عقدتى [عمدى A, عمدى N, עקדה H   4   ن ف [ر و N   4   موضع [موصعى :NH A

5   موضع تقاطع [موصعى نقاطعى :NH   وهما [A   5   من [م N   6   هر [ه N   متساويتين [مساوى A,

om. N, מהסאויהאן H   7   ميل متشابه [ملا متشابها A   8   ا ل ... ولنرسم [marg. A   فيبيّن [فسى A

marg. A   وبواحدة [وكل واحدة A   10   نقطتى [نقطه A   10   ل ن ب [ل ن 10

Sint itaque duo arcus qui producuntur a duobus nodis N et F,   5
scilicet a locis contactus, ad loca sectionum circuli ABG et duorum
circulorum $MN\acute{Y}$ QFC, duo arcus MN FC equales. Dico igitur quod

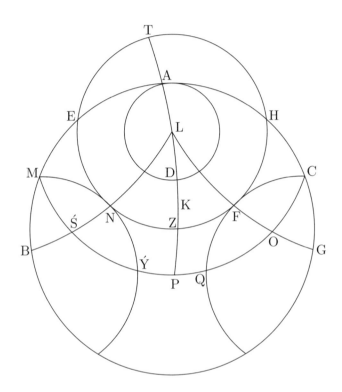

inclinatio duorum circulorum $MN\acute{Y}$ QFC super circulum ABG est
inclinatio similis. Notabo itaque polum duorum circulorum equidistantium AD $EZHT$, qui sit punctum L. Et signabo circulum maiorem   10
transeuntem per duo puncta A L, qui sit circulus $TALZP$. Ipse
igitur transibit etiam per punctum K, quod est polus circuli ABG. Et
signabo duos circulos maiores quorum quisque transeat per punctum L
et per unum duorum punctorum N et F, qui sint duo circuli LNB LFG.

5 duo arcus] *post* producuntur **BPsVa**   5 arcus] *om.* **V**   5 nodis] modis **Fi**   5 N] qui
sunt *add. supra* **Ps**, *in textu* **Va**   5 et] etiam **V**   6 scilicet] videlicet **PRVFi**, **Z** (*in corr.*)
6 locis] loco **Fi**   6 circuli ABG] *tr.* **Fi**   7 FC] ST **Fi**   7 igitur] ergo **BPsVa**   8 inclinatio] in
circulo **Fi** (*hic et saepius*), **Va**   8–9 circulorum ... similis] *om.* **Va**   8 QFC] QFT **Fi**
9 polum] polos **OBPsVa**   9–10 circulorum equidistantium] *tr.* **PRV**   10 sit] sint **PsVa**
10 punctum] punctus **OZBPsVa**   10–11 Et ... A L] *marg.* **R**   10 signabo] significabo **Va**, *hic
et saepius*   11 A] *supra* **BPs**; et *add.* **OBPsVa**   12 transibit] transit **OPsVa**   12 etiam] *om.* **Va**
12 quod] qui **FiZ**   13 quisque] uterque **MKgOBPsVa**   14 per] punctum *add.* **FiZ**   14 et] *om.*
BPsVa, etiam **V**   14 sint] sunt **OFiZBPsVa**   14 LNB] LNH **Kg**

ولأن دائرتى ه‍ز‍ح‍ط م‍ن‍س فى كرة وإحداهما تماس الأخرى وقد رسمت

دائرة عظيمة تمر بقطب إحداهما وبموضع الماسة وهى دائرة ل‍ن‍ب فإن

دائرة ل‍ن‍ب تمر أيضاً بقطبى دائرة م‍ن‍س وتكون قائمة عليها على زوايا

قائمة وكذلك نبين أن دائرة ل‍ف‍ج أيضاً تمر بقطبى دائرة ع‍ف‍ق وتكون

15   قائمة عليها على زوايا قائمة ولأنه قد عمل فى دوائر متساوية على أقطارها

التى تخرج من نقطتى ن‍ ف‍ قطعتان من دائرتين متساويتان قائمتان عليها

على زوايا قائمة وهما قطعتا ن‍ل ف‍ل مع القطع المتصلة بها وفصل منهما

قوسان متساويتان وهما قوسا ن‍ل ف‍ل وكانتا أصغر من نصف كل واحدة

من القوسين وفصل من الدوائر الأولى قوسان متساويتان وهما قوسا م‍ن

20   ف‍ق فيكون الخط المستقيم الذى يصل بين نقطة ل‍ ونقطة م‍ مساوياً للخط

المستقيم الذى يصل بين نقطة ل‍ ونقطة ق‍ فالدائرة التى ترسم على قطب ل‍

وببعد ل‍م‍ تمر بنقطة ق‍ أيضاً فلتمر ولتكن مثل دائرة م‍س‍ع‍ق وهى موازية

لدائرتى آد ه‍ز‍ح‍ط وذلك أنها على أقطاب بأعيانها ولأن دائرتى ا‍ب‍ج

م‍س‍ع‍ق فى كرة وإحداهما تقطع الأخرى وقد رسمت دائرة عظيمة تمر

25   بأقطابهما وهى دائرة ط‍د‍ك‍ز‍ت صارت دائرة ط‍د‍ك‍ز‍ت تقسم القطع التى

فصلت من الدوائر بنصفين نصفين فقوس م‍ت مساوية لقوس ت‍ق وأيضاً

لأن دائرتى م‍ن‍س م‍س‍ت‍ق تقطع إحداهما الأخرى وقد رسمت دائرة

11 om. A [2دائرة   13 om. A [أيضاً   13 الداره [1دائرة A   13 دائرة, supra A I [واحدهما,   [وإحداهما
14 دائرة [2 corr. ex N دائرتى [ولأنه   15 ملانه A [أقطارها   15 اقطابهما, اعطابهما marg. A   16 ن‍] أ
N ف‍ل] 17 marg. A   18 نصف] N   18 وىصل :AH [وفصل   19 supra A [الأولى   الاول N
20 فيكون [in corr. A   20-21 ل‍ ... نقطة] marg. A   22 م‍س‍ع‍ق [متسعق N   24 كرة [واحده N
N طلدكب, in corr. A, ط‍د‍ك‍ز‍ت [2   25 ط‍د‍ك‍ز‍ت [طلدكب N   25 supra A   24 تمر [add. N
26 فقوس [illeg. A   26 مساوية [مساوبا A   27 م‍س‍ت‍ق [in corr. A

Et quia duo circuli *EZHT MNÝ* sunt in spera et unus eorum contingit   15
alterum et iam signatus est circulus maior transiens per polos unius
eorum et per locum contactus, qui est circulus *LNB*, ergo circulus *LNB*
transit per polos circuli *MNÝ* et est super eum orthogonaliter erectus.
Simili quoque modo monstratur quod circulus *LFG* transit per polos
circuli *QFC* et est super eum orthogonaliter erectus. Et quia in   20
circulis equalibus iam facte sunt super eorum diametros, que a duobus
punctis *N* et *F* protrahuntur, due porciones circuli equales super eas
orthogonaliter erecte, que sunt due porciones *NL FL* cum porcionibus
eis continuis, et separantur ex eis duo arcus equales qui sunt *NL FL*,
et sunt minores medietate cuiusque duorum arcuum, et separantur ex   25
duobus circulis primis duo arcus equales, qui sunt duo arcus *MN FC*:
ergo linea recta que separat inter punctum *L* et punctum *M* est equalis
linee recte que separat inter punctum *L* et punctum *C*. Circulus igitur
qui super polum *L* cum spacio *LM* signatur transit etiam per punctum
C. Transeat ergo et sit sicut circulus *MÝQC*; ipse vero est equidistans   30
duobus circulis *AD EZHT* Quod ideo est quoniam ipsi sunt super
eosdem polos. Et quia duorum circulorum *ABG MÝQC* in spera unus
alterum secat et etiam signatus est circulus maior transiens per polos
eorum, qui est circulus *TDKZP*, ergo fit circulus *TDKZP* dividens
porciones, quas separat, in duo media et duo media: ergo arcus *MP* est   35
equalis arcui *PC*. Et etiam quia duorum circulorum *MNÝ MÝPC* unus

---

15 *MNÝ*] *in corr.* **O**   15 contingit] contingat **O**   15–16 contingit alterum] *tr.* **Z**   16 est] *om.*
FiZ   17 locum] loca **OBPs**   17 contactus] eius ab altero *add.* **MSS**   17 circulus²] *om.*
Fi   18 eum] *om.* **Fi**   19–20 Simili ... erectus] *om.* **B**, *marg.* **Ps**   20 *QFC*] *in corr.*
Ps   20 erectus] *om.* **OFi**   21 circulis] circulus **Fi**   22 *N* et *F*] *NZF* **Kg**   22 due] *in
corr.* **P**   23 *NL*] *NS* **R**   24 et] *om.* **Va**   24 separantur] separatur **V**, separant **Va**
25 cuiusque] utriusque **MOBPsVa**   26 duo] duos **V**   26 *MN*] *in corr.* **Ps**   27 *L*] *M* **MKgOZB**
PsVa   27 *M*] *L* **MKgOZBPsVa**   28 *L* et punctum] *om.* **V**   29 signatur] *post L* **OFiZBPsVa**,
om. **Kg**   30 ergo] igitur **FiZ**   30 equidistans] *in corr.* **BPs**   31 *EZHT*] *EŹHT* **V**, *EHT* **Va**
31 ipsi] *om.* **FiZ**, *in corr.* **B**   32 *MÝQC*] *corr. ex MNSQC* **V**, *MSC* (*in corr.*) **B**, *in corr.*
Ps; cum *add.* **OBPsVa**   32 unus] unius **B**   33 etiam] iam **VMFiZ**   33 signatus] signifa ua-
tus **B**   33 signatus est] significatus est (*in corr.*) **Ps**   33 est] *om.* **B**   33 circulus] circulis **Fi**
34 ergo ... *TDKZP*] *om.* **Fi**   34 fit] *om.* **OBPsVa**   34 *TDKZP²*] *in corr.* **B**; est *add.* **BPsVa**
34 dividens] dividet **OFiZ**   36 *MÝPC*] *in corr.* **Ps**

عظيمة تمر بأقطابهما وهى دائرة لنب صارت دائرة لنب تقسم القطع

التى فصلت من الدوائر بنصفين نصفين فقوس من مساوية لقوس نس

وقوس مس مساوية لقوس رس وكذلك نبين أن قوس عف أيضاً مساوية   30

لقوس فق وقوس عش مساوية لقوس شق فلأن قوس من مساوية

لقوس فق وقوس منس ضعف قوس من وقوس عفق ضعف قوس

فق تكون قوس منس أيضاً مساوية لقوس عفق والدائرتان متساويتان

فالخط الذى يوتر قوس منس مساوٍ للخط الذى يوتر قوس عفق ولكن

الخط الذى يوتر قوس منس يوتر أيضاً قوس مرس والخط الذى يوتر قوس   35

عفق يوتر أيضاً قوس عشق وقوسا مرس عشق من دائرة واحدة

بعينها فقوس مرس مساوية لقوس عشق وقوس مر نصف قوس مرس

وقوس قش نصف قوس عشق فقوس مر مساوية لقوس قش وكل

قوس مرست مساوية لكل قوس تعشق فقوس رست الباقية مساوية

لقوس تعش الباقية وهى من دائرة واحدة بعينها فقوس رست شبيهة   40

بقوس تعش ولكن قوس رست شبيهة بقوس نز وقوس تعش شبيهة

بقوس زف فقوس نز شبيهة بقوس زف وهى من دائرة بعينها

فقوس نز مساوية لقوس زف فبعد دائرتى منس عفق من نقطة

نش [رس N 30   مت [م رس N 30   ]الدوائر 29   repet. et del. A   30   A, באקטבהא H   ]اقطابها [بأقطابهما 28

N   31 الوضع فى add. A   ]مساوية 31   marg. A   31]لقوس فق ...مساوية [²   34–

N مثس [م رس N 36   ]عفق ولكن ...مثس [marg. A   35]م رس [in corr. A,   35]قوس ¹...ولكن

مث [م ر 38 add. A   38]داره [نصف 38   مثس N 37]مر [²   N مثس 37   ]م رس [²   N مرس 37   ]م ر [¹ 37 م رس

N   39]م رست [مثست N   39]تعشق [العاشق N   40–41 ¹...تعش [فقوس ...تعش [om. N

41]رست [رست N   41]ثست [نك N 42]نز [زف N 42]كف [²...فقوس [om. N 42]من [supra

A   43]نز [نك N 43]زف [كف N

alterum secat et iam signatus est circulus maior transiens per polos
eorum, qui est circulus LNB, fit circulus LNB dividens porciones,
quas ex duobus circulis separat, in duo media et duo media: ergo arcus
MN est equalis arcui $N\acute{Y}$ et arcus $M\acute{Y}$ est equalis arcui $\acute{S}\acute{Y}$. Similiter
quoque monstratur quod arcus QF equatur arcui FC, et arcus QO 40
est equalis arcui OC. Et quia arcus MN est equalis arcui FC et arcus
$MN\acute{Y}$ est duplus arcus MN et arcus QFC est duplus arcus FC, ergo
arcus $MN\acute{Y}$ est etiam equalis arcui QFC. Sed duo circuli sunt equales:
ergo linea que subtenditur arcui $MN\acute{Y}$ est equalis linee que subtendi-   45
tur arcui QFC. Linea vero que subtenditur arcui $MN\acute{Y}$ subtenditur
etiam arcui $M\acute{S}\acute{Y}$ et linea que subtenditur arcui QFC subtenditur
etiam arcui QOC et duo arcus $M\acute{S}\acute{Y}$ QOC sunt unius et eiusdem
circuli: ergo arcus $M\acute{S}\acute{Y}$ est equalis arcui QOC. Sed arcus $M\acute{S}$ est
medietas arcus $M\acute{S}\acute{Y}$ et arcus CO est medietas arcus QOC: ergo arcus   50
$M\acute{S}$ est equalis arcui CO. Totus vero arcus $M\acute{S}\acute{Y}P$ est equalis toti arcui
$PQOC$: reliquus igitur arcus $\acute{S}\acute{Y}P$ est equalis reliquo arcui PQO, et ipsi
sunt unius et eiusdem circuli. Sed arcus $\acute{S}\acute{Y}P$ est similis arcui NZ et arcus PQO est similis arcui ZF, et ipsi sunt unius et eiusdem circuli: ergo

37 transiens] *om.* **FiZ**   38 fit circulus LNB] *om.* **KgVa**   38 LNB] *in corr.* **Ps**
38 porciones] proporciones **V**   40 $N\acute{Y}$] *in corr.* **KgO**, NO **Va**   40 $M\acute{Y}$] MG **Kg**; MS, *corr. ex*
MG **Ps**; MSG **Va**   42 OC] PC **Fi**, DE **Va**   42 FC] *in corr.* **Ps**   42 et] ergo **O**   43 arcus[1]] ad
add. supra **BPs**, *in textu* **Va**   43 arcus[3]] arcui **BPsVa**   43 arcus FC] FO **Fi**   44 QFC] QFE
Fi   45 arcui] arcus **Va**   45 $MN\acute{Y}$] *in corr.* **Fi**   46–48 Linea ... QOC] *marg.* **BPs**, *om.* **Va**
47 etiam] et **Fi**, *om.* **BPsVa**   47 arcui QFC] *tr.* **Fi**   48 QOC^2] QCO **Fi**   48 unius] eius
Fi   49–50 $M\acute{S}$ ... arcus] *supra* **Ps**   49–50 est[2] ... $M\acute{S}\acute{Y}$] *marg.* **B**   50 $M\acute{S}\acute{Y}$] MS **B**
52 $\acute{S}\acute{Y}P$] *in corr.* **BPs**   52 PQO] *in corr.* **BPs**, QOP **Va**   53–54 et ... ZF] *primum false,*
deinde recte **R**   54 ZF] *corr. ex* FZF **P**

النصف من إحدى القوسين اللتين قسمت بهما دائرة هزحط بعد متساوٍ

والدوائر التى يكون بعدها من نقطة النصف من إحدى هاتين القوسين أيهما   45

كان بعداً متساوياً متشابهة الميل فميل دائرتى منس عفق على دائرة

ابجد متشابه ، وذلك ما أردنا أن نبين ·

[A] تمت المقالة الثانية من كتاب ثاودوسيوس فى الأكر وهى اثنان وعشرون

شكلاً والحمد لله رب العالمين ·

[N] تمت المقالة الثانية من كتاب ثاودوسيوس فى الكرات وهى ثلثة وعشرون

شكلاً إصلاح أبى الحسن ثابت بن قرة الحرانى بحمد الله ومنّه ·

[H] تמת   תמת אלמקאלה אלב מן כתאב תאודוסיוס פי אלאשכאל

אלכריה ·

מתסאויאן N, مساوية ‏מתסאויא 44 [هزحط] هر supra A   46 [الميل] احـ على داره احـ add. A   47 [متشابه] in corr. A, مساوية‏
H

arcus *NZ* est equalis arcui *ZF*. Ergo spacium duorum circulorum *MNÝ*   55
QFC a puncto medietatis unius duorum arcuum qui separantur ex circulo *EZHT* est spacium equale. Circuli vero quorum spacium a puncto
medietatis unius horum duorum arcuum, quicumque fuerit, est equale,
sunt similis inclinationis super circulum *ABG*: ergo inclinatio duorum
circulorum *MNÝ QFC* super circulum *ABG* est similis. Et illud est   60
quod demonstrare voluimus.

Pars secunda libri theodosii de speris expleta est.

---

56 unius] eorum horum *add.* **R**   56–58 qui … duorum arcuum] *repet.* **R**   56–57 circulo] cui
add. marg. **R**   57 vero] *supra* **M**   58 unius] unus **Va**   58 horum] *om.* **FiZBPsVa**
58 quicumque] quecumque **OFiZVa**; quicumque, *corr in* quecumque **BPs**   59 duorum] *in corr.*
Ps   60 *MNÝ*] *MNX* **Kg**   60 *QFC*] *om.* **R**   60 illud] iam **Fi**   61 demonstrare] *om.* **O**
61 demonstrare voluimus] volumus demonstrare **BPsVa**   61 voluimus] volumus **KgO**

المقالة الثالثة من كتاب ثاودوسوس فى الأكر

آ   إذا خُط فى دائرة خط ما مستقيم يقسم الدائرة بقسمين غير متساويين

وعمل عليه قطعة من دائرة ليست بأعظم من نصفها وكانت قائمة على

الدائرة على زوايا قائمة وقسمت قوس القطعة التى عملت على الخط

بقسمين غير متساويـيـن فإن الخط الذى يوتر القوس الصغرى يكون أقصر   5

جميع الخطوط المستقيمة التى تخرج من تلك النقطة التى انقسمت القوس

عليها إلى القوس العظمى من الدائرة الأولى وكذلك أيضاً إن كان الخط

المخرج قطر الدائرة وكانت سائر الأشياء التى كانت للقطعة التى ليست

بأعظم من نصف دائرة المعمولة على الخط على حالها فإن الخط المخرج الذى

تقدم ذكره أقصر جميع الخطوط المستقيمة التى تخرج من تلك النقطة بعينها   10

وتلقى الخط المحيط بالدائرة الأولى ويكون أعظمها الخط الذى يوتر القوس

العظمى

،

فليخط فى دائرة ا‏ب‏ج‏د خط ما مستقيم وهو خط ب‏د يقسم الدائرة

بقسمين غير متساويين ولتكن قوس ب‏ج‏د أعظم من قوس ب‏ا‏د ولتعمل

على خط ب‏د قطعة من دائرة ليست بأعظم من نصف دائرة قائمة على   15

المقالة 1 [الأكر ... من [*deest in*   N *add.* رت اعن ;AN *add. antea* بسم الله الرحمن الرحيم
H 1 [ثاودوسوس [N ثاودوسيوس   الأكر] 1 الكراب ,N وره الحرانى س بانت الحسن اى اصلاح
add. N   5 أقصر [H *add.* ٮٯ   8 ليست [A *add.* هى   9 بأعظم [A اعظم   15 ب‏د [A *supra*
marg. A] ليست ... على دائر‹ة› >   15–16

Incipit pars tercia libri theodosii de speris

1 Cum in circulo aliqua descripta fuerit linea secans circulum in duas partes inequales et super eam constituta fuerit portio circuli non maior medietate eius, que super circulum orthogonaliter sit erecta, et divisus fuerit arcus portionis qui super lineam constitutus fuit in duas partes 5 inequales, linea que minori arcui subtendetur erit minor omnibus rectis lineis que ab illo puncto protrahuntur super quod arcus divisus fuit ad maiorem arcum primi circuli. Et similiter etiam si fuerit linea protracta diametrus circuli et fuerint reliqua, que sunt porcionis que non est maior semicirculo facte super lineam, secundum habitudinem suam, 10 erit linea protracta quam prediximus minor omnibus rectis lineis que producuntur ab illis eisdem punctis et occurrunt linee primum circulum continenti, et erit maior eis linea que maiori arcui subtenditur.

In circulo itaque *ABG* linea recta describatur, que sit linea *BD* dividens circulum in duas partes inequales, que sint duo arcus *BGD* 15 *BAD*, et sit arcus *BGD* maior arcu *BAD*; et faciam super lineam *BD* porcionem circuli non maiorem medietate circuli orthogonaliter erec-

---

2 fuerit] *in corr.* **Ps** 2 secans circulum] *tr.* **OBPsVa** 2 in²] *supra* **Va** 3 partes inequales] *tr.*
PRVFiZ   3 eam] eum **Kg**   3 constituta] cumstituta **V**, *hic et saepius*   5–
6 partes inequales] *tr.* **PRVFiZ**   6 que] *supra* **Ps**   6 minori] *in corr.* **BPs**, minor **Kg**
6 subtendetur] subtenditur **RKgBPsVa**   8 etiam] *om.* **BPsVa**   8–9 protracta diametrus] *tr.*
Va (diametrus: diametro, *hic et saepius*)   9 circuli] *om.* **Va**   9 non] *om.* **PRVM** (**O:** *post*
est)   10 lineam, secundum] *om.* **Z**   11 prediximus] *in corr.* **B**   11 minor] *supra* **BPs**
11 minor omnibus] *tr.* **Va** 11 rectis lineis] *tr.* **Va** 13 eis] *in corr.* **Fi** 14 linea recta] *tr.* **KgZ**
(recta: *corr. ex* maior **Kg**)   14 describatur] desc'pbat̄ **V**   15 partes inequales] *tr.* **PRVMKgO**
FiZ   15 inequales] *in corr.* **Ps**, equales **Fi**   15 sint] *in corr.* **Kg**, sunt **Z**   16 *BAD*¹] *supra*
BPs, *om.* **Va**   16 maior arcu *BAD*] *om.* **V**   16 *BAD*²] fuerit *add.* **Va**   16 super lineam] *in*
corr. **Ps**   17 erectam] erecti **PRVM**

دائرة اٜبٜجد على زوايا قائمة وهى قطعة بٜهٜد ولتقسم قوس بٜهٜد
بقسمين غير متساويين على نقطة هٜ ولتكن قوس بٜهٜ أصغر من قوس هٜد

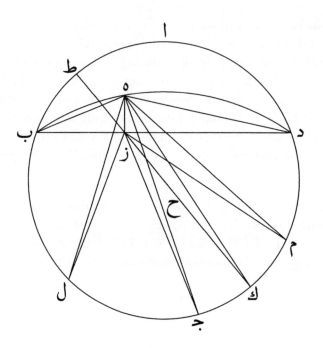

وليوصل خط هٜبٜ ، فأقول إن خط بٜهٜ أقصر جميع الخطوط المستقيمة التى
تخرج من نقطة هٜ إلى قوس بٜجد فليخرج من نقطة هٜ إلى سطح دائرة
اٜبٜجد عمود هٜز فهو يقع على الفصل المشترك لسطحى اٜبٜجد بٜهٜد الذى
هو خط بٜد لأن قطعة بٜهٜد قائمة على دائرة اٜبٜجد على زوايا قائمة
ولتتعلم مركز دائرة اٜبٜجد وليكن نقطة حٜ وليوصل خط زحٜ وليخرج فى
الجهتين إلى نقطتى طٜ كٜ وليخرج من نقطة هٜ إلى قوس بٜجد خط هٜلٜ

‏‫17 هٜد ... بٜهٜ [هب قوس من اعطم هٜد A 18 المستقيمة] om. N 21 قطعة] in corr. A
A بلحد [بٜجد 23 om. N 23 بٜجد] om. N 22 فى [من A 23 إلى [من 22 على ... اٜبٜجد 21-22

tam super circulum *ABGD*, que sit porcio *BED*; et dividam arcum *BED* in duas partes inequales super punctum *E*; et sit arcus *BE* minor

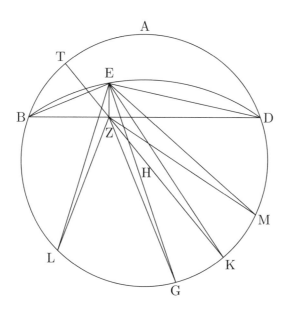

arcu *ED*; et producam lineam *BE*. Dico igitur quod linea *BE* est minor 20
omnibus rectis lineis que protrahuntur a puncto *E* ad arcum *BGD*. Protraham ergo a puncto *E* ad superficiem circuli *ABGD* perpendicularem
EZ: ipsa igitur cadit super differentiam communem *ABGD BED*, que
est linea *BD*, quoniam porcio *BED* erecta est super circulum *ABGD*
orthogonaliter. Et notabo centrum circuli *ABGD*, quod sit punctum 25
H, et producam lineam *ZH*, quam protraham in duas partes usque ad
duo puncta *T* et *K*. Et protraham a puncto *E* ad arcum *BGD* lineam

18 arcum] *om.* **Fi**   20 arcu ... minor] *marg.* **BPs**   20 BE^1] *KE* **Fi**   20 igitur] *om.* **Kg**,
ergo **FiZ**   20 BE^2] *KE* **Fi**   21 protrahuntur] producuntur **KgBPsVa**   21 *E*] *om.* **Va**   21–
22 ad ... *E*] *marg. inf.* **B**, *supra* **Ps**   22 ergo] autem **Fi**   22–23 perpendicularem *EZ*] *tr.* **Fi**
23 igitur] etiam **Kg**   23 *ABGD*] et *add. supra* **B**   24 *BED*] *BDE* **BPsVa**   24 erecta] *om.*
MBPsVa   24 *ABGD*] erecta *add. supra* **M**   25 notabo] vocabo **Fi**   26 *ZH*] *ZA* **Va**
26 in] *supra* **R**   27 et] *om.* **FiZ**   27 a puncto] *supra* **BPs**   27 *BGD*] *in corr.* **Kg**

وليوصل خط زلّ فخط هزّ عمود على سطح دائرة اب‌جد فهو يحدث مع

25 جميع الخطوط التى تخرج من طرفه وتكون فى سطح دائرة اب‌جد زوايا

قائمة وكل واحد من خطى زب زلّ اللذين هما فى سطح دائرة اب‌جد

يخرج من طرف خط هزّ فكل واحدة من زاويتى بزه لزه قائمة ولأن

خط زب أقصر من خط زلّ يصير المربع أيضاً الكائن من خط زب أقل من

المربع الكائن من خط زلّ ويجعل المربع الكائن من خط هزّ مشتركاً فالمربعان

30 الكائنان من خطى هزّ زب أقل من المربعين الكائنين من خطى هزّ زلّ ولكن

المربع الكائن من خط به مساوٍ للمربعين الكائنين من خطى هزّ زب

والمربع الكائن من خط له مساوٍ للمربعين الكائنين من خطى لز زه فالمربع

الكائن من خط به أيضاً أقل من المربع الكائن من خط له فخط هب أقصر

من خط هل ،

35 وكذلك نبين أنه أقصر من جميع الخطوط المستقيمة التى تخرج من

نقطة ه وتلقى قوس ب‌جد فخط به أقصر من جميع الخطوط المستقيمة التى

تخرج من نقطة ه وتلقى قوس ب‌جد ، وأقول إن الخط الذى بقرب منه

من الخطوط المستقيمة التى تخرج من نقطة ه فيما بين نقطتى كب أبداً

أقصر من الذى هو أبعد منه فليخرج خط جه وليوصل خط زج فلأن خط

24 وليوصل] فلنوصل N 25 طرفه] نقطه ز 〔زٓ in corr.〕A 27 ولأن] لان N 28 أيضاً] om. A 28–
24 〔... زب] marg. 29 〔زب أقل ... خط[1] marg. A 〔المربع pro مربع] 29 〔ويجعل] in corr. A 30–31 زب ... أقل[
A 〔للمربعين الكائنين: للمربع الكاس〕31 〔زب] رل A 33 من[2] om. A 36 〔به] ب‌جد] add.
A الخطوه 〔الخطوط] A 38 يحك 〔ب‌جد] A 37 المستقيمة] om. N 36 النH 〔من] et del. A
38 ه] om. A 38 〔كب] اقصر من در خط من فلان خط هم وليوصل خط هم وهو خط اخر مستقيم انضا فليحرج
خط add. 〔et del.?〕A 38–39 أبداً ... منه] repet. marg. A

EL, et producam lineam *ZL*. Linea autem *EZ* est perpendicularis super superficiem circuli *ABGD*: ergo ex ipsa cum omnibus lineis que ab eius extremitate protrahuntur et sunt in superficie circuli *ABGD* proveniunt anguli recti. Sed unaqueque duarum linearum *ZB ZL*, que sunt in superficie circuli *AGBD*, protrahitur ab extremitate linee *EZ*: ergo unusquisque duorum angulorum *BZE LZE* est rectus. Et quia linea *ZB* est brevior linea *ZL*, fit etiam quadratum factum ex linea *ZB* minus quadrato facto ex linea *ZL*. Ponam autem quadratum factum ex linea *EZ* commune: ergo duo quadrata facta ex duabus lineis *EZ ZB* sunt minora duobus quadratis factis ex duabus lineis *EZ ZL*. Quadratum autem factum ex linea *BE* est equale duobus quadratis factis ex duabus lineis *EZ ZB*, et quadratum factum ex linea *LE* est equale duobus quadratis factis ex duabus lineis *LZ ZE*: ergo quadratum factum ex linea *BE* est etiam minus quadrato facto ex linea *LE*. Ergo linea *EB* est brevior linea *EL*.

Et similiter ostenditur quod ipsa est minor omnibus rectis lineis que protrahuntur a puncto *E* et occurrunt arcui *BGD*. Ergo linea *BE* est brevior omnibus rectis lineis que producuntur a puncto *E* et occurrunt arcui *BGD*. Dico igitur quod linea, que est ei magis propinqua ex lineis rectis que producuntur a puncto *E* inter duo puncta *B K*, est minor ea que ab ea magis est remota. Protraham ergo lineam *GE* et coniungam lineam *ZG*. Et quia linea *LZ* est brevior linea *ZG*, erit

30

35

40

45

28 *ZL*] *Z* et **Fi**   29 ergo] *supra* **BPs**   29 cum] cā **Va**   29 lineis] rectis *add.* **FiZ**   30 eius] eis **Va**   30 circuli] *supra* **BPs**   31 unaqueque] utraque **KgBPsVa**, queque **Fi**   31 *ZB*] et *add.* **O**   31 *ZB ZL*] *BZL* **Fi**   32 protrahitur] protrahuntur **KgOBPsVa**   32 ab] *om.* **Fi**   32 linee] *om.* **OFiBPsVa**   33 unusquisque] uterque **MKgOB**, utrique **Va**   33 quia] *in corr.* **M**   34 est] *om.* **V**   34 quadratum factum] quadratus factus **PsVa**   34 ex linea] *ante* factum **V**   34–35 minus ... linea[1]] *repet.* **Fi**   35 *ZL*] *Z* **FiZ**   36 commune] *in corr.* **B**   36 *ZB*] *ZL* **FiZ**   37 minora] maiora **Z**   37 quadratis] *supra* **BPs**   37 duabus] *om.* **FiZ**   37 *EZ ZL*] *LZ ZE* **OBPs**   37 *ZL*] *BZ* **FiZ**   37–40 Quadratum ... *LZ ZE*] *om.* **BPsVa**   38 *BE*] *in corr.* **KgO**   38–39 duabus] duobus **Fi**   41 est] *om.* **Fi**   41 est etiam] *tr.* **MKgB PsVa**   41 quadrato facto] *tr.* **BVa**, *in corr.* **Ps**   42 linea] *om.* **R**   43 ostenditur] ostendam **OBPsVa**   43 ipsa] ipse **Va**   44 protrahuntur] *in corr.* **R**   44 occurrunt] occurrant **R**   44–46 Ergo ... arcui *BGD*] *om.* **OBPsVa**   45 que] *om.* **M**   45 producuntur] protrahuntur **R**   46 *BGD*] *om.* **M**   46 igitur] ergo **KgBPsVa**   46 que] est *BE add.* **Fi**   46 ei magis] *tr.* **Fi**   47 *B*] *G* **Fi**   48 ea] etiam *add.* **Va**   48 magis est] *tr.* **OZBPs**   48 est] *om.* **Va**   48 Protraham] Protrahatur **OBPs**, Protrahitur **Va**   48 ergo] igitur **Fi**   48 lineam] linea **OB PsVa**   49 erit] est **Kg**

لَزَ أقصر من خط زجَ يكون أيضاً المربع الكائن من خط لَزَ أقل من المربع ٤٠

الكائن من خط زجَ ويجعل المربع الكائن من خط زهَ مشتركاً فالمربعان

الكائنان من خطى زلَ زهَ أقل من المربعين الكائنين من خطى هزَ زجَ ولكن

المربعين الكائنين من خطى لَزَ زهَ مساويان للمربع الكائن من خط لَه

والمربعان الكائنان من خطى جزَ زهَ مساويان للمربع الكائن من خط هجَ

فالمربع الكائن من خط لَه أقل من المربع الكائن من خط هجَ فخط لَه أقصر ٤٥

من خط هجَ ،

وكذلك نبين أن ما قرُب من خط هبَ من الخطوط المستقيمة التى

تخرج من نقطة هَ فيما بين نقطتى بَ كَ أقصر مما بعُد وأيضاً فإنا نصل

خطى هكَ هد ، فأقول إن خط هكَ أطول جميع الخطوط المستقيمة التى

تخرج من نقطة هَ وتلقى قوس بَكد وإن خط هد أقصر من جميع الخطوط ٥٠

المستقيمة التى تخرج من نقطة هَ فيما بين نقطتى دَ كَ فلأن خط كَزَ أطول

من خط زجَ يكون المربع الكائن من خط كَزَ أعظم من المربع الكائن من

خط زجَ ويجعل المربع الكائن من خط زهَ مشتركاً فالمربعان الكائنان من

خطى كَزَ زهَ وهو المربع الكائن من خط هكَ أعظم من المربعين الكائنين من

خطى هزَ زجَ وهو المربع الكائن من خط هجَ فخط هكَ أطول من خط هجَ ٥٥

وكذلك نبين أن خط هكَ أيضاً أطول من جميع الخطوط المستقيمة التى تخرج

من نقطة هَ وتلقى قوس كَد فخط هكَ أطول جميع الخطوط المستقيمة التى

40 خط [1 om. N   40 خط [2 زلَ] حَ add. et del. N   42 زهَ] illeg. A   42 أقل] marg. A

N هزَ [هد   44 جزَ زهَ] illeg. A   47 نبين [ابضا add. A   48–50 بَكد ... فيما] repet. et del. A   49 أطول [H: من add. AN

A ذلك [هكَ   57 مشترك [مسترك A   53 وتلقى قوس بَكد] marg. A   50 أطول [من add. AN   49

57–58 قوس ... كَد] repet. N   57 أطول [من add. AN

etiam quadratum factum ex linea *LZ* minus quadrato facto ex linea 50
ZG. Ponam autem quadratum factum ex linea *ZE* commune: duo igitur
quadrata facta ex duabus lineis *ZL ZE* sunt minora duobus quadratis
factis ex duabus lineis *EZ ZG*. Sed duo quadrata facta ex duabus lineis
LZ ZE sunt equalia quadrato facto ex linea *LE* et duo quadrata facta
ex duabus lineis *GZ ZE* sunt equalia quadrato facto ex linea *EG*: ergo 55
quadratum factum ex linea *LE* est minus quadrato facto ex linea *EG*.
Ergo linea *LE* est minor linea *EG*.

Simili quoque modo monstratur quod quecumque rectarum linearum, que a puncto *E* inter duo puncta *B* et *K* protrahuntur, linee *EB*
existit propinquior, est minor ea que ab ea magis elongatur. Producam 60
etiam duas lineas *EK ED*. Dico igitur quod linea *EK* est longior omnibus rectis lineis que a puncto *E* protrahuntur et occurrunt arcui *BKD*
et quod linea *ED* est brevior omnibus rectis lineis que protrahuntur a
puncto *E* inter duo puncta *D K*. Et quia linea *KZ* est longior linea *ZG*
ergo quadratum factum ex linea *KZ* est maius quadrato facto ex linea 65
ZG. Ponam autem quadratum factum ex linea *ZE* commune: ergo duo
quadrata facta ex duabus lineis *KZ ZE*, que sunt quadratum factum ex
linea *EK*, sunt maius duobus quadratis factis ex duabus lineis *EZ ZG*,
que sunt quadratum factum ex linea *EG*. Ergo linea *EK* est longior
linea *EG*. Et similiter ostenditur quod linea *EK* est etiam longior 70
omnibus rectis lineis que producuntur a puncto *E* et occurrunt arcui *KD*. Ergo linea *EK* est longior omnibus rectis lineis que a puncto *E*

51 autem] *supra* **Ps**   51 *ZE*] *Z, supra E* **B**   51 duo] nunc **BPsVa**   51 duo igitur] dico
igitur quod **R**   52–54 sunt ... *LZ ZE*] *marg.* **M**   53 duabus[1]] *om.* **M**   53 duabus[2]] *om.*
M   54–55 *LZ* ... lineis] *om.* **Z**   54 *LE*] *BE* **Fi**   54–55 *LE* ... ex linea] *marg.* **BPs**,
om. **Va**   55 duabus lineis] *om.* **OBPs**   55 facto] quod *add.* **Fi**   56 *LE*] *in corr.*
Ps   57 minor] *in corr.* **B**, minus **PsVa**   58 quoque] *om.* **BPsVa**   58 modo] *om.* **P**
59 inter] intus **Fi**, *hic et saepius*   59 *B* et *K* protrahuntur] protrahuntur scilicet *B K* **MKgB**
PsVa   60 propinquior] propinquiorum **Fi**   60 elongatur] elongantur **Z**   62 a puncto *E*] ab *E*
BPsVa   62 protrahuntur] *post* que **MOBPsVa**   62–64 et ... *D K*] *marg.* **BPs**   64 linea[1]] *om.*
R   64 *KZ*] *in corr.* **B**, *ZN* **Va**   64 longior] *supra* **BPs**   64–65 longior ... *KZ* est] *om.*
Fi   64 linea[2]] *om.* **Va**   65–66 ergo ... *ZG*] *marg.* **BPsVa**   68 *EK*] *E* **Fi**   68–
69 *EK* ... ex linea] *om.* **Va**   68 duobus] *om.* **BPs**   69 linea[2]] *om.* **BPsVa**   69 est] *om.*
Va   70 ostenditur] ostendetur **FiZ**   70 etiam] *om.* **O**   70–72 etiam ... *EK* est] *om.* **FiZ**
71 producuntur] protrahuntur **OBPsVa**   71–75 *E* ... et *D*] *marg.* **BPs**   71 occurrunt] *in
corr.* **B**   72–74 que ... lineis] *marg. inf.* **M**

تخرج من نقطة هـ وتلقى قوس بـكـد ،

وأقول أيضاً إن خط هـد أيضاً أقصر من جميع الخطوط المستقيمة التى

تخرج من نقطة هـ فيما بين نقطتى كـد فليخرج أيضاً خط آخر مستقيم وهو

خط هـم وليوصل خط مـز فلأن خط دز أقصر من خط زم يكون المربع

الكائن من خط دز أقل من المربع الكائن من خط زم ويجعل المربع الكائن

من خط زه مشتركاً فالمربعان الكائنان من خطى هـز زد اللذان هما مثل

المربع الكائن من خط هـد أقل من المربعين الكائنين من خطى هـز زم الذين

هما مثل المربع الكائن من خط هـم فخط ده أقصر من خط مه ،

وكذلك نبين أن خط هـد أقصر من جميع الخطوط المستقيمة التى تخرج

من نقطة هـ وتلقى قوس كـد فيما بين نقطتى كـد فخط هـد أقصر من جميع

الخطوط المستقيمة التى تخرج من نقطة هـ وتلقى قوس كـد فيما بين نقطتى

كـد والخط الذى بقرب منه من الخطوط التى تخرج فيما بين نقطتى كـد

أقصر من الذى هو أبعد منه ولأن خط هـد أطول من خط هـب إذ كانت

قوس ده أيضاً أعظم من قوس هـب يكون خط هـب أقصر كثيراً من جميع

الخطوط المستقيمة التى تخرج من نقطة هـ إلى قوس كـد ،

وقد تبين أنه أقصر من جميع الخطوط المستقيمة التى تخرج إلى قوس

كـب أيضاً فخط بـه أقصر من جميع الخطوط المستقيمة التى تخرج من نقطة

60

65

70

59 marg. [خط زم ويجعل ‖ N أيضاً 59 ‖ 2 om. A ‖ 62 [دز] illeg. A ‖ 62 فاقول [وأقول] 59
A مشرك [مشتركاً 63 ‖ om. A المربع الكائن من [62-63 ‖ om. N ويجعل ... خط زه [62-63
A فيها [فيما 63 ‖ hic et saepius A الكران [اللذان 64 ‖ اللرر [اللذين 64 ‖ 67 من [2 om. H ‖ 68 فيها [
70 وحط هد اقصر من حميع [1 هـب] A 71 اذا [إذ] A 70 اعظم [أطول] A فه [منه] N 70 [هو] om. A ‖ 70
add. NH كثيراً [كثيراً] 71 om. NH ‖ 72 [كـد] add. A الخطوط المستقيمه الى تحرح من نقطه ه الى فوس كـد
73 جميع [om. A

producuntur et occurrunt arcui *BKD*.

Et dico etiam quod linea *ED* est brevior omnibus rectis lineis que protrahuntur a puncto *E* inter duo puncta *K* et *D*. Protraham igitur etiam aliam rectam lineam, que sit linea *EM*, et producam lineam *MZ*. Et quia linea *DZ* est brevior linea *ZM*, ergo quadratum factum ex linea *DZ* est minus quadrato facto ex linea *ZM*. Ponam autem quadratum factum ex linea *ZE* commune: ergo duo quadrata facta ex duabus lineis *EZ ZD*, que sunt equalia quadrato facto ex linea *ED*, sunt minora duobus quadratis factis ex duabus lineis *EZ ZM*, que sunt equalia quadrato facto ex linea *EM*. Ergo linea *DE* est brevior linea *ME*.

Et similiter ostenditur quod linea *ED* est brevior omnibus lineis rectis que protrahuntur a puncto *E* et occurrunt arcui *KD* inter duo puncta *K* et *D*: ergo linea *ED* est brevior omnibus rectis lineis que producuntur a puncto *E* et occurrunt arcui *KD* inter duo puncta *K* et *D*. Et quecumque rectarum linearum que protrahuntur inter duo puncta *K* et *D* ei magis propinquat est brevior ea que ab ea magis est remota. Et quia linea *ED* est longior linea *BE*, quoniam arcus *DE* etiam est maior arcu *EB*, ergo linea *EB* est multo brevior omnibus rectis lineis que a puncto *E* ad arcum *KD* protrahuntur.

Iam autem ostensum fuit quod ipsa etiam est brevior omnibus rectis lineis que ad arcum *KB* protrahuntur: ergo linea *BE* est brevior omnibus lineis rectis que a puncto *E* ad arcum *BKD* producuntur.

73 producuntur] protrahuntur, *post* que **KgOFiZ** 73–75 et ... puncto *E*] *om.* **Kg** 74 *ED*] *om.* **Va** 74 lineis] *om.* **OBPs** 75 protrahuntur] *corr. ex* producuntur **P**, producuntur vel protrahuntur **R** 75 igitur] ergo **Kg**, *om.* **OBPsVa** 76 etiam] *om.* **Kg** 76 linea] *om.* **OBPsVa** 76 *EM*] *in corr.* **Z** 76 *MZ*] *om.* **Fi** 77 Et quia] quia *Z* et quoniam **Fi** 77 quia] quoniam **Z** 77 linea[1]] *om.* **PRVFiZ** 77 linea[2]] *om.* **O** 80 *ZD*] *ZB* **Kg**, *in corr.* **B** 80–81 que ... *ZM*] *et in marg. et in textu* **B**, *marg.* **Ps** 80–82 sunt ... *EM*] *marg.* **M** 81 factis ex duabus lineis] *om.* **BPsVa** (**B** *in textu*) 82 facto] *supra* **B** 82–83 linea[2] ... brevior] *om.* **BPsVa** 83 quod] quo **Fi** 83 lineis] *om.* **R** 83–84 lineis rectis] *tr.* **KgOFi** 84 protrahuntur] producuntur **Kg** 84 et occurrunt] *illeg.* **Fi** 85–87 ergo ... *K* et *D*] *om.* **O** 85–87 linea ... *K* et *D*] *om.* **Va** 85 rectis lineis] *tr.* **Z** 85 lineis] *om.* **Fi** 87 Et] ergo **Va** 87–88 Et ... et *D*] *repet.* **B** 87 rectarum] istarum **Va** 87 protrahuntur] *corr. ex* producuntur **P** 88 propinquat] appropinquat **VMOZ**, eppropinquat **Fi**, appropinquatur **KgB**; et *add.* **Fi** 88 est] et **Va** 88 brevior] est *add.* **Va** 88 ab ea] ei **OBPsVa** 89 est remota] *tr.* **OBPsVa** 89–90 *DE* etiam] *tr.* **OBPsVa** 90 maior] minor **Z** 90 *EB*[1]] *in corr.* **B** 90 linea] *in corr. supra* **B**, *om.* **Ps** 90 multo] *om.* **BPsVa** 91 a puncto *E*] *om.* **BPsVa** 91 arcum] *om.* **Ps** 92 autem] etiam **KgOBPsVa** 92 ipsa etiam] *tr.* **OBPsVa** 92 etiam] autem **Fi** 93 rectis lineis] *tr.* **FiZ** 93 que] a puncto *add.* **V** 93 arcum] *om.* **BPsVa** 94 lineis rectis] *tr.* **PRVMKgFiZ** 94 *E*] *supra* **BPs** 94 *BKD*] *KD* **Fi**

ه‍ إلى قوس ب‍ك‍د وليكن خط ب‍د المخرج قطر الدائرة ولتكن سائر الأشياء   75

الباقية على حالها ، فأقول إن خط ه‍ب أقصر من جميع الخطوط التى تخرج

من نقطة ه‍ وتلقى الخط المحيط بدائرة اب‍ج‍د وإن خط ه‍د أطولها فإذا

عملت الأشياء التى وصفنا بأعيانها فإن قوس ده‍ إذ كانت أعظم من قوس

ه‍ب وقد أخرج عمود ه‍ز يكون خط د‍ز أطول من خط ز‍ب وخط ب‍د

قطر دائرة اب‍ج‍د فمركز الدائرة على خط ز‍د فخط ز‍د أطول من خط ز‍ج   80

وخط ز‍ج أطول من خط ز‍ب فالمربع الكائن من خط ز‍د أعظم من المربع

الكائن من خط ج‍ز والمربع الكائن من خط ز‍ج أعظم من المربع الكائن من

خط ز‍ب ويجعل المربع الكائن من خط ز‍ه‍ مشتركاً فالمربعان الكائنان من

خطى د‍ز ز‍ه‍ اللذان هما مثل المربع الكائن من خط ده‍ أعظم من المربعين

الكائنين من خطى ج‍ز ز‍ه‍ اللذين هما مثل المربع الكائن من خط ج‍ه‍   85

والمربعان الكائنان من خطى ج‍ز ز‍ه‍ اللذان هما مثل المربع الكائن من خط

ج‍ه‍ أعظم من المربعين الكائنين من خطى ب‍ز ز‍ه‍ اللذين هما مثل المربع

الكائن من خط ب‍ه‍ فخط ده‍ أطول من خط ه‍ج وخط ه‍ج أطول من خط

ه‍ب وكذلك نبين أن خط ه‍د أطول جميع الخطوط التى تخرج من نقطة ه‍

وتلقى الخط المحيط بدائرة اب‍ج‍د وخط ه‍ب أقصرها فخط ه‍د أطول جميع   90

الخطوط التى تخرج من نقطة ه‍ إلى الخط المحيط بدائرة اب‍ج‍د وخط ه‍ب

أقصرها ، وذلك ما أردنا أن نبين ·

رہ [خطى A 85   مشترك [مشتركاً A 83   داره [الدائرة 80   supra A [الخط 77   om. N [من 76

A واطول [²أطول A 88   om. A [وخط ه‍ج 88   om. A [والمربعان ... خط ج‍ه‍ 86–87   add. et del.

وكذلك [وذلك A 89   أطول [من خط ه‍ل add. et del. N; من add. NH 90   من [أطول add. AN 89

Sit autem linea *BD* protracta diametrus circuli et sint reliqua omnia 95
secundum suam habitudinem. Dico igitur quod linea *EB* est brevior
omnibus lineis que producuntur a puncto *E* et occurrunt linee continenti circulum *ABGD* et quod linea *ED* est eis longior. Cum ergo
fecerimus eadem que prediximus, quia arcus *DE* est maior arcu *EB* et
iam producta est perpendicularis *EZ* erit linea *DZ* longior linea *ZB*. 100
Linea vero *BD* est diametrus circuli *ABGD*: ergo centrum circuli est
super lineam *ZD*. Ergo linea *ZD* est longior linea *ZG* et linea *ZG* est
longior linea *ZB*: ergo quadratum factum ex linea *ZD* est maius quadrato facto ex linea *GZ*. Sed quadratum factum ex linea *ZG* est maius
quadrato facto ex linea *ZB* Ponam autem quadratum factum ex linea 105
ZE commune: duo igitur quadrata facta ex duabus lineis *DZ ZE*, que
sunt equalia quadrato facto ex linea *DE*, sunt maiora duobus quadratis
factis ex duabus lineis *GZ ZE*, que sunt equalia quadrato facto ex linea
GE. Et duo quadrata facta ex duabus lineis *GZ ZE*, que sunt equalia
quadrato facto ex linea *GE*, sunt maiora duobus quadratis factis ex 110
duabus lineis *BZ ZE*, que sunt equalia quadrato facto ex linea *BE*.
Ergo linea *DE* est longior linea *EG* et linea *EG* est longior linea *EB*.
Et similiter demonstratur quod linea *ED* est longior omnibus lineis que
producuntur a puncto *E* et occurrunt linee continenti circulum *ABGD*
et quod linea *EB* est eis brevior. Ergo linea *ED* est longior omnibus 115
lineis, que a puncto *E* ad circumferentiam circuli *ABGD* protrahuntur,
et linea *EB* est eis brevior. Et illud est quod demonstrare voluimus.

95 *BD*] *in corr.* **BPs**   95 diametrus] *in corr.* **BPs**   95 sint] *in corr.* **BPs**, sunt **FiVa**
95 reliqua omnia] *tr.* **OBPsVa**   96 Dico] Dato **Z**   96 *EB*] E, B *supra* **B**   96 est brevior] *tr.*
OBPsVa   97 omnibus] rectis *add.* **R**   97 linee] *in corr.* **Va**   98 quod] *om.* **Kg**
99 eadem] *illeg.* **M**, ea **KgOBPsVa**   99 *EB*] *ZB* (*in corr.*) **Kg**   100 iam] *om.* **BPsVa**
100 producta] productam **V**   100 *DZ*] *DE* **Kg**; *DZ, corr. ex DE* **B**   100 *ZB*] *EB*
Kg, *ZDL* **Fi**; *ZB, corr. ex EB* **B**   101 *BD*] *D* **Fi**   102 *ZG*] *ZB* **B**   102–
103 *ZG* est longior linea] *om.* **V**   103 *ZB*] *corr. ex ZD* **P**   103 factum ex linea] *om.*
BPsVa   104 facto ex linea] *om.* **BPsVa**   104 *GZ*] linee *add.* **PsVa**   104 factum ex linea] linee
O, *om.* **BPsVa**   105 facto] *om.* **BPsVa**   105 ex[1]] *supra* **Ps**, *XG* **Va**   105 linea[1]] *om.*
BPsVa   105 factum ex linea] linee **O**, *om.* **BPsVa**   106 facta] *om.* **BPsVa**   107 equalia] *post*
facto **Va**   107–109 facto ... linea *GE*] *marg.* **BPs**   108 duabus] *om.* **OVa**   109–
111 Et ... linea *BE*] *om.* **Va**   109 quadrata facta] *illeg.* **M**   109 facta ex duabus lineis] *om.*
BPs   109 *ZE*] linearum *add.* **BPs**   109 que] *om.* **R**   109 sunt equalia] *repet.* **Fi**
110 *GE*] *in corr.* **B**   110 maiora] maior **R**   110 ex[2]] *supra* **BPs**   111 duabus] duobus
Fi   111 duabus lineis] *om.* **BPs**   111 *ZE*] *in corr.* **P**   111 quadrato facto] *marg.* **B**, *supra* **Ps**
111 *BE*] *supra* **BPs**, *EA* **Kg**   112 *EG* ... linea[4]] *om.* **BPsVa**   113 demonstratur] monstratur
BPsVa, dividitur **Fi**   113 omnibus] rectis *add.* **O**, *add. et del.* **R**   114 continenti] continentis **V**
115 quod] quia **BPsVa**   116 *ABGD*] *ABG* **Fi**   116 protrahuntur] producuntur **KgOBPsVa**
117 illud] hoc **KgOBPsVa**   117 demonstrare voluimus] *om.* **Kg**, volumus monstrare **OBPsVa**

ب

إذا خُطّ فى دائرة خط ما مستقيم يفصل منها قطعة ليست بأصغر من نصف
دائرة ثم عمل عليه قطعة دائرة ليست بأعظم من نصف دائرة مائلة على
القطعة التى ليست بأعظم من نصف دائرة وقُسمت قوس القطعة التى
عملت بقسمين غير متساويين فإن الخط الذى يوتر القوس الصغرى أقصر
من جميع الخطوط المستقيمة التى تخرج من تلك النقطة التى انقسمت عليها ٥
إلى قوس القطعة التى ليست بأصغر من نصف دائرة ·

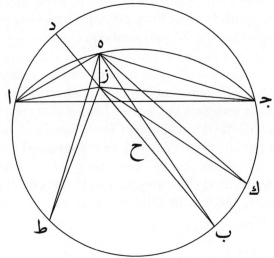

فليخط فى دائرة اب‍ج‍د خط ما وهو خط اج‍ يفصل من الدائرة قطعة
ليست بأصغر من نصف دائرة وهى قطعة اب‍ج‍ ولتُعمل على خط اج‍ قطعة
من دائرة ليست بأعظم من نصف دائرة وهى قطعة اه‍ج‍ مائلة على قطعة
اد‍ج‍ التى ليست بأعظم من نصف دائرة ولتقسم قوس اه‍ج‍ بقسمين غير ١٠
متساويين على نقطة ه‍ ولتكن قوس ج‍ه‍ أعظم من قوس ه‍ا‍ وليوصل خط
ه‍ا‍ ، فأقول إن خط ه‍ا‍ أقصر من جميع الخطوط المستقيمة التى تخرج من

2-3 دائرة ... مائلة [repet. N 4 الصغرى [marg. A 4 أقصر [اصغر A 5 من [om. NH -8

9 قطعة ... اب‍ج‍ [قطعه ادح طعه وبعمل على قطعه ادح A; اد‍ج‍ [اب‍ج‍ N 9 دائرة ... ليست [om. N :H

9 اه‍ج‍ [اه‍ج‍ , هد‍ , supra A 10 ولتقسم [illeg. N 12 من [om. NH

2   Cum in circulo describitur linea recta separans ex eo portionem semicirculo non minorem, deinde fit super eam portio circuli non maior
semicirculo inclinata super portionem que non est maior semicirculo,
et dividitur arcus portionis que facta est in partes inequales, linea que
subtenditur minori arcui est minor omnibus rectis lineis que ab illo   5
puncto producuntur super quod dividitur ad arcum portionis que non
est minor semicirculo.

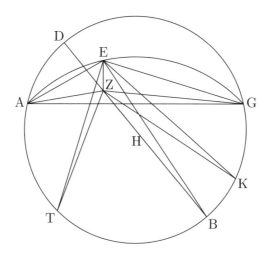

In circulo itaque *ABGD* aliqua describatur linea, que sit linea
AG, separans ex circulo portionem non minorem semicirculo, que sit
portio *ABG*. Et faciam super lineam *AG* portionem circuli, que sit   10
portio *AEG*, inclinatam super portionem *ADG*, que non est maior
semicirculo. Et dividam arcum *AEG* in duas partes inequales super
punctum *E*, et sit arcus *GE* maior arcu *EA*, et producam lineam *EA*.
Dico igitur quod linea *EA* est brevior omnibus rectis lineis que a puncto

---

1 separans] *in corr.* **Ps**   2 deinde] autem *add.* **FiZ**   2 fit] sit **RKgOFiZ**   2 eam] eum **R**
2 circuli] *om.* **FiZ**   3 inclinata] inclinato **Va**   3 semicirculo] *om.* **Va**   4 dividitur] dividatur
OFiZ   5 minori] maiori **Kg**   6 producuntur] protrahuntur **Z**   8 *ABGD*] *ABG, D supra* **BPs**
8 describatur] *in corr.* **Ps**, describitur **RFi**   9 separans] *in corr.* **Ps**   9 minorem] minor est
Fi   10 *ABG*] *BGA* **FiZ**   10 lineam *AG*] *tr.* **BPsVa**   11 *AEG*] *AGE* **V**; *EG, A supra* **BPs**
11 inclinatam] inclinata **PsVa**, declinatam **OZ**, declinati **Fi**, erectam **PRVM**   11 *ADG*] *AGD*
FiZ   12 partes] *supra* **BPs**   12 partes inequales] *tr.* **Va**   13 *EA*[1]] *ED* **Kg**, *EA (A corr. ex
D)* **BPs**   13 lineam *EA*] *tr.* **BPsVa**   13 *EA*[2]] *illeg.* **M**, *in corr.* **Ps**   14 igitur] ergo **OBPs**
14 linea *EA*] *tr.* **BPsVa**

نقطة ه إلى قوس اب‍ج فليخرج من نقطة ه إلى سطح دائرة اب‍ج‍د عمود

فهو يقع فيما بين خط اج‍ وقوس اد‍ج‍ لأن قطعة اه‍ج‍ مائلة على قطعة

اد‍ج‍ فليخرج وليكن خط ه‍ز يلق سطح الدائرة على نقطة ز ولنتعلم مركز

دائرة اب‍ج‍د فإن مركزها إما أن يكون على خط اج‍ وإما فيما بين خط اج‍

وقوس اب‍ج‍ لأنا كنا وضعنا أن قطعة اب‍ج‍ ليست بأصغر من نصف

دائرة ،

وليكن أولاً فيما بين خط اج‍ وبين قوس اب‍ج‍ وليكن نقطة ح

وليوصل خط زح وليخرج فى الجهتين إلى نقطتى د ب وليخرج من نقطة ه

إلى قوس اب‍ج‍ خط مستقيم يلقاها وهو خط ه‍ط وليوصل خطا از ز‍ط

ولأن خط ه‍ز عمود على سطح دائرة اب‍ج‍د فهو يحدث مع جميع الخطوط

التى تلقاه وهى فى سطح دائرة اب‍ج‍د زوايا قائمة وكل واحد من خطى از

ز‍ط اللذين هما فى سطح دائرة اب‍ج‍د يلقى خط ه‍ز فكل واحدة من

زاويتى از‍ه ط‍زه قائمة فلأن خط از أقصر من خط ز‍ط يكون المربع الكائن

من خط از أقل من المربع الكائن من خط ز‍ط ويجعل المربع الكائن من خط

ز‍ه مشتركاً فالمربعان الكائنان من خطى از ز‍ه اللذان هما مثل المربع الكائن

من خط اه‍ أقل من المربعين الكائنين من خطى ط‍ز ز‍ه اللذين هما مثل

المربع الكائن من خط ط‍ه فخط اه‍ أقصر من خط ط‍ه وكذلك نبين أنه أقصر

ولنتعلم [ولنتعلم N 15   وليقى [وليقى N 15   ولان N: AH   لأن [14 om. N   خط [14 N فلط] [فليخرج 13

نقط [نقطة 19   add. et del. A   وبين [19 خط   N فلكى [وليكن 19   A يح [اب‍ج‍] 17   A

A يلقاها هو [تلقاه وهى فى 23   A حطا [خط 22   ילאקאת H   يلقاها [21   A الى [إلى 20   A

24 واحدة [واحد A 25   ط‍زه [ط‍ز A 26   خط [1 om. A 29   ط‍ه [corr. ex د ه A

E ad arcum ABG producuntur. Protraham ergo a puncto E ad super-   15
ficiem circuli $ABGD$ perpendicularem. Ipsa igitur cadet inter lineam
AG et arcum ADG, quoniam portio AEG est inclinata super portionem ADG. Protrahatur ergo et sit linea EZ, et occurrat superficiei
circuli super punctum Z. Et notabo centrum circuli $ABGD$, quod aut
erit super lineam AG aut inter lineam AG et arcum ABG, quoniam   20
nos posuimus ut portio ABG non sit minor semicirculo.

Sit ergo primum inter lineam AG et arcum ABG sitque punctum
H; et describam lineam ZH, quam in utramque partem producam
usque ad duo puncta D et B. Et protraham a puncto E ad arcum ABG
lineam rectam occurrentem ei, que sit linea ET, et producam duas   25
lineas AZ ZT. Et quia linea EZ est perpendicularis super superficiem
circuli $ABGD$, ergo ex ipsa cum omnibus rectis lineis que ei occurrunt
et sunt in superficie circuli $ABGD$, recti proveniunt anguli. Unaqueque
vero duarum linearum AZ ZT, que sunt in superficie circuli $ABGD$,
occurrit linee EZ: ergo unusquisque duorum angulorum AZE TZE est   30
rectus. Et quia linea AZ est brevior linea ZT, ergo quadratum factum
ex linea AZ est minus quadrato facto ex linea ZT. Ponam autem
quadratum factum ex linea ZE commune: duo igitur quadrata facta
ex duabus lineis AZ ZE, que sunt equalia quadrato facto ex linea AE,
sunt minora duobus quadratis factis ex duabus lineis TZ ZE, que sunt   35
equalia quadrato facto ex linea TE. Ergo linea AE est brevior linea
ET. Et similiter ostenditur quod ipsa est brevior etiam omnibus rectis

---

15 arcum] circumferentiam **FiZ**   15 producuntur] protrahuntur **BPsVa**   15 ergo] *illeg.* **M**, igi-
tur **FiZ**   17 *ADG*] *ABG* **Fi**   17 *AEG*] *LEG* **R**   17 inclinata] inclinatio **Kg**   18 *ADG*] *AGD*
BPsVa   18 ergo] igitur **FiZ**   19 aut] autem **VFiZ**   20 *ABG*] *in corr.* **Ps**   21 nos] *om.*
KgOBPsVa   21 posuimus] possumus **Fi**; primo *add.* **Va**   21 portio] arcus **O**, arcus (*in corr.*)
BPs, porcio arcus **Va**   21 sit] *supra* **BPs**   22 primum] primo **MKgOBPsVa** (*supra* **BPs**)
22 et arcum *ABG*] *om.* **Fi**   22 *ABG*] *AGD* **Z**   23 utramque partem] utraque parte **PsVa**
24 ad] a **Fi**   24 *D* et *B*] *B* et *D* **OBPsVa**   24 arcum] *repet.* **Fi**   25 que] qui **V**   25 *ET*] *om.*
Va   25–26 duas lineas] *tr.* **Kg**   26 quia] *om.* **BPsVa**   26 *EZ*] *E T* **Fi**   27 *ABGD*] *ABG*
BPsVa   27 ex] *om.* **BPs**   28 in superficie] supra superficiem **R**   28 Unaqueque] Utraque
MKgBPsVa   29 in] *om.* **Fi**   30 occurrit] occurrunt **BPsVa**   30 unusquisque] uterque **MKgOB**
PsVa   30 angulorum] supra *add.* **PsVa**, *add. et del.* **B**   30 *AZE*] *AEZ* **Kg**; *AZE, corr. ex*
AEZ **BPs**   30 *AZE TZE*] *QZ et ZE* **Fi**   31 rectus] *in corr.* **Ps**; t *add.* **Va**   31 linea[1]] *om.*
MOBPsVa   31 ergo] *om.* **BPsVa**   31–32 factum ex linea] *om.* **BPsVa**   32 minus] *illeg.* **M**
32 facto ex linea] *om.* **BPsVa**   33 igitur] ergo **FiZ**   33–34 facta ex duabus lineis] duarum li-
nearum **BPsVa**   34 ex[2]] *repet.* **Kg**   35 minora] maiora **Fi**   35 *TZ ZE*] *ZE ZT* **OBPsVa**, *ZE*
TZ **Kg**   36 *TE*] *illeg.* **M**   37 etiam] *post* similiter **KgBPsVa**, *post* ipsa **Fi**, *om.* **Z**   37 rectis] *om.*
PVM, *post* lineis **Z**

30 أيضاً من جميع الخطوط المستقيمة التى تخرج من نقطة ﻫ إلى قوس اطٮ

فخط اﻫ أقصر من جميع الخطوط المستقيمة التى تخرج من نقطة ﻫ إلى قوس

اطٮ فيما بين نقطتى آ ٮ ،

وكذلك أيضاً نبين أن ما قرُب منه من الخطوط المستقيمة التى تخرج

من نقطة ﻫ إلى قوس اطٮ التى فيما بين نقطتى آ ٮ أقصر مما بعُد

35 وليوصل خط ﺑﻫ ، فأقول إن خط ﺑﻫ أطول جميع الخطوط التى تخرج

من نقطة ﻫ إلى قوس اٮﺟ فلأن خط ﺑﺰ أطول من خط زﻁ

يكون المربع الكائن من خط ﺑﺰ أعظم من المربع الكائن من خط زﻁ

ويجعل المربع الكائن من خط زﻫ مشتركاً فالمربعان الكائنان من خطى ﻫﺰ

زٮ اللذان هما مثل المربع الكائن من خط ﻫٮ أعظم من المربعين الكائنين

40 من خطى ﻫﺰ زﻁ اللذين هما مثل المربع الكائن من خط ﻫﻁ فخط ﺑﻫ

أطول من خط ﻫﻁ ،

وكذلك نبين أنه أطول أيضاً من جميع الخطوط المستقيمة التى تخرج

من نقطة ﻫ إلى قوس اٮﺟ فخط ﻫٮ أطول جميع الخطوط التى تخرج من

نقطة ﻫ إلى قوس اٮﺟ فليوصل ﻫﺟ ، فأقول إن خط ﻫﺟ أقصر جميع

45 الخطوط التى تخرج من نقطة ﻫ إلى قوس ٮﺟ فيما بين نقطتى ٮ ﺟ

فليخرج أيضاً خط آخر مستقيم وهو خط ﻫﻙ وليوصل خطا زﻙ زﺟ فلأن

خط زﺟ أقصر من خط زﻙ يكون المربع الكائن من خط زﺟ أقل من المربع

30 جميع [الجميع A   30 المستقيمة] om. N   31-32 اطٮ ... فخط] om. A   31 من¹] om. N
رآ [زٮ] 39 in corr. A   36 قوس [أطول] marg. A   36 اعظم [يكون المربع] 37 فالمربع A   37 بﺰ] in corr. A
A   40 الكائن [in corr. N   42 نبين] N سس   43 ﻫ د A, ﺟ N   43-45 ﻫﺟ ... فخط] om. A
N خطا [خط] 46 من add. H   43 أطول [من

lineis que producuntur a puncto E ad arcum ATB inter duo puncta A
et B. Ergo linea AE est brevior omnibus rectis lineis que producuntur
a puncto E ad arcum ATB.   40

Et similiter etiam ostenditur quod illa que ex rectis lineis, que a
puncto E producuntur ad arcum ATB que sunt inter duo puncta A et
B, ei est propinquior est brevior ea que est ab ea magis remota. Producam ergo lineam BE. Dico igitur quod linea BE est longior omnibus
rectis lineis que protrahuntur a puncto E ad arcum ABG. Et quia li-   45
nea BZ est longior linea ZT, ergo quadratum factum ex linea BZ est
maius quadrato facto ex linea ZT. Ponam autem quadratum factum
ex linea ZE commune: duo igitur quadrata facta ex duabus lineis EZ
ZB, que sunt equalia quadrato facto ex linea EB, sunt maiora duobus
quadratis factis ex duabus lineis EZ ZT, que sunt equalia quadrato   50
facto ex linea ET. Ergo linea BE est longior linea ET.

Et similiter ostenditur quod ipsa est longior etiam omnibus rectis
lineis que producuntur a puncto E ad arcum BG. Ergo linea EB est
longior omnibus lineis que producuntur a puncto E ad arcum ABG.
Describam ergo lineam EG. Dico igitur quod linea EG est brevior omni-   55
bus lineis que producuntur a puncto E ad arcum BG inter duo puncta
B G. Protraham ergo etiam aliam rectam lineam, que sit linea EK;
et producam duas lineas ZK ZG. Et quia linea ZG est brevior linea
ZK, ergo quadratum factum ex linea ZG est minus quadrato facto ex

---

38 inter] *om.* **M**   *38–40* inter ... *ATB*] *om.* **Kg**, *repet. et del.* **B**   *38–40* duo ... *ATB*] *marg.*
M   *39* *AE*] *in corr.* **BPs**   *39* brevior] minor **Z**   *40* *ATB*] *illeg.* **R**   *41* etiam ostenditur] *tr.*
Kg   *41* illa] ille **R**   *41* illa que] *om.* **FiZ**   *41* rectis lineis] *tr.* **O**   *42* producuntur] *post* que
PsVa   *43* *B*] et *add.* **FiZ**   *43* est ab ea magis] ab ea est **Fi**, ab ea magis est **Z**   *44* ergo] igitur
FiZ   *44* igitur] ergo **OBPsVa**   *45* *ABG*] *BGA* **FiZ**   *45* Et] *om.* **OBPsVa**   *46* ergo] *om.*
B, *supra* **Ps**   *47* facto ex linea] *om.* **BPsVa**   *48* duo] dico **Va**   *48* igitur] ergo **OBPsVa**
48 facta ex duabus lineis] *om.* **BPsVa**   *49–50* que ... *EZ ZT*] *marg.* **BPs**   *49* linea] *om.*
BPsVa   *50* factis ... lineis] *om.* **BPsVa**   *50* duabus lineis] *om.* **O**   *50* *EZ ZT*] *TEZ* **Va**
51 ex linea] *supra* **Ps**   *52* ostenditur] monstratur **MKgOBPsVa**   *52* etiam] *post* similiter
KgOBPsVa   *53* *BG*] *ABG* **KgBPs**   *53–54* *BG* ... arcum] *om.* **O**   *53–54* Ergo ... *ABG*] *om.*
KgBPs   *53–56* Ergo ... arcum *BG*] *om.* **Va**   *54* omnibus] rectis *add.* **MFiZ**   *55* ergo] igitur
OFiZ   *55* igitur] ergo **OBPs**   *55* linea *EG*] *tr.* **O**   *55* *EG*] *supra* **BPs**, *AEG* **Fi**   *55–*
56 omnibus] quibus **Fi**; rectis *add.* **KgFiZBPs**   *57* *B*] *G* **BPsVa**; et *add.* **OFiZ**   *57* *G*] *B* **BPsVa**
57 ergo] *om.* **KgOBPsVa**, igitur **FiZ**   *57* rectam lineam] *tr.* **Kg**   *58* est brevior linea] *ante* Et
V   *58–59* linea *ZK*] *tr.* **OBPsVa**   *59* ergo] *om.* **OBPsVa**   *59* factum ex linea] *om.* **OBPsVa**
59–60 facto ex linea] *om.* **OBPsVa**

الكائن من خط زك وليكن المربع الكائن من خط زه مشتركاً فالمربعان

الكائنان من خطى زج زه اللذان هما مثل المربع الكائن من خط هج أصغر

50 من المربعين الكائنين من خطى كز زه اللذين هما مثل المربع الكائن من

خط هك فخط جه أقصر من خط هك وكذلك نبين أنه أقصر أيضاً من جميع

الخطوط المستقيمة التى تخرج من نقطة ه إلى قوس ب ك ج فيما بين نقطتى

ب ج فخط هج أقصر جميع الخطوط التى تخرج من نقطة ه إلى قوس ب ك ج

فيما بين نقطتى ب ج ،

55 وكذلك أيضاً نبين أن ما قرُب منه من الخطوط المستقيمة التى تخرج

من نقطة ه إلى قوس ب ج فيما بين نقطتى ب ج أقصر مما بعُد فلأن خط

اه أقصر جميع الخطوط المستقيمة التى تخرج من نقطة ه إلى قوس ا ط ب

وخط هج أقصر الخطوط المستقيمة التى تخرج من نقطة ه إلى قوس ب ك ج

وخط اه أقصر من خط هج لأن القوس أيضاً أصغر من القوس يكون خط

60 اه أقصر كثيراً من الخطوط التى تخرج من نقطة ه إلى قوس ب ج فخط هاا

أقصر جميع الخطوط المستقيمة التى تخرج من نقطة ه إلى قوس ا ب ج

وكذلك أيضاً تبين أنه إن كانت قطعة ا ب ج نصف دائرة فإن خط اه أقصر

جميع الخطوط المستقيمة التى تخرج من نقطة ه إلى قوس ا ب ج ، وذلك ما

أردنا أن نبين .

المربع [المربعين 50 אעושם H | أصغر [49 marg. A | زج ... زج [om. A 49-51 ه ا ك ... ه ا [خطى 49-50

A 52 المستقيمة [om. A | وكذلك أيضاً [55 illeg. A | 55 من [om. A; جميع add. A 56-

om. [أقصر ... أيضاً N 59 اه ه [هج 58 illeg. A | اطب وخط [اطب 57-58 om. A | ب ج ... إلى قوس [57

A 60 كثيراً [كدرا A 62 أيضاً [om. A

linea *ZK*. Sit autem quadratum factum ex linea *ZE* commune: ergo   60
duo quadrata facta ex duabus lineis *ZG ZE*, que sunt equalia quadrato
facto ex linea *EG*, sunt minora duobus quadratis factis ex duabus lineis
KZ ZE, que sunt equalia quadrato facto ex linea *EK*. Ergo linea *GE*
est brevior *EK*. Et similiter monstratur quod ipsa est brevior etiam
omnibus rectis lineis que a puncto *E* producuntur ad arcum *BKG*   65
inter duo puncta *B* et *G*: ergo linea *EG* est brevior omnibus lineis que
protrahuntur a puncto *E* ad arcum *BKG* inter duo puncta *B* et *G*.

Et similiter etiam ostenditur quod quecumque rectarum linearum
est ei propinquior, que producuntur a puncto *E* ad arcum *ABG* inter
duo puncta *B* et *G*, est brevior ea que ab ea magis elongatur. Et quia   70
linea *AE* est brevior omnibus rectis lineis que producuntur a puncto
E ad arcum *ATB* et linea *EG* est brevior omnibus rectis lineis que a
puncto *E* ad arcum *BKG* protrahuntur et linea *AE* est brevior linea
EG quoniam arcus etiam est minor arcu: ergo linea *AE* est multo
minor omnibus lineis que protrahuntur a puncto *E* ad arcum *BG*.   75
Ergo linea *EA* est brevior omnibus rectis lineis que producuntur a
puncto *E* ad arcum *ABG*. Et similiter etiam monstratur, si fuerit portio
ABG semicirculus, quod linea *AE* est brevior omnibus rectis lineis que
protrahuntur a puncto *E* ad arcum *ABG*. Et illud est quod demonstrare
voluimus.   80

60–62 *ZK* ... ex linea] *om.* **Fi**   60 factum ex linea] *om.* **OBPsVa**   60 *ZE* commune] *illeg.*
M   61 lineis] zineis **Va**   62 facto] *om.* **RV**   62 facto ex linea] *om.* **BPsVa**
62 factis ex duabus lineis] *om.* **OBPsVa**   63 *ZE*] *in corr.* **P**   63 facto ex linea] *om.* **OBPsVa**
63 linea *GE*] *tr.* **BPsVa**   63 *GE*] *in corr.* **BPs**   64 brevior] linea *add.* **R**   64 ipsa] *EG*
add. supra **Ps**, *in textu* **Va**   64 etiam] *om.* **OBPsVa**   65 rectis] *post* lineis **FiZ**, *om.*
PRVMKg   65 *BKG*] *BKD* **Fi**   66–67 ergo ... et *G*] *om.* **Kg**   67 *BKG*] *in corr.* **M**, *BKD* **Fi**
68 etiam] *om.* **OBPsVa**, et **Fi**   68 rectarum linearum] *illeg.* **M**   69 producuntur] producitur
OBPsVa   69 a] ad **V**   69 arcum] *om.* **BPsVa**   69 *ABG*] *BKG* **OFiZ**   71 *AE*] *in corr.* **Kg**
71 producuntur] *post ATB* **BPs**   72–73 *ATB* ... arcum] *marg.* **BPs**   73 *BKG*] *BKD* **Fi**
73 protrahuntur] *post* que **BPs**; *post BKG* producuntur *add.* **B**, *add. et del.* **Ps**; producuntur
KgO (*post* que **O**)   74 etiam] *supra* **P**, *om.* **FiZ**   74 est] *om.* **O**   74 arcu] arcui **V**
74 *AE*] *supra* **BPs**   74 multo] *om.* **KgBPsVa**   75 omnibus] rectis *add.* **KgOBVa**   75 *E*] *om.*
Va   75–77 *BG* ... arcum] *om.* **FiZ**   76 *EA*] *in corr.* **BPs**   76 producuntur] protrahuntur
Kg   77 etiam] *om.* **OFiZBPsVa**   77 monstratur] ostenditur **KgBPsVa**   78 quod] ergo
Va   78 lineis] *om.* **BPsVa**   79 a puncto *E*] *om.* **BPsVa**   79 illud] hoc **O**, istud **BPsVa**
79 quod ... voluimus] *om.* **O**   79–80 demonstrare voluimus] dicere voluimus **Fi**, dicere
volumus **Z**, volumus probare **BPsVa**

إذا كانت فى كرة دائرتان عظيمتان قطعت إحداهما الأخرى وفصل من كل ج
واحدة منهما قوسان متساويتان متصلتان إحداهما بالأخرى عن جنبتى إحدى
النقطتين اللتين تتقاطعان عليهما فإن الخطوط المستقيمة التى تصل فيما بين
أطراف القسى التى تفصل فى جهة واحدة بعينها مساوٍ بعضها لبعض ·

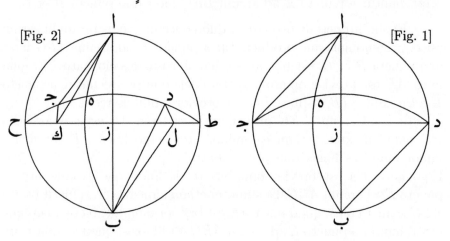

فلتكن فى كرة دائرتان عظيمتان وهما دائرتا اب جد تقطع إحداهما   5
الأخرى على نقطة ه ولنفصل من كل واحدة منهما قوسين متساويتين
متصلتين إحداهما بالأخرى عن جنبتى نقطة ه وهى قسى اه هب جه هد
ولتكن قوس اه مساوية لقوس هب وقوس جه مساوية لقوس هد وليوصل
خطا جا بد ،

فأقول إن خط جا مساوٍ لخط بد وذلك أن الدائرة التى ترسم على   10
قطب ه وببعد هآ تمر أيضاً بنقطة ب وأما بنقطة جـ فهى إما أن تمر وإما أن

om. A] مساوية 8   add. A بعمها] واحدة   6   in corr. A واحدة]   4   N تفصل supra ,بصل مصل] تفصل   4
افضا سقطه ب وأما]   2 تمر   11   A فاما] فهى إما   11   N فاما] وأما   11   A داره] الدائرة   10   N ط جا] جا   10
N الا] أن لا   11–12   add. A سقطه جـ كما مر

ح وذلك ان موس جـ اما ان نكون مساويه لقوس هجـ او غير مساويه فان كانت مساويه مرت الداره جـ وان هد   8
marg. N لم نكن مساويه لم نمر

3 Cum in spera fuerint duo maiores circuli, quorum unus alterum secet, et ab unoquoque eorum duo arcus equales separantur, quorum unus alteri continuetur ab utraque parte unius duorum punctorum super que se secant, linee recte que coniungunt inter extremitates arcuum qui separantur in una et eadem parte ad invicem sunt equales.   5

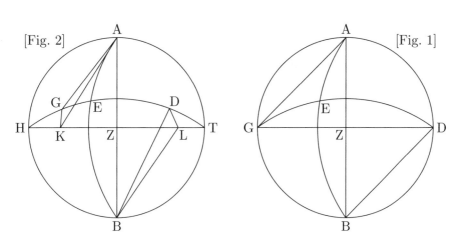

Sint ergo in spera duo circuli maiores, qui sint duo circuli *AB GD*, quorum unus alterum secet super punctum *E*, et separabo ex unoquoque eorum duos arcus equales, quorum unus alteri sit continuus ab utraque parte puncti *E*, qui sint arcus *AE EB GE ED,* et sit arcus *AE* equalis arcui *EB* et arcus *GE* sit equalis arcui *ED*; et producam   10 duas lineas *GA BD.*

Dico igitur quod linea *GA* est equalis linee *BD.* Quod ideo est quoniam circulus, qui signatur super polum *E* secundum longitudinem *EA*, transit etiam per punctum *B*, per punctum vero *G* aut transit

1 fuerint] *post* circuli **PRVMFiZ**   *2* arcus equales] *tr.* **FiZ**   *2* separantur] sperantur **Ps**
3 alteri] ad alterum **FiZ** *4* se] *om.* **OFiBPsVa** *4* recte] *om.* **Va** *5* qui] que **FiZ** *6* ergo] *post*
in spera **FiZ**   *6* sint] sunt **RMFiZVa**   *7* ex] ei **Fi**   *8* unoquoque] utroque **KgOBPsVa**
8 continuus] communis **Va**   *9* parte] *om.* **Va**   *9* sint] sunt **Fi**   *10* equalis] *repet.* **V**
10 EB] *in corr.* **BPs**   *10–12* et^2 ... linee] *om.* **Fi**   *11* duas lineas] *tr.* **Va**   *12* igitur] ergo
OBPsVa   *12* GA] GD **R**   *13* signatur] significatur **Ps**   *13* polum] punctum **PRVMFiZ**
14 EA] *in corr.* **B**

لا تمر فلتمر أولاً بنقطة جَـ كما فى الصورة الأولى فهى تمر بنقطة دَ أيضاً

وذلك أن قوس جَهَـ مساوية لقوس هَدَ فلتُعمل هذه الدائرة وهى دائرة

اجـبـد وليكن الفصل المشترك لدائرتى اجـبـد اهـب خط اَب والفصل

المشترك لدائرتى اجـبـد جَهَـد خط جَـد ولأن دائرة اهـب العظمى التى فى   15

كرة تقطع دائرة ما من الدوائر التى فى الكرة وتمر بقطبيها وهى دائرة

اجـبـد فهى تقطعها بنصفين وعلى زوايا قائمة فخط اَب قطر دائرة اجـبـد

وكذلك نبين أن خط جَـد أيضاً قطر دائرة اجـبـد فنقطة زَ مركز دائرة

اجـبـد فخطوط زاَ زب زجـ زد الأربعة مساوٍ بعضها لبعض فلأن خطى زاَ

زجـ مساويان لخطى زب زد كل واحد منهما لنظيره وزاوية ازجـ مساوية   20

لزاوية دزب وذلك أنهما متقابلتان تكون قاعدة اجـ مساوية لقاعدة دب ،

وأيضاً فلا تمر الدائرة التى ترسم على قطب هَ وببعد هاَ بنقطة جَـ لكن

تقع أبعد منها كما فى الصورة الثانية فهى تمر بنقطة بَ وتقع أبعد من نقطة

دَ فلتعمل ولتكن مثل دائرة احـبـط ولتتمم دائرة جَهَـد ولتلق دائرة احـبـط

على نقطتى حَ طَ وليكن الفصل المشترك لدائرتى احـبـط اهـب خط اَب   25

والفصل المشترك لدائرتى احـبـط حَهَـط خط حَط ونبين كما بينا آنفاً أن

نقطة زَ مركز دائرة احـبـط وأن كل واحدة من دائرتى اهـب حَهَـط قائمة

على دائرة احـبـط على زوايا قائمة فليخرج من نقطتى جَـ دَ إلى سطح

فلان [ولأن 15   *om.* A] اَب ... خط 14–15   A مساوٍ لقوس [مساوية لقوس 13   *om.* A] دَ 12

A   16 فى ... كرة [*om.* A   18 زَ] *add.* N ادا [فلا 22   فلان N   تمر 22   يمران A,

N   23 أبعد منها [أحدهما A   24 احـبـط ... ولتتمم [*om.* A   25–26 اهـب ... اَب [*om.* N

26 آنفاً [ايضا A   27 زَ] بَ A

aut non transit. Transeat igitur primo per punctum *G*, sicut est in   15
prima figura. Ipse igitur etiam transit per punctum *D*. Quod ideo est
quoniam arcus *GE* est equalis arcui *ED*. Hunc igitur notabo circulum,
qui sit circulus *AGBD*. Sit autem differentia communis duobus circulis
AGBD AEB linea *AB* et differentia communis duobus circulis *ABGD*
GED sit linea *GD*. Et quia circulus *AEB* maior qui est in spera secat   20
aliquem circulorum qui sunt in spera et transit per duos polos eius,
qui est circulus *AGBD*, ergo ipse secat eum in duo media et orthogonaliter. Linea igitur *AB* est diametrus circuli *AGBD*. Simili quoque
modo ostenditur quod etiam linea *GD* est diametrus circuli *AGBD*:
ergo punctum *Z* est centrum circuli *AGBD*. Quattuor igitur linee *ZA*   25
ZB ZG ZD ad invicem sunt equales. Et quia due linee *ZA ZG* sunt
equales duabus lineis *ZB ZD*, queque videlicet earum sue relative, et
angulus *AZG* est equalis angulo *DZB* – et hoc ideo quoniam ipsi sunt
oppositi –, ergo basis *AG* equatur basi *DB*.

Sit etiam circulus, qui signatur super polum *E* et secundum lon-   30
gitudinem *EA*, non transiens per punctum *G*, sed cadat longe ab eo,
sicut in secunda apparet figura. Ipse igitur transit per punctum *B* et
cadit longius a puncto *D*. Signetur ergo et sit sicut circulus *AHBT*, et
compleatur circulus *GED* et occurrat circulo *AHBT* super duo puncta
H et *T*. Et sit differentia communis duobus circulis *AHBT AEB* linea   35
AB et differentia communis duobus circulis *AHBT HET* sit linea *HT*.
Ostendam autem sicut etiam ostendi quod punctum *Z* est centrum
circuli *AHBT* et quod unusquisque duorum circulorum *AEB HET* est
orthogonaliter erectus super circulum *AHBT*. Protraham ergo a duobus punctis *G* et *D* ad superficiem circuli *AHBT* duas perpendiculares   40

15 igitur] ergo **Kg**, *om.* **BPsVa**   15 primo] primi **V**   16 prima figura] *tr.* **Va**   16 igitur] ergo
Ps, *om.* **BVa**   16 etiam transit] *tr.* **RVMKgOFiZ**   17 Hunc] hinc **PsVa**   17 igitur] ergo
OFiZBPsVa   18 *AGBD*] *ABGD* **KgOB**   19 *AGBD*] *in corr.* **Ps**, *ABGD* **KgOFiZBVa**
19 communis] cummunis **V**   19 duobus] *om.* **BPsVa**   19 circulis] *om.* **Va**   19 *ABGD*] *AGBD*
(*in corr.*) **Ps**, *GABD* **Va**   20 *AEB*] *ABE* **KgB**; est *add.* **OBPsVa**   20 qui] quia **O**   20 est] *om.*
O   20 secat] *in corr.* **Ps**, secant **B**   21 aliquem] alium **Fi**   22 *AGBD*] *ABGD* **KgOBPsVa**
22 secat] secacat **Fi**   23 est] *supra* **V**   23 *AGBD*] *ABGD* **KgOBPsVa**   23 Simili] Similiter
FiZ   23 quoque] *om.* **BPsVa**   24 modo] *om.* **PFiZ**   24 ostenditur] monstratur **RKgOB**
PsVa   24 etiam] *om.* **ROFiZ**   24 *AGBD*] *AGD* **V**, *ABGD* **KgOBPsVa**   25 *Z*] etiam
V   25 *AGBD*] *ABGD* **KgOFiZBVa**   25 linee] line **Fi**   25 *ZA*] *ZL* **Fi**   26–
27 Et ... equales] *repet.* **R**   27 queque] *illeg.* **M**, utraque **KgOBPsVa**   27 videlicet] *om.*
OBVa, set *scr. et del.* **Ps**, scilicet **MKg**   27 earum] *om.* **KgOBPsVa**   27 sue] *in corr.*
B   28 *AZG*] *AZGE* **Fi**   28 *DZB*] *ZDB* **Kg**, *ZBD* **B**, *BZD* **OPsVa**   28 hoc] est *add.* **Va**
28 ideo] est *add.* **KgOBPs**   28 ipsi] *om.* **OFiZBPsVa**   29 oppositi] *in corr.* **BPs**, composi-
ti **Fi**, oppositum **Z**   29 equatur] est equalis **R**   29 *DB*] *EB* **R**   30 signatur] significatur
Va   30 et] *om.* **VOFiZVa**   32 sicut] sunt **Fi**   32 Ipse] ipsi **Va**   32 punctum] *om.* **FiZ**
33 Signetur] Significetur **BPs**   33 ergo] igitur **OPsVa**; circulus *add.* **OBPsVa**   33 *AHBT*] *in
corr.* **B**   35 et] *om.* **O**, *supra* **Ps**   36 *AHBT*] *AHT* **V**   36 *HET*] *in corr.* **B**
37 autem] *om.* **BPsVa**   37 sicut etiam] *tr.* **BPsVa**   37 etiam] et **R**, *om.* **OFiZ**   38 quod] quia
BPsVa   38 unusquisque] uterque **MKgOBPsVa**   39 erectus] *om.* **FiZ**   39 ergo] autem **M**
40 *G* et *D*] *D* et *G* **OBPsVa**   40–42 duas ... circuli *AHBT*] *marg.* **BPs**

دائرة اح ب ط عمودا جك دل وليوصل خطا اك ل ب فلأن قوس ه ح

مساوية لقوس ه ط وذلك أن قطب دائرة اح ب ط هو نقطة ه وقوس ج ه ‏ 30

من إحداهما مساوية لقوس ه د من الأخرى تصير قوس ج ح الباقية مساوية

لقوس د ط الباقية فلأن قوس ح ه ط قطعة دائرة وقد فصل منها قوسان

متساويتان وهما قوسا ح ج د ط وقد أخرج عمودا ك ج دل يصير عمود ك ج

مساويا لعمود د ل ويصير خط ح ك مساويا لخط ط ل وكل خط ح ز مساو

لكل خط ز ط فخط ز ل الباقي مساو لخط ك ز الباقي وخط از مساو لخط ز ب ‏ 35

فخطا اك ل ب متساويان متوازيان فلأن خط اك مساو لخط ل ب وخط ك ج

مساو لخط د ل يكون خطا اك ك ج جميعا مساويين لخطي ب ل ل د جميعا

كل واحد منهما لنظيره وزاوية جك ا مساوية لزاوية دل ب وذلك أن كل

واحدة منهما قائمة فقاعدة اج ج مساوية لقائدة د ب ، وذلك ما أردنا أن

نبين . ‏ 40

د   إذا كانت فى كرة دائرتان عظيمتان تقطع إحداهما الأخرى وفصل من

إحداهما قوسان متساويتان متصلتان إحداهما بالأخرى كل واحدة منهما من

ناحيتى إحدى النقطتين اللتين تتقاطعان عليهما ثم أخرج سطحان متوازيان

يمران بالنقطتين الحادثتين وأحدهما يلقى الفصل المشترك لسطحى الدائرتين

خارج بسيط الكرة من جهة النقطة التى ذكرنا وكانت كل واحدة من القوسين ‏ 5

29 جك [هك A 29 [ه ح A 31 إحداهما] احداهما;A ورص add. A 32 [د ط A زط 32 قطعة [N من

add. A 33 كج ... كج [كدلك يكون عمود كم A 34 [ح ز A 35 حب [لكل خط A لخط] زل [رط

A لخطا [N 37 حط A حط 2 خطا [Hadd. A 2 منهما] في [بالأخرى A 3 ناحيتى] A illeg. 36 [خطا

3 عليهما [H: عليها AN الدائرتين] NH: الراوسن A 5 بسيط [لسط A

GK DL, et producam duas lineas *AK LB*. Et quia arcus *EH* est equalis arcui *ET*, et hoc ideo quoniam polus circuli *AHBT* est punctum *E* et arcus *GE* unius eorum est equalis arcui *ED* alterius, fit reliquus arcus *GH* equalis reliquo arcui *DT*. Et quia arcus *HET* est portio circuli ex qua duo arcus equales sunt separati, qui sunt duo arcus *GH DT*, et due perpendiculares *KG DL* iam sunt producte, fit perpendicularis *KG* equalis perpendiculari *DL* et fit linea *HK* equalis linee *TL*. Tota autem linea *HZ* est equalis toti linee *ZT*. Ergo linea *ZL* reliqua equatur relique linee *KZ*. Sed linea *AZ* est equalis linee *ZB*: ergo due linee *AK LB* sunt equales et equidistantes. Et quia linea *AK* est equalis linee *LB* et linea *KG* est equalis linee *DL*, ergo due linee *AK KG* sunt equales duabus lineis *BL LD*, queque scilicet earum sue relative, et angulus *AKG* est equalis angulo *DLB*. Quod ideo est quoniam unusquisque eorum est rectus. Ergo basis *AG* est equalis basi *DB*. Et illud est quod demonstrare voluimus.

4   Cum in spera duo circuli maiores fuerint quorum unus alterum secet et separabuntur ex uno eorum duo arcus equales, quorum unus alteri continuetur in unaquaque duarum partium unius duorum punctorum super que se secant, postea producentur due superficies equidistantes transeuntes per duo puncta que accidunt, quarum una occurrat communi differentie superficierum duorum circulorum extra superficiem spere a parte puncti quod prediximus, et fuerit unusquisque duorum

45

50

55

5

41 producam] protraham **FiZBPsVa**   41 lineas] *om.* **Va**   42 ideo] est *add.* **OBPsVa**   43 unius] unus **R**   43 arcui] *supra* **OPs**, *om.* **B**   43 arcui *ED*] *tr.* **Va**   43 fit] sit **VFiZB**   43 reliquus arcus] *tr.* **MKgOBPsVa**   44 *GH*] *EH* **P**   44 *DT*] *in corr.* **B**   45 qua] *supra* **Ps**   45 arcus equales] *tr.* **PV**   45 equales] *om.* **FiZ**   45 equales ... arcus²] *marg.* **M**   45 separati] *in corr.* **Ps**   45 duo²] *om.* **M**   46 due] duo **Fi**   46 fit] sit **KgFiZ**   47 fit] *om.* **OBPsVa**, sit **VKgFiZ**   47 *HK*] *GHK* **V**   47 linee] *om.* **BPsVa**   48 equalis] totalis **Fi**   49 *KZ*] *in corr.* **Ps**   49 linee²] *om.* **Ps**   50–51 et ... sunt equales] *marg.* **M**   52 *BL LD*] *in corr.* **B**   52 queque] utraque **MKgOBPs**   52 scilicet] videlicet **R**, *om.* **PsVa**   52 earum] *om.* **MKgOBPsVa**   52 relative] *in corr.* **BPs**   53 Quod ideo est] *om.* **Va**   53 unusquisque] uterque **MKgOBPsVa**   54 est¹] *om.* **FiZ**   54 illud] hoc **O**, istud **BPsVa**   54 quod ... voluimus] *om.* **O** 55 demonstrare voluimus] dicere voluimus **Fi**, dicere volumus **Z**, volumus probare **BPsVa**   1 Cum] Cc **V**   3 in] *om.* **Z**   3 unaquaque] utraque **MKgOB PsVa**   3 partium] parium **Fi**, *in corr.* **Z**   4 se] *om.* **Z**   4 producentur] producantur **O**, producuntur **FiZ**   4 equidistantes] equedistantes **R**   5 accidunt] ad iungunt **Va**   7 et] *marg.* **R**   7 fuerit] fuit **Va**   7 unusquisque] -que *supra* **BPs**, uterque **MKg**

المتساويتين أعظم من كل واحدة من القوسين اللتين فُصلتا من الدائرة
الأخرى العظيمة بالسطحين المخرجين مما يلى تلك النقطة بعينها فإن القوس
التى بين النقطة التى تقاطعت عليها الدائرتان العظيمتان وبين السطح الذى
لا يلقى الفصل المشترك أعظم من القوس التى بين تلك النقطة وبين سطح
الدائرة الذى يلقى الفصل المشترك ·

10

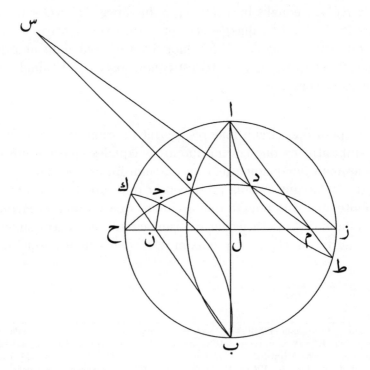

فلتكن فى كرة دائرتان عظيمتان وهما دائرتا اهب جهد تقطع إحداهما
الأخرى على نقطة ه ولتفصل من دائرة اهب منهما قوسان متساويتان وهما
قوسا اه هب المتصلتان إحداهما بالأخرى فى كل واحدة من ناحيتى نقطة ه
وليخرج سطحان متوازيان يمران بنقطتى آ ب وهما سطحا اد جب وليكن

---

arcuum equalium maior unoquoque duorum arcuum qui ex altero separantur circulo maiore secundum duas superficies productas ab ea
parte qua sequitur idem punctum, arcus, qui est inter punctum super   10
quod duo maiores circuli se secant et inter superficiem que non occurrit
communi differentie, est maior arcu qui est inter illud punctum et inter
superficiem circuli que occurrit communi differentie.

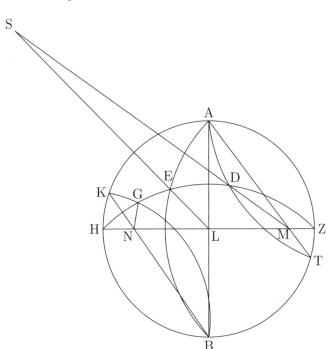

Sint ergo in spera duo maiores circuli, qui sint duo circuli *AEB*
GED, quorum unus alterum secet super punctum *E*, et separentur ex   15
circulo *AEB*, qui est unus eorum, duo arcus equales, qui sint arcus *AE*
EB, quorum unus alteri sit continuus in unaquaque duarum partium
puncti *E*. Et producantur due superficies equidistantes transeuntes
per duo puncta *A* et *B*, que sint superficies *AD GB*, et sit superfi-

8 maior] in *add.* **P**   8 unoquoque] utroque **MKgOBPsVa**   9 ab] ex **KgOBPsVa**
10 parte qua] *supra* **B**, *marg.* **Ps**   11 que] *in corr.* **Ps**   12–13 est … differentie] *marg.* **B**,
add. manu rec. **Ps**   12 inter[1]] *om.* **B**, *supra* **Ps**   12 illud punctum] *tr.* **MKg**   12 inter[2]] *om.*
OBVa, *supra* **Ps**   14 maiores circuli] *tr.* **OPsVa** (**Ps** *supra*)   14 circuli[1]] *om.* **FiB**, *su-*
pra **Ps**   14 sint] sunt **FiZVa**   14 duo circuli] *om.* **MKgOBPsVa**   14 *AEB*] *in corr.* **B**
15 secet] secat **R**   15 *E*] *supra* **BPs**   15 separentur] *in corr.* **BPs**   15 ex] a **KgOBPsVa**
16 *AEB*] *supra* **BPs**   16 arcus equales] *tr.* **PRVMFiZ**   16 sint] *in corr.* **Ps**, sunt **FiZ**
16 arcus[2]] *om.* **OBPsVa**   17 continuus] *corr. ex* communis **B**   17 unaquaque] utraque
MKgOBPsVa   18 producantur] producentur **B**, *corr. ex* producentur **Ps**   19 duo] *om.* **BPsVa**
19 sint] sunt **ZVa**   19 sit] fit **Ps**

سطح آد منهما ملاقياً للفصل المشترك لسطحى اه‌ب جه‌د خارج بسيط

الكرة من جهة نقطة ه‌ ولتكن كل واحدة من قوسى اه ه‌ب المتساويتين

أعظم من كل واحدة من قوسى جه ه‌د ، فأقول إن قوس جه أعظم من

قوس ه‌د وذلك أن الدائرة التى ترسم على قطب ه‌ وببعد ه‌آ تمر بنقطة ب

وتقع أبعد من نقطتى جه د لأن كل واحدة من قوسى اه ه‌ب أعظم من كل

واحدة من قوسى جه ه‌د فلتخرج ولتكن مثل دائرة اح‌ب‌ز ولتتمم الدائرتين

ولتلق دائرة جه‌د دائرة اح‌ب‌ز على نقطتى ح ز ولتلق دائرة آد دائرة

اح‌ب‌ز على نقطة ط ودائرة ب‌ج دائرة اح‌ب‌ز على نقطة ك وليكن

الفصل المشترك لدائرتى اه‌ب اح‌ب‌ز خط اب والفصل المشترك لدائرتى

ح‌ه‌ز اح‌ب‌ز خط ح‌ز والفصل المشترك لدائرتى ادط اح‌ب‌ز خط اط

والفصل المشترك لدائرتى كه‌ج‌ب اح‌ب‌ز خط كب والفصل المشترك لدائرتى

ح‌ه‌ز اه‌ب خط ه‌ل والفصل المشترك لدائرتى ح‌ه‌ز ادط خط م‌د والفصل

المشترك لدائرتى كه‌ج‌ب ح‌ه‌ز خط ج‌ن وسطح آد يلقى الفصل المشترك

لسطحى ح‌ه‌ز اه‌ب الذى هو خط ه‌ل خارج بسيط الكرة من جهة

نقطة ه‌ فليقه على نقطة س فنقطة س فى سطح ادط ولكنها فى سطح

ح‌ه‌ز أيضاً ونقطتا د م فى كلى سطحى ادط ح‌ه‌ز خط م‌د يلقى خط له‌

خارج بسيط الكرة من جهة نقطة ه‌ وهما يلتقيان على نقطة س فليلتقيا عليها

حـ [ح 21   A الدايره [الدائرتين 20   قوسى جه ه‌د [illeg. A   17 آد حـ [اه ه‌ب 16

N   اح [¹اح‌ب‌ز 21-22   حـ [آد ... ¹اد 21   H: om. AN   ⁴دائرة 21   اح‌ب‌ز [²اح‌ب‌ز 22

A   23-24 والفصل ... حز   om. A   ²اح‌ب‌ز [اح‌ب‌ط 24   AN   ¹اط [اط supra A

A فسطح [وسطح   ه‌د [ه‌ل A 27   والفصل [¹والفصل om. A 25   كه‌ج‌ب [H: كح sub حـ N 26   كب ... والفصل 25

كل [كلى N 30   A فلملماه, فلملاه   فليقه [فليلقه   om. N 29   ¹نقطة A 29 خارج العمود لبسط [خارج بسيط 28

N   A فلملمان, فلملما   فليلتقيا [فليلتقيا   om. A 31   يلتقيان [يلتقيان A 31 سط [بسيط A

cies AD, que est una earum, occurrens communi differentie duarum   20
superficierum $AEB\ GED$ extra superficiem spere a parte puncti E. Et
sit unusquisque duorum arcuum equalium $AE\ EB$ maior unoquoque
duorum arcuum $GE\ ED$. Dico igitur quod arcus GE est maior arcu ED.
Quod ideo est quoniam circulus qui signatur super polum E secundum
spacium EA transit per punctum B et cadit longe a duobus punctis   25
G et D, quoniam unusquisque duorum arcuum $AE\ EB$ est maior
unoquoque duorum arcuum $GE\ ED$. Protrahatur ergo et sit sicut circulus $AHBZ$; et complebo circulos. Et occurrat circulus GED circulo
$AHBZ$ super duo puncta H et Z, et circulus AD occurrat circulo $AHBZ$
super punctum T, et circulus BG occurrat circulo $AHBZ$ super punc-   30
tum K. Et sit differentia communis duorum circulorum $AEB\ AHBZ$
linea AB, et differentia communis duorum circulorum $HEZ\ AHBZ$
linea HZ, et differentia communis duorum circulorum $ADT\ AHBT$ est
linea AT, et communis differentia duorum circulorum $KGB\ AHBZ$ est
linea KB, et differentia communis duorum circulorum $HEZ\ AEB$ est   35
linea EL, et communis differentia duorum circulorum $HEZ\ ADT$ est
linea DM, et communis differentia duorum circulorum $KGB\ HEZ$ est
linea GN. Superficies igitur AD occurrit communi differentie duarum
superficierum $HEZ\ AEB$, que est linea EL, extra superficiem spere
a parte puncti E. Occurrat ergo ei super punctum S. Punctum ergo   40
S est in superficie ADT. Ipsum vero est etiam in superficie HEZ et duo
puncta $D\ M$ sunt in unaquaque duarum superficierum $ADT\ HEZ$:
ergo linea MD occurrit linee LE extra superficiem spere a parte puncti
E, et ipse concurrunt super punctum S. Concurrant ergo super ipsum.

---

20 AD] *om.* **Kg**; A, *supra* D **BPs**   20 una] superficierum *add.* **Va**   20 earum] superficierum
[...] *add. supra* **Ps**; equidem *add.* **Va**   20 occurrens] *in corr.* **Va**   21 superficiem spere] *tr.*
PRVMFiZ   22 unoquoque] utroque   **MKgOBPsVa**   23 ED^1] *in   corr.*   **Va**
24 signatur] significatur **Va**   25 EA] E **R**   26 quoniam] quod, *supra* id est quia
BPs, quia **Va**   26 unusquisque] uterque **MKgOBPsVa**   27 unoquoque] utroque **MKgOBPsVa**
27 Protrahatur] protrahantur **VZ**   28 $AHBZ$] *corr. ex* $AHZB$ **BPsVa**, $AHZB$ **PRVMKgOFiZ**
28 occurrat] occurret **BPsVa**   29 $AHBZ$] $AHZB$ **FiZ**   29–30 H ... punctum^1] *om.* **Fi**   31–
34 AEB ... circulorum] *om.* **BPsVa**   31 $AHBZ$] AHB **V**, $AH\ BH$ **FiZ**   33 $AHBT$] $AHBZ$
OFiZ   33 est] *om.* **OFiZ**   34–35 et ... KB] *marg.* **R**   34 communis differentia] *tr.*
Kg   34 KGB] KBG **Kg**   34 $AHBZ$] HBZ **FiZ**   34 est] *om.* **OFiZ**   35 linea] line **V**
35 differentia communis] *tr.* **R**   35 duorum] *om.* **OFiZ**   35–37 HEZ ... circulorum] *om.*
BPsVa   35 AEB] AFB **R**   35 est] *om.* **OBPsVa**   36 communis] cummunis **V**
36 communis differentia] *tr.* **MKg**   36 est] *om.* **OFiZ**   37 KGB] *in corr.* **Ps**, KBG
KgB   37 HEZ] KEZ **Kg**   37 est] *om.* **OFiZ**   38–39 duarum superficierum] duorum
circulorum **KgBPsVa**   39 que] *supra* **BPs**   39 superficiem] superficie **Fi**   40 Punctum] *om.*
Va   41 est] *post* superficie **BPsVa**   41 ADT] ATD **PRVMKgOFiZ**   41 vero] *om.* **OBPsVa**
41 est etiam] *tr.* **RBPsVa**   41 etiam] *om.* **KgZ**   41 superficie] circuli *add.* **MKgOBPs**
42 $D\ M$] *in corr.* **B**, D et M **FiZ**   42 in] *om.* **V**   42 unaquaque] utraque **MKgOBPsVa**
42 superficierum] equidistantium *add.* **KgOB**, *add. et del.* **Ps**   42 ADT] *in corr.* **B**,
ABT **Kg**   43 ergo] igitur **Va**   43 linee] line **V**   43 LE] *in corr.* **Kg**   44 ipse] ipsi
FiZ   44 concurrunt] *in corr.* **PB**, occurrunt **RPsVa**   44 super ... Concurrant] *om.* **Fi**
44 Concurrant] *in corr.* **B**, Occurrant **R**   44 ipsum] punctum **Kg**; ipsum, *corr. ex* punctum
B; punctum, *supra* ipsum **Ps**

ودائرة اهب العظمى فى كرة تقطع دائرة احبز من الدوائر التى تكون فى

الكرة وتمر بقطبيها فهى تقطعها بنصفين وعلى زوايا قائمة لخط اب قطر دائرة

احبز وكذلك نبين أن خط حز أيضاً قطر دائرة احبز فنقطة ل مركز

الدائرة ،

35

ولأن سطحى كجب ادط المتوازيين قد قُطعا بسطح احبز يكون

الفصلان المشتركان لهما متوازيين لخط كب موازٍ لخط اط وأيضاً فلأن سطحى

كجب ادط المتوازيين قد قُطعا بسطح حهز فيكون الفصلان المشتركان

لهما متوازيين لخط جن موازٍ لخط دم ولأن كل واحد من سطحى اهب حهز

قائم على سطح احبز على زوايا قائمة يكون الفصل المشترك لهما أيضاً

عموداً على سطح احبز والفصل المشترك لهما هو خط هل لخط هل عمود

على سطح احبز فهو يحدث مع جميع الخطوط التى تلقاه فى سطح

احبز زوايا قائمة وكل واحد من خطى اب حز اللذين هما فى سطح

احبز يلقى خط هل لخط هل عمود على كل واحد من خطى اب حز ،

ولأن زاوية سلن خارجة من مثلث سلم تكون أعظم من زاوية

سمل الداخلة المقابلة لها وزاوية سلن قائمة فزاوية سمل حادّة فزاوية

سمز منفرجة ولأن خط جن موازٍ لخط دم وقد وقع عليهما خط حز

تكون زاوية جنح مساوية لزاوية سمل وزاوية سمل حادة فزاوية

جنح حادة ولأن خط اط موازٍ لخط كنب وقد أخرج فيما بينهما خطا

A احب [احبز 41 A اب [احبز 40 A اب [اهب 39 om. N 39 اهب] سطحى 39 A فلان [ولأن 36
A لب [لكنب 49 A سمل [سمز 47 A فلان [ولأن 45 A واحدة [واحد 44 A حب [حز 43

Circulus autem *AEB* maior in spera secat circulum *AHBZ*, qui est   45
unus ex circulis qui sunt in spera, et transit per duos polos eius: ergo
secat eum in duo media et orthogonaliter. Ergo linea *AB* est diametrus circuli *AHBZ*. Et similiter monstratur quod linea *HZ* etiam est
diametrus circuli *AHBZ*: ergo punctum *L* est centrum circuli.

Et quia duas superficies equidistantes *KGB ADT* iam secuit su-   50
perficies *AHBZ*, due communes earum differentie sunt equidistantes:
ergo linea *KB* equidistat linee *AT*. Et etiam quia duas superficies equidistantes *KGB ADT* iam secuit superficies *HEZ*, ergo due communes
differentie earum sunt equidistantes: ergo linea *GN* equidistat linee
DM. Et quia unaqueque duarum superficierum *AEB HEZ* orthogo-   55
naliter est erecta super superficiem *AHBZ*, ergo communis differentia
earum etiam est perpendicularis super superficiem *AHBZ*. Communis
autem earum differentia est linea *EL*: ergo linea *EL* est perpendicularis super superficiem *AHBZ*. Ergo ex ipsa cum omnibus lineis, que ei
occurrunt et sunt in superficie *AHBZ*, recti proveniunt anguli. Una-   60
queque autem duarum linearum *AB HZ*, que sunt in superficie *AHBZ*,
occurrunt linee *EL*: ergo linea *EL* est perpendicularis super unamquamque duarum linearum *AB HZ*.

Et quia angulus *SMZ* extrinsecus est trianguli *SLM*, ergo est
maior angulo *SLM* intrinseco sibi opposito. Angulus autem *SLM* est   65
rectus: ergo angulus *SLM* est acutus. Ergo angulus *SMZ* est expansus.
Et quia linea *GN* equidistat linee *DM* et iam cecidit super eas linea
HZ, ergo angulus *GNH* est equalis angulo *SML*. Angulus autem *SML*
est acutus: ergo angulus *GNH* est acutus. Et quia linea *AT* equidistat
linee *KNB* et iam producte sunt inter eas due linee *AB MN* et linea   70

45 *AHBZ*] *HABZ* **KgBPs**   45 qui] quoniam **Fi**   45–48 qui ... *AHBZ*] *om.*
Va   46 unus ex circulis] *permut.* **BPs**   46 sunt] *post* spera **OPs**   47 et] *supra*
Ps   48 etiam] et **Fi**   48 etiam est] *tr.* **OBPsVa**   49 *AHBZ*] *HABZ* **BPs**
50 equidistantes] *om.* **V**   51 *AHBZ*] ergo *add.* **OFiZBPs**   51–53 *AHBZ* ... superficies] *om.*
Va   51 earum differentie] *tr.* **MKgFiZ**   52–54 ergo ... equidistantes] *marg.* **BPs**   52–
54 *KB* ... ergo linea] *om.* **Fi**   54 differentie earum] *tr.* **OBPsVa**   55 *DM*] *AM* **Kg**;
AM, supra D   55 unaqueque] utraque **MKgOBPsVa**   56 est erecta] *tr.* **BPsVa**   56–
57 ergo ... *AHBZ*] *om.* **BPsVa**   56 communis] omnis **Fi**   56–57 differentia earum] *tr.* **R**
57 earum] eorum **OFiZ**   58 earum] eorum **OFiZ**   58 est[1]] *om.* **Kg**, *supra* **BPs**   58 linea] *om.*
OFiZBPsVa   58 ergo linea *EL*] *om.* **Fi**   59 ex] *om.* **B**, *supra* **Ps**   59 omnibus] rectis *add.*
MKgOFiZBPsVa   60 *AHBZ*] *in corr.* **Ps**, *AKBZ (in corr.)* **KgB**   60–61 Unaqueque] Utraque
MKgOBPsVa   61 que ... *AHBZ*] *om.* **Fi**   62 occurrunt] *supra* **BPs**   62 *EL*[1]] *L, supra E* **Ps**
62 *EL*[2]] *in corr.* **BPs**   62–63 unamquamque] utramque **MKgBPsVa**   63 duarum] *in corr.*
B   64 angulus] anagulus **V**   64 *SMZ*] *FMZ* **PsVa**   64 extrinsecus est] *tr.* **MKgOBPsVa**
64 trianguli] *in corr.* **B**, circuli **Kg** 64–65 ergo ... *SLM*] *om.* **B**, *marg.* **Ps** 65 maior] *o.t. add.*
Va 65 angulo] circulo **Va** 65 *SLM*[1]] *LSM* **Kg** 65 Angulus autem] Sed angulus **MKgOBPsVa**
65 *SLM*[2]] *LSM* **KgB**   66 ergo ... acutus] *om.* **O**, *repet.* **B**, *hanc sententiam post* obtusus
(*sic pro* expansus) **Va**   66 *SLM*] *SML* **RMBPsVa** (**M** *in corr.*)   66 *SLM* ... angulus*[2]*] *om.*
OFiZ   66 Ergo ... expansus] *marg.* **Ps**   66–69 Ergo ... acutus*[2]*] *om.* **B**   66 *SMZ*] *in corr.*
O, *SML* **Kg**   66 expansus] actutus *supra* **O**, obtusus **PsVa**   67–69 Et ... acutus*[1]*] *om.* **PsVa**
68 *GNH*] *ENH* **Z**   68 autem] etiam **Kg**   69 *GNH*] *in corr.* **Ps**, *LNGH* **Va**   69 *AT*] *CT* **Kg**
70 *KNB*] *AND* **Kg**; *KNB, corr. ex AND* **B**   70 producte] protracte **R**   70 sunt] *supra* **R**

اب مَن وخط ال مساوٍ لخط لٮب يكون خط نل مساوياً لخط لَم وكل خط   50

حَل مساوٍ لكل خط لز فخط حَن الباقى مساوٍ لخط مَز الباقى ولأن حَز

قطعة دائرة وقد فصل من وترها خطان متساويان وهما خطا حَن مَز وقد

أخرج خطا جَن دَم متوازيين وزاوية جَنح حادة وزاوية دمز منفرجة

تكون قوس حَج أصغر من قوس دَز فلأن كل قوس حَه مساوية لكل

قوس هَز وقوس جَح أصغر من قوس دَز تصير قوس جَه الباقية أعظم من   55

قوس هَد الباقية ،   وذلك ما أردنا أن نبين .

ه   إذا كان قطب الدوائر المتوازية التى فى كرة على الخط المحيط بدائرة عظمى

من دوائرها وقطعت هذه الدائرة دائرتان عظيمتان على زوايا قائمة إحداهما

من الدوائر المتوازية والأخرى مائلة على الدوائر المتوازية وفصل من الدائرة

المائلة قوسان متساويتان متصلتان إحداهما بالأخرى فى جهة واحدة بعينها

عن الدائرة العظمى من الدوائر المتوازية ثم رسمت دوائر من الدوائر المتوازية   5

تمر بالنقط الحادثة فإنها تفصل من الدائرة الأولى العظمى قسياً غير متساوية

فيما بينهما وما كان من هذه القسى أقرب إلى الدائرة العظمى من الدوائر

المتوازية فهو أعظم من القوس التى هى أبعد منها .

50 من] A مَ 50 خط [وخط A 50 يكون] A مكون 50 نل] رل, N 51 لم] A, رل [repet. مساوٍ ... الباقى [2
A 51 ولأن] فلان N 51 ولأن] N 52 قطعة [من add. A 53 حادة ... وزاوية [1 tr. post منفرجة A 53-
54 منفرجة ... وزاوية [2 A, om. NH 54 فلأن] فلان A ولان 55 الباقية [الباقى A 56 الباقية [الباقى A, om. NH
الاولى [العظمى A 5 فى [عن N 5 الدوائر [corr. ex الدائرة N 3 وفصل [وقطعت N 2 كاب N 1 كان [
هو [هى N 8 فهو [فهى N 8 منها [بينهما N 7 قسياً [قسى N 6 om. N 5 من الدوائر المتوازية [add. N
N

AL equalis est linee *LB*, ergo linea *NL* est equalis linee *LM*. Tota autem linea *HL* est equalis toti linee *LZ*, reliqua ergo linea *HN* relique linee *MZ* equatur. Et quia *HEZ* est portio circuli et iam separate sunt ex corda eius due linee equales, que sunt *HN MZ*, et iam protracte sunt due linee equidistantes *GN DM* et angulus *GNH* est acutus et 75 angulus *DMZ* est expansus, ergo arcus *GH* est minor arcu *DZ*. Et quia totus arcus *HE* est equalis toti arcui *EZ*, sed arcus *GH* est minor arcu *DZ*, fit reliquus arcus *GE* maior reliquo arcu *ED*. Et illud est quod demonstrare voluimus.

5 Cum fuerit polus circulorum equidistantium qui sunt in spera super lineam continentem circulum maiorem ex circulis eius, et secuerint hunc circulum duo circuli maiores orthogonaliter, quorum unus sit ex circulis equidistantibus et alter sit inclinatus super circulos equidistantes, et separati fuerint ex circulo inclinato duo arcus equales, quorum 5 unus alteri continuetur in eadem parte a circulo maiore primo, deinde signati fuerint circuli qui sint ex circulis equidistantibus transeuntes per puncta provenientia, ipsi separabunt ex circulo primo maiore arcus inter se inequales. Et quicumque horum arcuum fuerit propinquior circulo maiori qui est ex circulis equidistantibus, erit maior arcu qui 10 ab eo fuerit magis remotus.

71 *AL*] *in corr.* **BVa**, *AB* **V**   71 equalis est] *tr.* **MOZBPsVa**   71 linee²] *om.* **Fi** 72 est equalis] equatur **OBPsVa**   72 *LZ*] et *add.* **Va**   72 ergo] igitur **BPs**   72 *HN*] *KN* **V**   73 equatur] est equalis (*post HN*) **FiZ**   73–74 Et … et] *om.* **FiZ**   74– 76 et … angulus] *marg.* **R**   75 linee equidistantes] *tr.* **R**   75 *GNH*] *GAH* **Va**   75– 76 *GNH* … angulus] *om.* **Fi**   76 minor] maior (*in corr.*) **Kg**   76–78 Et … *DZ*] *om.* **O**   77 est equalis] equatur **BPsVa**   77 toti arcui] *tr.* **Fi**   77 sed] si **Z**   78 fit] sit **Kg**   78 fit … arcu *ED*] *primum false, deinde recte* **R**   78 illud] istud **OBPsVa** 78 quod … voluimus] *om.* **O**   79 demonstrare voluimus] volumus demonstrare **BPsVa** 1 Cum] in spera *add.* **OBPsVa**   1 polus circulorum] *tr.* **OBPsVa**   2 secuerint] secuerunt **Z** 3 circuli] *om.* **M**   3 unus sit] *tr.* **V**   3 sit] fit **BVa**   4 super] duos *add.* **R**   5 arcus] *repet.* **Va**   6 parte] parate **M**   7 sint] sunt **OB+ PsVa**   8 ipsi] *in corr.* **B**   8 ex] a **OBPsVa** 8 a (*sc.* ex) circulo] *supra* **BPs**   9 inequales] *in corr.* **BPs**, equales **Va**   9 Et] *repet.* **Fi** 10 circulo maiori] *tr.* **R**   10 erit] *in corr.* **Fi**, est **Z**   11 ab … remotus] est remotior ab eo **OBPsVa**   11 fuerit] *post* remotus **Kg**   11 fuerit magis] *tr.* **RM**

فليكن على خط ا ب ج المحيط بالدائرة العظمى قطب الدوائر المتوازية

10   وهو نقطة آ ولتقطع هذه الدائرة دائرتان عظيمتان هما دائرتا ب ز ج د ز ه

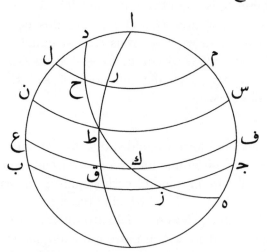

على زوايا قائمة إحداهما من الدوائر المتوازية والأخرى مائلة على الدوائر

المتوازية ولتفصل من الدائرة المائلة قوسان متساويتان وهما قوسا ك ط ط ح

على الولاء فى جهة واحدة بعينها عن دائرة ب ز ج العظمى من الدوائر

المتوازية ولترسم دوائر من الدوائر المتوازية تمر بنقط ك ط ح وهى دوائر

15   ع ك ف ن ط س ل ح م ، فأقول إنها تفصل من دائرة ا ب ج الأولى العظمى

قسياً غير متساوية وتكون القوس التى هى أقرب من الدوائر العظمى من

الدائرة المتوازية أبداً أعظم من القوس التى هى أبعد منها ، فأقول إن قوس

ع ن أعظم من قوس ن ل فلترسم دائرة عظيمة تمر بنقطتى آ ط وهى دائرة

ا ط ق وأيضاً فلأن نقطة آ قطب دائرة ع ك ف تكون قوس ان ع مساوية

المتوازية [11   A   ولتقطع [ولقطع   10   add. A   وهى دائره اح [العظمى   9   A   الحط [خط ا ب ج   9

om.] هى   16   A   ومر [تمر   14   A   وهى دائره دزه [المتوازية   12   add. A   وهى دائره رح

A   وأيضاً [19   illeg. A   نل [18   (الداره :الدوائر)   marg. A   من الدوائر العظمى [16   om. N

19-20   دائرة ... ع ك ف [om. A

Sit ergo super lineam continentem circulum maiorem ABG polus circulorum equidistantium, qui sit punctum A, et secent hunc circulum duo circuli maiores, qui sint circuli BZG DZE, orthogonaliter, quorum

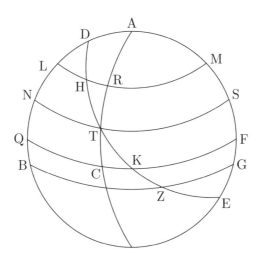

unus sit ex circulis equidistantibus et alter sit declinatus a circulis   15
equidistantibus. Et separentur ex circulo declinato duo arcus equales,
qui sint arcus KT TH continui in una et eadem parte a circulo maiore
BZG, qui est ex circulis equidistantibus. Et signentur circuli qui sint ex
equidistantibus transeuntes per puncta K T H, qui sint circuli QKF
NTS LHM. Dico igitur quod ipsi separant ex circulo primo maiore   20
ABG arcus inequales, et arcus qui est propinquior circulo maiori qui
est ex circulis equidistantibus semper est maior arcu qui magis est ab eo
remotus: dico igitur quod arcus QN est maior arcu NL. Signabo itaque
circulum maiorem transeuntem per duo puncta A et T, qui sit circulus
ATC. Et quia punctum A est polus circuli QKF, ergo arcus ANQ est   25

12 ergo] quoque **OBPsVa**   12 continentem] *om.* **Z**   12 circulum maiorem] *tr.* **OBPsVa**
12 *ABG*] *ALG* **M**   13 punctum] punctus **BPsVa**   13 hunc] cuus **Fi**   14 sint] sunt **BPsVa**
14 circuli²] *om.* **OBPsVa**   14 *DZE*] *DEZ* **FiZ**   15 unus] *repet.* **Kg**   15 sit] *om.* **Va**
15 equidistantibus] equidistantium **Fi**   15 sit declinatus] declinetur **OBPsVa**   16 ex] a **KgB**
PsVa   16 declinato] declinacio **Fi**   17 *KT TH*] *HT KT* **OBPsVa**, *H T* **V**   17 *TH*] qui
sint *add.* **OKgBPsVa** (sint: sunt **BPsVa**)   17 in] ea *add.* **Fi**   17 circulo maiore] *tr.* **RV**
18 *BZG*] *BZD* **Kg**   18 ex] *supra* **Va**   18 signentur] signetur **V**   18 sint] sunt **FiZ**
19 equidistantibus] et *add.* **BPs**   19 puncta] *in corr.* **Ps**   19 sint] sunt **OFiZ**   20 *LHM*] *LKM*
V   20 ipsi] *in corr.* **B**   20 separant] separent **B**; separant, *corr. ex* separent **Ps**; secant **FiZ**   20 primo] *om.* **B**, *in corr.* **Ps**   20 primo maiore] *tr.* **OPsVa**   22 semper] *om.*
FiZ   22 est²] cum **Va**   22 magis] *post* ab eo **FiZ**   23 igitur] ergo **O**   23 *QN*] *AL* **Fi**
23 arcu] quam arcus **KgBPsVa**   23 Signabo] *post* igitur **O**, Significabo **BPs**   23 itaque] igitur
MKgOBPsVa (**O** *post* circulum); a *add.* **V**   24 maiorem] *supra* **O**, *om.* **BPsVa**   24 duo] *supra*
V   25 *ATC*] *ATE* **ROB**, *ATR* **Va**   25 quia] *supra* **M**, *in corr.* **Ps**   25 punctum] punctus **Z**
25 *QKF*] quia *KF* **Fi**   25–26 est equalis] equatur **OBPsVa**, *hic et saepius*

قوس اطق وأيضاً لأن نقطة آ قطب دائرة نطس تكون قوس الن 20

مساوية لقوس ارط فقوس نع الباقية مساوية لقوس طق الباقية وكذلك

نبين أن قوس نل أيضاً مساوية لقوس رط فقوس نع مساوية لقوس طق

وقوس لن مساوية لقوس رط ودائرة اطق العظمى فى كرة تقطع دائرة من

الدوائر التى تكون فى الكرة وهى دائرة عقف وتمر بقطبيها فهى تقطعها

بنصفين وعلى زوايا قائمة فدائرة اطق قائمة على دائرة عقف على زوايا 25

قائمة ،

فقد عمل على قطر دائرة عقف الذى يخرج من نقطة ق قطعة من

دائرة قائمة على دائرة عقف على زوايا قائمة وهى قطعة قط معما يتصل

بها وقد فصل منها قوس هى أصغر من نصف القطعة التى عملت وهى قوس

طق فالخط المستقيم الذى يصل بين نقطة ق ونقطة ط أقصر جميع الخطوط 30

المستقيمة التى تخرج من نقطة ط إلى دائرة عقف فالخط المستقيم الذى

يصل بين نقطة ق ونقطة ط أقصر من الخط المستقيم الذى يصل بين نقطة

ط ونقطة ك ودائرتا ده اق متساويتان وذلك أنهما عظيمتان فقوس طق

أصغر من قوس طك وكذلك نبين أن قوس طر أصغر من قوس طح لأنه

قد عمل على قطر دائرة لحم قطعة من دائرة قائمة عليها على زوايا قائمة 35

وهى قطعة رط معما يتصل بها وفصل قوس رط أصغر من نصف القطعة

التى عملت وأيضاً فإن خط كط مساوٍ لخط طح فكل واحد من خطى كط

_____

22 فقوس [وقوس N   24 تكون [N om.   24 عقف [A, hic et saepius   24 illeg. A   27 دائرة [om. N

28 قائمة على دائرة [om. A   32 ط ... أقصر [om. A   33 ده اق [33 אק ואשׁ אטק H   اق [اٯ A

34 طك [طل A   36 وهى قطعة رط [om. AN   37–38 طر ... كط [خط كط مساوه لقوس طح كط

وس كط مساوه لقوس طح فكل واحده من موسى قط طر اصغر من كل واحدة من موسى كط طح A

equalis arcui *ATC*. Et etiam quia punctum *A* est polus circuli *NTS*, ergo arcus *ALN* est equalis arcui *ART*: ergo reliquus arcus *NQ* est equalis reliquo arcui *TC*. Simili quoque modo ostenditur quod arcus *NL* est equalis arcui *RT*: ergo arcus *NQ* est equalis arcui *TC* et arcus *LN* est equalis arcui *RT*. Circulus autem *ATC* maior in spera secat 30 unum ex circulis qui sunt in spera, qui est circulus *QCF*, et transit per polos eius: ergo ipse secat eum in duo media et orthogonaliter. Ergo circulus *ATC* orthogonaliter est erectus super circulum *QCF*.

Iam igitur super diametrum circuli *QCF*, que protrahitur a puncto *C*, constituta est portio circuli orthogonaliter erecti super 35 circulum *QCF*, que est portio *CT* cum eo quod ei adiungitur, et separatus est ex ea arcus qui est minor medietate portionis que facta est, qui est arcus *TC*. Ergo linea recta que coniungit inter punctum *C* et punctum *T* est brevior omnibus rectis lineis que a puncto *T* producuntur ad circulum *QCF*. Ergo linea recta que 40 coniungit inter punctum *C* et punctum *T* est brevior linea recta que coniungit inter punctum *T* et punctum *K*. Duo autem circuli *DE AC* sunt equales. Quod ideo est quoniam ipsi sunt maiores, ergo arcus *TC* est minor arcu *TK*. Et similiter ostenditur quod arcus *TR* est minor arcu *TH*, quoniam super diametrum circuli *LHM* 45 iam facta est portio circuli erecta super eam orthogonaliter, que est portio *RT* cum eo quod ei coniungitur, et separatus est arcus *RT* minor medietate portionis que facta est. Et etiam quia linea *KT* est

---

26 arcui] *om.* **BPs**   26 *ATC*] *ATE* **RO**; *ATC, supra CQF* **B**; *ATC, supra nota illeg.* **Ps**   26–27 Et … *ART*] *primum false, deinde recte* **BPs**   26 polus] est *add.* **Va**   26 *NTS*] *NT* **KgB**, *corr. ex MF* **O**   27 ergo] *om.* **OBPsVa**   27 arcui] *om.* **B**, *marg.* **Ps**   27 *ART*] *in corr.* **B** 27 arcus] *om.* **Fi** 27 *NQ*] *in corr.* **Ps** 28 reliquo] *om.* **BPs** 28 *TC*] *TE* **OVa**; *in corr., supra F* **B**   28 Simili quoque] Similique **BPsVa**   28 ostenditur] monstratur **OBPsVa**   29 *RT*] *KT, supra RT* **B**   29–31 ergo … arcui *RT*] *marg.* **R**   29 *TC*] *TE* **OZVa**   30 arcui] *om.* **BPs** 30 *ATC*] *ATE* **OVa**; qui *add.* **O**; qui *supra* **B**   30 maior] est *add.* **PsVa** 31 *QCF*] *QEF* **OVa**   32 et] *om.* **OBPsVa**   33 *ATC*] *ARE* **O**, *ATE* **Fi**; *AKC, supra R* **BPs**; *ARC* **Va**   33 orthogonaliter est] *tr.* **Kg**   33 est erectus] erigitur **OBPsVa**   33 *QCF*] *QEF* **OB**   34 igitur] ergo **R**   34 diametrum circuli *QCF*] *permut.* **BPsVa**   34 *QCF*] *QEF* **O**   34–36 que … *QCF*] *marg.* **BPs**   34 protrahitur] protrahuntur **Fi**, producitur **B**   35 *C*] *E* **O**, *T* **Va**   35 circuli] circulis **Fi**   35 erecti] ex recta **Fi**   36 *QCF*] *QEF* **O**   36 *CT*] *ET* **OB**, *om.* **Va** 36 cum] in, *supra* cum **B**; et cum **ZVa**   36 quod ei] *supra* **B**, quidem eidem **Va**   36 ei] eidem **Ps**   38 *TC*] *TE* **OZ**   38 recta] *om.* **Va**   38 coniungit] coniungitur **M**   39 *C*] *E* **OFiVa**   39–41 omnibus … brevior] *om.* **Kg**   40 *T*] *supra* **BPs**   40 *QCF*] *in corr.* **B**, *QEF* **O**, *QFC* **PsVa** 41 coniungit] coniungitur **M**   41 *C*] *E* **O**, *T* **FiZ**   41 *T*] *om.* **V**, *E* **FiZ**   41 brevior] longior **Z**   42 *T*] *C* **Fi**   43 *DE*] *in corr.* **O**   43 *DE AC*] *DE AE* **FiB**, *DEA* et **R**   43 est] *supra* **M**   44 arcus] diametrum *add.* **V**   44 *TC*] *AG* **M**, *TE* **OVa**; *TE, supra C* **B**   44 *TK*] *TK, supra N* **B**   45 *TR*] *TK, supra R* **B**   45 arcu] arcui **V**, *om.* **Z**   45 circuli] *TCS false add.* **Kg** 45 *LHM*] *in corr.* **BPs**   46 iam facta est] *om.* **OFiZ**   47 *RT*[1]] *DY* **KgB** (*supra RT* **B**), *DT* **PsVa**   47 quod] *om.* **V**   47 ei coniungitur] *tr.* **BPsVa**   47 coniungitur] adiungitur **KgOFiZ** 47 *RT*[2]] *supra* **B**, *corr. ex HT* **O**; *BY* (*in corr.*) **Kg**; *BY, supra HC* **Ps**; *HT* **Va**   48 etiam] *om.* **Va**   48 quia] *om.* **Kg**   48–49 est equalis] equatur **OFiZ**

طح أطول من كل واحد من خطى قط طر ،

فلأن دائرة بزج موازية لدائرة لحم ودائرة بزج تلقى الفصل

40 المشترك لدائرتى حطك ارق داخلاً أعنى على مركز الكرة صارت دائرة لحم

تلقى الفصل المشترك لدائرتى حطك اطق خارج بسيط الكرة من جهة نقطة

ط فلأن دائرتى حطك رطق العظيمتين تقطع إحداهما الأخرى وقد فصل

من دائرة حطك منهما قوسان متساويتان وهما قوسا كط طح على الولاء فى

كل واحدة من ناحيتى نقطة ط وقد عمل سطحان متوازيان يمران بنقطتى

45 ح ك وهما سطحا لحم عقف وسطح لحم منهما يلقى الفصل المشترك

لسطحى حطك رطق خارج بسيط الكرة من جهة نقطة ط وكل واحدة من

قوسى كط طح المتساويتين أعظم من كل واحدة من قوسى قط طر

تكون قوس قط أعظم من قوس طر ولكن قوس قط مساوية لقوس عن

وقوس طر مساوية لقوس نل فقوس عن أعظم من قوس نل ، وذلك ما

50 أردنا أن نبين ·

و   إذا كان قطب الدوائر المتوازية التى فى كرة على الخط المحيط بدائرة من

الدوائر الكبار وقطعت هذه الدائرة دائرتان عظيمتان على زوايا قائمة وكانت

إحدى الدائرتين من الدوائر المتوازية وكانت الأخرى من الدوائر المائلة على

لحم [احم N   40 داخل [داخلاً N   40 ازق [ارق N   40 متواربه [موازية N   39 ولان [فلأن N   39

طز [طر N   47 زطق [رطق N   42 فصل [فضل N   43 الوالى [الولاء :NH   46 طق [رطق A   42 رطق A 42

49 من [quod sequitur non videtur N   49 نل [نل² A   1 دد [A   1 الدوائر [الداره A   1 على ... ية [.invis

N 2 كبار رٹ [وقطعت [مفطعب A   2 الدائرة [الدوار N   2 عظيمتان [ان ... ان [.invis N

3 الدائرتين ى [.invis N   3 الأخرى [.invis N

equalis linee *TH*, ergo unaqueque duarum linearum *KT TH* est longior
unaquaque duarum linearum *CT TR*.　　　　　　　　　　　　　　　50

　　Et quia circulus *BZG* equidistat circulo *LHM* et circulus *BZG*
occurrit communi differentie duorum circulorum *HTK ARC* intus, scilicet super centrum spere, fit circulus *LHM* occurrens communi differentie duorum circulorum *HTK ATC* extra superficiem spere a parte
puncti *T*. Et quia duorum circulorum *HTK RTC* maiorum unus alte-　55
rum secat et iam separati sunt ex circulo *HTK*, qui est unus eorum,
duo arcus equales, qui sunt arcus *KT TH* secundum continuitatem in
unaquaque duarum partium puncti *T*, et iam facte sunt due superficies
equidistantes transeuntes per duo puncta *H K*, que sunt due superficies *LHM QCF*, et superficies *LHM*, que est una earum, occurrit　60
communi differentie duarum superficierum *HTK RTC* extra superficiem spere a parte puncti *T* et unusquisque duorum arcuum *KT TH*
equalium est maior unoquoque duorum arcuum *CT TR*: ergo arcus
CT est maior arcu *TR*. Sed arcus *CT* est equalis arcui *QN* et arcus
TR est equalis arcui *NL*: ergo arcus *QN* est maior arcu *NL*. Et illud　65
est quod demonstrare voluimus.

6　Si polus circulorum equidistantium qui sunt in spera super lineam continentem aliquem circulorum maiorum fuerit, quem duo maiores circuli
orthogonaliter secent, quorum unus sit ex circulis equidistantibus
et alter sit ex circulis inclinatis super circulos equidistantes, et separati

49 unaqueque] utraque **MKgOFiZBPsVa**　　　49 duarum linearum] *tr.* **Va**　　49 *KT*] *AT*
Z　　　49 longior] maior　**OFiZBPsVa**　　　　50 unaquaque] utraque　　**MKgOFiZBPsVa**
50 duarum linearum] *om.* **OFiZBPsVa**　　　50 *CT*] *in corr.* **B**, *AT* **R**, *ET* **OVa**　　51–
54 Et … puncti *T*] *om.* **Va** 51–54 circulus … Et quia] *om.* **FiZ** 51 et circulus *BZG*] *marg.*
BPs　　51 *BZG*] *BZH* **O**　　52 occurrit] *in corr.* **BPs**　　52 differentie duorum] *tr.* **OBPs**
52 *HTK*] *ATK* **Kg**　　52 *ARC*] *in corr.* **BPs**　　52–53 *ARC* … *HTK*] *om.* **O**　　52–
53 intus … *ATC*] *om.* **B** 53 fit] sit **Kg** 53 *LHM*] *LKM* **V** 53–54 differentie duorum] *tr.*
Kg, *in corr.* **M**　54 *ATC*] *RTE* **O**　55 *RTC*] *in corr.* **B**, *RTE* **O**　55 maiorum] maior **RB**
57 *KT TH*] *KTH* **Fi**　57 continuitatem] *in corr.* **V**　58 unaquaque] utraque **MKgOFiZ**
58 iam] *om.* **Z**　　59 equidistantes … due superficies] *repet.* **B**　　59 *H K*] *K H* **OZBVa**;
que sunt due superficies equidistantes transeuntes per duo puncta *K H false add. et del.* **Ps**
59–60 superficies] equidistantes *add.* **O**　60 *LHM*[1]] *SHM* **Va**　60 *LHM QCF*] *in corr.* **B**
60 *QCF*] *QEF* **O**, **QCT Fi**　61 *HTK*] *BTK* **Fi**　61 *RTC*] *RCT* **Kg**, *RTE* **O**, *RCE* (*supra*) **B**
62 *T*] *in corr.* **O** 62 unusquisque] uterque **MKgOBPsVa** 62 *TH*] *CH* **R** 63 equalium] *post*
arcuum **MKgOBPsVa**　63 unoquoque] utroque **KgOBPsVa**　63 arcuum] *corr. ex* circulorum
O, circulorum **Kg**　63 *CT*] *ET* **O**　63 *CT TR*] *in corr.* **B**　64 *CT*[1]] *TE* **O**, *TC* **KgFiPsVa**,
illeg. **B**　64 *TR*] *TK, supra R* **B**　64 *CT*[2]] *TR, corr. ex CT* **R**; *lac.* **Kg**; ille, *supra TE* **O**;
ille, *supra scilicet CE* **B**; *TC* **FiPsVa**　64 est equalis] equatur **O**　64 *QN*] ergo arcus *add.* **Fi**
65 *TR*] *in corr.* **B**　65 est equalis] equatur **O**　65 illud] istud **BPsVa** 66 demonstrare] *om.*
B　66 voluimus] volumus **ZBPsVa**　1 equidistantium] *corr. ex* equalium **Ps**　1 lineam] linea
V　2 maiorum] maior **Kg**　2 circuli] *om.* **FiZ**　3 secent] *in corr.* **B**　3 sit] fit **BPsVa**
4 sit] fit **PsVa**

الدوائر المتوازية وفصل من الدائرة المائلة قسى متساوية متصلة على الولاء فى

ناحية واحدة عن الدائرة التى هى أعظم الدوائر المتوازية ورسمت دوائر   5

عظيمة تمر بالنقط الحادثة وبالقطب فهى تفصل من الدائرة العظيمة من

الدوائر المتوازية فيما بينها قسياً غير متساوية والقوس التى تقرب من الدائرة

الأولى العظمى أبداً أعظم من القوس التى هى أبعد منها ·

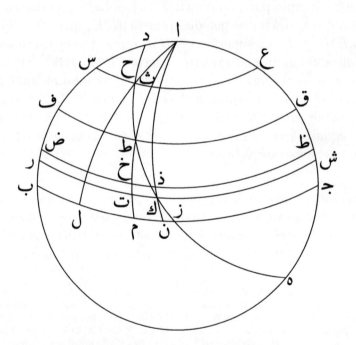

فليكن على خط اَبَجَ المحيط بالدائرة العظمى قطب الدوائر المتوازية

وهو نقطة اَ ولتقطع دائرة اَبَجَ دائرتا بَزَجَ دزه العظيمتان على زوايا   10

قائمة فلتكن دائرة بَزَجَ أعظم الدوائر المتوازية ودائرة دزه مائلة على

الدوائر المتوازية ولتفصل من دائرة دزه قوسان متساويتان وهما قوسا كَطَ

invis. [واحدة ... الدوا 5 التوالى A [الولاء invis. N 4 [قسى متساوية invis. N 4 [اثر ... المتوازية 4

N 6 العظيمة ... بالنقط [invis. N 7 الدوائر [om. A, אלדאירה H 7 غير ... من [invis. N

8–9 على ... م [invis. N, sequitur قطب [invis. N 9–10 اَ ... لتقطع [دا]رتان عطمتان N 10 ولتقطع[

A [العظيمتان 10 A وكف قطع [تا ... invis. N 11 فلتكن[وليكن A

fuerint ex circulo inclinato arcus equales coniuncti secundum continui-   5
tatem in parte una a circulo qui est maior ex circulis equidistantibus,
et signati fuerint circuli maiores transeuntes per puncta que proveniunt et per polum, ipsi separabunt ex circulo maiore qui est ex circulis
equidistantibus inter se arcus inequales, et arcus qui fuerit propinqui-
or maiori circulo primo semper erit maior arcu qui ab eo magis erit   10
remotus.

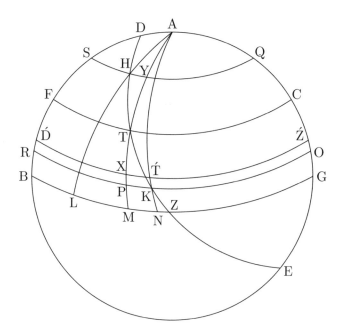

Sit itaque super lineam *ABG* continentem circulum maiorem
polus circulorum equidistantium, qui sit punctum *A*, et circulum *ABG*
secent duo circuli *BZG DZE* maiores orthogonaliter. Sitque circulus
BZG maior ex circulis equidistantibus, et circulus *DZE* sit inclinatus   15
super circulos equidistantes, et separentur ex circulo *DZE* duo arcus
equales, qui sint arcus *KT TH*, secundum continuitatem in parte una a

---

6 in] ex **FiZ**   6 a] *om.* **Z**   7 fuerint] sunt **V**   8 separabunt] *in corr.* **Ps**, *hic et saepi-*
us   8 circulis] *om.* **PRMKgBPsVa**   9 inequales] *in corr.* **B**   10 maiori] *om.* **R**, *in corr.*
B   10 primo] *post* propinquior **OVa**, *om.* **BPs**   10 semper erit] *in corr. supra* **B**, *in corr.* **Ps**
10 magis erit] *tr.* **R**   10 erit] est **FiZ**   10 erit remotus] *tr.* **OBPsVa**   12 continentem] *in*
corr. **B**   14 secent] secant **Fi**   15 *BZG*] B et G **Fi**   15 ex circulis] *om.* **Kg**   15 *DZE*] *DEZ*
Kg   16 *DZE*] *in corr.* **Ps**   17 sint] sunt **FiBPsVa**   17 *KT*] *HT* **Fi**   17 parte] parate **M**, *hic*
et saepius

ط ح على الولاء فى ناحية واحدة عن دائرة ب ز ج التى هى أعظم الدوائر

المتوازية ولترسم دوائر عظيمة تمر كل واحدة منها بنقطة آ وبواحدة واحدة

من نقط ح ك ط ط وهى دوائر اح ل اطم اكن ، فأقول إن قوس ل م أعظم

من قوس م ن فلترسم دوائر من الدوائر المتوازية تمر بنقط ح ط ك وهى

دوائر س ح ع ف ط ق رك ش فقوس رف أعظم من قوس ف س لما بيّنا فيما

تقدم ولكن قوس رف مساوية لقوس ت ط وقوس ف س مساوية لقوس

ط ث فقوس ت ط أعظم من قوس ط ث فلتوضع قوس ط خ مساوية لقوس

ث ط وقوس ح ط مساوية لقوس ط ك فالخط المستقيم الذى يصل بين نقطة

ح ونقطة ث مساوٍ للخط المستقيم الذى يصل بين نقطة خ ونقطة ك فلترسم

دائرة موازية للدوائر الأول تمر بنقطة خ وهى دائرة خ ذ ض فلأن دائرة

اذكن العظيمة التى فى كرة تقطع دائرة من الدوائر التى تكون فى الكرة

وهى دائرة خ ذ ض وتمر بقطيها فهى تقطعها بنصفين وعلى زوايا قائمة

فدائرة اذكن قائمة على دائرة خ ذ ض على زوايا قائمة ،

فلأن سطحى ب ز ج خ ذ ض المتوازيين قد قُطعا بسطح اذكن صارت

الفصول المشتركة لهما متوازية فالفصل المشترك لسطحى اذكن ب ز ج موازٍ

للفصل المشترك لسطحى اذكن خ ذ ض والفصل المشترك لسطحى ب ز ج

اذكن هو قطر دائرة اذكن الذى خرج من نقطة ن فالفصل المشترك

لسطحى اذكن خ ذ ض موازٍ لقطر دائرة اذكن الذى يخرج من نقطة ن

15

20

25

30

نكس [ركش 17   A ونمر [تمر 16   N وبواحد [وبواحدة واحدة 14   NH منهما [منها 14

N   [المستقيم 18   om. N   21 المستقيم] om.N   N طد [طك 20   supra A 20   ط ط [ت ط 19   A فوس [قوس 18

N مهو [هو A, و del. وهو [هو 29   A على [وعلى 24   A متوارنة [موازية 22

29 اذكن] om. A   30 ن ... خ ذ ض] marg., ن ... الذى in textu etiam N   29 اذكن[¹¹اذكن] supra A

circulo *BZG* maiore, qui est ex circulis equidistantibus. Signabo autem circulos maiores quorum quisque transeat per punctum *A* et per unum punctorum *H T K*, qui sunt circuli *AHL ATM AKN*. Dico igitur quod arcus *LM* est maior arcu *MN*. Signabo ergo circulos qui sunt ex equidistantibus transeuntes per puncta *H T K*, qui sint circuli *SHQ FTC RKO*. Arcus ergo *RF* est maior arcu *FS* propter hoc quod in precedentibus est ostensum; arcus vero *RF* est equalis arcui *PT* et arcus *FS* est equalis arcui *TY*: ergo arcus *PT* est maior arcu *TY*. Ponam autem arcum *TX* equalem arcui *YT*. Sed arcus *HT* est equalis arcui *TK*: ergo linea que coniungit inter punctum *H* et punctum *Y* est equalis linee que coniungit inter punctum *X* et punctum *K*. Signabo itaque circulum equidistantem circulis primis transeuntem per punctum *X*, qui sit circulus *XT́D́*. Et quia circulus *AT́KN* qui est in spera secat unum ex circulis qui sunt in spera, qui est circulus *XT́D́*, et transit per polos eius, ergo secat eum in duo media et orthogonaliter. Ergo circulus *AT́KN* orthogonaliter est erectus super circulum *XT́D́*.

Et quia duas superficies *BZG XT́D́* equidistantes superficies *AT́KN* iam secuit, fiunt differentie earum communes equidistantes. Differentia ergo communis duabus superficiebus *AT́KN BZG* equidistat differentie communi duabus superficiebus *AT́KN XT́D́*. Differentia vero communis duabus superficiebus *AT́KN BZG* est diametrus circuli *AT́KN*, que a puncto *N* protrahitur: differentia ergo communis duabus superficiebus *AT́KN XT́D́* equidistat diametro circuli *AT́KN*, que protrahitur a puncto *N*. Iam ergo protracta est in circulo *AT́KN*

20

25

30

35

40

18 BZG] *in corr.* **R**   *18* maiore] *post* circulo **R**, *om.* **BPsVa**   *18* Signabo] Significabo **BPs**, *hic et saepius*   *19* transeat] transit (*post A*) **OBPsVa**   *19* punctum *A* et per] *om.* **Va**   *20* sunt] sint **MVZ**, sit **PsVa**   *20* circuli] *om.* **Z**, circulus **PsVa**   *20 AHL*] *HAL* **KgB**, *ALH* **Z**   *20 AKN*] *om.* **V**   *20* igitur] ergo **BPsVa**   *21 MN*] *in corr.* **Kg** *21* circulos] alios **B**   *21* ex] circulis *add.* **FiZ**   *22* sint] sunt **FiPsVa**   *22 SHQ*] *in corr.* **B**   *22 FTC*] *FET* **FiZ**   *23 RKO*] *in corr.* **B**   *23* ergo] igitur **R**   *23 RF*] *TF* **Z**, *in corr.* **BPsVa**   *23–24* maior ... *RF* est] *om.* **Ps**   *23* in] *supra* **Va**   *24* est¹] *om.* **Fi** *24* ostensum] ostensus **Kg**   *24 RF*] *in corr.* **RB**   *24–25* est equalis] equatur **OBPsVa**, *hic et saepius*   *25 TY¹*] *RZ* **R**   *25 TY²*] *in corr.* **Ps**   *26 TX*] *om.* **Va**   *26 YT*] *in corr.* **Ps**   *26 TK*] *in corr.* **Kg**   *27–28 H* ... punctum¹] *om.* **Z**   *28* linee] *in corr.* **BPs**   *28* itaque] autem **KgOBPsVa**   *30* sit] *in corr.* **Va**   *30 XT́D́*] *XEFi* **Kg**, *XRFi* **B**   *30* circulus²] maior *add. marg.* **Ps**, *add. in textu* **Va**   *30 AT́KN*] *vide* **B** *in lin.* 35 *30* secat] circulum *add.* **BPsVa**   *31* ex circulis qui sunt] *om.* **BPsVa**   *31* circulus] *in corr.* **B**   *32* polos eius] eos (*in corr.*) **R**   *32* eius] *om.* **Z**   *32* in ... orthogonaliter] *permut.* **M**   *33* est erectus] erigitur **BPsVa**   *33* super circulum] *om.* **Va**   *33 XT́D́*] *in corr.* **PsVa**   *34* superficies²] *om.* **BPsVa**   *35* iam ... equidistantes] *false post AT́KN* (*l. 30*) **B**   *35* fiunt] fuerint **Fi**   *35* differentie earum] *tr.* **OFiZ**   *35* earum] eorum **KgBPsVa** *36* Differentia] *in corr.* **Z**; vero *add. supra* **Ps**   *36* ergo] vero **KgVa**   *36* ergo communis] *tr.* **BPs** (*in corr.* **Ps**)   *36–37* equidistat] equidistant **R**   *36–38* equidistat ... *BZG*] *om.* **KgPsVa** *37* duabus superficiebus] *om.* **B**   *37–38 XT́D́* ... *BZG*] *om.* **Fi**   *39* protrahitur] *post* que **KgBPsVa** (**B** *in corr.*)   *39–41* differentia ... puncto *N*] *om.* **Va**   *41–44* Iam ... puncto *N*] *marg.* **BPs**   *41* protracta] producta **FiZ**   *41* in] a **Va**

فقد أخرج فى دائرة اذكن خط ما وهو الفصل المشترك لدائرتى اذكن

خذص يقسم الدائرة بأقسام غير متساوية موازياً لقطر دائرة اذكن الذى

يخرج من نقطة ن وقد عملت عليه قطعة من دائرة قائمة على دائرة اذكن

على زوايا قائمة وهى قطعة خذ مع القطعة المتصلة بهذه وقُسمت قوس

35 القطعة القائمة بأقسام غير متساوية على نقطة خ وقوس خذ أصغر من

نصف القطعة المعمولة فالخط المستقيم الذى يصل بين نقطة خ ونقطة ذ

أقصر جميع الخطوط المستقيمة التى تخرج من نقطة خ إلى قوس ذكن

فالخط المستقيم الذى يصل بين نقطة خ ونقطة ذ أقصر من الخط المستقيم

الذى يصل بين نقطة خ ونقطة كـ والخط الذى يصل بين نقطة خ ونقطة كـ

40 أطول من الخط الذى يصل بين نقطة خ ونقطة ذ والخط الذى يصل بين

نقطة خ ونقطة كـ مساوٍ للخط الذى يصل بين نقطة ح ونقطة ث فالخط

الذى يصل بين نقطة ح ونقطة ث أطول من الخط الذى يصل بين نقطة خ

ونقطة ذ فلأن دائرة خذص أقرب إلى مركز الكرة من دائرة سحع تكون

دائرة خذص أعظم من دائرة سحع فلأن دائرتى سحع خذص غير

45 متساويتين ودائرة سحع أصغرهما وقد خرج فى دائرة سحع الخط الذى

يصل بين نقطة ح ونقطة ث وفى دائرة خذص الخط الذى يصل بين نقطة

خ ونقطة ذ وكان الخط الذى يصل بين نقطة ح ونقطة ث أطول من الخط

الذى يصل بين نقطة خ ونقطة ذ تكون قوس حث أعظم من القوس

ذ] N 38   موار [موازياً 32   om. A (add. supra رح) 32–33 اذكن ... يقسم [om. AH 32   خذص] in 32
corr. A   38 فالخط [والخط N   يصل] 39   om. A   والخط [فالخط 39   المستقيم [om. A   أقصر من [marg. A 38
N مكان [وكان 47   om. A سحع ...²] om. A   فلأن [ولأن N   45–46 دائرة ... سحع N   أطول ... ذ [40–41

aliqua linea que est communis differentia duorum circulorum $A\acute{T}KN$ $X\acute{T}\acute{D}$ dividens circulum in partes inequales equidistans diametro circuli $A\acute{T}KN$, que protrahitur a puncto N, super quam portio circuli facta est super circulum $A\acute{T}KN$ orthogonaliter erecta, que est portio $X\acute{T}$ cum portione que cum hac continuatur, et divisit arcum portionis erecte in partes inequales super punctum X. Arcus autem $X\acute{T}$ est minor medietate portionis date; linea igitur recta que coniungit inter punctum X et punctum $\acute{T}$ est brevior omnibus rectis lineis que a puncto X ad arcum $\acute{T}KN$ producuntur. Ergo linea recta que coniungit inter punctum X et punctum $\acute{T}$ est brevior linea recta que coniungit inter punctum X et punctum K. Linea autem que coniungit inter punctum X et punctum K est longior linea que coniungit inter punctum X et punctum $\acute{T}$. Linea vero que coniungit inter punctum X et punctum K est equalis linee que coniungit inter punctum H et punctum Y: ergo linea que coniungit inter punctum H et punctum Y est longior linea que coniungit inter punctum X et punctum $\acute{T}$. Et quia circulus $X\acute{T}\acute{D}$ est centro spere propinquior circulo SHQ, ergo circulus $X\acute{T}\acute{D}$ est maior circulo SHQ. Et quia duo circuli SHQ $X\acute{T}\acute{D}$ sunt inequales et circulus SHQ est minor ex eis et iam protracta est in circulo SHQ linea, que coniungit inter punctum H et punctum Y, et in circulo $X\acute{T}\acute{D}$ linea que coniungit inter punctum X et punctum $\acute{T}$, sed linea que coniungit inter punctum H et punctum Y est longior linea que coniungit inter punctum X et punctum $\acute{T}$: ergo arcus HY est maior arcu qui est similis arcui $X\acute{T}$

45

50

55

60

43 circulum] *marg.* **R**; $A\acute{T}KN$ *add.* **OFiZ** ($A\acute{T}HN$ **Fi**)   43 equidistans] *om.* **Va**   45 $X\acute{T}$] ZXI **Kg**   46 divisit] divisimus **OFiZBPsVa** (**B** *in corr.*)   46 portionis] in duo media *add.* **Kg**, *add. et del.* **Ps**   46 erecte] *marg.* **M**   47 inequales] *in corr.* **BPs**   47 X] L **Kg**   47 $X\acute{T}$] X **Va**   48 portionis] *om.* **V**, *in corr.* **Ps**   48 date] *in corr.* **BPs** 48 igitur] ergo **KgOZBPsVa**   48 punctum] puncta **Fi**   49 $\acute{T}$] L **R**   50 $\acute{T}KN$] LKN **R**, KN **Va**   50 producuntur] protrahuntur (*post* que) **BPsVa**   50–51 inter … coniungit] *marg.* **Ps** 50 X] *om.* **Fi**   51 $\acute{T}$] L **R**   51 linea recta] *tr.* **OFiZ**   52–53 Linea … punctum K] *marg.* **B**   52–53 Linea … punctum $\acute{T}$] *scr. et del.* **M**, *om.* **OFiZ**   52 autem] vero **KgOBPsVa** 52 X] H **PRVMZ**   53–54 est … punctum K] *om.* **B**   53–54 longior… est] *om.* **KgOPsVa** 53 $\acute{T}$] K **MSS**; ergo ipsa est longior linea que coniungit inter punctum X et punctum I *add.* **PRV** (**R** *post* punctum K, *l.* 52)   55 H] *in corr.* **B**   55 Y] *in corr.* **B**   55–56 ergo … punctum Y] *marg.* **BPs**   55 linea] recta *add.* **OBPsVa**   57 punctum X] puncta quod **Fi**   57–58 centro … circulus $X\acute{T}\acute{D}$] maior circulo **Fi**   58 circulo[1]] *om.* **R** 58 SHQ] sed HQ **R**, $SHYQ$ **Z**, *in corr.* **B**   58–59 ergo … circulo SHQ] *marg.* **BPs**   58– 59 ergo … circuli SHQ] *marg.* **M**   59 SHQ^2] *in corr.* **B**   59 sunt] fuerint **V**   60 ex] *supra* **BPs**   60 in] a *add.* **V**   61 Y] *supra* **B**   62 inter[1]] *supra* **O**   62 X et punctum] *om.* **Fi** 62 sed] et **OFiBPs**, *om.* **Va**   62–64 sed … punctum $\acute{T}$] *marg.* **BPs**   62 punctum[3]] puncta **OBPsVa**   63 H] hec **V**   63 H et punctum Y] H Y **OFiZBPsVa**   63 punctum[2]] puncta **OB** **PsVa**   63–64 X etpunctum $\acute{T}$] X $\acute{T}$ **OBPsVa**

الشبيهة بقوس خَذ من دائرتها ولكن قوس حَث شبيهة بقوس لَم وقوس

خَذ شبيهة بقوس مَن فقوس لَم أعظم من القوس الشبيهة بقوس مَن من ٥٠

دائرتها وهما من دائرة واحدة فقوس لَم أعظم من قوس مَن ، وذلك ما

أردنا أن نبين ·

ز   إذا كانت فى كرة دائرة عظيمة تماس دائرة من الدوائر المتوازية وكانت دائرة

أخرى عظيمة مائلة على الدوائر المتوازية تماس دائرتين أعظم من الدائرتين

اللتين كانت الدائرة الأولى تماسها وكانت مواضع المماسة أيضاً على الدائرة

الأولى العظمى وفصلت من الدائرة المائلة قسى متساوية متصلة على الولاء

فى جهة واحدة بعينها من الدائرة العظمى من المتوازية ورسمت دوائر متوازية ٥

تمر بالنقط الحادثة فإنها تفصل فيما بينها من الدائرة الأولى العظمى قسياً غير

متساوية والقوس القريبة من أعظم الدوائر المتوازية أعظم من القوس التى

هى أبعد منها ·

فلتماس فى كرة دائرة اَبَجَ العظمى دائرة ما من الدوائر التى تكون

فى الكرة وهى دائرة اَدَ على نقطة آ ولتماس دائرة أخرى عظيمة مائلة على ١٠

الدوائر المتوازية وهى دائرة هزحَ دائرتين أعظم من الدائرتين اللتين كانت

تماسهما دائرة اَبَجَ ولتكن مواضع المماسة أيضاً على دائرة اَبَجَ على نقطتى

هَ حَ ولتكن أعظم الدوائر المتوازية دائرة بَزجَ ولتفصل من الدائرة المائلة

---

50 الشبيه [الشيهة A   1 دائرة³] om. N   5 من المتوازية] متوازية [add. N غير مساوبه
AN اَبَ :H in corr.   10 اَد] add. AH المتوازية [الدوائر N   9 وابها [فإنها :H بالنقطة AN وإنها [بالنقط 6
N 11 وهى دائرة هزحَ] om. NH   12 اَبَجَ¹] اَحَد N   13 أعظم [من add. A الدائرة [الدوائر A, داره
N

ex suo circulo. Arcus autem *HY* est similis arcui *LM* et arcus *XT̄* est   65
similis arcui *MN*: ergo arcus *LM* est maior arcu simili arcui *MN* ex
suo circulo; ipsi autem sunt ex uno circulo: ergo arcus *LM* est maior
arcu *MN*. Et illud est quod demonstrare voluimus.

7   Si in spera fuerit circulus maior contingens aliquem ex circulis equidistantibus et fuerit alius circulus maior inclinatus super circulos equidistantes contingens duos circulos maiores circulis quos primus circulus
fuit contingens fuerintque etiam loci contactus super circulum primum
maiorem et separati fuerint ex circulo inclinato duo arcus equales con-   5
iuncti secundum continuitatem in parte una a circulo qui est maior ex
circulis equidistantibus, et signati fuerint circuli equidistantes inequales
transeuntes per puncta que provenient, ipsi inter se separabunt ex circulo primo maiore arcus inequales; et arcus propinquior maiori circulo
qui est ex equidistantibus est maior arcu ab eo remotiore.   10

     In spera itaque circulus maior *ABG* aliquem contingat ex circulis
qui sunt in spera super punctum *A*, qui sit *AD*, et contingat alius
circulus maior inclinatus super circulos equidistantes duos circulos
maiores circulis quos circulus *ABG* contingit. Et sint etiam loci
contactus super circulum *ABG* super duo puncta *E* et *H*; et sit maior   15
ex circulis equidistantibus circulus *BZG*, et separentur ex circulo *EZH*

66 arcu] *in corr.* **P**   67 circulo¹] *supra* **R**   68 illud] hoc **O**, istud **PsVa**
68 quod … voluimus] *marg.* **V**, *om.* **O**   68 demonstrare] monstrare **V**, dicere **FiZ**
68 demonstrare voluimus] volumus demonstrare **BPsVa**   1 Si] Cum **VFiZ**   1 fuerit] *post*
maior **PRVMKg**   2 et] cum **PsVa**   3 maiores] qui sunt ex *add.* **PRVMFiZ** (sunt: sint
PRV); ex *add.* **OVa**, *add. supra* **BPs**   3 circulus] *om.* **Va**   4 fuit] fuerit **OBPsVa**, erat
Fi   4 fuerintque] fueritque **PRV**, fuerint quod **Fi**   4 etiam loci] *tr.* **O**   4 loci] *om.* **Fi**
4 primum] primo **Kg**   5 et] etiam **M**   5 fuerint] fuerit **V**   6 parte] parate **M**, *hic et
saepius*   6 maior] maximus **FiZ**   7 equidistantibus] duo circuli *DTK ZE AIKN* (*I supra*) in puncto *K* se secare debent *add.* **RV** (**R**: *marg., ad l. 24*)   7 signati] significati
Z   7 circuli equidistantes] *supra* **B**, *marg.* **Ps**   7 equidistantes inequales] *tr.* **FiZ**
7 inequales] equidistantes *add.* **O**   7 inequales transeuntes] *tr.* **BPsVa**   8 puncta] *in corr.*
BPs   10 est²] *om.* **Fi**   10 arcu] *supra* **R**   10 remotiore] remotior **R**   12 *A*, qui sit *AD*] *AD*
autem qui sit **V**   12 qui sit *AD*] *om.* **MKgOFiZBPsVa**   13 equidistantes] *in add.*
V   13 duos circulos] *tr.* **M**   14 maiores] ex *add.* **PRVMOFiZBPsVa** (**BPs** *supra*)   14 quos] *supra* **Ps**   14 circulus] circulis **V**   14 circulus *ABG*] *tr.* **MKgOBVa**
14 sint] sit **PRVMKg**   15 *E* et *H*] *in corr.* **BPs**; circuli *EZH* **FiZ** (*EZH: EH* **Fi**)   16–
17 circulus … equidistantibus] *marg.* **Ps**

على الدوائر المتوازية وهى دائرة هزح قوسان متساويتان وهما قوسا لك كط

15   على الولاء فى جهة واحدة من الدائرة العظمى من الدوائر المتوازية ولترسم

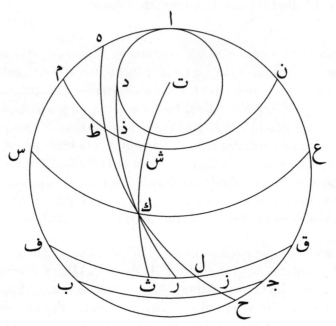

دوائر متوازية تمر بنقط طا كا لا وهى دوائر مطن سك ع فلق ، فأقول

إن قوس ف س أعظم من قوس م س فلترسم دائرة عظيمة تمر بنقطة كـ

وتكون مماسة لدائرة آد وهى دائرة ركد فليس يلقى نصف الدائرة الذى

يخرج من نقطة آ إلى ناحية ب نصف الدائرة الذى يخرج من نقطة د إلى

20   ناحية ر وليتعلم قطب الدوائر المتوازية وليكن نقطة ت ولترسم دائرة عظيمة

تمر بنقطتى ت كـ وهى دائرة تكث فدائرة تكث العظمى التى فى كرة

تقطع دائرة ما من الدوائر التى تكون فى الكرة وهى دائرة فلق وتمر

داره [الدائرة 18 A هاسه [مماسة 18 N داره [دوائر 16 *om.* NH 16 [وهى دائرة هزح 14 N عى [على 14
الى [الذى 19 A داره [الدائرة 19 N سصف AH, [نصف AH, [لصف [نصف 19 A آب [ب 19 A والى [الذى 18 A
N 20 [ر (.*ed*): در A, درد بد N, ا H 21 [فدائرة ودارة N 21 [كرة A الكره A 22 الكره ... دائرة] *om.*
A

inclinato a circulis equidistantibus duo arcus equales, qui sint arcus *LK*
KT, secundum continuitatem in parte circuli maioris qui est ex circulis

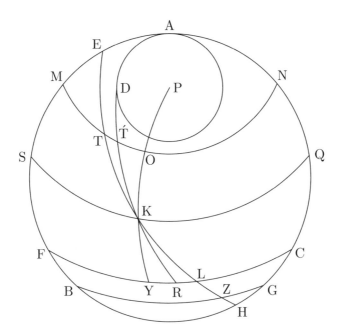

equidistantibus. Signabo autem circulos equidistantes transeuntes per
puncta *T K L*, qui sint circuli *MTN SKQ FLC*. Dico igitur quod arcus   20
FS est maior arcu *SM*. Signabo igitur circulum maiorem transeuntem
per punctum *K*, qui sit contingens circulum *AD*, qui sit circulus *RKD*.
Medietas ergo circuli que producitur a puncto *A* ad partem *B* non
occurrit medietati circuli que protrahitur a puncto *D* ad partem *RD*.
Signabo autem polum circulorum equidistantium, qui sit punctum *P*,   25
et describam circulum maiorem transeuntem per duo puncta *P* et *K*,
qui sit circulus *PKY*. Circulus autem maior *PKY* qui est in spera secat aliquem ex circulis qui sunt in spera, qui est circulus *FLC*, et transit

17 inclinato] *in corr.* **V**   17 sint] sunt **Fi**   19 transeuntes] *om.* **Fi**   20 sint] sunt **FiZ**
20 *MTN*] *MN, supra T* **O**   20 *FLC*] *FCL* **R**, *FLT* **Va**   21 *FS*] *FL* **Fi**, *FC* **Z**   21 arcu] *om.*
Fi   22 sit contingens] contingat **OBPsVa**, contingit **FiZ**   22 *RKD*] *RKD RK* **V**, *RKID* **Fi**,
TKLD **Z**   23 ergo] igitur **PsVa**   23 producitur] *in corr.* **P**   23 *B*] *supra* **BPs**, *LX* **Z**
24 occurrit] concurrit **BPsVa**   24 *D*] *A* **Fi**   24 partem] punctum **Z**   24 *RD*] *K* **OBPs**
Va (**BPs**: *corr. ex KD*), *R* **FiZ**   25 *P*] p⁹ **Fi**   26 maiorem] *om.* **BPsVa**   26 duo] *om.*
BPsVa   27 maior] *om.* **KgBPsVa**   27 spera] *in corr.* **B**   28 ex circulis] circulorum **MKgBPsVa**
28 *FLC*] *FLE* **Va**   28–30 et … *FLC*] *marg.* **P**

بقطريها فهى تقطعها بنصفين وعلى زوايا قائمة فدائرة ت‌ك‌ث قائمة على
دائرة ف‌ل‌ق ،

٢٥   فقد عمل على قطر دائرة ف‌ل‌ق الذى يخرج من نقطة ث قطعة من
دائرة قائمة عليها على زوايا قائمة وهى قطعة ت‌ث مع القطعة المتصلة بها
وقد قُسمت قوس القطعة التى عملت بأقسام غير متساوية على نقطة ك
وقوس ك‌ث هى الصغرى من القسمين فيكون الخط المستقيم الذى يصل بين
نقطة ك ونقطة ث أقصر جميع الخطوط المستقيمة التى تخرج من نقطة ك
٣٠   إلى الخط المحيط بدائرة ف‌ل‌ق والخط القريب منها أبداً أصغر من الذى هو
أبعد فالخط الذى يصل بين نقطة ك ونقطة ر أقصر من الخط الذى يصل
بين نقطة ك ونقطة ل <...> أطول من الخط الذى يصل بين نقطة ك
ونقطة ر ودائرتا د‌ر ه‌ل‌ح متساويتان وذلك أنهما عظيمتان فقوس ك‌ل أعظم
من قوس ك‌ر وكذلك نبين أن قوس ط‌ك أيضاً أعظم من قوس ك‌ذ وقوس
٣٥   ط‌ك مساوية لقوس ك‌ل فكل واحدة من قوسى ط‌ك ك‌ل أعظم من كل
واحدة من قوسى ك‌ذ ك‌ر ولأن دائرة ب‌ز‌ج موازية لدائرة م‌ط‌ن ودائرة
ب‌ز‌ج تلقى الفصل المشترك لدائرتى ط‌ك‌ل ذ‌ك‌ر داخل بسيط الكرة صارت
دائرة م‌ط‌ن تلقى الفصل المشترك لدائرتى ط‌ك‌ل ذ‌ك‌ر خارج بسيط الكرة من

الى تصل [المتصلة A   ٢٦ قطعه [القطعة A   ٢٦ فقد ... om. N   فـ‌ل‌ق; AH] وقد [فقد ٢٥
كون [فيكون N   ٢٨ الموسى AH: [القسمين N   ٢٨ وقسمت بقسمين مختلفين [وقد ... متساوية ٢٧
A مقطه [ونقطة A   ٣٠ حط [الخط A   ٣١-٣٢ ر] ... ونقطة [om. H   ٣٢ بين¹ [من A   ٣٢ مقطه
N أيضاً¹ [أيضاً N   ٣٤ ك‌ر] كن N   ٣٤ ن] ر N   ٣٣ ر] om. H   ٣٣ ونقطة] om. H   ٣٢-٣٣ ر... أطول] om. NH
واحد [واحدة A وكل [فكل A   ٣٥ مساو [مساوية A   ٣٥ وبه add. A   ٣٥ ط‌ك‌ك] om. A   ٣٥ ك‌ذ وقوس ٣٤
A من¹ و, supra N   ٣٨ من] و N   ٣٨ لدا] لدائ N   ٣٨ [الدائرى A خارج [خارج A   ٣٧ ك‌ر‌ك A داخل] داخل ٣٦ ك‌ذ ك‌ر] ك‌ذ‌ك‌ر A

per polos eius: ergo ipse secat eum in duo media et orthogonaliter.
Ergo circulus PKY est erectus super circulum FLC. 30

Iam autem facta est super diametrum circuli FLC, que protrahitur a puncto Y, portio circuli orthogonaliter super eam erecta, que est
portio PY cum portione ei continua, et secatur arcus portionis que
facta est in partes inequales super punctum K. Et arcus KY est minor
duabus sectionibus: ergo linea recta que coniungit inter punctum K 35
et punctum Y est brevior omnibus rectis lineis que a puncto K producuntur ad lineam continentem circulum FLC. Et linea recta que ei
propinquior existit semper est minor ea que ab ea magis est remota: ergo linea que coniungit inter punctum K et punctum R est brevior linea
que coniungit inter punctum K et punctum L. Ergo linea que coniungit 40
inter punctum K et punctum L est longior linea que coniungit inter
punctum K et punctum R. Duo autem circuli DR ELH sunt equales.
Quod ideo est quoniam ipsi sunt maiores. Ergo arcus KL est maior
arcu KR. Et similiter ostenditur quod arcus TK est maior arcu $K\acute{T}$.
Sed arcus TK est equalis arcui KL: ergo unusquisque duorum arcuum 45
TK KL est maior unoquoque arcuum $K\acute{T}$ KR. Et quia circulus BZG
equidistat circulo MTN et circulus BZG occurrit communi differentie
duorum circulorum TKL $\acute{T}KR$ intra superficiem spere, fit circulus
MTN occurrens communi differentie duorum circulorum TKL $\acute{T}KR$

29 eum] *corr. ex* ipsum **Z**   30 *PKY*] *in corr.* **BPs**   30 *FLC*] orthogonaliter *add.* **OFiZ**
31 Iam autem facta est] et **Va**   31 autem] ergo **OFiBPs**   31 super] *om.* **V**   31 circuli] *om.*
Kg   31 que] qui **Va**   32 portio] *in corr.* **B**   33 *PY*] *PKY* **Kg**   33 secatur] secant
R   33 arcus portionis] portio **OBPsVa** (*corr. ex* portionis **BPs**)   35 duabus] duobus **P**
35 coniungit] contingit **Va**   36 et] inter *add.* **Z**   36 rectis lineis] *tr.* **B**   36 a puncto *K*] *post*
FLC **OBPsVa**   36 *K*] *A* **Kg**   36–37 producuntur] protrahuntur (*post* que **OPs**)
37 *FLC*] *in corr.* **B**, *FCL* **FiZ**, *FLE* **Va**   37–38 que ... existit] propinquior ei existens
OBPsVa   38 propinquior] proprior **R**   39 *R*] *K* **Fi**   39–40 est ... punctum *L*] *marg.* **BPs**
39 brevior] recta *add.* **Va**   40 punctum[1]] *om.* **R**   40–42 Ergo ... punctum *R*] *om.* **OB**
PsVa   42 *DR*] *DK* **PsVa**   42 *ELH*] *in corr.* **O**   44 *K\acute{T}*] *KY* **Kg**   45 *TK*] *om.* **Va**
45 unusquisque] uterque **MKgOBPsVa**   46 *TK*] *CK* **Fi**   46 unoquoque] utroque **MKgOZB**
PsVa   46 unoquoque arcu] *tr.* **V**   46 arcuum] arcu **PRVMFiZ**   46 circulus] arcus **R**
46 *BZG*] *FLC* **FiZ**   47 equidistat ... *BZG*] *om.* **V**   47 *MTN*] *MT* **Fi**   47 *BZG*] *in corr.*
O, *MTN* **FiZ**   48 *\acute{T}KR*] *IKT* **P**   48 intra] *in corr.* **B**, extra **FiZ**   48–49 intra ... *\acute{T}KR*] *om.*
OFiPsVa   48 fit] sit **KgB**   48–50 fit ... spere] *om.* **Z**   49 *TKL* *\acute{T}KR*] *ZKL* et *KR* **B**

جهة نقطة كَ فلأن دائرتى طاكـل ذكـر العظيمتين فى كرة تقطع إحداهما

الأخرى على نقطة كَ وقد فصل من إحداهما قوسان متساويتان وهما قوسا

طاكـ كَل متصلتان على الولاء فى كل واحدة من ناحيتى النقطة التى

تتقاطعان عليها وقد مرّ بنقطتى طا لَ سطحان متوازيان وهما سطحا فـلـق

مـطـن وسطح مـطـن منهما يلقى الفصل المشترك لسطحى طاكـل ذكـر خارج

بسيط الكرة من جهة نقطة كَ وكان كل واحدة من قوسى طاكـ كَل أعظم من

كل واحدة من قوسى راكـ كـذ تكون قوس راكـ أعظم من قوس كـذ ولكن

قوس راكـ مساوية لقوس فـس وقوس كـذ مساوية لقوس مـس فقوس

فـس أعظم من قوس مـس ، وذلك ما أردنا أن نبين ٠

ح

إذا كانت فى كرة دائرة عظيمة تماس دائرة من الدوائر المتوازية وكانت فيها

دائرة أخرى عظيمة مائلة على الدوائر المتوازية تماس دائرتين أعظم من التى

كانت تماسهما الدائرة الأولى وكانت مواضع المماسة أيضاً على الدائرة الأولى

العظمى وفصل من الدائرة المائلة قسى متساوية متصلة على الولاء فى جهة

واحدة بعينها من الدائرة التى هى أعظم الدوائر المتوازية ورسمت دوائر

عظام تمر بالنقط الحادثة وتفصل من الدوائر المتوازية فيما بينها قسياً متشابهة

فهى تفصل من الدائرة التى هى أعظم الدوائر المتوازية فيما بينها قسياً غير

om. A [إحداهما الأخرى 39–40   A داره [دائرتى 39   A ولكن [فلأن 39   om. A [جهة 39

ال سطحان طَل [طا لَ سطحان 42   A سقط [بنقطتى 42   A الى [التى 41   A وصل [وقد فصل 40

NH تماسها [تماسهما 3   DIPS H [وقوس 46   om. A [فـس ... لقوس 46   N زدك [راك 46   A

3 موضع [مواضع 3   A تمر 6   om. N

extra superficiem spere a parte puncti K. Et quia duorum circulorum   50
TKL $\acute{T}KR$ maiorum in spera unus alterum secat super punctum K et
iam separati sunt ex uno eorum duo arcus equales, qui sunt arcus TK
KL coniuncti secundum continuitatem in unaquaque duarum partium
puncti super quod se secant, et iam producte sunt a duobus punctis
T et L due superficies equidistantes, que sunt superficies FLC MTN,   55
et superficies MTN, que est una earum, occurrit differentie communi
duabus superficiebus TKL $\acute{T}KR$ extra superficiem spere a parte puncti
K, et unusquisque duorum arcuum TK KL est maior unoquoque arcu
RK $K\acute{T}$, ergo arcus RK est maior arcu $K\acute{T}$. Sed arcus RK est equalis
arcui FS et arcus $K\acute{T}$ est equalis arcui SM: ergo arcus FS est maior   60
arcu SM. Et illud est quod demonstrare voluimus.

8   Cum in spera fuerit circulus maior contingens aliquem ex circulis equidistantibus et fuerit in ea alius circulus maior inclinatus super circulos
equidistantes contingens circulos qui sint maiores eo quem circulus
primus fuit contingens fueritque etiam loci contactus super circulum
primum maiorem et separati fuerint ex circulo inclinato arcus equales   5
secundum continuitatem coniuncti in parte una et eadem a circulo qui
est maior ex circulis equidistantibus et signati fuerint circuli maiores
transeuntes per puncta provenientia et separaverint inter se ex circulis
equidistantibus arcus similes, ipsi separabunt inter se ex circulo qui est

50–51 extra … $\acute{T}KR$] *marg.* **R**   51 *TKL*] *supra* **BPs**   51 $\acute{T}KR$] *DKR* **PRVMKgOFiZPsVa**
52 iam] *om.* **BPs**   52 duo] duorum **O**   52 arcus] circulis **Fi**   53 coniuncti] *om.* **OB**
PsVa   53 unaquaque] utraque **MKgOBPsVa**   54 puncti] *K add.* **OZBPsVa** (**BPs** *supra*)
54 quod se] *tr.* **Ps**   54 se] *om.* **Va**   55 *T*] *R* **Z**   55 et] *om.* **KgOBPsVa**   55 *MTN*] *corr.*
in MCN **Z**   56 et superficies *MTN*] *om.* **VFi**   56 differentie communi] *tr.* **MKgOFiZBPsVa**
57 duarum superficierum] **OFiZBPsVa** (superficiebus: super sinistrum **Fi**)   57 $\acute{T}KR$] *DKR*
OFiPsVa, *DKI* **Z**   57 spere] *om.* **Fi**   58 arcuum] *in corr.* **BPs**   58 unoquoque] utroque
MKgOZBPsVa   59 $K\acute{T}$] *KL* **Kg**   59 arcu] *om.* **Va**, arcui **Fi**   59 $K\acute{T}$] *KL* **Kg**
60 est equalis] equatur **OBPsVa**   60 arcui] *om.* **OBPsVa**   60 *SM*] *FM* **Ps**   60 ergo] *in*
corr. **B**   61 illud] hoc **OBPsVa**   61 quod … voluimus] *om.* **OBPs**   61 demonstrare] *om.* **Va**
61 voluimus] volumus **Va**   1 maior] *et add.* **FiZ**   3 contingens] et duo *add.* **V**   3 sint] sunt
KgOFiZBPsVa   3 quem] quoniam **Fi**   4 primus] *supra* **B**, *marg.* **Ps**   4 fuit] fuerit **Fi**
4 fuerintque] fueritque **PVM**, fuerit **Kg**   5 circulo] illo *add.* **Z**   8–9 circulis … ex] *marg.*
sup. **M**   9 separabunt] *in corr.* **Ps**

متساوية والقوس التى تقرب منها من الدائرة الأولى العظمى أعظم من التى

تبعد منها .

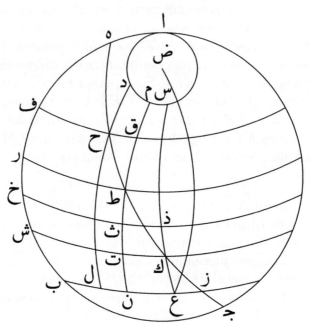

10   فلتكن فى كرة دائرة ا ب جـ العظمى ولتماس دائرة من الدوائر المتوازية

التى تكون فى الكرة وهى دائرة ا د على نقطة آ ولتكن دائرة أخرى عظيمة

مائلة على الدوائر المتوازية وهى دائرة ه ز جـ تماس دائرتين أعظم من الدائرتين

المتوازيتين اللتين كانت تماسهما دائرة ا ب جـ الأولى العظمى ولتكن أيضاً

مواضع الماسة على دائرة ا ب جـ على نقطتى ه جـ ولتكن أعظم الدوائر

15   المتوازية دائرة ب ز ولتفصل من دائرة ه ز جـ المائلة قوسان متساويتان وهما

قوسا ح ط ط ك المتصلتان على الولاء فى جهة واحدة من دائرة ب ز العظمى

من الدوائر المتوازية ولترسم دوائر عظيمة تمر بنقط ح ط ك وهى دوائر د ح ل

---

maior ex circulis equidistantibus arcus inequales et arcus primi circuli 10
maioris, qui est ei magis propinquus, est maior eo qui ab eo magis
remotus est.

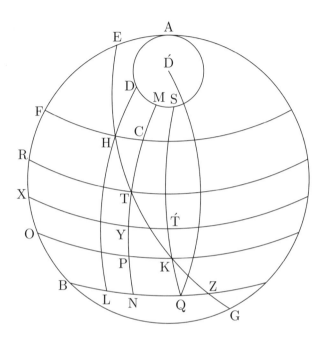

In spera itaque sit circulus maior *ABG*, qui contingat aliquem
ex circulis equidistantibus qui sunt in spera, qui sit circulus *AD*, super
punctum *A*. Et sit circulus alius maior inclinatus super circulos equi- 15
distantes, qui sit circulus *EZG*, contingens duos circulos maiores qui
sunt ex circulis equidistantibus, quos circulus primus maior *ABG* fuit
contingens, et sint etiam loci contactus super duo puncta circuli *ABG*.
Et sit circulus maior qui est ex circulis equidistantibus circulus *BZ*. Et
separabo ex circulo inclinato *EZG* arcus equales coniunctos secundum 20
continuitatem in parte una a circulo *BZ* maiore qui est ex circulis
equidistantibus, *HT TK*. Et signabo circulos maiores transeuntes per
puncta *H T K*, qui sint circuli *DHL MTN SKQ*, occurrentes circulo

11 qui ab eo] *om.* **Fi**   12 remotus] retus **V**   12 remotus est] *tr.* **KgOFiZ**
13 contingat] contingit **FiZ**   15 sit] sic **Fi**, fit **Ps**   15 alius maior] *tr.* **Kg**
16 contingens] continges **Fi**   16 duos] *supra* **O**, *om.* **PFiZ**   16 qui²] *repet.* **Fi**
19 circulus *BZ*] *tr.* **O**   20 separabo] *in corr.* **Ps**   22 *HT*] *KT* **V**   22 *TK*] *TH* **Fi**
23 *K*] et per punctum *Fi* qui est polus circulorum equidistantium *add.* **PsVa**   23 sint] sunt
FiZ   23 sint circuli] sit circulus **BPsVa**   23 *SKQ*] *SQK* **KgBPsVa**; *SKQ, corr. ex SQK* **O**

م ط‍ن س‍ك‍ع تلقى دائرة اد على نقط د م س وتفصل من الدوائر المتوازية

فيما بينها قسياً متشابهة ، فأقول إن قوس ل‍ن أعظم من قوس ن‍ع فلترسم

20   دوائر متوازية تمر بنقط ح ط ك وهى دوائر ف‍ح‍ق رط ش‍ت‍ك فقوس

رش أعظم من قوس ر‍ف ولكن قوس رش مساوية لقوس ط‍ت وقوس

رف مساوية لقوس ط‍ق فقوس ت‍ط أعظم من قوس ط‍ق فلتكن قوس

ط‍ث مساوية لقوس ط‍ق وقوس ح‍ط مساوية لقوس ط‍ك فالخط المستقيم

الذى يصل بين نقطة ح ونقطة ق مساوٍ للخط المستقيم الذى يصل بين

25   نقطة ث ونقطة ك فلترسم دائرة موازية لأى دائرة كانت من دوائر ف‍ح‍ق

رط ش‍ت‍ك ب‍ز تمر بنقطة ث وهى دائرة خ‍ث‍ذ وليتعلم قطب الدوائر

المتوازية وليكن نقطة ض ولترسم دائرة عظيمة تمر بنقطتى ض ع وهى دائرة

ض‍ع ،

فدائرة ض‍ع العظمى فى كرة تقطع دائرة من الدوائر التى تكون فى

30   الكرة وهى دائرة ب‍ز وتمر بقطبيها فهى تقطعها بنصفين وعلى زوايا قائمة

فدائرة ض‍ع قائمة على دائرة ب‍ز على زوايا قائمة فدائرة س‍ع مائلة على

دائرة ب‍ز إلى ناحية اه ب فدائرة ب‍ز مائلة على دائرة س‍ع إلى ناحية س

ودائرة ب‍ز موازية لدائرة خ‍ث‍ذ فدائرة خ‍ث‍ذ مائلة على دائرة س‍ع إلى

ناحية س ولأن سطحى ب‍ز خ‍ث‍ذ المتوازيين قد قُطعا بسطح س‍ع يكون

18 نقط [نقطة AN   19 بينها [سنهما N   19 متشابهة :H مشابها A, مساوه [مساويه N   19 ل‍ن [ربه   A [فقوس [وقوس A   20 فقوس … طق 22 repet. A (فقوس: قوس)   22 فقوس … طق A   22 فلتكن [ولكن   A [وقوس ح‍ط [وهو حط A   23 مساوية [2لخط add. A   24–25 نقطة … ح [om. A   23   27 بنقطتى [بنقطى N   29 العظمى [العطى A   31 زوايا قائمة [om. A   32 اه ب [34–   hic et saepius A   34 سطحى [سطح H   om. H [ودائرة … ناحية س   33–34 نقطתי H نקطה אם   35 بسطح … متوازيين [om. A

AD super puncta *D M S*, et separent inter se ex circulis equidistantibus arcus similes. Dico igitur quod arcus *LN* est maior arcu *NQ*. 25
Signabo igitur circulos equidistantes transeuntes per puncta *H T K*,
qui sint circuli *FHC RT OPK*. Arcus igitur *RO* est maior arcu *RF*;
arcus autem *RO* est equalis arcui *TP* et arcus *RF* est equalis arcui
TC: ergo arcus *PT* est maior arcu *TC*. Sit autem arcus *TY* equalis arcui *TC*. Sed arcus *HT* est equalis arcui *TK*: linea igitur recta 30
que coniungit inter punctum *H* et punctum *C* est equalis linee recte
que coniungit inter punctum *Y* et punctum *K*. Describam ergo circulum cuilibet horum circulorum equidistantem – *FHC RT OPK BZ* –
transeuntem per punctum *Y*, qui sit circulus *XYT́*. Et signabo polum
circulorum equidistantium, qui sit punctum *D́*. Et describam circulum 35
maiorem transeuntem per duo puncta *D́ Q*, qui sit circulus *D́Q*.

Circulus igitur *D́Q* maior in spera secat aliquem ex circulis qui
sunt in spera, qui est circulus *BZ*, et transit per polos eius: ergo ipse
secat eum in duo media et orthogonaliter. Ergo circulus *D́Q* est erectus
super circulum *BZ* orthogonaliter. Circulus igitur *SQ* est inclinatus 40
super circulum *BZ* ad partem *A E B*; ergo circulus *BZ* est inclinatus
super circulum *SQ* ad partem *S*. Circulus vero *BZ* equidistat circulo
XYT́: ergo circulus *XYT́* est inclinatus super circulum *SQ* ad partem
S. Et quia duas superficies *BZ XYT́* equidistantes iam secuit superfi-

27 sint] sunt **FiZVa**   27 *FHC*] *FAC* **Kg**, *SHC* **Fi**, *FH* **B**   27 arcu] *om.* **PsVa**   27 arcu *RF*] *tr.*
OB   28 est equalis[1]] equatur **OBPsVa**   28 et] *om.* **Va**   28 est equalis[2]] equatur (*post*
TC) **BPsVa**   29 *PT*] *in corr.* **Ps**   29 *TC*] *TP* **R**, *IE* **M**, *RO* **FiVa**   30 *HT*] *KT* **R**
30 est equalis] equatur **OBPsVa**   30 *TK*] *TH* **Fi**   30 igitur] ergo **BPsVa**   30 recta] *marg.*
R   31 punctum[1]] *in corr.* **B**   31 est equalis] equatur **OPsVa**   31 linee recte] *tr.* **Kg**
32 ergo] igitur **FiZ**   33 equidistantem] equidistans **PRV**   33 *FHC*] *in corr.* **B**, *FHT* **R**, *FKC*
V   34 transeuntem] *om.* **Z**   34 punctum *Y*] puncta *Y* **BPsVa** (*Y corr. ex TY* **B**), puncta *I Y*
PRVMKg   34 qui] quid **Fi**   34 *XYT́*] *XIY* **Kg**   35 punctum] punctus **OBPsVa**   35 *D́*] *FY*
R   37 Circulus] *in corr.* **V**   37 Circulus igitur *D́Q*] *om.* **Z**, *supra* **BPs**   38 et transit] *om.* **Fi**
39 erectus] *in corr.* **V**   40 circulum *BZ*] punctum *B* **BPs**   40 orthogonaliter] *post* erectus
MKgOBPsVa   40–41 Circulus . . . *A E B*] *om.* **OFiZVa**   40 igitur] *om.* **BPs**   40 *SQ*] *supra* **Kg**,
in corr. **B**, *FSQ* **Ps**   40 est inclinatus] inclinatur **BPs**   41 *BZ*[1]] orthogonaliter *add.* **Kg**, *add.*
et del. **MBPs**   41 est] cum **Fi**   43 *XYT́*[1]] *in corr.* **B**, *CXYI* **Fi**   43 ergo circulus *XYT́*] *marg.*
B, *supra* **Ps**   43 est inclinatus] inclinatur **OBPsVa**   44 secuit] secant **MFi**

الفصلان المشتركان لها متوازيين فالفصل المشترك لسطحى س ع خ ث ذ مواز ٣٥

للفصل المشترك لسطحى ب ز س ع والفصل المشترك لسطحى ب ز س ع هو

قطر دائرة س ع الذى خرج من نقطة ع فالفصل المشترك لسطحى س ع

خ ث ذ مواز لقطر دائرة س ع الذى يخرج من نقطة ع فقد خرج فى دائرة

س ع خط ما وهو الفصل المشترك لسطحى س ع خ ث ذ يقسم الدائرة بقسمين

غير متساويين وذلك أنه مواز لقطر دائرة س ع وقد عمل عليه قطعة دائرة ٤٠

وهى قطعة ث ذ معما يتصل بها مائلة على القطعة التى ليست بأعظم من

نصف دائرة وقد قُسم قوس القطعة القائمة بقسمين غير متساويين على

نقطة ث وقوس ث ذ أصغر من نصف القطعة التى عملت فالخط المستقيم

الذى يصل بين نقطة ث ونقطة ذ أقصر جميع الخطوط المستقيمة التى تخرج

من ث إلى القوس التى ليست بأصغر من نصف دائرة فالخط الذى يصل بين ٤٥

نقطة ث ونقطة ذ أقصر من الخط الذى يصل بين نقطة ث ونقطة ك والخط

الذى يصل بين نقطة ث ونقطة ك قد بيّنّا أنه مساوٍ للخط الذى يصل بين

نقطة ح وبين نقطة ق فالخط الذى يصل بين نقطة ث ونقطة ذ أقصر من

الخط الذى يصل بين نقطة ح ونقطة ق فالخط الذى يصل بين نقطة ح

ونقطة ق أعظم من الخط الذى يصل بين نقطة ث ونقطة ذ ولأن دائرة ٥٠

خ ث ذ أقرب إلى مركز الكرة من دائرة ف ح ق تكون دائرة خ ث ذ أعظم

٣٥ لها [להםא H لمعلع, supra ٣٨-٣٩ مواز ... خ ث ذ] .om A ٣٩ بقسمين [قسمين AN ٤٠ لقطر [لمعلع

AN مختلفين :H ٤٠ غير متساويين [المعطة A ٤١ القطعة [.om A, :H داره N ٤١ قطعة [لقطر N

وسطة [وبين نقطة .om N ٤٨ التى [A موس ٤٥ القوس [A والخط ٤٣ فالخط [A الصف A الصف ٤٣ نصف [A

A (l. ٥١) أقرب ante [أعظم ... خ ث ذ .om A ٥١-٥٢ تكون دائرة خ ث ذ ٥١ A والخط [فالخط N ٤٨

cies SQ, erunt differentie eis communes equidistantes. Differentia ergo 45
duabus superficiebus SQ $XY\acute{T}$ communis equidistat differentie communi
muni duabus superficiebus BZ SQ. Sed differentia communis duabus
superficiebus BZ SQ est diametrus circuli SQ, que producitur a puncto
Q. Communis ergo differentia duarum superficierum SQ $XY\acute{T}$ equidistat
stat diametro circuli SQ que producitur a puncto Q. Iam ergo protracta 50
est in circulo SQ linea aliqua que est differentia communis duabus
superficiebus SQ $XY\acute{T}$ secans circulum in duas sectiones inequales.
Quod ideo est quoniam ipsa equidistat diametro circuli SQ. Iam autem
super ipsam constituta est portio circuli, que est $Y\acute{T}$ cum eo quod ei
coniunctum est, inclinata super portionem que non est maior semicir- 55
culo super punctum Y. Et arcus $Y\acute{T}$ est minor medietate portionis
que est constituta; linea igitur recta que coniungit inter punctum Y et
punctum $\acute{T}$ est brevior omnibus rectis lineis que a puncto Y ad arcum
qui non est minor semicirculo protrahuntur. Ergo linea que coniungit
inter punctum Y et punctum $\acute{T}$ est brevior linea que coniungit inter 60
punctum Y et punctum K. Sed linea que coniungit inter punctum
H et punctum C est equalis linee que coniungit inter punctum Y et
punctum K: ergo linea que coniungit inter punctum Y et punctum $\acute{T}$
est brevior linea que coniungit inter punctum H et punctum C. Linea
igitur que coniungit inter punctum H et punctum C est maior linea que 65
coniungit inter punctum Y et punctum $\acute{T}$. Et quia circulus $XY\acute{T}$ est
magis propinquus centro spere quam circulus FHC, ergo circulus $XY\acute{T}$

45 equidistantes] distantes **Va**   45 Differentia] Differentie **KgOBPsVa**   45 ergo] *repet.*
Fi   46 duabus superficiebus] *om.*   **FiZ**   46 communis] communi   **OBPsVa**
46 differentie] differentia **OBPsVa** (*corr. ex* differentie **Ps**)   46–47 communi] communis
OBPsVa   47–48 Sed ... SQ] *om.*   **Fi**   47 differentia communis] *tr.*   **KgBPsVa**
48 producitur] producuntur **Z**   49–50 Communis ... puncto Q] *om.* **BPsVa**   49 ergo] igitur
OFiZ   49 differentia] *in corr.* **Fi**   49 duarum] *om.* **OFiZ**   49 SQ $XY\acute{T}$] *tr.* **FiZ**
50 que ... Q] *om.* **OFiZ**   50 ergo] igitur **VKgBPsVa**   51 aliqua] *in corr.* **B**, alia **PsVa**
51 differentia communis] *tr.* **MKgBPsVa**   52 SQ] RQ **Va**   53–54 Iam autem] Et iam
FiZ   53 autem] igitur **OBPsVa**   54 que] qui **OFiZBPsVa**   54 est] circulus *add.* **KgO**
FiZBPsVa   55 portionem] portionum **Fi**   55 non] *supra* **V**   56 Y] I **MZ**, YI **Fi**
56 minor] maior **V**   57 est constituta] *tr.* **FiZ**   57 igitur] ergo **KgOBPsVa**   57 recta] *om.*
BPsVa   57 inter punctum] *repet.* **Va**   57 Y] Q **Kg**; Y, *corr. ex* Q **B**   58 que] est *add.* **Fi**
58 Y] R **M**; Y, *corr. ex* **BPs**   60 Y] X **R**   61 Y] I **Kg**; Y, *corr, ex* Z **B**   62 C] IC **Fi**
62 est equalis] equatur **OBPsVa**   63 que] *supra* **BPs**   63 Y] *illeg.* **M**; Y, *corr. ex* K **Kg**
64–65 Linea ... punctum C] *om.* **V**   65 igitur] ergo **PBPsVa**   67 propinquus] propinquior
Va   67 FHC] FKC **V**, FAC **Kg**   67–68 ergo ... circulo FHC] *om.* **Va**

من دائرة ف ح ق ولأن دائرتى خ ث ذ ف ح ق غير متساويتين ودائرة ف ح ق

أصغرهما وقد أخرج فى دائرة ف ح ق منهما الخط الذى يصل بين نقطة ح

ونقطة ق وأخرج فى دائرة خ ث ذ الخط الذى يصل بين نقطة ث ونقطة ذ

وكان الخط الذى أخرج فى الدائرة الصغرى أطول من الخط الذى أخرج فى   55

الدائرة العظمى لأن الخط الذى بين نقطة ح ونقطة ق أطول من الخط الذى

بين نقطة ث ونقطة ذ تكون قوس ح ق أعظم من القوس الشبيهة بقوس

ث ذ من دائرتها ولكن قوس ح ق شبيهة بقوس ل ن وقوس ث ذ شبيهة

بقوس ن ع فقوس ل ن أعظم من القوس الشبيهة بقوس ن ع من دائرتها

وهما من دائرة واحدة بعينها فقوس ل ن أعظم من قوس ن ع ، وذلك ما   60

أردنا أن نبين .

ط   إذا كان قطب الدوائر المتوازية على الخط المحيط بالدائرة العظمى وقطعت

هذه الدائرة دائرتان عظيمتان على زوايا قائمة إحداهما من الدوائر المتوازية

والأخرى مائلة على الدوائر المتوازية وفصلت من الدائرة المائلة قوسان

متساويتان غير متصلتين على الولاء فى جهة واحدة بعينها من الدائرة التى

هى أعظم الدوائر المتوازية ثم رسمت دوائر عظيمة تمر بالنقط الحادثة   5

وبالقطب فإنها تفصل من أعظم الدوائر المتوازية فيما بينها قسياً غير متساوية

والقوس القريبة من الدائرة الأولى العظمى أبداً أعظم من التى هى أبعد

منها .

52 ف ح ق[2] om. A   53-54 بين ... ح[om. A   55 أخرج[2] ح رح A   56-

دائرتا[وهما add. N 59-60   دائرتها[او شبهه ها A موس القوس[om. N 59   بين نقطة ح ... add. N 57

add. A 1 كان[كاب N 1 وقطعت[وقد قطعت A   فى ... التى 4 om. N

est maior circulo *FHC*. Et quia duo circuli *XYT́ FHC* sunt inequales et circulus *FHC* est minor ex eis et in circulo *FHC*, qui est unus eorum, iam protracta est linea que coniungit inter punctum *H* et punctum *C* 70 et in circulo *XYT́* protracta est linea que coniungit inter punctum *Y* et punctum *T́* et linea que protracta est in circulo minore fuit longior linea que protracta est in circulo maiore, eo quod linea que coniungit inter punctum *H* et punctum *C* est longior linea que coniungit inter punctum *Y* et punctum *T́*: ergo arcus *HC* est maior arcu simili arcui 75 *YT́* ex suo circulo. Arcus autem *HC* est similis arcui *LN* et arcus *YT́* est similis arcui *NQ*: ergo arcus *LN* est maior arcu simili arcui *NQ* ex suo circulo. Ipsi autem sunt ex uno et eodem circulo: ergo arcus *LN* est maior arcu *NQ*. Et illud est quod demonstrare voluimus.

9 Si in spera fuerit polus equidistantium circulorum super lineam continentem circulum maiorem et secuerint hunc circulum duo maiores circuli orthogonaliter, quorum unus sit ex circulis equidistantibus et alter sit inclinatus super circulos equidistantes, et ex circulo inclinato fuerint separati duo arcus equales non continue coniuncti in una et 5 eadem parte a circulo qui est maior ex circulis equidistantibus, et post signati fuerint circuli maiores transeuntes per puncta que proveniunt et per polum, ipsi separabunt ex circulo maiore qui est ex circulis equidistantibus inter se arcus inequales, et arcus circuli maioris primi, qui ei fuerit propinquior, semper erit maior eo qui ab eo erit magis remotus. 10

68 FHC[1]] *HFC* **M**, *FQC* **Kg**   *68* duo] *om.* **Va**   *69* ex] *supra* **BPs**   *69* in] etiam **BPsVa** *69 FHC*] *THC* **Kg**   *69* qui ... eorum] *om.* **OBPsVa**   *70–71* punctum[1] ... inter] *marg.* **Ps**, *et in textu et in marg.* **B**   *71–74* ... punctum *C*] *om.* **Fi**   *71 XYT́*] iam *add. supra* **B**, *in textu* **PsVa**   *72* est] fuit **M**   *72–73* minore ... circulo] *om.* **R**   *72–73* fuit ... maiore] *marg.* **Z** *75 Y*] *H* **Fi**   *75* arcui] *om.* **FiZ**   *77* simili] sm° **Va**   *77* simili arcui] *om.* **Fi**   *78* circulo] aut eo *add. et del.* **M**, autem eo *add.* **Kg**   *79* illud] hoc **OFiZBPsVa**   *79* quod ... voluimus] *om.* **OBPsVa**   *1* Si] Cum **Z**; circulus per *add. et del.* **V**   *1* fuerit] *om.* **Fi**, *post* maiorem **PRVM** *1* polus] *ante* in spera **V**; circulus **Kg**   *2* maiorem et] maioremque **Kg**   *2* secuerint] secuerit **R**   *2* hunc circulum] *tr.* **R**   *5* equales] *supra* **M**, *post* coniuncti **FiZ**   *6* a] ex **R**   *7* signati] *om.* **BPsVa**   *7* et] *om.* **Kg**   *8* qui est] *om.* **Fi**   *8* est] *om.* **Va**   *8* ex] illis *add.* **KgOBPsVa**, *add. et del.* **M**   *9* arcus[1]] *illeg.* **Z**   *9* arcus[2]] arcum, *in corr.* **Z**   *9* maioris] *in corr.* **V**   *9* primi] vel primo *add.* **Va**   *9* ei] *om.* **FiZ**   *10* fuerit] fuit **FiBPsVa**   *10* semper] *om.* **FiZ**   *10* erit[2]] est **FiZ**   *10* erit magis] *tr.* **PMKgOFiZBPsVa**

فليكن فى كرة على الخط المحيط بدائرة ا‌ب‌ج قطب الدوائر المتوازية

وهو نقطة آ ولتقطع دائرة ا‌ب‌ج دائرتان عظيمتان وهما د‌ه‌ج ب‌ه على زوايا 10

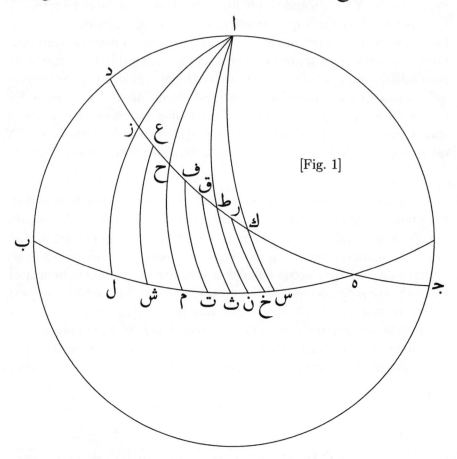

[Fig. 1]

قائمة ولتكن دائرة ب‌ه من الدوائر المتوازية ودائرة د‌ه‌ج مائلة على الدوائر

المتوازية ولتفصل من دائرة د‌ه‌ج قوسان متساويتان وهما قوسا ز‌ح ط‌ك غير

متصلتين على الولاء فى جهة واحدة بعينها من الدائرة التى هى أعظم الدوائر

المتوازية ولترسم دوائر عظيمة تمر بنقط ز‌ح ط‌ك وبقطب آ وهى دوائر

ازل ا‌ح‌م ا‌ط‌ن ا‌ك‌س ، فأقول إن قوس ل‌م أعظم من قوس ن‌س وذلك 15

س‌ا [ن‌س ‌ add. N‌ 16‌ ‌ من [أعظم 14‌ om. A‌ د‌ه‌ج] 11‌ هما A‌ وهما] 11‌ N‌ وهى [وهى 11‌ وهو] 11

Sit ergo in spera super lineam continentem circulum ABG polus circulorum equidistantium, qui sit punctum A, et secent circulum ABG
duo circuli maiores, qui sint $DEG\ BE$, orthogonaliter. Sitque circulus

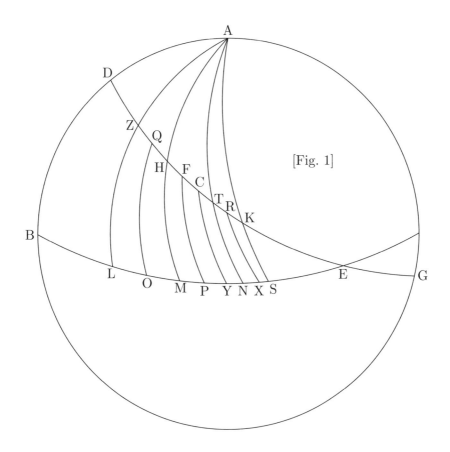

[Fig. 1]

BE ex circulis equidistantibus, et circulus DEG sit inclinatus super
circulos equidistantes. Et separentur ex circulo DEG duo arcus equales,   15
qui sint arcus $ZH\ TK$, non coniuncti secundum continuitatem in una
et eadem parte a circulo qui est maior ex circulis equidistantibus. Et
describam circulos maiores transeuntes per puncta $Z\ H\ T\ K$ et per
polum A, qui sint circuli $AZL\ AHM\ ATN\ AKS$. Dico igitur quod
arcus LM est maior arcu NS. Quod ideo est quoniam arcus HT aut   20

12 secent] secet **B**   13 circuli] *illeg.* **M**   13 sint] sunt **FiBPsVa**   13 *DEG*] *supra*
BPs   13 circulus] *om.* **OBPsVa**   14 *BE*] *om.* **Fi**, *in corr.* **B**   14 sit] *supra* **M**   14–
15 sit … circulo *DEG*] *marg.* **BPs**   14 super] *in corr.* **Fi**   15 separentur] separantur **MVa**
15 ex circulo] *om.* **Fi**   16 sint] sunt **FiVa**   17 a] alicui **FiZ**   19 sint] sunt **Fi**   19 *AZL*] *in
corr.* **B**, *ALZ* **R**   19 *AZL … AKS*] *AZ LA HM AT NA K S* **Fi**   19 *ATN*] *om.* **Z**, *in corr.* **B**
20 *LM*] *illeg.* **Z**   20 quoniam] quia **B**

أن قوس حط إما أن تكون مشاركة فى المقدار لقوسى زح طك وإما أن لا

تكون مشاركة لهما ،

فلتكن أولاً فى الصورة الأولى قوس حط مشاركة فى المقدار لقوسى

زح طك ولتُقسم قسى زح حط طك بذلك المقدار الذى تشترك فيه على

نقط ع ف ق ر ولترسم دوائر عظيمة تمر بنقط ع ف ق ر وبقطب آ وهى 20

دوائر ع ش ف ت ق ث رخ فلأن قسى زع عح حف فق قط طر رك

متصلة متوالية مساوٍ بعضها لبعض تكون قسى لش شم مت تث ثن

نخ خس متصلة متوالية غير مساوٍ بعضها لبعض وأعظمها قوس لش وما

بعد ذلك منها على الولاء فلأن قوس لش أعظم من قوس نخ وقوس شم

أعظم من قوس خس تكون كل قوس لم أعظم من كل قوس نس ، 25

ثم لا تكون قوس حط مشاركة فى المقدار لقوسى زح طك ،

فأقول إن قوس لم أيضاً أعظم من قوس نس وذلك أنه إن لم تكن قوس

لم أعظم من قوس نس فإنها إما أن تكون أصغر منها أو مساوية لها فلتكن

أولاً إن أمكن قوس لم أصغر من قوس نس كما فى الصورة الثانية ولتكن

قوس لم مساوية لقوس نع ولترسم دائرة عظيمة تمر بقطب آ وبنقطة ع 30

وهى دائرة عف فلما كانت قسى كط طف حط ثلثاً تعلمنا قوساً ما وهى

قوس طق أعظم من قوس طف وأصغر من قوس طك مشاركة فى المقدار

20 حط] om. A   20 تشترك] مشرك N   21 بنقط] سقطه A   22 عف] om. N   27 لا] ثم 20
الا   om. A   28 مساوٍ] مشاركة N   27 مشاركة (مساوية: supra)] ثم ... نس repet. N   27-28 A
28 نس ... وذلك] om. A   30 لم] مساوه add. A   32 ثلثاً] لب A

32 عف] حط لقوس [sic] ومباده وأصغر من قوس طك وأعظم من قوس طف قوس طق ولنفرض add. N

erit communicans in quantitate duobus arcubus *ZH TK* aut erit non communicans eis.

Sit itaque primum sicut in prima continetur figura arcus *HT* communicans in quantitate duobus arcubus *ZH TK*. Et dividantur arcus *ZH HT TK* secundum illam quantitatem in qua communicant super 25 puncta *Q F C R*, et describantur circuli maiores transeuntes per puncta *Q F C R* et per polum *A*, qui sint circuli *QO FP CY RX*. Et quia arcus *ZQ QH HF FC CT TR RK* continue coniuncti ad invicem sunt equales, ergo arcus *LO OM MP PY YN NX XS* continue coniuncti sunt ad invicem inequales. Et maior eorum est arcus *LO* et qui sunt 30 post eum minores secundum continuitatem. Et quia arcus *LO* est maior arcu *NX* et arcus *OM* est maior arcu *XS*, ergo totus arcus *LM* est maior toto arcu *NS*.

Post hec ponatur ut arcus *HT* non sit communicans in quantitate duobus arcubus *ZH TK*. Dico igitur quod arcus *LM* est etiam maior 35 arcu *NS*. Quod ideo est quoniam si arcus *LM* non fuerit maior arcu *NS*, ergo vel ipse erit minor eo aut equalis ei. Sit ergo, si est possibile, arcus *LM* minor arcu *NS*, sicut in secunda continetur figura, et sit arcus *LM* equalis arcui *NQ*. Describam igitur circulum maiorem transeuntem per punctum *Q* et per polum *A*, qui sit circulus *QF*. Propter hoc igitur 40 quod fuerunt tres arcus *KT TF HT*, notabo aliquem arcum, qui sit arcus *TC*, maiorem arcu *TF* et minorem arcu *TK* communicantem in

21 non] *om.* **Z**   23 primum] prius **OFiZBPsVa**   23 continetur] *in corr.* **V**   23–
24 communicans] *post* quantitate **PRVFiZ**   24 arcubus] *marg.* **R**   24 Et] *om.* **O**   24–
25 Et . . . *TK*] *om.* **B**, *marg.* **Ps**   24–25 dividantur . . . *TK*] *post Q F C R* **O**   25 *TK*] *supra*
M, *CK* **Fi**, *om.* **V**   25 secundum] *et in marg. et in textu* **Ps**   26 per] ea *add.* **FiZ**
27 *Q F C R*] *om.* **FiZ**   27 per] *om.* **M**   27 *QO*] *in corr.* **B**, *supra* **Ps**, *QC* **Kg**   27 *FP*] *SP* **Va**
27 *RX*] *IX* **Fi**   28 *FC*] *FO* **Fi**   28 *CT*] *C, supra T* **B**   28 *RK*] *IK* **B**   28 coniuncti] sunt
equales *add.* **Va**   28 coniuncti] *post* invicem **Z**   29–30 equales . . . inequales] *marg.* **BPs**   29–
30 ergo . . . inequales] *marg.* **R**   29 *OM*] *AM* **Kg**   29 *PY*] *P* **Fi**, *PX* **Va**   29 *YN*] *in corr.*
Kg   30 sunt] *post* invicem **KgFiZ**   30 ad invicem] *supra* **O**   30 inequales] minores *add. et del.*
BPs   31 minores] *om.* **Kg**, *marg.* **V**, *supra* **M**, *post* secundum **BPsVa**, *post* continuitatem **OFiZ**
31 continuitatem] *scr. et del. hic, add. supra* minores **B**   32 arcus[1]] *om.* **BPsVa**   33 toto] *om.*
V   33 toto arcu] *tr.* **Fi**   33 *NS*] *NC* **Fi**   34 hec] hoc **Fi**   34 ponatur] *in corr.* **BPs**, po-
nam **M**   35 etiam] *om.* **FiZ**   37 erit] *supra* **B**   37 ei] *om.* **FiZ**   37 est possibile] *tr.* **MKgO**
FiZBPsVa (possibilis **Fi**)   38 minor] *in corr.* **B**   38 *NS*] *MS* **Fi**   38 continetur figura] *tr.*
PsVa   38 figura] *supra* **BPs**   38 *LM*] *AM* **Fi**   39 igitur] ergo **KgOBPs**   40 per] *om.* **RKg**
40 polum *A*] *tr.* **BPsVa** (*in corr.* **BPs**)   41 fuerunt] fuerint **KgOFiBPsVa**   41 tres arcus] *tr.*
OBPsVa   41 arcus . . . *HT*] *HT TF* continui *HT* **Kg**, continui *KT HT* (*TH:* **OB**) *HZ*
incommunicantes **OFiZBPsVa** (incommunicantes: *marg.* **B**, *supra* **Ps**)   41 qui] *in corr.* **B**
42 *TF* . . . arcu] *om.* **VFi**   42 *TK*] *in corr.* **B**

لقوس حط ولتكن قوس حر مساوية لقوس طق ولترسم دائرتان عظيمتان
تمران بنقطتى ر ق وبقطب آ وهما دائرتا رش قت فلأن قوس رح مساوية

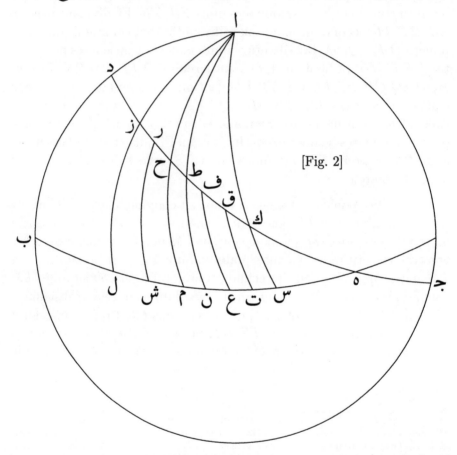

[Fig. 2]

لقوس طق وقوس حط مشاركة فى المقدار لكل واحدة من قوسى رح 35
طق تكون قوس مش أعظم من قوس نت وتكون قوس لم أعظم من
قوس شم فقوس لم أعظم من قوس نع كثيراً وقد كانت مساوية لها
أيضاً هذا غير ممكن فليس قوس لم بأصغر من قوس نس ،

A ووس [وتكون قوس 37   A مشاركا [مشاركة 36   N ولان [فلأن 35   34 حط] *in corr.* A
38 قوس²... شم [*om.* A 38   A كثير [كثيراً 39   A اصعر [بأصغر 39

38 شم [نع وقوس نث أعظم من قوس *add.* NH

quantitate arcui *HT*. Et sit arcus *HR* equalis arcui *TC*. Signabo autem
duos circulos maiores transeuntes per duo puncta *R* et *C* et per polum
A, qui sint duo circuli *RO CÝ*. Et quia arcus *RH* est equalis arcui *TC*   45

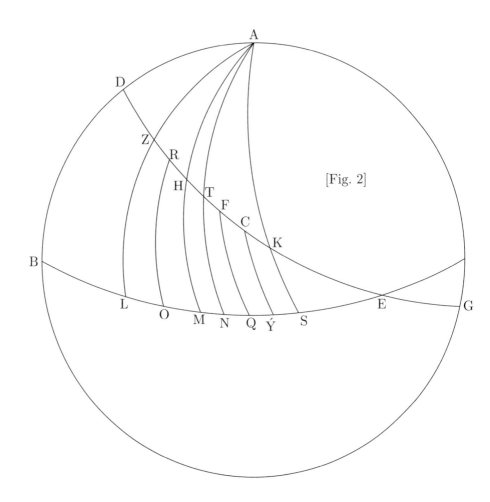

[Fig. 2]

et arcus *HT* communicat in quantitate unicuique duorum arcuum *RH*
TC, ergo arcus *MO* est maior arcu *NÝ*. Arcus vero *LM* est maior arcu
OM et arcus *NÝ* est maior arcu *NQ*: ergo arcus *LM* est multo maior
arcu *NQ*. Iam autem fuit ei etiam equalis, quod est impossibile: arcus
ergo *LM* non est minor arcu *NS*.   50

43 *HT*] *HO*   **Fi**   45 *RH*] *in   corr.*   **BPs**   45 est equalis] equatur   **OBPsVa**
46 unicuique] utrique   **MKgOBPsVa**   46 *RH*] *in   corr.*   **BPs**   47 *NÝ*] *in   corr.*   **Kg**
47 vero] *om.* **M**   48 *NÝ*] *NP* **Kg**   48–49 ergo ... impossibile] *marg.* **BPs**   48 multo] *supra*
P   49 arcu] *om.* **FiZ**   49 autem] *om.* **R**   49 fuit] fuerit **Va**   49 ei etiam] *tr.* **MKg**
49 etiam] *om.* **OBVa**   49 est impossibile] *tr.* **M**   49–50 arcus ergo] *tr.* **OFiZBPsVa**
50 ergo] igitur **R**   50 est] *om.* **Fi**

فأقول إنها ليست بمساوية لها أيضاً فإن أمكن فلتكن مساوية لها كما

فى الصورة الثالثة ولتقسم قوسا زح طك بنصفين نصفين على نقطتى ع ف   40

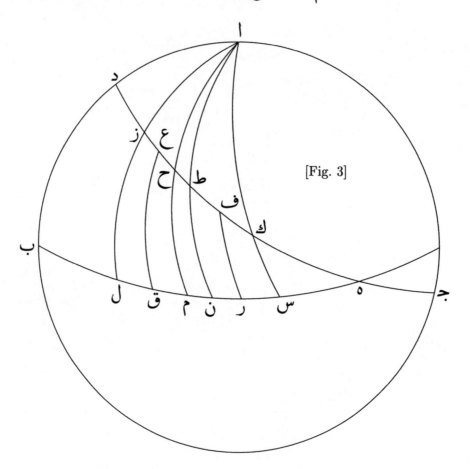

[Fig. 3]

ولترسم دائرتان عظيمتان تمران بنقطتى ع ف وبقطب أ وهما دائرتا عق فر

فلأن قوس زع مساوية لقوس عح تكون قوس لق أعظم من قوس قم

فقوس لم أكبر من ضعف قوس مق وأيضاً فلأن قوس طف مساوية

لقوس فك تكون قوس نر أعظم من قوس رس فقوس سر أصغر من

Dico igitur quod nec etiam equalis ei existit. Quod si possibile fuerit, sit ei equalis quemadmodum in tertia ponitur figura. Dividam autem duos arcus *ZH* et *TK* in duo media super duo puncta *Q F* et

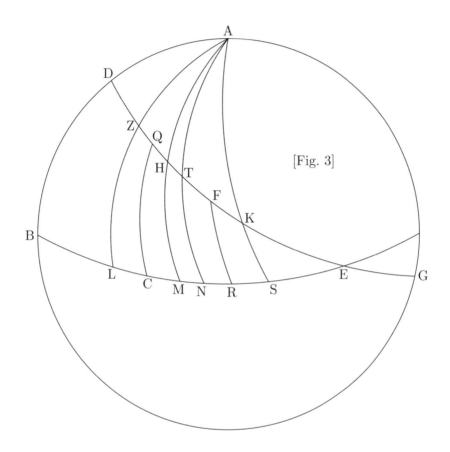

[Fig. 3]

signabo duos circulos maiores transeuntes per duo puncta *Q F* et per polum *A*, qui sint duo circuli *QC FR*. Et quia arcus *ZQ* est equalis arcui *QH*, erit arcus *LC* maior arcu *CM*. Arcus igitur *LM* est maior duplo arcus *MC*. Et etiam quia arcus *TF* est equalis arcui *FK*, erit arcus *NR* maior arcu *RS*. Ergo arcus *SR* est minor arcu *RN*; ergo

55

51 igitur] ergo **M**, etiam **PsVa**   51 nec] neque **OBPsVa**   51 etiam] *om.* **OBPsVa**   51 possibile] possibilis **Fi**   52 ei] *om.* **VOBPsVa**   52 quemadmodum] quod **Kg**   52 ponitur] *om.* **BPsVa**, portio **Fi**   53 autem] ergo **KgOFiZBPsVa**   53 et[1]] *om.* **KgFi ZPsVa**   53 *Q*] *et add.* **OFiZ**   54 transeuntes] *om.* **Fi**   54 *Q*] *et add.* **OFiZ**   55 sint] sunt **FiZ**   55 *FR*] *in corr.* **B**   55 est equalis] equatur **OBPsVa**   56 *QH*] *QNH* **Fi**   56 Arcus] *om.* **Fi**   57 arcus [1]] arcu **BPsVa**   57 etiam quia] *tr.* **FiZ**   57 *TF*] *CF* **B**   57 est equalis] equator **OBPsVa**   57 equalis] qualis **Fi**   57 *FK*] *FiK* **V**   58 *NR*] etiam *add.* **V**   58 *RS*] *TS* **R**   58 *SR*] *in corr.* **B**

قوس نَرَ فقوس نَسَ أصغر من ضعف قوس نَرَ فلأن قوس لَمَ مساوية 45

لقوس نَسَ وقوس لَمَ منهما أعظم من ضعف قوس مَقَ وقوس نَسَ

أصغر من ضعف قوس نَرَ تكون قوس قَمَ أصغر من قوس نَرَ وقد كنا

وضعنا أن قوسى حَعَ طَفَ متساويتان وذلك غير ممكن للذى تبين فى

الصورة الثانية من هذا الشكل فليس قوس لَمَ بمساوية لقوس نَسَ وقد

كان تبين أنها ليست بأصغر منها فقوس لَمَ أعظم من قوس نَسَ ، وذلك 50

ما أردنا أن نبين .

إذا كان قطب الدوائر المتوازية على الخط المحيط بدائرة عظيمة وقطعت هذه

الدائرة دائرتان عظيمتان على زوايا قائمة وكانت إحداهما من الدوائر المتوازية

وكانت الأخرى مائلة على الدوائر المتوازية وتعلمت على الدائرة المائلة

نقطتان كيف ما وقعتا فى جهة واحدة بعينها عن الدائرة العظمى من الدوائر

المتوازية ورسمت دوائر عظيمة تمر بالنقط الحادثة وبالقطب فإن نسبة 5

القوس من الدائرة العظمى من الدوائر المتوازية التى تقع فيما بين الدائرة

الأولى العظمى وبين الدائرة العظمى التى رسمت من بَعد فمرت بالأقطاب

إلى القوس من الدائرة المائلة التى تقع فيما بين هاتين الدائرتين بأعيانهما

كنسبة القوس من الدائرة العظمى من الدوائر المتوازية التى تقع فيما بين

الدوائر العظيمة التى تمر بقطب الدوائر المتوازية وبالنقط التى تعلمت إلى 10

ى

مَقَ] H: وتر N 47 أعظم [قوس 46 om. A فلأن [... 1 نَسَ 46 س [نَسَ A 46 نَسَ] A
مساوية ... بـ [marg. 50 لَمَ 50 A للسركل [الشكل 50 قوس [قوسى 49 om. A قوس نَرَ] 1 نَسَ 48
الدوار [الدائرة 2 add. N من [وقطعت 1 مولق [قوس نَس 51 (مساوه : بمساوية) A
om. A التى ... الدثرة 8-7 A والعظمى [العظمى 7 داره [دوائر 5 A واحداهما [وكانت إحداهما 2
هذا אלدואיر N, هده الدوار ٮاعٮانها [هاتين الدائرتين بأعيانهما 8 om. A بين [بين 8 A وقع [تقع 8
H בעינהא

arcus *NS* est minor duplo arcus *RN*. Et quia arcus *LM* est equalis
arcui *NS* et arcus *LM*, qui est unus eorum, est maior duplo arcus *MC* 60
et arcus *NS* est minor duplo arcus *NR*, erit arcus *CM* minor arcu *NR*.
Iam autem posuimus quod duo arcus *HQ* et *TF* sunt equales quod est
impossibile propter hoc quod in secunda forma huius figure ostensum
est. Non est igitur arcus *LM* equalis arcui *NS*, et iam fuit ostensum
quod non est eo minor. Arcus igitur *LM* est maior arcu *NS*. Et illud 65
est quod demonstrare voluimus.

10 Si polus circulorum equidistantium super lineam continentem circulum maiorem fuerit et secuerint hunc circulum duo circuli maiores
orthogonaliter, quorum unus sit ex circulis equidistantibus et alter sit
inclinatus super circulos equidistantes, et signata fuerint super circulum inclinatum duo puncta, qualitercumque contingat, in una et eadem 5
parte a circulo maiore qui est ex circulis equidistantibus, et signati fuerint circuli maiores transeuntes per puncta provenientia et per polum,
erit proportio arcus circuli maioris qui est ex circulis equidistantibus,
qui cadit inter circulum primum maiorem et inter circulum maiorem
qui postea signatus fuit et transivit per polos, ad arcum circuli inclinati 10
qui cadit inter hos eosdem circulos sicut proportio arcus circuli maioris
qui est ex circulis equidistantibus, qui cadit inter circulos maiores

59 *NS*] *CNS* **Fi**   59 arcus²] arcu **BPsVa**   59 *RN*] *IN* **Fi**   60 *LM*] *LN* **Fi**   60 qui] *supra*
Kg   60 maior] *supra* **BPs**   61 est] *om.* **Va**   61 arcus²] arcu **Fi**   61 arcu] *om.* **M**   61 *NR*] *in*
corr. **B**   62 posuimus] possumus **Va**   62 *HQ*] *BQ* **Fi**   62 et] *om.* **KgBPsVa**   62 est] isti
Fi   63 hoc] *om.* **FiZ**   64 est] *post* igitur **R**, *post LM* **OPsVa**   64 arcus] *om.* **KgOB**
PsVa   65 eo minor] enntiorum **V**   65 illud] istud **OPsVa**   66 quod ... voluimus] *om.* **OB**
66 demonstrare] *om.* **Va**   66 voluimus] volumus **Va**   1 continentem] *in corr.* **B**; coniun-
gentem, *supra* contingentem **M**; contingentem **Ps**   2 circulum duo] *supra* **Ps**   3 sit¹] *supra*
O, *om.* **B**, *marg.* **Ps**   3 sit²] *om.* **R**   4 circulos] *in corr.* **P**   4 signata] signati **FiZ**
6–7 signati ... transeuntes] signatus fuerit circulus maior transiens **PRVM**   7 per²] *supra*
M   8 proportio] porcio **FiVa**   8 arcus] *om.* **R**   8 est] *om.* **Z**   9 cadit] *in corr.* **BPs**
9 primum] *supra* **R**   9 et] *om.* **R**   9 et ... maiorem] *marg.* **M**   9 inter²] *om.* **BPsVa**
10 arcum] *post* inclinati **BPsVa**   11 inter] *supra* **M**   11 hos] os **Va**   12 est] *post* circulis
PV

قوس ما هى أصغر من القوس من الدائرة المائلة فيما بين النقط التى

تعلمت ·

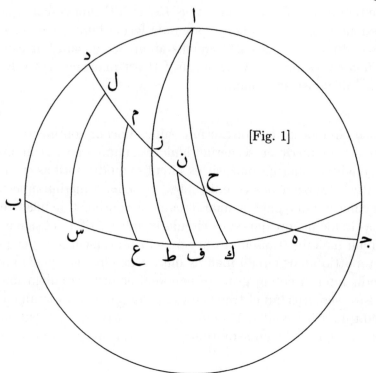

[Fig. 1]

فليكن على الخط المحيط بدائرة اب‌جَ العظمى قطب الدوائر المتوازية

وهو قطب آ ولتقطع دائرة اب‌جَ دائرتان عظيمتان على زوايا قائمة وهما

دائرتا دهجَ به ودائرة به من الدوائر المتوازية ودائرة دهجَ مائلة على

الدوائر المتوازية ولتتعلم على دائرة دهجَ نقطتان كيف ما وقعتا وهما نقطتا زَ

حَ فى جهة واحدة بعينها من دائرة به العظمى من الدوائر المتوازية ولترسم

على نقطتى زَ حَ وعلى قطب آ دائرتان عظيمتان وهما دائرتا ازطَ احكَ ،

فأقول إن نسبة قوس به‌طَ إلى قوس دزَ كنسبة قوس طهكَ إلى قوس ما هى

_____

qui transeunt per polos circulorum equidistantium et per puncta que
fuerint notata, ad aliquem arcum qui est minor arcu circuli inclinati
qui est inter puncta que fuerint signata.   15

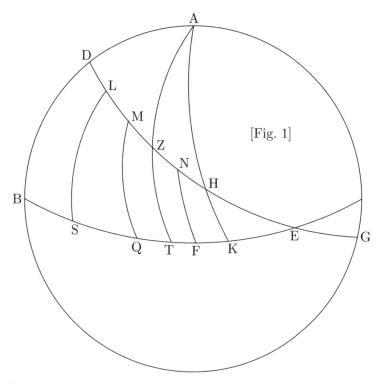

[Fig. 1]

Super lineam itaque continentem circulum *ABG* maiorem sit polus circulorum equidistantium, qui sit polus *A*, et circulum *ABG* secent
duo circuli maiores orthogonaliter, qui sint duo circuli *DEG BE*. Et sit
circulus *BE* ex circulis equidistantibus et circulus *DEG* sit inclinatus super circulos equidistantes, et signentur super circulum *DEG* duo   20
puncta, qualitercumque contingat, que sint puncta *Z H*, in una et eadem parte a circulo *BE* maiore qui est ex circulis equidistantibus. Et
signentur super duo puncta *Z H* et super polum *A* duo circuli maiores,
qui sint duo circuli *AZT AHK*. Dico igitur quod proportio arcus *BT*
ad arcum *DZ* est sicut proportio arcus *TK* ad aliquem arcum qui est   25

13 qui transeunt] transeuntes **BPsVa**   14 fuerint] fuerunt **P**   14 est minor] *tr.* **OFiZBPsVa**
14 minor] maior **V**   14 circuli] simili **Va**   15 que] *in corr.* **Fi**   15 fuerint] fuerunt **PRV**
15 signata] notata **V**, significata **Z**   17 polus] punctum **OFiZ**   18 sint] sunt **Fi**   18 duo] *om.*
PRVMFiZ   18 sit] si **Va**   20 signentur] *in corr.* **V**, signetur **R**; significentur **B**, *hic et saepi-*
us   21 que] qui **OFiZBPsVa**   21 sint puncta] *tr.* **V**   22 *BE* maiore] *tr.* **FiZ**   22 est] *supra*
R   23 signentur] signetur **Fi**, significentur **O**   23 *Z*] *et add.* **PMKgBPsVa**   23 *A*] *H A*
Fi   24 sint] sunt **FiVa**   24 *AZT*] *AET* **Kg**, *AZC* **Fi**   24 igitur] ergo **O**   24 *BT*] *BC*
Fi   25 *DZ*] *BZ* **Z**   25 sicut] *repet.* **Kg**   25 arcus] *om.* **BPsVa**   25 arcum^2] *in corr.* **P**
25 qui] qua **Fi**

أصغر من قوس زح وذلك أن قوس زح إما أن تكون مشاركة فى المقدار 20

لقوس زد وإما أن لا تكون كذلك ،

فلتكن أولاً فى الصورة الأولى مشاركة لها ولتقسم قوسا زد زح بذلك

المقدار على نقط ل م ن ولترسم دوائر عظيمة تمر بنقط ل م ن وبقطب آ

وهى دوائر لس م ع ن ف فلأن قسى دل لم مز زن نح متصلة على

الولاء مساوٍ بعضها بعض تكون قسى ب س س ع ط ف ك بعضها 25

أعظم من بعض على الولاء إذا ابتدئ من قوس ب س العظمى فلأن قسى

ب س س ع ط ف ك متوالية بعضها أعظم من بعض وقسى دل لم

مز زن نح متوالية مساوٍ بعضها لبعض وعدد قسى ب س س ع ط مساوٍ

لعدد قسى دل لم مز وعدد قوسى ط ف ك مساوٍ لعدد قوسى زن نح

تكون نسبة قوس ب ط إلى قوس دز أعظم من نسبة قوس طك إلى قوس 30

زح وذلك أنه لما كانت قوس ب س أعظم من قوس ط ف وقوس دل

مساوية لقوس زن وإذا كانت أقدار غير متساوية فنسبة القدر الأعظم منها

إلى قدر واحد بعينه أعظم من نسبة القدر الأصغر إليه ونسبة قدر جميع

المقدمات إلى جميع التوالى أعظم من نسبة جميع المقدمات إلى جميع التالية

فإن نحن صيّرنا نسبة قوس ب ط إلى قوس دز كنسبة قوس طك إلى قوس 35

ما صارت تلك القوس أصغر من قوس زح ،

ثم لا تكون قوس حز مشاركة فى المقدار لقوس دز ، فأقول إن

28–N اصغر H, אכבר [أعظم 26   om. A [لم 24   om. A [وبقطب 23   فقط A 23   [ن² om. A 23

المقدمه [¹مقدمات 34   AH: om. N 34   غير [تكون 30   add. A قوس 32   [بس ... ¹قوسى 29

N 34 نسبة [مقدمات ²المقدمه N 34 التالية [الباقه A 35 نسبة [om. N   مقدمات 34 om. A

minor arcu *ZH*. Quod ideo est quoniam arcus *ZH* aut est communicans in quantitate arcui *ZD* aut non est ita.

Sit igitur primum in forma prima communicans ei. Et dividantur duo arcus *DZ ZH* secundum illam quantitatem super puncta *L M N* et signentur circuli maiores transeuntes per puncta *L M N* et per polum *A*, qui sint circuli *LS MQ NF*. Et quia arcus *DL LM MZ ZN NH* coniuncti secundum continuitatem ad invicem sunt equales, erunt arcus *BS SQ QT TF FK* inter se vicissim minores secundum continuitatem cum inceptum fuerit ab arcu *BS* maiore. Et quia arcuum *BS SQ QT TF FK* alii sunt aliis maiores et arcus *DL LM MZ ZN NH* continui sunt ad invicem equales et numerus arcuum *BS SQ QT* est equalis numero arcuum *DL LM MZ* et numerus arcuum *TF FK* est equalis numero arcuum *ZN NH*, erit porportio arcus *BT* ad arcum *DZ* maior proportione arcus *TK* ad arcum *ZH*. Quod ideo est quoniam propter hoc quod arcus *BS* est maior arcu *TF* et arcus *DL* est equalis arcui *ZN* et cum fuerint quantitates inequales, proportio quantitatis maioris earum ad quantitatem unam et eandem erit maior proportione minoris quantitatis ad eam, et proportio quantitatis omnium antecedentium ad omnes consequentes est maior proportione omnium antecedentium ad omnes consequentes. Et si nos proposuerimus proportionem arcus *BT* ad arcum *DZ* sicut proportionem arcus *TK* ad aliquem arcum, fiet arcus ille minor arcu *ZH*.

26 *ZH*] *in corr.* **Fi**   27 *ZD*] *ZA* **B**   28 igitur] ergo **VMFiZB**   28 primum] *om.* **BPsVa**
28 prima] primo **BPsVa**   29 puncta] *in corr.* **Ps**, *post L M N* **O**   29–30 et ... *L M N*] *marg.*
BPs   31 *DL LM*] *in corr.* **Ps**   31 *LM*] *in corr.* **B**   32 secundum continuitatem] *supra*
Kg   32 erunt] *in corr.* **Ps**   32 arcus] *marg.* **R**   33–35 inter ... *FK*] *marg. inf.*
B, *marg.* **Ps**   34 cum] in **Va**   34 fuerit] fuerint **Va**   34 arcuum] arcum **R**
34 *BS SQ*] *BSQ* **Fi**   35 *MZ ZN*] *MZN* **Fi**   36 sunt] *post* invicem **RMKgO(B)PsVa**
36 invicem] *hinc usque ad finem prop. deest* **B**   36 arcuum] arcum **R**   36 *BS SQ*] *BSQ*
Fi   36 *QT*] *TF TK add.* **Kg**   36 equalis] *post* arcuum **Va**   37 est equalis] equatur
OPsVa   38 *ZN*] *in corr.* **Ps**   38 arcus] *om.* **PsVa**   39 arcus] *om.* **PsVa**   39 arcus *TK*] *tr.*
O   39 *ZH*] *in corr.* **V**   40 hoc quod] *marg.* **P**   40 est equalis] equatur **OPsVa**
41 proportio] porcio **Va**   41–43 maioris ... quantitatis[2]] *om.* **Va**   42 earum] *in corr.* **M**
42 maior] *supra* **Ps**   44 omnium antecedentium] *tr.* **PsVa**   44 antecedentium] adcedentium
V   45 proposuerimus] posuerimus **RMKgOFiZPsVa** (*post BT* **Va**)   45 arcus] *marg.* **R**
46 proportionem] proportioni **Fi**, proportio **Ps**   46 aliquem arcum] *tr.* **Kg**   46 fiet] fi et **PsVa**

نسبة قوس بط إلى قوس دز كنسبة طك إلى قوس ما أصغر من قوس

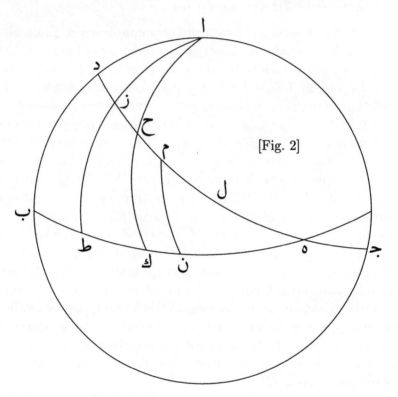

[Fig. 2]

زح فإن لم يكن ذلك كذلك فإنه إما أن تكون نسبتها إليها كنسبة طك إلى

قوس هى أعظم من قوس زح وإما أن تكون كنسبتها إلى قوس زح فلتكن   40

أولاً إن أمكن كنسبة طك إلى قوس هى أعظم من قوس زح وهى قوس

زل كما فى الصورة الثانية ولما كانت قسى لز زح زد ثلثاً فصلنا قوساً

أخرى أصغر من قوس لز وأعظم من قوس زح مشاركة فى المقدار لقوس

زد وهى قوس زم ولترسم دائرة عظيمة تمر بنقطة م وبقطب آ وهى دائرة

من فلأن قوس زم مشاركة فى المقدار لقوس دز تكون نسبة قوس بط   45

إلى قوس دز كنسبة قوس طن إلى قوس ما أصغر من قوس زم ونسبة

Post hec non sit arcus *HZ* communicans in quantitate arcui *ZD*. Dico igitur quod proportio arcus *BT* ad arcum *DZ* est sicut proportio

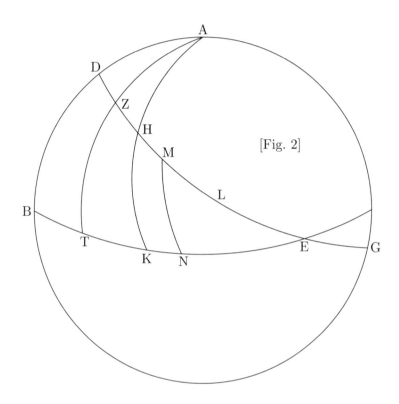

[Fig. 2]

arcus *TK* ad aliquem arcum minorem arcu *ZH*. Quod si hoc non ita   50
fuerit, ergo aut erit proportio eius ad ipsum sicut proportio arcus *TK*
ad arcum qui sit maior arcu *ZH* aut erit sicut proportio eius ad arcum
ZH. Si ergo possibile est, sit primum sicut proportio *TK* ad arcum qui
sit maior arcu *ZH*, qui sit arcus *ZL*, sicut in secunda constat forma. Et
quia arcus *LZ ZH* sunt continui, ponemus alium arcum minorem arcu   55
LZ et maiorem arcu *ZH*, communicantem in quantitate arcui *ZD*, qui
sit arcus *ZM*. Et signabo circulum maiorem transeuntem per punctum
M et per polum *A*, qui est circulus *MN*. Et quia arcus *ZM* communicat in quantitate arcui *DZ*, erit proportio arcus *BT* ad arcum *DZ* sicut

48 *HZ*] *in corr.* **Z**   50 arcu] arcum **Fi**   50 non ita] *tr.* **OPsVa**   50 ita] facta **Fi**
51 fuerit] fuit **Fi**   51 arcus] *om.* **KgOPsVa**   53 possibile est] *tr.* **FiZVa**   53 est] *om.* **Ps**
53 sit] *in corr.* **Ps**, sic **Va**, non sit **Kg**   53 proportio] arcus *add.* **Fi**   55 continui] continua **R**
(*marg.*)   56 quantitate] quantitatem **Fi**   57 transeuntem] *om.* **PsVa**   57 per punctum] *repet.*
Va   58 *A*] *supra* **Ps**   58 est] sit **OFiZ**   59 arcus *BT*] *tr.* **PsVa**

قوس بط إلى قوس دز كنسبة قوس طك إلى قوس زل فنسبة قوس طك

إلى قوس لز كنسبة قوس طن إلى قوس ما هى أصغر من قوس زم

وقوس نط أعظم من قوس طك فالقوس التى هى أصغر من قوس زم هى

50   أعظم من قوس زل ولكنها أصغر منها وذلك غير ممكن فليس نسبة قوس

بط إلى قوس زد كنسبة قوس طك إلى قوس هى أعظم من قوس زح ،

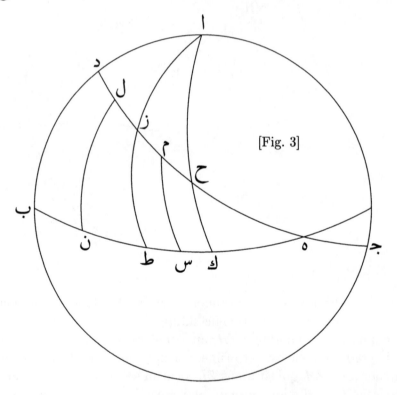

[Fig. 3]

فأقول إنه ليس نسبتها إليها كنسبة طك إلى زح فإن أمكن فلتكن

نسبة قوس بط إلى قوس زد كنسبة قوس طك إلى قوس زح كما فى

الصورة الثالثة ولتقسم كل واحدة من قوسى دز زح بنصفين نصفين على

48 طن] H دزا [زل primum | أصغر ... من قوس N 50-51 وهى [هى N 49 repet. A [طك ... زم] هى

N طك رح [زح 53 N رح [زح 52 N رح , supra مح [زح 52 N وامول [فأقول | false, deinde recte A 52

proportio arcus *TN* ad aliquem arcum minorem arcu *ZM*. Proportio   60
autem arcus *BT* ad arcum *DZ* est sicut proportio arcus *TK* ad arcum
ZL: ergo proportio arcus *TK* ad arcum *LZ* est sicut proportio arcus *TN*
ad arcum aliquem minorem arcu *ZM*. Arcus vero *NT* est maior arcu
TK: ergo arcus qui est minor arcu *ZM* est maior arcu *ZL*. Sed ipse est
minor eo, quod quidem est impossibile. Non est igitur proportio arcus   65
BT ad arcum *ZD* sicut proportio arcus *TK* ad arcum qui est maior
arcu *ZH*.

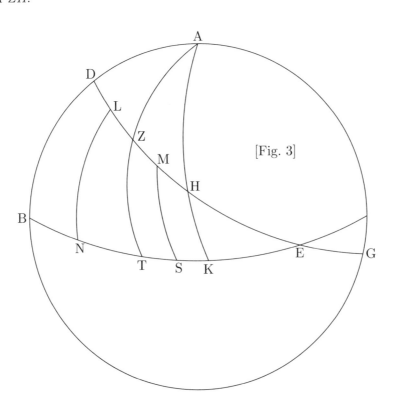

[Fig. 3]

Et dico etiam quod non est proportio eius ad ipsum sicut
proportio *TK* ad *ZH*. Quod si possibile fuerit, sit proportio arcus *BT*
ad arcum *ZD* sicut proportio arcus *TK* ad arcum *ZH*, sicut in ter-   70
70

60 arcum] *hic desinit prop. 10, sequitur textus prop. 11, l. 60; reliquum prop. 10 add. manu
rec. marg. inf.* **Ps**   61 arcus[1]] *marg.* **R**   61 est] sit **O** (*supra*)   63 arcum] qui est *add.*
V   63 arcum aliquem] *tr.* **O**   64 *ZL*] *LX* **Kg**, *LZ* **OPsVa**   65 quidem] *om.* **Va**, quid **V**
65 impossibile] possibile **M**   65 arcus] *marg.* **R**   66 *ZD*] *ZB* **Fi**   66 proportio] ad *add.*
Fi   68 dico] dicam **Kg**   69 proportio] arcus *add.* **Z**   69 *TK*] arcus *add.* **Fi**   69 ad] arcum
add. **FiZ**   69 *ZH*] arcum *add.* **O**   69 fuerit] fuit **Fi**   69 sit] si **R**   70 *ZD*] *ZT* **Fi**   70–
71 tertia] secunda *add.* **V**

نقطتى لَ مَ ولترسم دائرتان عظيمتان تمر كل واحدة منهما بقطب آ وبواحدة ٥٥

من نقطتى لَ مَ وهما دائرتا لنَ مسَ فلأن قوس دلَ مساوية لقوس لزَ

تكون قوس بنَ أعظم من قوس نطَ وتكون قوس بطَ أعظم من مثلى

قوس طنَ وكذلك نبين أيضاً أن قوس كطَ أصغر من مثلى قوس طسَ فلأن

قوس بطَ أعظم من مثلى قوس طنَ وقوس طكَ أصغر من مثلى قوس

طسَ تكون نسبة قوس بطَ إلى قوس طكَ أعظم من نسبة قوس نطَ إلى ٦٠

قوس طسَ فنسبة قوس نطَ إلى قوس طسَ أصغر من نسبة قوس بطَ

إلى قوس طكَ ونسبة قوس بطَ إلى قوس طكَ كنسبة قوس دزَ إلى قوس

زحَ فنسبة قوس نطَ إلى قوس طسَ أصغر من نسبة قوس دزَ إلى قوس

زحَ ونسبة قوس دزَ إلى قوس زحَ كنسبة قوس لزَ إلى قوس زمَ فنسبة

قوس نطَ إلى قوس طسَ أصغر من نسبة قوس لزَ إلى قوس زمَ وإذا ٦٥

بدلنا تكون نسبة قوس نطَ إلى قوس لزَ أصغر من نسبة قوس طسَ إلى

قوس زمَ فإن صيّرنا نسبة قوس نطَ إلى قوس لزَ كنسبة قوس طسَ إلى

قوس ما صارت تلك القوس أعظم من قوس زمَ وقد كان تبين فى الصورة

الثانية أن ذلك غير ممكن فليس نسبة قوس بطَ إلى قوس دزَ كنسبة قوس

طكَ إلى قوس زحَ وقد كان تبين أنه ليس نسبتها إليها كنسبة قوس طكَ ٧٠

إلى قوس هى أعظم من قوس زحَ فهى إذاً كنسبتها إلى قوس هى أصغر منها

٥٥ كل [بكل A   ٥٥ وبواحدة ... [واحدة] om. A   ٥٧ قوس³] om. A   ٥٨ قوس³] طن وقوس لطا

H: ٥٨–٦١ أصغر من مثلى قوس ... طسَ¹] add. A   ٥٩–٦٠ فلأن ... [ ] om. A   ٦٠ وقوس ... قوس بطَ

٦٢ طكَ] om. N   ٦٣ قوس¹] om. A   ٦٤–٦٥ ونسبة ... قوس زمَ] repet. A   ٦٣ قوس¹] repet. A

٦٥ وإذا] ادا A   ٦٨ القوس] الصورة N   ٦٩ الثانية] الثالثه A   ٧٠ تبين] add. N   ٧٠ قوس²] om. A

tia videtur forma. Dividam autem unumquemque duorum arcuum *DZ*
ZH in duo media et duo media super duo puncta *L M*, et signabo
duos circulos maiores quorum quisque transeat per polum *A* et unum
duorum punctorum *L* et *M*, qui sint circuli *LN MS*. Et quia arcus *DL*
est equalis arcui *LZ*, erit arcus *BN* maior arcu *NT* et erit arcus *BT* 75
maior duplo arcus *TN*. Et similiter etiam monstratur quod arcus *KT*
est minor duplo arcus *TS*. Et quia arcus *BT* est maior duplo arcus *TN*
et arcus *KT* est minor duplo arcus *TS*, erit proportio arcus *BT* ad
arcum *TK* maior proportione arcus *NT* ad arcum *TS*. Ergo proportio
arcus *NT* ad arcum *TS* est minor proportione arcus *BT* ad arcum *TK*. 80
Proportio vero arcus *BT* ad arcum *TK* est sicut proportio arcus
DZ ad arcum *ZH*: ergo proportio arcus *NT* ad arcum *TS* est minor
proportione arcus *DZ* ad arcum *ZH*. Sed proportio arcus *DZ* ad arcum
ZH est sicut proportio arcus *LZ* ad arcum *ZM*: ergo proportio arcus
NT ad arcum *TS* est minor proportione arcus *LZ* ad arcum *ZM*. Cum 85
autem permutaverimus, erit proportio arcus *NT* ad arcum *LZ* minor
proportione arcus *TS* ad arcum *ZM*. Si ergo posuerimus ut sit proportio arcus *NT* ad arcum *LZ* sicut proportio arcus *TS* ad aliquem arcum,
fiet arcus ille maior arcu *ZM*. In secunda vero forma fuit ostensum
illud fore impossibile. Non est ergo proportio arcus *BT* ad arcum *DZ* 90
sicut proportio arcus *TK* ad arcum *ZH*; et iam fuit ostensum etiam quod non est eius proportio ad ipsum sicut proportio *TK* ad arcum

71 unumquemque] utrumque **MKg**   73 duos circulos] *tr.* **R**   73 quisque] uterque **MKg**
73 et] per *add.* **VMKgOFiZPsVa**   74 et] *om.* **OPsVa**   74 sint] sunt **FiZ**   75 est] *om.*
Fi   75 arcui] arcum **Fi**   75 arcus[1]] *in corr.* **R**   75 *BN*] *LM* **R**   76 arcus[1]] arcu
Va   76 etiam] *post* monstratur **R**, *post* quod **PV**; *iterum post* quod **OFiZ**   76 *KT*] *K* **Fi**
77 arcus] arcu **Va**   77 *TS*] erit proportio arcus *BT hic false add.* **M**   78 est] *post* et
O   78 *BT*] *BD* **Fi**, *BG* **Ps**   79–80 *NT* … arcus[2]] *om.* **R**   80–83 *BT* … arcus[1]] *om.* **Va**
81 vero arcus] *tr.* **V**   81 arcus[2]] *om.* **Kg**   82 *ZH*] *ZB* **Fi**   83 *ZH*] *ZB* **Fi**   84 *ZH*] est pro-
porcio arcus *DZ* ad arcum *ZH false add.* **Va**   84 sicut] *om.* **Va**   85 *LZ*] *BZ* **Va**   86 *LZ*] *ZB* **Z**
87 Si] li **Va**   87 ergo] autem **FiZ**   88 arcum[1]] *om.* **PsVa**   88 arcus[2]] *om.* **Kg**   90 illud] istud
Fi   90 *BT*] *HT* **Kg**   91–92 etiam] *om.* **OFiZ**   92 est] fuit **OFiZPsVa**

فنسبة قوس ب‍ط إلى قوس دز كنسبة قوس ط‍ك إلى قوس ما هى أصغر

من قوس ز‍ح ، وذلك ما أردنا أن نبين .

يا   إذا كان قطب الدوائر المتوازية على الخط المحيط بدائرة عظيمة وقطعت هذه

الدائرة دائرتان عظيمتان على زوايا قائمة وكانت إحداهما من الدوائر المتوازية

وكانت الأخرى مائلة على الدوائر المتوازية وكانت دائرة أخرى عظيمة تمر

بقطبى الدوائر المتوازية وتقطع الدائرة المائلة فيما بين الدائرة العظمى من

الدوائر المتوازية وبين الدائرة التى تماسها الدائرة المائلة من الدوائر المتوازية   5

فإن نسبة قطر الكرة إلى قطر الدائرة التى تماسها الدائرة المائلة أعظم من

نسبة القوس من الدائرة العظمى من الدوائر المتوازية التى تقع فيما بين

الدائرة الأولى العظمى وبين الدائرة التى تتلوها وتمر بقطبى الدوائر المتوازية

إلى القوس من الدائرة المائلة التى تقع فيما بين تلك الدوائر بأعيانها .

فليكن على الخط المحيط بدائرة أ‍ب‍ج العظمى قطب الدوائر المتوازية   10

وهو نقطة آ ولتقطع دائرة أ‍ب‍ج دائرتان عظيمتان على زوايا قائمة وهما

دائرتا ب‍ه‍ج د‍ه‍ز ولتكن دائرة ب‍ه‍ج الدائرة العظمى من الدوائر المتوازية

ودائرة د‍ه‍ز مائلة على الدوائر المتوازية ولتكن دائرة أخرى عظيمة وهى دائرة

أ‍ح‍ك تقطع دائرة د‍ه‍ز وتمر بقطبى الدوائر المتوازية فيما بين دائرة ب‍ه‍ج

وبين الدائرة التى تماسها دائرة د‍ه‍ز وهى دائرة د‍ل‍م ، فأقول إن نسبة قطر   15

ورضت [وكانت 3   om. H [المتوازية 3   الأخرى] 3 om. H [المحيط 1   المحيطه A   كاس N [كان 1
A   4 بقطبى] بقطب A   وتقطع الدائرة المائلة 4 [ante تمر (l. 3, cum transpositione) و NH
om.   12 الدائرة] A موس [القوس 9   موس A   نسبة القوس من 7 [ممdaها H تماسها 5
A   المتوازية ... ودائرة [om. A 14 أ‍ح‍ك [om. A 15 وبين الدائرة] H: والدار A, والداره
A   قطر] repet. A 15

qui sit maior arcu *ZH*: ergo est sicut proportio eius ad arcum qui est
minor eo. Ergo proportio arcus *BT* ad arcum *DZ* est sicut proportio
arcus *TK* ad aliquem arcum qui est minor arcu *ZH*. Et illud est quod   95
demonstrare voluimus.

11 Si polus circulorum equidistantium fuerit super lineam continentem
circulum maiorem et secuerint hunc circulum duo circuli maiores orthogonaliter, quorum unus sit ex circulis equidistantibus et alter sit
declinatus a circulis equidistantibus, et fuerit alius circulus maior secans circulum declinatum – et transiens per duos polos circulorum   5
equidistantium – inter circulum maiorem qui est ex circulis equidistantibus et inter circulum quem circulus declinatus contingit qui est
ex circulis equidistantibus, erit proportio diametri spere ad diametrum
circuli quem contingit circulus declinatus maior proportione arcus circuli maioris qui est ex circulis equidistantibus, qui cadit inter circulum   10
primum maiorem et inter circulum qui eum sequitur et transit per
polos circulorum equidistantium, ad arcum circuli declinati qui cadit
inter illos eosdem circulos.

Super lineam itaque continentem circulum *ABG* maiorem sit
polus circulorum equidistantium, qui sit punctum *A*, et secent circulum   15
ABG duo circuli maiores orthogonaliter, qui sint circuli *BEG DEZ*.
Sitque circulus *BEG* circulus maior qui est ex circulis equidistantibus,
et circulus *DEZ* sit inclinatus super circulos equidistantes, et sit
alius maior circulus, qui sit circulus *AHK*, secans circulum *DEZ* – et
transiens per duos polos circulorum equidistantium – inter circulum   20
BEG et circulum quem contingit circulus *DEZ*, qui est circulus *DLM*.
Dico ergo quod proportio diametri spere ad diametrum circuli *DLM*

93 arcu] *om.* **Va**   93 arcu *ZH*] *tr.* **O**   93 proportio eius] *tr.* **MKg**   95 arcus] *supra*
R   95 arcu] *om.* **Va**   95–96 quod … voluimus] *om.* **O**   96 voluimus] volumus
Va   *1* Si] *hanc prop. usque ad HF, l. 60, add. in fine libri eadem manu* **B**, *alia*
manu **Ps**   *1* super lineam] lnreueam **Fi**   *1–2* lineam … maiorem] circulum lineam maiorem continentem **Kg**   *2* circulum maiorem] *tr.* **OBPsVa**   *5* duos polos] *tr.*
R   *7–8* et … equidistantibus] *marg.* **M**   *7* circulum] equidistantem *add.* **Kg**
7 contingit] *post* quem **Kg**   *7–8* qui … equidistantibus] *om.* **Kg**   *8* proportio] dia
add. **Va**   *11* primum maiorem] *tr.* **BPsVa**   *14* maiorem] *in corr.* **V**   *15* secent] hunc
add. **FiZBPsVa**   *16* *ABG*] *om.* **OFiZBPsVa**   *16* *BEG*] *BT EG* **O**, *B TEG* **B**, *BTEG* **PsVa**
16 *DEZ*] *om.* **Fi**   *17* Sitque circulus *BEG*] *om.* **Fi**   *17* *BEG*] *DEG* **R**   *19* alius] *in corr.*
Va   *19* qui sit circulus] *om.* **MVa**   *19* circulus²] *om.* **Fi**   *19* et] *om.* **R**   *21* *DEZ*] *in corr.*
Va   *22* ergo] igitur **MKgOFiZPsVa**   *22* diametri] *supra* **R**

الكرة إلى قطر دائرة د ل م أعظم من نسبة قوس ب ط إلى قوس د ح فلترسم

دائرة من الدوائر المتوازية تمر بنقطة ح وهى دائرة ن ح س ولتكن الفصول

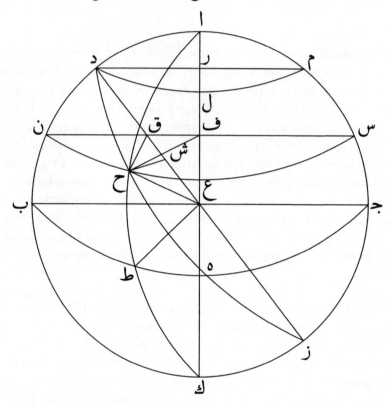

المشتركة لهذه السطوح خطوط ا ك د ز ب ج ن س د م ط ع ح ف ح ق ع ح

فدائرة ا ب ج العظمى فى كرة تقطع دوائر من الدوائر التى فى الكرة وهى

دوائر د ل م ن ح س ب ه ج وتمر بأقطابها فهى تقطعها بنصفين وعلى زوايا

قائمة وتكون خطوط د م ن س ب ج أقطار دوائر د ل م ن ح س ب ه ج ودائرة

ا ب ج قائمة على كل واحدة من دوائر د ل م ن ح س ب ه ج على زوايا قائمة

فلأن دوائر د ل م ن ح س ب ه ج متوازية فى كرة وقد أخرج خط مستقيم يمر

16 [د ل م] لم A   17 [ن ح س] محمر N   18 [ع ح] om. A, نع N   19-50 مساو N   [فدائرة ا ب ج ...] om. A
23 [ب ه ج] H: om. N

est maior proportione arcus *BT* ad arcum *DH*. Signabo itaque
circulum qui sit ex circulis equidistantibus transeuntem per punctum
H, qui sit circulus *NHS*. Sintque differentie communes his superficiebus   25

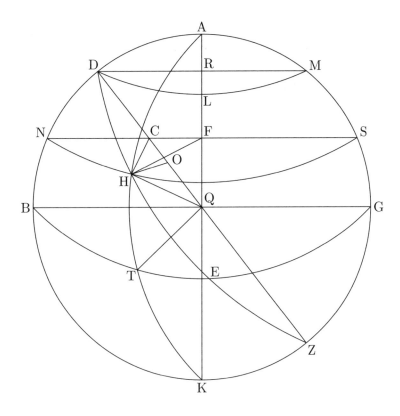

linee *AK DZ BG NS DM TQ HC HF*. Circulus ergo *ABG* maior
in spera secat circulos qui sunt ex circulis qui sunt in spera, qui sunt
circuli *DLM NHS BEG*, et transit per polos eorum. Ergo ipse secat
eos in duo media et orthogonaliter, et linee *DM NS BG* sunt diametri
circulorum *DLM NHS BEG*, et circulus *ABG* erigitur super unum-   30
quemque circulorum *DLM NHS BEG* orthogonaliter. Et quia circuli
DLM NHS BEG equidistantes sunt in spera et iam producta est linea

23 Signabo] Significabo **Ps**   *24* sit] est **O**   *25* his] in **Va**   *26* linee] *in corr.* **Ps**   *26 NS*] *NL*
FiZ   *26 DM*] *HQ add.* **KgBPsVa**   *26 TQ*] *TL* **Fi**   *26 HC*] *BC* **Fi**   *26 HF*] *in corr.* **V**, *BF*
Z   *26* ergo] igitur **BPs**   *30 DLM NHS*] *DL MN HS* **Fi**   *30–31* et ... orthogonaliter] *marg.*
R   *30* erigitur] erigit **Fi**   *30–31* erigitur ... *BEG*] *om.* **V**   *31 NHS*] *NH* **Fi**   *31 BEG*] *GED*
O

بقطبيها وهو خط اك يكون خط اك قائماً على كل واحدة من دوائر دلم

نحس بهج على زوايا قائمة ويمر بمراكزها وبمركز الكرة أيضاً فنقط ر ٢٥

ف ع مراكز دوائر دلم نحس بهج ولأن سطوح دلم نحس بهج

المتوازية قُطعت بسطح ابج تكون الفصول المشتركة لها متوازية فخطوط دم

نس بج مواز بعضها لبعض وأيضاً لأن سطحى نحس بهج المتوازيين

قد قُطعا بسطح احك يكون الفصلان المشتركان لهما متوازيين فخط حف

مواز لخط طع وخط نف يلقى خط فح وهما موازيان لخطى بع طع ٣٠

اللذين يلقى أحدهما الآخر وليست هذه الخطوط فى سطح واحد فهى تحيط

بزاويتين متساويتين وتكون زاوية نفح مساوية لزاوية بعط ولأن

دائرتى نحس دهز قائمتان على دائرة ابج على زوايا قائمة يكون الفصل

المشترك لدائرتى نحس دهز أيضاً عموداً على دائرة ابج والفصل المشترك

لهما هو خط حق فخط حق عمود على دائرة ابج وهو يحدث مع جميع ٣٥

الخطوط المستقيمة التى تخرج منه فى سطح دائرة ابج زوايا قائمة وكل

واحد من خطى فق عق اللذين هما فى سطح دائرة ابج يخرج من خط

حق فكل واحدة من زاويتى حقف حقع قائمة ولأن خط اك قائم على

خط نس على زوايا قائمة تكون زاوية قفع قائمة ولأن زاوية قفع

قائمة تكون زاوية فعق حادّة فخط عق أطول من خط قف وليجعل خط ٤٠

قش مساوياً لخط فق وليوصل خط حش ،

فلأن خط فق مساو لخط قش وخط حق مشترك يكون خطا فق

---

٢٨ بج] هح N   ٢٨ مواز] موازنه N   ٣٨ واحدة] واحد N   ٤٢ وخط حق] وحق, *supra* خط N

recta transiens per polos eorum, que est linea AK, ergo linea AK est orthogonaliter erecta super unumquemque circulorum DLM NHS BEG
et transit per centra eorum et per centrum spere etiam. Ergo puncta   35
R F Q sunt centra circulorum DLM NHS BEG. Et quia superficies
DLM NHS BEG equidistantes secat superficies ABG, erunt differentie
communes eis equidistantes; ergo linee DM NS BG ad invicem sunt
equidistantes. Et etiam quia duas superficies NHS BEG equidistantes
iam secuit superficies AHK, erunt due differentie eis communes equidi-   40
stantes; ergo linea HF equidistat linee TQ. Et quia linea NF occurrit
linee FH et sunt equidistantes duabus lineis BQ QT, quarum una alteri
concurrit, et he linee non sunt in una superficie, ergo ipse continent angulos equales: ergo angulus NFH est equalis angulo BQT. Et quia duo
circuli NHS DEZ sunt orthogonaliter erecti super circulum ABG, erit   45
etiam differentia communis circulis NHS DEZ perpendicularis super
circulum ABG. Differentia vero eis communis est linea HC: ergo linea
HC est perpendicularis super circulum ABG et ex ea cum omnibus
rectis lineis, que ab ea in superficie circuli ABG protrahuntur, proveniunt anguli recti. Sed unaqueque duarum linearum FC QC, que sunt   50
in superficie circuli ABG, protrahitur a linea HC: ergo unusquisque
duorum angulorum HCF HCQ est rectus. Et quia linea AK orthogonaliter est erecta super superficiem NS, erit angulus CFQ rectus; et
quia angulus CFQ est rectus, erit angulus FQC acutus. Ergo linea QC
est longior linea FC. Ponam autem lineam CO equalem linee CF et   55
producam lineam HO.

   Et quia linea CF est equalis linee CO, si posuero lineam HC
communem, erunt due linee FC CH equales duabus lineis CO CH,

33 que] qui **Fi** 33 ergo linea AK] *marg.* **M**, *om.* **Z** 34 unumquemque] unum quem **V**, unumquodque **Fi** 34–35 DLM ... per centrum] *om.* **Fi** 35 et^2] *supra* **Va** 35 etiam] *om.* **OFiZB**
PsVa 36 Q] qui **Fi** 36 NHS] NHG **Kg** 36 BEG] $SBEG$ **Fi** 37 DLM ... superficies] *om.*
Fi 37 equidistantes secat] *tr.* **MO** 37 secat] secant **Z** 37–38 secat ... equidistantes] *om.* **R**
38 eis equidistantes] *tr.* **Va** 39 Et ... equidistantes] una **V** 39 etiam] *om.* **Va** 40 eis] *post*
erunt **Kg** 40 eis communes] *tr.* **Fi** 41 ergo] *om.* **Fi** 41 linea] *in corr.* **V** 41 HF] HK **Fi**
41 occurrit] concurrit **PKg** 42 FH] *in corr.* **Fi** 42 et sunt] *marg.* **P** 43 concurrit] occurrit
Va 43 una superficie] *tr.* **MKgO** 43 ipse] non *add.* **R** 46 circulis] *in corr.* **Ps**, circulorum **Va**
47 Differentia] *illeg.* **Fi** 47 eis] *om.* **OBPsVa** 47 HC] HO **Fi** 47–48 ergo linea HC] *om.*
VVa 48 HC] HO **Fi** 49 circuli] *supra* **B** 50 Sed] et **OFiZBPsVa** 50 unaqueque] utraque
MKg 51 a] *supra* **M** 51 unusquisque] uterque **MKg** 52 HCF] HOF **Kg** 52 rectus] erectus
Va 53 est erecta] *tr.* **OBPsVa** 53 NS] NF **RKg** 53 erit] EQT **Va** 53 CFQ] EFQ
Va 54 rectus] erectus **Fi** 54 FQC] FAC **Z** 54 linea] *marg.* **B** 55 lineam] *in corr.* **V**
55 CO] *om.* **Fi** 56 HO] HC **Fi** 57 HC] HT **Va** 58 CO] TO **Fi** 58 CH] TH **Va**

قح مساويين لخطى قش قح كل واحد لنظيره وزاوية فقح القائمة

مساوية لزاوية شقح القائمة فقاعدة حش مساوية لقاعدة حف ومثلث

فقح مساوٍ لمثلث شقح والزوايا الباقية مساوية للزوايا الباقية كل واحدة

لنظيرتها وهى الزوايا التى توترها الأضلاع المتساوية فزاوية حفق مساوية

لزاوية حشق لكن زاوية حفق مساوية لزاوية طعب فزاوية حشق

مساوية لزاوية طعب ولأن حعق مثلث والزاوية التى على نقطة ق منه

قائمة وقد أخرج فيه خط حش تكون نسبة خط عق إلى خط قش أعظم

من نسبة زاوية قشح إلى زاوية قعح وخط قف مساوٍ لخط قش وزاوية

قشح مساوية لزاوية طعب فنسبة خط عق إلى خط قف أعظم من

نسبة زاوية بعط إلى زاوية قعح ولكن نسبة خط قع إلى خط قف

كنسبة خط عد إلى خط در وهى نسبة خط دز إلى خط دم ونسبة زاوية

بعط إلى زاوية قعح كنسبة قوس بط إلى قوس دح فنسبة خط زد

أيضاً إلى خط دم أعظم من نسبة قوس بط إلى قوس دح وخط دز قطر

الكرة وخط دم قطر دائرة دلم فنسبة قطر الكرة إلى قطر دائرة دلم أعظم

من نسبة قوس بط إلى قوس دح ، وذلك ما أردنا أن نبين .

يب   إذا كانت فى كرة دائرتان عظيمتان تماسان دائرة واحدة بعينها من الدوائر

المتوازية وتفصلان فيما بينهما من الدوائر المتوازية قسياً متشابهة وكانت دائرة

أخرى عظيمة مائلة على الدوائر المتوازية تماس دائرتين أعظم من الدائرتين

---

om. [من 51 A .om | نسبة 51 A ماالحط [الحط 50 N مق [قف 50 H: om. N [حشق ... زاوية 47
A 54 [بعط A [زد 54 A .om [دلم 2 A [دل 1 دائرتان عظيمتان تماسان [marg.,
N دارىان [دائرتين 3 A تماسان pro

queque scilicet earum sue relative, et angulus *FCH* rectus est equalis angulo *OCH* recto: ergo basis *HO* est equalis basi *HF* et triangulus *FCH* est equalis triangulo *OCH* et reliqui anguli sunt equales reliquis angulis, quisque videlicet suo relativo. Et ipsi sunt anguli quibus latera subtenduntur equalia: ergo angulus *HFC* est equalis angulo *HOC*. Angulus autem *HFC* est equalis angulo *TQB*: ergo angulus *HOC* est equalis angulo *TQB*. Et quia *HQC* est triangulus et angulus eius qui est super punctum *C* est rectus et iam protracta est in eo linea *HO*, erit proportio linee *QC* ad lineam *CO* maior proportione anguli *COH* ad angulum *CQH*. Linea autem *FC* est equalis linee *CO* et angulus *COH* est equalis angulo *TQB*: ergo proportio linee *QC* ad lineam *FC* est maior proportione anguli *BQT* ad angulum *CQH*. Sed proportio linee *CQ* ad lineam *CF* est sicut proportio *QD* ad lineam *DR* – et ipsa est proportio linee *ZD* ad lineam *DM* – et proportio anguli *BQT* ad angulum *CQH* est sicut proportio arcus *BT* ad arcum *DH*. Sed proportio linee *ZD* etiam ad lineam *DM* est maior proportione arcus *BT* ad arcum *DH*. Linea autem *DZ* est diametrus spere et linea *DM* est diametrus circuli *DLM*. Ergo proportio diametri spere ad diametrum circuli *DLM* est maior proportione arcus *BT* ad arcum *DH*. Et illud est quod demonstrare voluimus.

12 Cum in spera fuerint duo circuli maiores contingentes unum et eundem circulum ex circulis equidistantibus et separaverint inter se ex circulis equidistantibus arcus similes et fuerit alius circulus maior inclinatus super circulos equidistantes contingens duos circulos maiores duobus

59 queque] utraque **MKg**   59 scilicet] *om.* **KgOFiBPsVa**   59 earum] *om.* **RM**, illarum **Kg**   59 est equalis] *tr.* **MKg**   60 *HF*] *hic desinit* **B**, *f. 112v, et* **Ps**, *f. 173v*   60 et] *hic incipit pars posterior prop. 11 in* **Ps**, *f. 170v*   60 triangulus] circulus **KgPsVa**   61 *FCH*] *FHC* **Kg**   61 est equalis] equatur **PsVa**, *hic et saepius*   61 triangulo] angulo **Kg**, circulo **PsVa**   62 quisque videlicet] scilicet uterque **MKgPsVa**   62 ipsi] *om.* **PsVa**   63 *HFC*] *HST* **Va**   63–64 *HOC ... angulo*] *om.* **V**   64 *HFC*] *AFC* **Z**, *HCT* **Va**   64–65 ergo ... *TQB*] *om.* **PsVa**   65 *TQB*] *QB* **Fi**   65 triangulus] circulus **KgPsVa**   68 *CQH*] *COH* **Fi**, *SQH* **PsVa**   68 autem] *om.* **R**   69 *COH*] *CHO* **KgFiZ**   70 *BQT*] *BT QT* **Fi**, *HTQ* **PsVa**   70 *CQH*] *CQB* **Z**, *TQH* **Va**   71 *CQ*] *TQ* **Va**   71 lineam *CF*] *tr.* **O**   72 *BQT*] *BQC* **PsVa**   73 Sed] ergo **KgOFiZPsVa**   74 etiam] est **Va**   75 *DH*] *DB* **Ps**   76 *DLM*] *DML* **Ps**   77 *BT*] *BR* **Va**   77 *DH*] *DB* **Z**, *ZB* **PsVa**   77 illud] hoc **OFiZ**, istud **PsVa**   78 quod ... voluimus] *om.* **O**   78 demonstrare] *om.* **PsVa**   78 voluimus] volumus **PsVa**   2 et separaverint] *om.* **Va**   2 separaverint] *in corr.* **Ps**   2 inter se] *post* equidistantibus (*l. 3*) **Ps**; unum *add.* **Kg**   2–3 ex circulis equidistantibus] *om.* **Va**   3 alius] alicuius **Z**   4 circulos[1]] *in corr.* **Ps**   4 duobus] duo **V**

اللتين كانت تماسهما الدائرتان الأوليان وتقطع الدائرتين اللتين تماسان دائرة

واحدة بعينها من الدوائر المتوازية فيما بين الدائرة العظمى من الدوائر

المتوازية وبين الدائرة التى ماستها الدائرتان الأوليان فإن نسبة ضعف قطر

الكرة إلى قطر الدائرة التى تماسها الدائرة المائلة أعظم من نسبة القوس من

الدائرة العظمى من الدوائر المتوازية التى تقع فيما بين الدائرتين اللتين تماسان

دائرة واحدة بعينها إلى القوس من الدائرة المائلة التى تقع فيما بين تلك

الدوائر بأعيانها ·

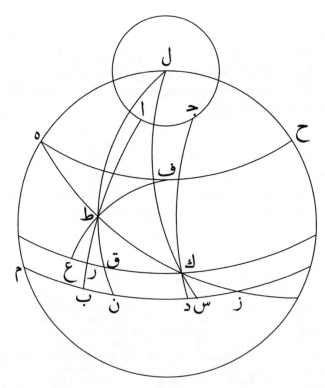

circulis, quos duo primi circuli fuerint contingentes, et secuerit duos cir-   5
culos qui contingunt unum et eundem ex circulis equidistantibus inter
circulum maiorem qui est ex circulis equidistantibus et circulum quem
contingunt duo primi circuli, proportio dupli diametri spere ad diametrum circuli quem contingit circulus inclinatus est maior proportione
arcus circuli maioris, qui est ex circulis equidistantibus et cadit inter   10
circulos qui contingunt eundem circulum, ad arcum circuli inclinati qui
cadit inter illos eosdem circulos.

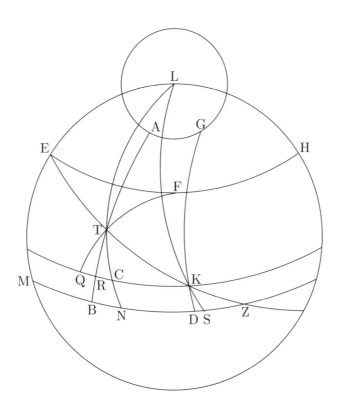

---

5 duo] duos **Va**   5 fuerint] fuerunt **Z**   5 duos] duo **Z**   6 eundem] *in corr.* **V**; circulum
add. **FiZPsVa**   6 ex circulis] *supra* **Ps**   7 est] *om.* **Fi**   7 quem] quod **Fi**   8 ad] *supra* **Ps**
9 quem] que **Fi**   12 cadit] cadet **Fi**   12 illos] *om.* **OFiZPsVa**

فلتماس فى كرة دائرتا اب جد العظيمتان دائرة واحدة بعينها من

الدوائر المتوازية وهى دائرة اج على نقطتى آ جـ ولتفصلا من الدوائر المتوازية

فيما بينهما قسياً متشابهة ولتماس دائرة أخرى عظيمة وهى دائرة هز المائلة

على الدوائر المتوازية دائرتين أعظم من الدائرتين اللتين تماسهما دائرتا اب

جد ولتقطع دائرتى اب جد فيما بين الدائرة العظمى من الدوائر المتوازية 15

وبين دائرة اج التى تماسها دائرتا اب جد ولتكن الدائرة العظمى من الدوائر

المتوازية دائرة م ب ز ولتكن الدائرة التى تماسها دائرة هز من الدوائر المتوازية

دائرة هح ، فأقول إن نسبة ضعف قطر الكرة إلى قطر دائرة هح أعظم من

نسبة قوس ب د إلى قوس طك وليكن قطب الدوائر المتوازية نقطة لـ

ولترسم دوائر عظام تمر بنقطة لـ وبواحدة واحدة من نقط ه ط كـ وهى 20

دوائر ه م ح لـ طن لكس ولترسم دائرة من الدوائر المتوازية تمر بنقطة كـ

وهى دائرة عك ولترسم دائرة عطف العظمى مارة بنقطة طـ ومماسة لدائرة

هح على نقطة فـ فلأن دائرتى عك هفح متوازيتان وقد رسمت دائرتان

عظيمتان وهما دائرتا هطكز عطف مماستين لدائرة هفح على نقطتى ه

فـ ورسمت دائرة عظمى تمر بقطب لـ وبنقطة طـ وهى دائرة لطق 25

العظمى تكون قوس عق مساوية لقوس قك فقوس رق أصغر من قوس

قك فقوس رك أقل من ضعف قوس كق ولكن قوس رك شبيهة بقوس

A ماله [المائلة 13 سها N [بينهما 13 N ولمعصل [ولتفصلا 12 A داره ما [دائرتا 11 om. N [فى كرة 11
A دالى [دائرتى 15 add. AH دائرة هز [ولتقطع 15 A اب [آب 14 om. A [أعظم من الدائرتين 14
عظاما [عظام 20 A الدوار [دوائر 20 N مس :AH [م ب ز 17 A داره [الدائرة 16 A ومر [وبين 16
AN 20 رك [² رك A 27 موس [فقوس 27 A موس [فقوس om. A 26 [لـ وبنقطة 25 om. A [طـ
A

In spera itaque duo maiores circuli *AB GD* eundem contingant circulum, qui sit ex circulis equidistantibus sitque circulus *AG*, super duo puncta *A* et *G* et separent inter se ex circulis equidistantibus arcus similes. Et contingat circulus alius maior, qui sit circulus *EZ*, inclinatus super circulos equidistantes duos circulos maiores duobus circulis, quos duo circuli *AB GD* contingunt, et secet circulus *EZ* duos circulos *AB GD* inter circulum maiorem qui est ex circulis equidistantibus et inter circulum *AG*, quem duo circuli *AB GD* contingunt. Et sit circulus maior qui est ex circulis equidistantibus circulus *MS*, sitque circulus, quem contingit circulus *EZ* qui est ex circulis equidistantibus, circulus *EH*. Dico igitur quod proportio dupli diametri spere ad diametrum circuli *EH* est maior proportione arcus *BD* ad arcum *TK*. Sit igitur polus circulorum equidistantium punctum *L*. Signabo autem circulos maiores transeuntes per punctum *L* et per unum et unum punctorum *E T K*, qui sint circuli *EMHL LTN LKS*, et signabo aliquem ex circulis equidistantibus transeuntem per punctum *K*, qui sit circulus *QK*, et signabo circulum maiorem *QTF* transeuntem per punctum *T* et contingentem circulum *EH* super punctum *F*. Et quia duo circuli *QK EFH* sunt equidistantes et iam signati sunt duo circuli maiores, qui sunt duo circuli *ETKZ QTF*, contingentes circulum *EFH* super duo puncta *E* et *F* et signatus est circulus maior transiens per polum *L* et per punctum *T* qui est circulus *LTC* maior, erit arcus *QC* equalis arcui *CK*. Arcus autem *RC* est minor arcu *CK*: ergo arcus *RK* est minor duplo arcus *KC*. Arcus vero *RK* est similis arcui *BD* et arcus *KC* est

15

20

25

30

35

13 contingant] *in   corr.* **P**   *13* contingant circulum] *tr.*   **PsVa**   *14–*
15 sitque … equidistantibus] *marg.* **M**   *15* separent] separant **Va**   *18* circulis] circulus **Fi**
20 quem] quod **Fi**   *21* sit circulus] *tr.* **FiZ**   *22* MS … circulus²] *marg.* **Ps**   *22* quem] que
Fi   *22* est] *om.* **Va**   *23* circulis] *om.* **PR**, *post* equidistantibus **VVa**   *23* igitur] ergo
PsVa   *24* spere] *supra* **Ps**   *25* igitur] ergo **KgOFiPsVa**   *27* et unum] *supra* **M**, *om.*
KgOPsVa   *27* E T K] T E K **KgOPsVa**   *27* sint] sunt **FiPsVa**   *27* EMHL] EMKL
R   *27* EMHL LTN LKS] EM   HL   LM   LSK   **Fi**   *28* et signabo aliquem] *marg.* **Ps**
30 EH] *in corr.* **Ps**   *31* F] S **Fi**   *32* duo] *om.* **FiZ**   *32* ETKZ] TKZ **R**, ET HZ **Kg**
32 ETKZ QTF] ZK ZQTF **Fi**   *34* L] *om.* **Va**   *34* per] *om.* **FiPsVa**   *35* QC] QT **Va**
36 RC] *in corr.* **Ps**   *37* arcus¹] *supra* **Kg**, arcu **Va**   *37* KC] KT **Va**   *37* vero] autem **O**,
om. **Fi**   *37* est similis] *repet.* **Va**

ب د وقوس كﻗ شيهة بقوس ن س فقوس ب د أقل من ضعف قوس ن س

ولأن نسبة قطر الكرة إلى قطر دائرة هح أعظم من نسبة قوس م ن إلى

30   قوس هط ونسبة قوس م ن أيضاً إلى قوس هط أعظم من نسبة قوس ن س

إلى قوس طك فنسبة قطر الكرة أيضاً إلى قطر دائرة هفح أعظم من نسبة

قوس ن س إلى قوس طك وإذا أخذت أضعاف المقدمات كانت نسبة ضعف

قطر الكرة إلى قطر دائرة هفح أعظم من نسبة القوس التى هى ضعف

قوس ن س إلى قوس طك ونسبة ضعف قوس ن س إلى قوس طك أعظم

35   من نسبة قوس ب د إلى قوس طك وذلك أن القوس التى هى ضعف قوس

ن س هى أعظم من قوس ب د فنسبة ضعف قطر الكرة إلى قطر دائرة

هفح أعظم كثيراً من نسبة قوس ب د إلى قوس طك ، وذلك ما أردنا أن

نبين .

يج ·   إذا كانت فى كرة دوائر متوازية تفصل من دائرة ما عظيمة قسياً متساوية مما

يلى الدائرة العظمى من الدوائر المتوازية و رسمت دوائر عظيمة تمر بالنقط

الحادثة وتكون إما مارة بأقطاب الدوائر المتوازية وإما مماسة لدائرة واحدة

بعينها من الدوائر المتوازية فإنها تفصل من الدائرة العظمى من الدوائر

5   المتوازية فيما بينها قسياً متساوية ·

32 أخذت [حد ب A   33 قطر [A   33 هو [هى A   36 نس [سه A   36 من [om. A

36 ضعف [H: om. AN   36 إلى [صعف add. A   37 ب د [ستة add. A   1 دوائر [داره A

1 مما [om. A   2 تمر [ما م A   3 واحدة [وحده A, واحد N   4 الدائرة [الدوار A   4 العظمى [om. A

similis arcui *NS*: ergo arcus *BD* est minor duplo arcus *NS*. Et quia proportio diametri spere ad diametrum circuli *EH* est maior proportione
arcus *MN* ad arcum *ET* et proportio arcus *MN* etiam ad arcum *ET*   40
est maior proportione arcus *NS* ad arcum *TK*: ergo proportio diametri spere etiam ad diametrum circuli *EFH* est maior proportione arcus
NS ad arcum *TK*. Et cum antecedentium multiplicia accepta fuerint,
erit proportio dupli diametri spere ad diametrum circuli *EFH* maior
proportione arcus qui est duplus arcus *NS* ad arcum *TK*. Proportio   45
autem dupli arcus *NS* ad arcum *TK* est maior proportione arcus *BD*
ad arcum *TK*. Quod ideo est quoniam arcus qui est duplus arcus *NS*
est maior arcu *BD*: ergo proportio dupli diametri spere ad diametrum
circuli *EFH* est valde maior proportione arcus *BD* ad arcum *TK*. Et
illud est quod demonstrare voluimus.   50

13 Cum circuli equidistantes in spera fuerint separantes ex aliquo circulo
maiore arcus equales ab ea parte qua sequitur circulus qui est maior
ex circulis equidistantibus et signati fuerint circuli maiores transeuntes per puncta provenientia, qui aut transeant per polos circulorum
equidistantium aut contingant unum et eundem circulum ex circulis   5
equidistantibus, ipsi separabunt ex circulo maiore qui est ex circulis
equidistantibus inter se arcus equales.

38 arcus²] arcu **PsVa**   39 ad] *om.* **Fi**   39 maior] minor **Va**   40 *MN*¹] *in corr.* **Z**
40 *ET*¹] *om.* **V**   40 *MN*²] *supra* **Ps**   40 etiam] *om.* **R**   41–42 ergo ... *TK*] *marg.*
Ps   42 etiam] *om.* **OZPsVa**   42 circuli] circulum **Fi**   42 *EFH*] *EF* **Va**   42–
45 est ... *TK* est] *om.* **Va**   44 proportio] proporcione **Fi**   45 qui ... arcus²] *om.* **Ps**
46 *BD*] *LD* **R**   47 ideo est] *tr.* **V**   47 qui est] *om.* **Va**   47 arcus²] arcu **Va**
48 arcu] *om.* **FiZ**   48 dupli] *om.* **OFiZPsVa**   48 spere] *supra* **Ps**   49 Et ... voluimus] *om.*
O   50 illud] hoc **FiZ**, istud **PsVa**   50 demonstrare voluimus] volumus probare **PsVa**
1 aliquo] alio **KgFi**   *2* sequitur] *in corr.* **V**   *3* circulis] *om.* **PRVMKg**   *4* qui] que **R**
4 aut] autem **Z**   *5* contingant] contingent **R**, contingat **Fi**   *6* ipsi ... maiore] *marg.* **P**   *6–*
7 ipsi ... equidistantibus] *marg.* **Ps**   *6–7* qui ... equidistantibus] *post* equidistantibus (*l. 6*)
V   *7* inter se] *post* maiore **MKg**, *om.* **PsVa**

فلتكن فى كرة دائرتا ا ب جد المتوازيتان ولتفصلا من دائرة اه د

العظمى قوسين متساويتين وهما قوسا اه ه د مما يلى دائرة زح العظمى من

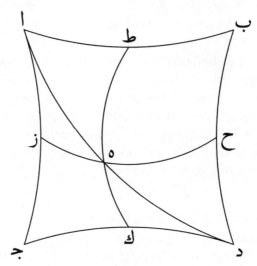

الدوائر المتوازية ولترسم دوائر عظيمة تمر بنقط ا ه د وهى دوائر ازج طه ك

ب ح د تكون إما مارة بقطب الدوائر المتوازية وإما مماسة لدائرة واحدة

بعينها من الدوائر المتوازية ، فأقول إن قوس زه مساوية لقوس ه ح وذلك   10

أنه لما كانت فى كرة دائرتان متوازيتان وهما دائرتا ا ب جد تفصلان من دائرة

ما عظيمة وهى دائرة اه د قوسين متساويتين وهما قوسا اه ه د مما يلى دائرة

زح العظمى من الدوائر المتوازية تكون دائرة ا ب مساوية لدائرة جد ولأن

دائرتى ا ب جد المتوازيتين المتساويتين تفصلان من دائرة كط العظمى قوسى

طه ه ك مما يلى دائرة زح العظمى من الدوائر المتوازية تكون قوس طه   15

مساوية لقوس ه ك وقوس اه مساوية لقوس ه د فالخط المستقيم الذى يصل

Sint itaque in spera duo circuli *AB GD* equidistantes et separantes ex circulo maiore duos arcus equales, qui sint duo arcus *AE ED*,
ab ea parte qua circulus *ZEH* maior qui est ex circulis equidistantibus   10

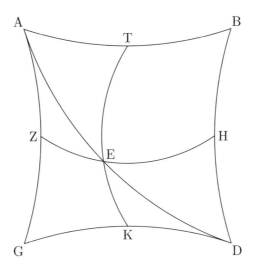

sequitur. Signabo autem circulos maiores transeuntes per puncta *A E*
D, qui sint circuli *AZG TEK BHD*, qui aut sint transeuntes per polos
circulorum equidistantium aut contingant unum et eundem ex circulis
equidistantibus. Dico igitur quod arcus *ZE* est equalis arcui *EH*. Quod
ideo est quoniam propter hoc quod in spera sunt duo circuli equidi-   15
stantes, qui sunt circuli *AB GD*, separantes ex aliquo circulo maiore,
qui est circulus *AED*, duos arcus equales, qui sunt arcus *AE ED*, ab ea
parte qua circulus *ZH* qui est maior ex circulis equidistantibus sequitur, erit circulus *AB* equalis circulo *GD*. Et quia duo circuli *AB GD*
equidistantes et equales separant ex circulo *KT* maiore arcus *TE EK*   20
ab ea parte qua sequitur circulus *ZH*, qui est maior ex circulis equidistantibus, erit arcus *TE* equalis arcui *EK*. Sed arcus *AE* est equalis
arcui *ED*: ergo linea recta que coniungit inter punctum *K* et punctum

9 duos] *om.* **FiZ**   9 sint] sunt **PsVa**   10 ab] *AB* **Kg**   10 ea] a **Kg**   10 *ZEH*] *EZH* **MKgO**
PsVa   10 *ZEH* maior] *tr.* **OPsVa**   10 circulis] *om.* **PRV**   12 sint] sunt **OFiPsVa**   12 *AZG*] *in*
corr. **Z**   12 *TEK BHD*] et *EKB HD* **Z**   12 sint] sunt **OFiPsVa**   13 contingant] contingunt
OPsVa   13 eundem] circulum *add.* **OPsVa**   14 est equalis] equatur **OPsVa**, *hic et saepi-*
us   15 est] *om.* **Va**   15 propter] *marg.* **Ps**   15 duo circuli] *tr.* **OFi**   16 sunt] sint **Kg**
16 *AB GD*] *ADG A* **Fi**   16 ex] *om.* **Va**   17 circulus] ex circulis **Z**   17 duos] duo **Fi**
17 sunt] sint **M**, *in corr.* **Ps**   18 *ZH*] *ZH, corr. ex EH* **P**; *ZH, E supra* **Ps**, *EZH* **Va**
18 maior] *om.* **Kg**, *supra* **Ps**   18–19 sequitur] *supra* **Ps**, *om.* **PRVMKg**   19 erit] *om.* **Fi**
19 circulus] circulis **Fi**   19 *AB*] *supra* **Va**   20 separant] *in corr.* **Ps**   21 ab ea parte] a
parte illa **OPsVa**   21 *ZH*] *ZB* **Fi**   21–22 equidistantibus] *supra* **Ps**   22 *EK*] *TK* **Fi**

بين نقطة آ ونقطة طٰ مساوٍ للخط المستقيم الذى يصل بين نقطة كـ ونقطة دٰ

فوتر قوس اطٰ مساوٍ لوتر قوس كدٰ والدوائر متساوية فقوس اطٰ شبيهة

بقوس كدٰ ولكن قوس اطٰ شبيهة بقوس زهٰ وقوس كدٰ شبيهة بقوس هحٰ

فقوس زهٰ شبيهة بقوس هحٰ وهى من دائرة واحدة فقوس زهٰ مساوية 20

لقوس هحٰ ، وذلك ما أردنا أن نبين ·

يد إذا ماست فى كرة دائرة عظيمة دائرة ما من الدوائر المتوازية التى فى الكرة

وكانت دائرة أخرى عظيمة مائلة على الدوائر المتوازية تماس دوائر أعظم من

الدوائر التى تماسها الدائرة الأولى فإن الدائرتين العظيمتين تفصلان من

الدوائر المتوازية فيما بينهما قسياً غير متشابهة وما قرُب من هذه القسى أحد

القطبين أيهما كان يكون أعظم من القوس من دائرته الشبيهة بما بعُد منها · 5

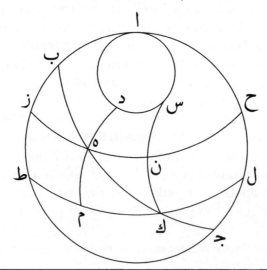

om.　تماس [2 ‏　*om.* N　 2　عظيمة [2 مموس [بقوس　 20　A شبيه [2شبيهة　 19　A ممد [كدٰ 18

add. A　من القطبين [القطبين　 5　N سها [بينهما　 4　A الدوائر [الدائر　 3　AN دائرة [الدائرة :H　دوائر [2　A

A دارىا [من دائرته 5

add. A وذلك أنهما فيما بين دائرتين إما مماستان لدائرة من الدوائر المتوازية أو تمران بأقطابهما [كدٰ1 19

D est equalis linee que coniungit inter punctum *A* et punctum *T*. Corda igitur arcus *AT* est equalis corde arcus *KD*, sed circuli sunt equales: ergo arcus *AT* est similis arcui *KD*. Arcus vero *AT* est similis arcui *ZE* et arcus *KD* est similis arcui *EH*: ergo arcus *ZE* est similis arcui *EH*; ipsi vero sunt ex uno circulo: ergo arcus *ZE* est equalis arcui *EH*. Et illud est quod demonstrare voluimus.

14 Cum in spera tetigerit circulus maior aliquem ex circulis equidistantibus qui sunt in spera et fuerit circulus alius inclinatus super circulos equidistantes contingens circulos maiores circulis quos primus circulus contingit, duo circuli maiores separabunt inter se ex circulis equidistantibus arcus dissimiles, et quicumque horum arcuum fuerit magis propinquus uni duorum polorum, quicumque fuerit, erit maior arcu sui circuli simili ei qui est magis remotus ab eo.

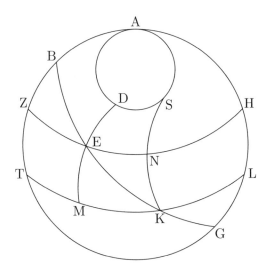

24 *A* et punctum] *om.* **Fi**   24 *T*] *in corr.* **Ps**   25 sed] fi **Fi**   26 *KD*] *KC* **Fi** 26 *KD* ... arcui] *om.* **Va**   27 *ZE*[1]] *in corr.* **Ps**   27 *EH*] *in corr.* **Ps**   27–28 ergo ... *EH*] *marg.* **P**   27 est similis] *corr. supra ex* equatur **Ps**   29 Et ... voluimus] *om.* **O**   29 illud] istud **PsVa**   29 quod ... voluimus] *om.* **Ps**   29 demonstrare] *om.* **Va** 29 demonstrare voluimus] *om.* **M**   29 voluimus] volumus **ZVa**   1 tetigerit] contigerit **FiZ** 1 maior] *supra* **Ps**   2 sunt] sit **FiZ**   2 alius] *marg.* **Ps**, *om.* **KgFi**   2 inclinatus] contingens *add.* **FiZ**   2 super] duo *add. et del.* **M**, duos *add.* **KgOPsVa**   3 equidistantes ... circulos] *om.* **OFiZ**   3 contingens] duos *add. et del.* **V**, *add.* **KgPsVa**   3 maiores] duobus *add.* **V** 3 primus] *hic redit* **B**   3–4 circulus contingit] *tr.* **BPsVa**   5 arcuum] *in corr.* **Fi**   5 fuerit] fuit **Z**   6 quicumque] quecumque **Fi**   6 sui] *om.* **PRVMKgVa**, *supra* **B**, *marg.* **Ps**

فلتكن فى كرة دائرة عظيمة وهى دائرة ا ب ج تماس دائرة ما من

الدوائر المتوازية التى فى الكرة وهى دائرة ا د س على نقطة آ ولتماس دائرة

أخرى عظيمة وهى دائرة ب ه ج المائلة على الدوائر المتوازية دوائر هى أعظم

من الدوائر التى ماستها دائرة ا ب ج الأولى ، فأقول إن دائرتى ا ب ج ب ه ج

١٠ تفصلان من الدوائر المتوازية فيما بينهما قسياً غير متشابهة وما قرب منها من

أحد القطبين أيهما كان يكون أعظم من القوس من دائرته الشيهة بما بعد

فليتعلم على دائرة ب ج المائلة نقطتا ه ك كيف ما وقعتا ولترسم على نقطتى

ه ك دائرتان موازيتان لدائرة ا د س وهما دائرتا ز ه ح ط ك ل ، فأقول إن

قوس ه ح أعظم من القوس من دائرتها الشيهة بقوس ك ل وإن قوس ط ك

١٥ أعظم من القوس من دائرتها الشيهة بقوس ز ه فلترسم دائرتان عظيمتان

تمران بنقطتى ه ك وهما دائرتا د ه م س ن ك مماستين لدائرة ا د س فنصف

الدائرة الذى يخرج من نقطة د إلى ناحية م لا يلقى نصف الدائرة الذى

يخرج من نقطة آ إلى ناحية ز ط ونصف الدائرة الذى يخرج من نقطة س

إلى ناحية ك لا يلقى نصف الدائرة الذى يخرج من نقطة آ إلى ناحية ل

٢٠ فلأن نصفى دائرتى ا ل س ك لا يلتقيان وفيما بينهما من الدوائر المتوازية قوسا

ن ح ك ل تكون قوس ن ح شيهة بقوس ك ل ولهذه الأسباب أيضاً تكون

سها [بينهما 10   *om.* A   [ا ب ج 9 NH الدائرة [الدائرة 9 H דאירה [دوائر 8   A الداره [الدوائر 8
AN مماستان H: [مماستين 16   A مواربان [موازيتان 13   A دارها [دائرته 11   N الى [²من 10
A من [س 18   N الى [الذى 18   N الى [²الذى 17   *om.* N 17   N نصف [نصف 17   N الى [¹الذى 17
A لفمان [يلتقيان 20   N داره [دائرتى 20   N الى [الذى 19   *om.* N 19   [لا 19

Sit ergo in spera circulus maior, qui sit circulus *ABG*, contingens aliquem ex circulis equidistantibus qui sunt in spera, qui sit circulus *ADS*, super punctum *A*. Et contingat circulus alius maior, qui sit circulus *BEG*, inclinatus super circulos equidistantes circulum qui sit maior circulo quem contingit circulus primus *ABG*. Dico igitur quod duo circuli *ABG BEG* separant ex circulis equidistantibus inter se arcus dissimiles, et quicumque eorum fuerit propinquior uni duorum polorum, quicumque fuerit, erit maior arcu sui circuli simili ei qui est magis remotus. Super circulum igitur *BG* inclinatum notabo duo puncta *E K* quocumque modo contingat et signabo super duo puncta *E* et *K* duos circulos equidistantes circulo *ADS*, qui sint circuli *ZEH TKL*. Dico igitur quod arcus *EH* est maior arcu sui circuli simili arcui *KL* et quod arcus *TK* est maior arcu sui circuli simili arcui *ZE*. Signabo ergo duos circulos maiores transeuntes per duo puncta *E* et *K*, qui sint circuli *DEM SNK*, contingentes circulum *ADS*. Ergo medietas circuli que producitur a puncto *D* ad finem *M* non occurrit medietati circuli que producitur a puncto *A* ad partem *ZT*, et medietas circuli que producitur a puncto *S* ad partem *K* non occurrit medietati circuli que producitur a puncto *A* ad partem *L*. Et quia due medietates circuli *AL SK* non concurrunt et inter eas sunt duo arcus – ex circulis equidistantibus – *NH KL*, erit arcus *NH* similis arcui *KL*. Et propter hec

قوس زه شبيهة بقوس طم ولأن قوس نح شبيهة بقوس كل تكون كل قوس

ه ح أعظم من القوس من دائرتها الشبيهة بقوس كل ومن قبل ذلك أيضاً

تكون قوس طك أعظم من القوس من دائرتها الشبيهة بقوس زه ، وذلك

25

ما أردنا أن نبين .

[A] تمت المقالة الثالثة من كتاب ثاودوسوس فى الأكر وهى أربعة عشر شكلاً

وبانقضائها كمل الكتاب ، والله أعلم .

[N] تمت المقاتة الثالثة من كتاب ثاوذوسيوس فى الكرات وبتمامها تم الكتاب

بأسره بحمد الله ومنّه وهى أربعة عشر شكلاً وعدد أشكال المقالات الثلثة

تسعة وخمسون شكلاً إصلاح ثابت بن قرة الحرانى الصابئ ، نقلتُ هذا

الكتاب من خط قرة بن سنان بن منصور بن سعيد بن ثابت بن سنان بن

ثابت بن قرة الحرانى الصابئ بمدينة الموصل حماها الله ، فى المدرسة

5

النظامية عمرها الله لست ليالٍ بقيت من جمادى الأولى لسنة أربع وخمسين

وخمس مائة هجرية على صاحبها أفضل السلام ، ووجدت فى آخر الكتاب

مكتوباً أن قد فرغ من تشكيل هذا الكتاب الحسن بن سعيد ولم يكن

الأصل الذى نقل منه الأشكال موثوقاً به بل كان فيه فساد ويجب أن يقابل

---

22 طم [طح A   23 من القوس [موس A   23 ومن قبل [ومل A   24 القوس [موس A

3 وخمسون [وحمسن MS   9 موثوقاً [موثوق MS *corr. ex*

---

22 طم [قوس حب هر قره [sic] من أحد القطبين وقوس لكمط قرية من القطب الآخر *add.* A

eadem etiam erit arcus *ZE* similis arcui *TM*. Et quia arcus *NH* est similis arcui *KL*, erit arcus *EH* maior arcu sui circuli simili arcui *KL*. 30 Et propter hoc etiam erit arcus *TK* maior arcu sui circuli simili arcui *ZE*. Et illud est quod demonstrare voluimus.

Explicit pars tertia libri theodosii de speris que est eius pars ultima, cum qua totus finitur liber.

---

29 erit] est **M**   30 erit] *H   add.* **Fi**   30 *EH*] *in   corr.* **BPs**   30 circuli simili] *tr.* **Fi**   30 simili] *in corr.* **V**   31 erit] *post TK* **O**, *om.* **BPsVa**   31 *TK*] est *add.* **BPsVa** 31 sui circuli] *om.* **PRVMFiZ**   32 illud] istud **OBPsVa**   32 demonstrare voluimus] volumus probare **BPsVa**   32 voluimus] volumus **O**

بالأشكال بنسخة أخرى وذلك فى ليلة الثلثاء لثمانى ليالٍ بقيت من ١٢ سنة   10

٤٢١ ، الحمد لله كثيراً وصلوته على محمد وآله أجمعين ·

[H] תמת אלמקאלה תאלתה מן כתאב תאודוסיוס ובתמאמהא תם

נמיע אלכתאב ואלחמד לאללה ·

Notes on the Text by al-Ḥasan b. Saʿīd

[end I 3, **N** p. 189 *marg.*]

قال الحسن بن سعيد وقد يجوز أن يقال فى تماسهما على نقطتين وقد بيّن

أقليدس أنه لا يمكن أن تماس دائرة [...]

[end II 13, **N** p. 223 *in textu*]

قال الحسن بن سعيد لهاتين الصورتين هما جميعاً للشكل ١٤ من هذه المقالة

وقد صححتهما بحسب الطاقة ودائرتى ب ز ط جه‍ا إما أن تماسان دائرة

ك ق ل (MS: كول) على طرفى قطرها أو على غير طرفى (sic) فإن ماستها

على طرفى (sic) على ما فى الصورة الثانية كانت الدائرة التى تمر بنقطتى م

ل تمر أيضاً بنقطة ك كما فى الصورة الثانية وإن لم تماسها على طرفى القطر

كانت الدائرة التى تمر بنقطة م وبنقطتى ك ل واحدة مما يفسد البرهان إذ

كانت أنصاف الدوائر كلها متشابهة أيضاً وهذه الصورة الثانية ليس توجد فى

شىء من النسخ وإنما خطرت ببالى فى وقت عملى لهذا الكتاب فأثبتها ·

[against II 13-14, **N** p. 223 *marg.*]

قال أبو الحسن بن سعيد إنما جعل العلامة ما بين هاتين الدائرتين لأنه لو

كانت فى داخل إحداهما لما أمكن أن تمر بها دائرة عظيمة تماس الدائرة

المفروضة ·

[end II 14, **N** p. 226, vertically written in lower half of page – the

ends of all the lines are cut off]

هذا مما نقلتُ أنا من خط الحسن بن سعيد ، قال الحسن بن سعيد قوس

ب‌ج لا تخلوا من أن تكون إما مساوية للقوس التى يوترها [...] المربع

وإما أعظم وإما أصغر وقد بُيّن كيف ترسم الدائرة التى تمر بنقطة ج

وت‌[...] دائرة ا‌ب متى كانت قوس ب‌ج مساوية للقوس التى يوترها ضلع

5   المربع ومتى كا[...] أيضاً أصغر وقد بقى أن نبين كيف ترسم هذه الدائرة

متى كانت قوس ب‌ج أع[...] القوس التى يوترها ضلع المربع وبيانه على

هذه الصفة معلوم أن القوس من ال[...] العظيمة التى فيما بين دائرة ا‌ب

والدائرة الموازية المساوية لها أقل من [...] فإذا كانت قوس ب‌ج أعظم

من ربع دائرة عظيمة تكون القوس التى فيما [...] نقطة ج وبين الدائرة

10   المساوية الموازية لدائرة ا‌ب أقل كثيراً من القوس [...] يوترها ضلع المربع

فإذا أخرجت بالعمل الذى تقدم فى هذا الشكل دائر[...] الدائرة المساوية

الموازية لدائرة ا‌ب وتمر بنقطة ج فمن البين أنها [...] دائرة ا‌ب إذ كانت

عظيمة ومماسة إحدى دائرتين متساويتين متوازيتين وهى الدائرة المساوية

الموازية لدائرة ا‌ب فهى تماس [...]

Translation

1. Al-Ḥasan b. Saʿīd said: it is possible to say about their touching
at two points: Euclid has shown that it is not possible that a circle
touches [...]

2. Al-Ḥasan b. Saʿīd said about these two figures: both belong
to Proposition 14 of this book. I have corrected them according to my

ability. Circles BZT and GEA touch circle KCL either at the ends of its diameter or not at its ends. If they touch at [its] ends, as in the second figure, the circle passing through points M L passes also through point K, as in the second figure. If they do not touch it at the ends of the diameter, the circle passing through point M and through points K L is the same – which spoils the proof, since all semicircles are also similar. This second figure is not found in any of the copies [of the text], but it came to my mind while working on this book. So I put it [here].

3. Abū al-Ḥasan b. Saʿīd said: he [Theodosius] put the point between these two circles because, if it were inside one of the two, it would not be possible that a great circle pass through it touching the given circle.

4. This is from what I have copied from the handwriting of al-Ḥasan b. Saʿīd. Al-Ḥasan b. Saʿīd said: arc BG does not need to be either equal to the arc subtended by [...] the square or greater or smaller. It has been shown how the circle is drawn passing through point G and pass[...] circle AB when arc BG is equal to the arc subtended by the side of the square also and when [...] also smaller. There remains for us to show how this circle is drawn when arc BG is gr[...] the arc subtended by the side of the square. The explanation of it is in the follwing way: it is known that the arc of [...] a great [circle] that is between circle AB and the circle parallel and equal to it is smaller than [...]. So when arc BG is greater than a quarter of a great circle, the arc between [...] point G and the circle equal and parallel to circle AB is much smaller than the arc subtended [...] by the side of the square. So when there is drawn in the preceding construction in this proposition cir[...] the circle equal and parallel to circle AB and passing through point G. So it is clear that it [...] circle AB, since it is great and touches one of the two equal and parallel circles – which is the circle equal and parallel to circle AB – so it touches [...].

Appendix

Lemmas to III 11 in **A**, 53r–v

وهذا هو الشكل الذى ذكرناه فى آخر الكتاب

[۱]   مثلث ا ب ج زاوية ب منه قائمة وقد أخرج خط ج د كيف ما وقع ،

فأقول إن نسبة ا ب إلى ب د أعظم من نسبة زاوية ب د ج إلى زاوية

ب ا ج ، ولننزل ذلك فتكون نسبة ا د إلى د ب على الفصل المشترك أعظم من

نسبة زاوية د ج ا إلى زاوية ب ا ج ، وندير على مثلث ا د ج دائرة ا ج   5

ونخرج د ح إلى ح يوازى ب ج ونخرجه على استقامة إلى ك ونصل ج ك

ا ك ، فأقول إن نسبة ا ح إلى ح ج أعظم من نسبة قوس ا د إلى د ج فندير

على نقطة ك وببعد ج ك قوس ج ز ط فيجب مما أنزلناه أن تكون نسبة ا ح

إلى ح ج أعظم من نسبة قطاع ط ز ك إلى قطاع ز ج ك وهكذا هو لما نذكره

وهو أن نصل ج ز ونخرجه على استقامة إلى م فنسبة قطاع ط ز ك إلى قطاع   10

ج ز ك كنسبة زاوية ز ك ط إلى زاوية ج ك ز وهى كنسبة زاوية د ج ا إلى زاوية

د ا ج ونسبة مثلث ك ز م إلى مثلث ك ج ز أعظم من نسبة القطاع إلى القطاع

فنسبة م ز إلى ز ج أعظم من نسبة القطاع إلى القطاع ونخرج ز ل يوازى

ا ك فنسبة ا ل ⟨إلى ل ج⟩ أعظم من نسبة القطاع ⟨إلى القطاع⟩ ونسبة ا ح

---

5 د جا [د ج MS   دائرة [داير MS 6 ب ج يوازى ح [ح يوارى MS 7 ا ك [ا ك MS   ح ج إلى ح ا

MS   د ح [add. MS 8 نسبة [سبه MS 9 ح ج [ح ج MS 11 كنسبة [ليست 1 MS 11 إلى [2 د ح MS

12 د ا ج [د ا ج MS 13 داح MS 14 ونخرج [ونحر MS 14 ا ح [ا ح MS

إلى حَجَّ أعظم من نسبة آلَ إلى لَجَّ فنسبة آحَ إلى حَجَّ أعظم ‹من نسبة› 15

القطاع إلى القطاع ،

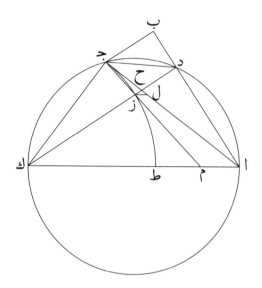

وتركيب البرهان أن نعمل ما عملنا فيحصل أن نسبة القطاع إلى

القطاع أصغر من نسبة آحَ إلى حَجَّ ونسبة آحَ إلى حَجَّ كنسبة آدَ إلى دَبَّ

ونسبة القطاع إلى القطاع كنسبة زاوية زكَطَ إلى زاوية جكَزَ وهى نسبة

زاوية دجآ إلى زاوية دآجَ وهى زاوية بآجَ فتكون نسبة آدَ إلى دَبَّ 20

أعظم من نسبة زاوية دجآ إلى زاوية دآجَ فإذا ركبنا كانت نسبة آبَ إلى

بَدَ أعظم من نسبة زاوية بَدجَ إلى زاوية بآجَ ، وذلك ما أردنا أن

نبين .

وهنا استبان أنه إذا فصل من محيط دائرة قوس أصغر من نصف

وأخرج دجكَ وترها وأخرج من أحد طرفى الوتر قطر الدائرة وأخرج من 25

الطرف الآخر من طرفى القطر خط ا‍ج‍ يقطع الوتر كيف ما اتفق وينتهى إلى

قوس د‍ج‍ك‍ المذكورة فإن نسبة القسم الذى يلى القطر من الوتر إلى قسم

ا‍د‍ب‍ الآخر أعظم من نسبة <٠٠٠> التى تلى القطر من القوس المفروضة

إلى القوس الأخرى ، وبالله التوفيق ·

[٢]   مثلث ا‍ب‍ج‍ زاوية ب‍ منه قائمة وأخرج خط ج‍د‍ كيف ما اتفق ، فأقول

إن نسبة خط ا‍ب‍ إلى ب‍د‍ أعظم من نسبة زاوية ب‍د‍ج‍ إلى ب‍ا‍ج‍ ،

برهان ذلك أنا نخرج من نقطة د‍ خط د‍ه‍ موازياً لخط ا‍ج‍ فتبين أن خط د‍ه‍

أعظم من خط د‍ب‍ وأصغر من خط د‍ج‍ فإذا جعلنا نقطة د‍ مركزاً وأدرنا

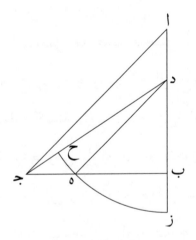

ببعد د‍ه‍ دائرة كانت تقع داخل مثلث ج‍ب‍د‍ وخارجه فلتكن مثل ز‍ه‍ح‍   5

فلأن ه‍د‍ مواز لـج‍ا‍ تكون نسبة ا‍د‍ إلى د‍ب‍ كنسبة ج‍ه‍ إلى ه‍ب‍ ونسبة ج‍ه‍

إلى ه‍ب‍ كنسبة مثلث د‍ه‍ج‍ إلى مثلث ب‍د‍ه‍ فنسبة خط ا‍د‍ إلى د‍ب‍ كنسبة

ح‍ [ب‍ا‍ج‍ 2 MS   د‍ [ب‍د‍ج‍ 2 *add.* MS   القسمى 2 من [2 28 MS   اعط [أعظم 28 MS   القسم [قسم 27

ره‍ [ز‍ه‍ح‍ 5 MS   ودرحاغه [وخارجه ج‍ب‍د‍ 5 MS   كان معطع [كانت تقع 5 MS   واردنا [وأدرنا 4 MS

MS 7 [د‍ه‍ج‍] *illeg.* MS

مثلث ده‌ج إلى مثلث د‌به ونسبة مثلث دجه إلى مثلث د‌به أعظم من

نسبة مثلث دجه بعينه إلى قطاع دزه لأنه أعظم من مثلث ب‌ده فنسبة

10 خط اد إلى د‌ب أعظم من نسبة مثلث ده‌ج ⟨إلى⟩ قطاع ده‌ز

⟨ومثلث ده‌ج أعظم من قطاع⟩ ده‌ح فنسبة اد إلى د‌ب أعظم كثيراً من

نسبة قطاع ده‌ح إلى قطاع دزه فإذا ركبنا كانت نسبة ا‌ب إلى ب‌د أعظم

من نسبة قطاع دزح إلى قطاع دزه ونسبة قطاع دزح إلى قطاع دزه

كنسبة زاوية ب‌دح إلى زاوية ب‌ده وزاوية ب‌ده مساوية لزاوية ب‌اج

15 فنسبة ا‌ب إلى ب‌د أعظم من نسبة زاوية ب‌دج إلى زاوية ب‌اج ، وذلك

ما أردنا أن نبين ·

[٣] مثلث ا‌ب‌ج زاوية ب منه قائمة وأخرج خط ج‌د كيف ما اتفق ، فأقول

إن نسبة خط ا‌ب إلى ب‌د أعظم من نسبة زاوية ب‌دج إلى زاوية ب‌اج

وبالتفصيل نسبة اد إلى د‌ب أعظم من نسبة زاوية دجا إلى زاوية دا‌ج ،

برهان ذلك أنا ندير على مثلث ا‌دج دائرة ونخرج دح‌ك موازياً ل‌ب‌ج

5 ونصل ا‌ك ك‌ج وندير ببعد ك‌ج قوس ج‌ز‌ط ونصل ج‌ز ونخرج جز إلى م

ونخرج ل‌ز موازياً ل‌ا‌م فلأن قطاع ك‌ط‌ز أصغر من مثلث م‌ز‌ك ومثلث

ك‌ز‌ج أصغر من قطاع ك‌ز‌ج تكون نسبة ⟨مثلث م‌ز‌ك إلى⟩ مثلث ك‌ز‌ج

أعني نسبة م‌ز إلى ز‌ج أعني نسبة ا‌ل إلى ج‌ل أعظم من نسبة قطاع ط‌ز‌ك

MS فزاوه [وزاوية 14   MS د‌به [دزه 12   A اعط [أعظم 11   MS د [د‌ب 10   MS ا‌ب [اد 10

MS ان [ل‌ز 6   MS م د [م 5   MS دح [جز 5   marg. MS [دجا إلى زاوية 3   MS ا [ب‌اج 15

إلى قطاع ك‍ز‍ج‍ لكن نسبة اح‍ إلى ح‍ج‍ أعظم من نسبة خط ا‍ل‍ إلى

ل‍ج‍ فنسبة اح‍ إلى ح‍ج‍ أعنى نسبة ا‍د‍ إلى د‍ب‍ أعظم كثيراً من نسبة قطاع 10

ط‍ز‍ك‍ إلى قطاع ك‍ز‍ج‍ أعنى نسبة زاوية ط‍ك‍ز‍ إلى زاوية ز‍ك‍ج‍ أعنى نسبة

<قوس> ا‍د‍ إلى قوس د‍ج‍ أعنى نسبة زاوية ا‍ج‍د‍ إلى زاوية د‍ا‍ج‍ وبالتركيب

نسبة ا‍ب‍ إلى د‍ب‍ أعظم من نسبة زاوية ط‍ك‍ج‍ إلى زاوية ج‍ك‍ز‍ أعنى نسبة

زاوية ب‍د‍ج‍ إلى زاوية د‍ا‍ج‍ ، وذلك ما أردنا أن نبين ·

[٤] وإن أخرج د‍ه‍ موازياً ل‍ج‍ا‍ وعمل <على> مركز د‍ وببعد د‍ه‍ دائرة قُطع

ب‍ج‍ وقُطع ب‍د‍ إذا أخرج على استقامة ز‍ ولتكن ز‍ح‍ وأكثر ما على ذلك ·

Lemma to III 11 in **H**, 53r

مقدمة احتاج إليها ثاودوسيوس فى الشكل الحادى عشر من المقالة الثالثة من

هذا الكتاب لثابت بن قرة الحرانى ·

قال ثابت ليكن مثلث عليه ا‍ ب‍ ج‍ وزاوية ب‍ منه ليست بأصغر من زاوية ا‍

وأخرج من زاوية ا‍ خط ا‍د‍ إلى خط ب‍ج‍ كيف ما وقع فى داخل المثلث ،

فأقول إن نسبة ج‍ب‍ إلى ب‍د‍ أعظم من نسبة زاوية ا‍د‍ب‍ إلى زاوية 5

ا‍ج‍ب‍ ،

9 خط [ط‍ MS   10 ل‍ج‍ [ا‍ح‍ MS   12 د‍ا‍ج‍ [د‍ح‍ا‍ MS   13 ا‍د‍ [ا‍ب‍ MS   13 ج‍ك‍ز‍ [ك‍ر‍ح‍ MS

1 د‍ه‍ [ا‍ه‍ 2 MS   استقامة [ا‍ه‍ 2 MS   احتاج [ا‍ه‍ مانمه 1 MS   اححن‍ MS 1 الثالثة [الثلثه MS 4 إلى [على‌ا‍
MS

برهان ذلك أن نخرج من خط جد خطاً موازياً لـاجـ وليلق اب على ه

ونجعل نقطة د مركزاً ونخط دائرة تمر بنقطة ه وتلقى اد على ح ونخرج

جب حتى يلقاه على ز فنسبة قطاع دهح إلى قطاع دهز كنسبة زاوية

حده إلى زاوية هدز فنسبة مثلث اده إلى قطاع هدز أعظم من نسبة زاوية

حده إلى زاوية هدز فنسبة مثلث اده إلى مثلث هدب أعظم بكثير من

نسبة زاوية اده إلى زاوية هدب فإذا ركبنا كانت نسبة مثلث ادب إلى

مثلث هدب أعظم من نسبة زاوية ادب إلى زاوية هدب وزاوية هدب مثل

زاوية اجب ونسبة مثلث ادب إلى مثلث هدب كنسبة خط اب إلى خط

هب فنسبة خط اب إلى خط هب أعظم من نسبة زاوية ادب إلى زاوية جـ

لكن نسبة اب إلى به كنسبة جب إلى بد فنسبة جب إلى بد أعظم

من نسبة زاوية ادب إلى زاوية جـ ، وذلك ما أردنا أن نبين ·

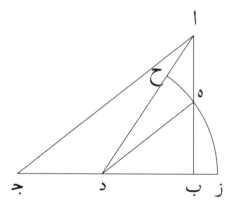

Lemma to III 11 in **P** etc.*

Quod proportio linee CQ ad lineam CO sit maior proportione anguli COH ad angulum CQH sic probatur: protraham a puncto Q lineam

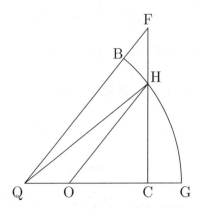

CQ equidistantem linee OH, que sit linea QF, que concurrat linee CH supra punctum F. Et ponam punctum Q centrum et secundum longi tudinem linee QH describam circulum GHB, et copulabo punctum C 5 puncto G. Et quia proportio sectoris BHQ ad sectorem HQG est sicut proportio anguli BQH ad angulum GQH et proportio sectoris BHQ ad triangulum HQC est maior proportione sectoris BHQ ad sectorem HQG, erit proportio sectoris BHQ ad triangulum CQH maior propor tione anguli HQB ad angulum HQG. Ergo, cum composuerimus, erit   10 proportio superficiei $BHCQ$ ad triangulum CHQ maior proportione anguli BQG ad angulum GQH. Et hoc ideo quia abscindam de sectore BQH aliquid cuius proportio ad triangulum HQC sit sicut proportio anguli BQH ad angulum HQG. Ergo proportio trianguli QFC ad tri-

*From **P**, 8r, *marg.*, almost opposite the inequality in III 11; **R**, 49v-50r, in text after III 11; **V**, 63va, in text after line 50 (above edition) of III 11; and **M**, 13r, at end of text.

3 equidistantem] equidistantium **V**   3 *OH* que sit linea] *marg.* **P**   3 *CH*] *COH* **R**
5 *GHB*] *GHD* **R**   6 *HQG*] quod probatur sicut probatur ultima figura sexti libri Euclidis *add.* **PRV**; (ultima figura sexti: sic probat ultima figura secti) **V**   7 *BQH*] et proportio
sectoris *BHA add. et del.* **R**   8 triangulum] angulum **R**   10 composuerimus] posuerimus
R   11 *BHCQ*] *BHOQ* **R**   11 *CHQ*] *corr. ex CQHQ OHQ* **R**   12 *BQG*] *BQC* **R**
12 abscindam] abscidam **P**   13 triangulum] angulum **R**

angulum HQC est multo maior proportione anguli BQC ad angulum   15
HQC. Sed proportio trianguli FQC ad triangulum HQC est sicut proportio linee FC ad lineam CH. Proportio linee FC ad lineam CH est
sicut proportio linee CQ ad lineam CO, quia linea OH est equidistans
basi QF: ergo proportio linee CQ ad lineam OC est maior proportione
anguli GQB ad angulum GQH. Sed angulus GQB est equalis angulo   20
COH: ergo proportio linee CQ ad lineam OC est maior proportione
anguli COH ad angulum CQH. Et illud est quod demonstrare voluimus.

Note to II 11 in **Kg** etc.[†]

Theodosius de speris in xia secundi proposuit perpendicularem HK ca-
dentem supra communem differentiam superficierum AHG ABG intra
circulum ABG et perpendicularem TL intra circulum DEZ. Possibile
vero est ut portiones erecte supra circulos ABG DEZ, que sunt portiones AHG DTZ, sint adeo magne et arcus qui separantur ex eis ab   5
ea parte qua diametrorum sequuntur extremitates adeo parvi ut perpendiculares ab eis protracte non cadant secundum hoc quod ostendit
Theodosius, scilicet intra circulos ABG DEZ, sed cadant extra eos
supra communes differentias si protrahantur, cadant ergo perpendiculares HK TL extra quemadmodum in subscriptis patet et diametri GA 10
ZD concurrant eis super duo puncta K L. Dico igitur quod si linea HB
fuerit equalis linee TE, arcus AB equabitur arcui DE et e converso.

Probatio: quoniam portio AHG equatur portioni DTZ et arcus DT
equalis arcui AH, remanet arcus HG equalis arcui TZ. Ergo corda
DT equatur corde AH et corda HG equatur corde TZ et linea DZ 15

[†]**Kg**, p. 258; **O**, 51v; **Fi**, 165v; **B** 112r; **Ps**, 172v; **Z**, 289r; **Pt**, 114r-v. In every
manuscript this note appears at the end of the text of the *Sphaerica*, though the
text of Prop. III 11 is placed after it in **BPs**.

---

15 est] maior *add. et del.* **P**   15 *BQC*] *HQO* **R**   16 Sed] et **R**   16 trianguli] *H add. et del.*
P   16 est sicut] *om.* **R**   17 *CH*[1]] *OH* **R**   17 Proportio] Et proportio **V**   17 linee[2]] *marg.* **P**,
f *add. et del.* **P**   17 lineam[2]] *OH* et proportio linee *FC* ad lineam *add.* **R**   18 sicut] linea *add.*
et del. **R**   18 linee] *CG add. et del.* **P**   18 *OH*] *CH* **P**   20–23 equalis … voluimus] *om.*
V   22–23 Et … voluimus] l'.h.e.q.d'.u. **P**, Et h.e.q.d.m.v. **M**   1 secundi] secundam **Ps**
1 proposuit] posuit **Kg**   2 *AHG*] *AGB* **B**   2–3 intra circulum *ABG*] *om.* **OFiBPsZPt**
3 et] *om.* **Fi**   5 *DTZ*] *in corr.* **B**   6 diametrorum] diametrum **Pt**   7 ab eis] **Kg**, ad eum
OBPs, ad ea **Pt**   7 cadant] cadat **Kg**   8 scilicet] *om.* **KgB**   8 intra] ultra **Fi**, inter **KgPs**   8–
9 *ABG … TL*] *om.* **Fi**   9 protrahantur] et *add.* **Pt**   11 igitur] ergo **OFiBPsZ**   12 *DE*] *om.*
Fi   12 et e converso] *om.* **FiZPt**   13 Probatio] *om.* **O**   13 *AHG*] *HAHG* **Pt**   13 arcus] arcum
Fi   13 *DT*] *DZ* **Ps**, est *add.* **Fi**   14 *AH*] *KH* **KgPs**, *HA in corr.* **Fi**   14 *HG*] *AG* **Z**, *in corr.* **Ps**
15 equatur[2]] equalis **O**   15 linea] *om.* **KgBPs**

est equalis linee *AG*; ergo triangulus *TDZ* equatur triangulo *AHG* et
anguli relativi sibi equantur; ergo angulus *TDZ* equatur angulo *HAG*;
ergo reliquus *TDL* equatur reliquo *HAK*. Et angulus *TLD* rectus et
etiam *AKH* quia *HK* et *TL* perpendiculares sunt et *LD* equatur *KA*.
Sed *DN MA* sunt equales: ergo *LN KM* sunt equales. Item quadratum   20
HB et quadratum *TE* equalia equantur quadratis *HK KB TL LE*;
et quadrata *HK TL* sunt equalia: ergo *KB* equatur *LE*. Sed *KM MB*
equantur *LN NE* et basis *KB* basi *LE*. Ergo angulus *KMB* equatur
angulo *LNE* et sunt supra centrum circulorum equalium; ergo arcus
AB DE sunt equales. Et hoc est quod subiecta figura declarat.   25

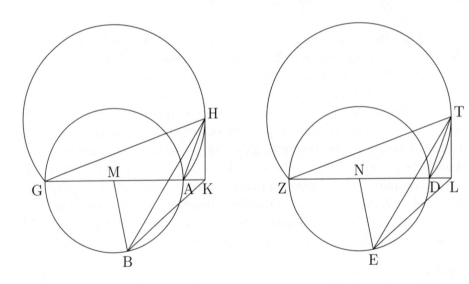

16 est] *om.* **Kg**   16 linee] *om.* **KgBPs**   16 linee *AG*] *tr.* **O**   16 triangulus ... *AHG*] circulus
equatur circulo *HAG* **Ps**   17 equantur] equales **KgOBPs**   17 *TDZ*] *DTZ* **Pt**
17 *HAG*] *AHG* **Fi**   18 *TDL*] *DTL* **Pt**   18 *TLD*] *TDL* **Fi**, *TDB* **Ps**   18–
19 *TLD* ... quia] *marg.* **B**   18–19 et etiam ... sunt] perpendicularis *KH* sue relative
Kg   19 *HK* et *TL*] *corrupt.* **B**   19 *TL*] *in corr.* **Fi**   19 sunt] est *KH* sue relative *in
corr.*   20–21 quadratum *HB* et quadratum *TE*] quadrata *HB TE* **OBZPt** (*HB TE corrupt.*
B)   21 *HK KB*] et quadratis *add.* **Kg**   21 *TL LE*] *om.* **B**   22 et quadrata *HK TL*] *om.* **Z**
22 ergo] linea *add.* **Fi**   22 equatur] est equalis **Kg**   23 basis] basi **Pt**   23 equatur] *om.*
B   24 et] ipsi *add.* **Kg**; propterea *add.* **Ps**   24 arcus] *in corr.* **B**; etiam *add.* **Kg**
25 declarat] conversa huius probatur ex predictis *add.* **Kg**

Note to II 11 in **M** etc. ‡

Hec probationes sunt necessarie in theoremate undecimo partis secunde
libri Theodosii de speris.

Perpendicularis protracta a puncto H ad superficiem circuli ABG
cadit super communem differentiam circulorum ABG AHG, quia protraham ab aliquo puncto signato in superficie circuli ABG perpen-   5
dicularem super superficiem circuli ABG. Ergo superficies protracta
a perpendiculari cum superficie circuli ABG angulum obtinebit rectum. Ergo due superficies, scilicet protracta a perpendiculari et superficies circuli AHG, sese secantes sunt super superficiem circuli ABG
perpendiculariter erecte. Ergo communis earum differentia est super   10
superficiem circuli ABG perpendiculariter erecta, que etiam cadit super lineam AG. A puncto igitur notato quocumque casu in linea AG
producam lineam equidistantem communi sectioni sitque linea KH. Ergo linea KH est super superficiem circuli ABG perpendicularis, quia
si fuerint due linee equidistantes et una super superficiem fuerit per-   15
pendicularis alia super illam superficiem erit perpendicularis. Si ergo
producetur a puncto H perpendicularis que cadat extra lineam AG et
cum coniungentur duo puncta perpendicularium proveniet triangulus
cuius duo anguli erunt recti, quod est impossibile.

Portiones equales AHG DTZ circulorum equalium que eriguntur   20
super diametros circulorum equalium possunt esse semicirculi et minores semicirculis et maiores. Et quando sunt semicirculi vel minores,
perpendiculares protracte cadent infra arcus; cum autem sunt maiores, quandoque cadent extra arcus super lineas protractas secundum
rectitudinem et coniunctionem diametrorum. Cum ergo protrahentur a   25
punctis H T perpendiculares cadentes extra arcus, et protraham lineas
AH HG DT TZ, erunt duo anguli GAH ZDT expansi, quia cadunt in
portionibus minoribus semicirculis. Et quia arcus AH est equalis arcui
DT et arcus AG est equalis arcui DZ, erit totus arcus HAG equalis
arcui TDZ. Et linea ZT linee GH equalis: ergo quadratum linee GH 30
est equale quadrato linee ZT. Sed quadratum GH est equale quadrato
GA et quadrato AH et duplo superficiei linee GA in lineam que sepa-

‡From **M**, 50v, and **Px** (= Paris, BnF, lat. 7377B), 38r-39r. Vatican Reg. 1268,
207v and 209r (the folios are out of order) also carries the text, but Tummers has
shown (Anaritius [1994], pp. xxi-xxv) that the codex is, in the text of Anaritius,
a copy from **M**, and this appears to go also for much of the rest of the codex.
Accordingly, we have not collated this manuscript.

3 ad] a **Px**   4–6 AHG ... ABG] om. **Px**   9 sunt] supra **M**   16 super] lineam add. et del.
M   22 Et] om. **Px**   28 portionibus minoribus] tr. **Px**   28 est] om. **Px**

rat perpendicularem ab angulo obtuso. Et quadratum ZT est equale
duobus quadratis ZD DT et duplo superficiei linee ZD in lineam que
separat perpendicularem ab angulo obtuso. Sed duo quadrata GA AH 35
sunt equalia duobus quadratis ZD DT: ergo duplum superficiei linee
GA in lineam que separat perpendicularem ab angulo obtuso est equale
duplo superficiei que fit ex ZD in lineam que separat perpendicularem
ab angulo obtuso. Ergo et superficies que fit ex GA in lineam que separat perpendicularem ab angulo obtuso est equalis superficiei que fit   40
ex ZD in lineam que separat perpendicularem ab angulo obtuso. Sed
linea GA est equalis linee ZD: ergo ea que separat perpendicularem
ab angulo obtuso est equalis ei que separat perpendicularem ab angulo
obtuso. Et hoc ex 14° 6$^\mathrm{i}$ libri Euclidis.

Ponam centrum circuli ABG punctum M et circuli DEZ punc-   45
tum N et protraham lineas HB TE MB NE BA ED et copulabo duo
puncta B E duobus punctis supra que cadunt due perpendiculares.
Et quia linea HA est equalis linee TD, ergo quadratum linee AH est
equale quadrato linee TD. Sed angulus quem facit perpendicularis cum
linea que separat perpendicularem ab angulo obtuso est equalis angulo   50
quem facit alia perpendicularis cum linea que ipsam separat ab angulo obtuso, quia uterque est rectus, et ea que separat perpendicularem
ab angulo obtuso est equalis ei que separat perpendicularem ab angulo obtuso. Ergo perpendicularis est equalis perpendiculari. Et linea
HB est equalis linee TE: ergo linea que coniungit inter punctum H 55
et punctum super quod cadit perpendicularis est equalis ei que coniungit inter punctum E et punctum super quem cadit perpendicularis.
Et quia linea que coniungit inter centra duorum circulorum et puncta
supra que cadunt perpendiculares sunt equales, et linea MB est equalis
linee NE, et linee que producuntur a punctis B E ad puncta supra que   60
cadunt perpendiculares sunt equales, erit angulus M equalis angulo N.
Sed proportio anguli ad angulum est sicut arcus ad arcum: ergo arcus
BA est equalis arcui ED.

Et dico quod si arcus AB AH sunt equales arcubus DE DT uterque suo relativo, erit linea HB equalis linee TE. Quia angulus BAG 65
est equalis angulo ZDE, erit angulus quem facit linea BA cum linea
que separat perpendicularem ab angulo obtuso equalis angulo quem
facit ED cum linea que separat perpendicularem ab angulo obtuso; et

33 est] *supra* **M**   38 duplo] *supra* **Px**   38 que fit ex] *om.* **M**   39 et] *supra* **M**   41 ab] *om.*
Px   41 angulo obtuso] *tr.* **Px**   42 linee] *marg.* **M**   43 equalis] *in corr.* **M**   46 *NE*] *MB*
add. **Px**   51 quem] que **M**   52 perpendicularem] *ad add. et del.* **M**   55 linea^2] *marg.* **M**
55 *H*] *B* **Px**   56 equalis ei] *tr.* **Px**   57 perpendicularis] ei est equalis *add.* **Px**   58 quia] a
add. et del. **M**   62 est] *om.* **M**   63 *BA*] *HA* **Px**   63 *ED*] unusquisque suo relativo *add. et del.*
M   64 *AH*] fuerit *add. et del.* **M**   64–65 uterque suo relativo] *marg.* **M**   65 angulus] angulo
Px

linea *BA* et linea que separat perpendicularem ab angulo obtuso sunt
equales linee *DE* et ei que separat perpendicularem ab angulo obtuso. 70
Ergo linee a duobus punctis *B E* protracte ad puncta supra que cadunt
perpendiculares sunt equales. Et perpendicularis est equalis perpendiculari, et angulus qui est in puncto super quod cadit perpendicularis
est equalis angulo qui est in puncto supra quod cadit alia perpendicularis. Ergo linea *HB* est equalis linee *TE*. Et hoc est quod demonstrare 75
voluimus.

Notes on the diagrams

The Arabic diagrams have been drawn so that they are as near as possible to those of the main tradition of the Arabic manuscripts. The Latin diagrams are derived from them by exchanging their Arabic letters for Gerard's equivalent letters. In at least one proposition (II 22) this means keeping to Gerard's letters in the diagram even if this means an unusual transcription of the Arabic.

Points and the letters labelling them are here normally referred to by their form in our Latin text. Only when there is a compelling reason (e.g. if the shape of the Arabic letters is important) do we use Arabic or Greek letters.

In almost every case the Arabic diagrams have been established by the agreement of two (or more) of the following witnesses: **Cz**, i.e. the Greek diagram (or one of the diagrams) reproduced by Czinczenheim in her edition of the text, mostly from her MS **A**, and our **A**, **N** and **P** (here **P** is counted as a witness to the Arabic tradition). Only seldom has **H** or any of the Latin manuscripts other than **P** been consulted. Where possible, the agreement of **Cz** with one (or more) of the Arabic witnesses **PAN** is taken as the basis of a drawing, since the agreement of any Greek manuscript and any Arabic manuscript ensures correctness, at least at the time of translation. In this matter it is very useful that Dr. Czinczenheim took care, in copying the figures, to copy the mistakes[1].

By a single point, say A, we mean the relevant point on the figure. By a letter with inverted commas, say "A", we mean the letter labelling the point; the inverted commas are sometimes omitted when the meaning is clear.

The following notes are included to indicate how the figures in the extant manuscripts look. Only mathematically essential differences and gross differences in shape and orientation are noticed. This is to facilitate judgement on possible copying of diagrams. Lack of specific remarks about a diagram is to be interpreted as agreement with the diagram as drawn.

The manuscript figures are described by reference to our figures: when a manuscript diagram is said, for instance, to be rotated, only a description is meant, not an account of how the orientation in the

---

[1]See the preface to her edition of the manuscript figures, p. 680.

manuscript figure arose and not a judgement of whether either orientation is correct. No account is taken of the points on a diagram letter: if, for instance, ت is represented by ث or by ب, or if خ is represented by ح, there is usually no comment.

The first figure in a proposition is sometimes referred to as Fig. (i), and so for subsequent figures in a proposition.

Book I

1. Our figure is based on **Cz**, Fig. (ii), and **AP**. **N** is quite different: in addition to its being rotated anti-clockwise through a right angle, we find ح difficult and ب impossible to read; "A" labels the intersection of DB with the circle.

2. Our figure is copied from **Cz**, based on our **P**, but similar to Greek MS **D**, and **P**. In **A** "Z" and "G" are switched and "B" is displaced to the highest point of the circle, "A" to the lowest. In **N** "B" is invisible in our copy; "G" labels te centre of the circle and "Z" apparently labels the same point as "A".

3. Our figure is copied from **CzP**. **AN** have ح at the top of the figure.

4. Based on **CzPA**(Fig. (ii))**N**. In **Cz** "D" labels the uppermost part of circle AGD.

A has another diagram, to the right of this, with the circles much closer together and with "Z" and "K" appearing to label, respectively, the intersections of line EA with circle AT and line LA with circle AG. In this diagram, too, "D" labels the upper intersection of the two circles.

5. Based on **N**. In **Cz**, which has an additional horizontal line through A, AB is vertical. In **A** AG is to the left of AB. **P** has AG horizontal and about equal to AB.

6. Based on **CzP**. **A** has the circles touching. **PN** have the circles the same size. In **N** "Z" is missing and "A" is cut off.

7. Based on **PN**. In **Cz** lines *EA* and *EG* are drawn. In **A** "*B*" and "*D*" are exchanged; *A* is on the right of the circle, not above it.

8. Based on **P**, though it has "*K*" for "*D*" (also in text). **Cz** has no lines *ZB*, *ZT*, *HB*, *HT*. The figure in **PN** is rotated 90° clockwise. In **A** *D* is on the *H*-side of *E*; in **N** "*D*" appears to label the intersection of *EH* with the circle.

9. The figure from the previous proposition is repeated in **PN**.

10. Based on **CzP**. **A** is rotated 90° clockwise. The figure in **N** is a left-right mirror image of ours.

11. Based on **PA**. The diagram in **Cz** is rotated 90° clockwise. In **N** the diagram is like ours rotated through 90° anticlockwise, but with *AHG* running from the top right quadrant to bottom left. "*Z*" is cut off in **N**.

12. Based on **CzPAN**.

13. Based on **CzA**. **PN** have *T* on $\odot GED$ and *K* on $\odot AEB$. In **N** the whole diagram is rotated 180°.

14, 15, 16. There is one diagram for these propositions, though in each manuscript three are drawn.

In **Cz**'s diagrams for Propositions 14 and 15, which are rotated clockwise through 90°, $\odot BD$ is replaced by two arcs *BED* and *BZD* of about 90° each; *BD* is approx. a diameter of $\odot ABG$. "*H*" appears to mark the pseudo-intersection of $B\widehat{E}D$ with the line *AG*. In Prop. 16 the figure is the same, except "*H*" is absent, and the two arcs *BED BZD* are much lower in the figure.

In **P** the figures for Props. 15 and 16 ae turned 90° clockwise. In 15 "*H*" appears to mark the pseudo-intersection of $B\widehat{E}D$ with line *GA*. In 16 "*E*" appears to mark this; "*H*" is missing.

In **A** all three figures are rotated through 180°.

In **N** the diagram for Prop. 14 is like ours, except for a large (extra) "ζ" in the top half of the left circle. The diagrams for Props. 15 and 16 are lettered as though they are to be read from the L.H.S. of the page. In both cases "*E*" appears to mark the pseudo-intersection of $B\widehat{D}Z$ and line *AG*. In 15 "*B*" and "*D*" are switched. In 16 a second "*A*" appears instead of "*B*".

17. Based on **PA**, though **P** has Z further to the right. In **Cz** the diagram is rotated anticlockwise by 90°; "B" and "D" are interchanged. In **N** the diagram is rotated so that Z is at the top; line ZB is added.

18. Based on **CzP**. In **AN** the diagram is rotated clockwise through 90°. In **A** the circles are much closer together; and in **N** "G" and the lower part of the figure has been cut off.

19. Based on **CzP**, though **P** has the two part-diagrams separated vertically, not horizontally. **A** has the rhombus-figure on the left. In **N** "A" is cut off; ∫ is put apparently to mark the centre of $\odot AGTB$. The quarters of the rhombus-figure are labelled, in a later, grosser, hand, with inexplicable letters. The top of the double figure, together with "A" and "D", are cut off.

20. Based on **CzP**. In **A** $EHTZ$ is rotated 180° and stands on the left; the other part of the figure is rotated through 90° clockwise. In **N** "L" appears to mark the pseudo-intersection of AK with $\odot BDG$.

21. Based on **N**. The figure in **Cz** is rotated, so that AB is horizontal, and then switched L-R. "D" marks the intersection of $\odot$s EGD and EZH. "Z" is between E and A.

In **A** the figure is rotated clockwise 120°. "D" marks the second point of intersection of $\odot$s EGD and AEZ (the first being E).

In **P** the figure is again rotated by 120° and the reflected L-R. "G" marks the $\odot$s GHZ and BGD (opp. E); "T" marks the intersection of $\odot$s BA and BGD. "D" marks the intersection of $\odot$s EBG and AB.

22. Based on **CzP** (though the circles are of equal size in **P** and T is not visible in the available copy).

In **A** the circles are of equal size and switched L-R. Instead of ⋧ a second ⸲ appears.
In **N** the figure is upside-down.

Book II

1 and 2. All four witnesses have two identical diagrams for the two propositions. Our diagram is based on **CzN**, though the inner circle in **Cz** is relatively smaller. In **P** the diagram is rotated anticlockwise by about 60°; in **A** it is rotated clockwise about 90°.

3. Our figure is from **PN**, though in **P** "*D*" is almost diametrically opposite on $\odot DGE$. In **A** the diagram is a mirror-image (top-to-bottom) of our figure rotated anticlockwise by 90°; $\odot ABG$ is a little larger than $\odot DGE$ and $\odot AGE$ is much bigger than either. **Cz**: here $\odot AGE$ is the largest and $\odot ABG$ the smallest circle; of the $\odot AGE$ only $A\overset{\frown}{G}E$ is drawn; "*B*" and "*D*" mark approximately opposite points to those in our diagram.

4. Based on **CzP**, though in **P** $B\overset{\frown}{Z}$ and $B\overset{\frown}{H}$ are both longer. In **AN**, in which $\odot GDE$ is drawn smaller and does not appear parallel to $\odot BTK$, the full $\odot ZBH$ (a little smaller than the other two circles) is drawn – to pass through T.

5. Based on **PA**. The figure in **Cz** is switched L-R; "*A*" is near the top of $\odot ABG$ and "*B*" is slightly below the leftmost point of $\odot ABG$; "*E*" and "*D*" are interchanged.

N is turned through 180°. H and T are the intersections of $\odot EDG$ with arcs NKG and NLG. B is the leftmost point and D the rightmost point of $\odot EDG$; "*A*" is missing.

6. Copied from **PCz**. **A** is rotated 90° anticlockwise. **N** is like **A**, only the curve of arc $GEDZH$ bends the other way.

7. Based on **PA**, though **A** is rotated 180°. In **Cz** the figure is rotated 90° clockwise; $\odot EZ$ is outside the big circle and $\odot GD$ is inside. In **N** the figure is rotated 90° anticlockwise; "*G*" and "*D*" are interchanged; $\odot AB$ and $\odot EZ$ are drawn so big that they cross.

8. Based on **PA**, though **A** is rotated 90° anticlockwise. **Cz** has $\odot ABG$ bigger than $\odot BDZ$. **N** has $A\overset{\frown}{E}H$ curved the other way; it has "*T*" above "*A*", labelling the upper circle.

9. Based on **CzP**, though in **Cz** G and D are lower in the figure. **A** is the same, but reflected L-R. In **N** the figure is rotated through 180°.

10. Based on **PAN**. In **Cz** L is not at the centre of $\odot ABGD$, but placed further to the right and a little lower.

11. Based on **CzPA**. **N** differs in that (1) each half is rotated 90° anticlockwise and (2) the diagram is drawn twice, repeated (with the two parts exchanged left for right) for the latter part of the proposition (the Greek texts makes this part into a separate proposition).

12. Taken from **A** and the diagram in **Cz** for the Greek Prop. 14. **P** is the same, but rotated 90° anticlockwise. In **N** the circles are the same size; "E" and "Z" are interchanged. In **CzPN** only arc EBZ of $\odot EBZ$ is drawn.

13. Copied from **P**. **Cz** has $R\ Z\ H$ as one point. In **A** the figure is rotated 90° clockwise; ت and ث (our "P" and "Y") are interchanged, and so are س and ش (our "S" and "O"). Where **PA** have "F" **Cz** has Φ ($=$ ث or "Y").

N is quite different and has two diagrams. The first, on p. 222, is labelled الصورة الثانية "the second figure". $\odot$s AEK and DTL are not drawn side by side, but aligned so that $\odot AEK$ touches $\odot KL$ at its lowest point, K, and $\odot DTL$ touches it at its highest point, L (opposite K). The intersections of $\odot$s AEK and DTL are at س and ص . $PQLM$ and $MKSN$ are drawn as two parts, upper and lower, of a vertical straight line, which continues to ث and ف (intersection with $\odot$s DTL and AEK, resp.). Instead of ل **N** has لع .

N's second diagram carries this note: الصورة الاولى التى البرهان بحسبها في هذه النسخة (the first figure, which the proof in this copy [of the text] supposes). It is not well drawn, $MLQP$ and MO are recognizable as two different arcs containing an angle, but $\odot$s $DTL\ HG\ TQ\ PGN$ all concur at G (also labelled "B"). Another circle is drawn among the parallel circles. Many letters are omitted.

14. Based on **PCzA**, though in **P** $H\widehat{B}E$ does not touch $\odot ANM$, the figure in **Cz** is a mirror-image of our diagram and in it KM and $\widehat{KG}$

are missing, and **A** is rotated through 180° and has an extra circle about midway between ⊙s *ANM* and *ZHGE*. On the extra circle are marked: و , ت and ث at the intersections with *GDAZ*, *GMQ* and *GH* resp.

N represents this diagram very poorly, most of the letters and some of the lines being missing. There is another diagram in **N**: in our copy a small sheet containing this figure has been pasted partly over the diagram for III 5 (q.v.) and partly over its text. Here " ن " stands for "ز " (despite the genuine " ن "); and $E\widehat{H}$ is extended at both ends to reach below *K* and *L*.

15. Our diagrams are copied from **A**. In **PA** the three figures are together, with an extra one at the end for **P**; in **N** the first two are on p. 228 and the third on p. 229. In **A** they are labelled "the first", etc. and in **N** "the first of the sixteenth proposition", etc. In **N** all arcs are made with complete circles.

(i) In **Cz** the diagram is like ours, only rotated clockwise by 45°. In **P** *G* is near *M*. In **N** *B* and *M* are closer and ⊙s *THE* and *AGD* are closer than they are in our diagram; *G* is lower and *A* higher in ⊙*AGD*; *K* is mislabelled "*T*", despite the real *T* and its label.

(ii) **P** approximates to our diagram turned clockwise by 45°, but "*B*" and "*K*" (here "*L*") are interchanged, "*D*" labels *S* and "*S*" is missing. The two concentric circles in **N** are labelled *TLKGBA* (outer) and *HZE* (inner) – so that *T* is on the outer circle. Here also "*B*" labels point *G* and ⊙*AEB* does not meet the two circles touching the inner parallel circle at *Z*; these two circles are *LZG* (for our *DZB*) and [*S*]*ZK*.

Cz is quite different: e.g. $A\widehat{E}G$ ("Γ" is missing) is there, only from L. to R., and apparently touching ⊙*TEZ* at *E*, and *DZB*, cutting ⊙*TEZ* at *T* and *Z* is from R. to L. "*H*" seems to mark the same point as "*Z*".

(iii) **P** is very similar to **A**, though it has *Q* only a little over half way up the R.H.S. of the diagram and **A** has "*K*" for "*L*" (also sometimes in the text) and an extra س , apparently labelling the same point as "*Z*". In both **PA** "*L*" ("*K*" in **A**) seems to label a conglomerate point of contact of ⊙s *MLS AEG BTD*. Our *L* is the meeting-point of ⊙s *AEG MLS*.

Cz is again different: $AE\widehat{L}HG$ runs from L. to R. and $GH\widehat{L}EA$ from R. to L. Our *QZN* appears to be represented by $M\widehat{Z}N$, but ⊙*MLS* is not touched.

In the last (4th) diagram in **P** L is no longer the apparent touching-place of $D\widehat{TZ}B$ with $\odot LMS$.

16. Copied from **CzP** (though **Cz** interchanges "C" and "F"). In **A** $LNHTFKM$ is curved to the right; MS is a short straight line, almost horizontal, with S to the left of M. In **N** the whole is rotated anticlockwise by 90°, and then reflected top-to-bottom; N appears to label the point of pseudo-intersection of $\overline{AB}$ and $LN\widehat{H}M$; there are some arcs, apparently cancelled, from E and Z to CQ and from C and Q to $\overline{GD}$.

17. Copied from **CzPA**. The diagram in **N** is reflected top-to-bottom; the bottom part of the diagram, including AB, is cut off.

18. Taken from **A**. In **CzPN** arc $KZNHMET$ does not extend below K and T; in **N** arc $KGLBT$ extends beyond K and T. In **P** K Z and D are not visible in the inner margin of the codex.

The diagram in **Cz** is reflected L-R. It does not have "L".

19. Our diagram is based on **PA**, though in both manuscripts KH and HT are drawn as one continuous arc. In **N** the diagram is turned anticlockwise through 90°: "T" is not visible – perhaps cut off in the top margin. Again, KH and HT are drawn continuously.

Cz has a diagram like the first of the two reproduced by Ver Eecke (see Mathematical Summary on II 19), only without $\widehat{HE}$ $\widehat{HZ}$, with HD HZ as straight lines, and with H on $\odot ABDG$. Ver Eecke's diagrams have $H\widehat{T}G$ and $H\widehat{K}D$ curved in the opposite sense to ours.

20. Here **P** is copied, though "N" had to be supplied. In **Cz** "B" and "D" are interchanged; also "T" and "Z". In **A** each of the circles is rotated 90° clockwise.

In **N** each of the two part-diagrams has been rotated 90° clockwise. In the left part-diagram (right in ours) $D\widehat{K}S$ is made into a complete circle and "G" is now shifted to the intersection (on the left) of this circle and line AS. Similar remarks apply to the right part-diagram; also "Z" and "N" are interchanged.

21. Copied from **P**, though "I" had to be supplied (from **A** and a rather questionable part of the text). **Cz** has $\widehat{PT}$ and $\widehat{RO}$ drawn as not intersecting; "R" is at the bottom of the curve. "D" is almost at the bottom of its circle. $X\widehat{L}F$ and $T\widehat{A}LZ$ are continued to $\odot AHGBE$.

In **A** the diagram is rotated 90° anticlockwise. $\widehat{PT}$ (drawn as a complete circle) $\widehat{RO}$ do not intersect; $M\widehat{N}S$ and $B\widehat{Z}G$ are drawn almost touching. Some of the letters in the middle of the diagram are not clear.

As for **N**, arcs $M\widehat{N}S$ $B\widehat{Z}G$ $Q\widehat{F}C$ $\widehat{PT}$ are formed with smaller radius than in our diagram and none intersects with any other; $\widehat{PT}$ is a full circle. There is an extra ع at the end of $\widehat{RO}$. Some of the letters in the middle of the diagram are displaced: e.g. "A", "U", "Y". On the top left is the note: هذا الشكل غير صحيح (this diagram is not correct).

22. Based on **Cz**, though this has $TLZP$ extending beyond P to ABG. **P** is almost the same, though rotated 180°; but for ر it has S (in our Latin text denoted by $\acute{S}$) and for س it has Y (here $\acute{Y}$ – also in the Latin text). The letters in our Latin diagram are Gerard's.

In **A**, which is rotated 90° anticlockwise, TLP, which is curved in the opposite sense, is extended to touch $\odot QFO$.

N is rotated through 180°. Here the intersection (not labelled) of $S\widehat{N}$ with $\odot AGB$ is separated from the intersection, labelled "M", of $C\widehat{O}QR$ with $\odot ABG$. $T\acute{A}LZ$ meets $C\widehat{O}QB$ at S and both ت and س label the point of intersection.

Book III

1. Based on **CzAN**. **P** is essentially the same, though it looks different: M and K are closer together and closer to D; G is distinctly on the left of the figure, so that B L G are also closer together. **AN** also have G on the left, and D M K G L B are more evenly spread than in **Cz**. There are lines in **N** joining G and M with the pseudo-intersection of BD and EK and subsequently erased.

2. Copied from **CzP**. **A** seems carelessly drawn: $\triangle ETZ$ is very thin, with Z almost on ET. B is at the bottom of the circle, D near the top.

In **N** B and T are on the L.H.S. of the diagram, B almost at the bottom. An arc is drawn through AEZ. There are also straight lines EL and LB, where L is on AG, just to the right of the centre. An uncertain letter (similar to ل or ك) marks the pseudo-intersection of AG and EK; it should be noted that both "K" (in the long form normal for diagram letters) and "L" both appear elsewhere in the diagram.

3. Based on **CzPA**, though **Cz** swaps the constituent figures, so that the left circle appears on the right, and in **P** the right circle is rotated 90° clockwise, $G\widehat{E}D$ being in the opposite sense, so that E is above left w.r.t. the centre. **A** has the right circle rotated 90° anticlockwise, and "A" and "B" are swapped. In **N** the right circle is rotated 180° and "A" and "B" are exchanged.

4. Copied from **CzPA**, though in **Cz** S is not so far to the left.

In **N** "K" is missing and "ر" is substituted for "A". The meeting of LE and MD is indicated by an angle in LE – S would otherwise be well off the page. "S" is cut off. LG and ME are drawn and produced to meet just outside the present page. Any letter there is cut off. "K" stands at the intersection of ME and the circle.

5. Copied from **CzP**, though in **P** $Q\widehat{F}$ and $B\widehat{G}$ are lower, and $AR\widehat{T}C$ stops at arc BG.

In **A** "E" is missing and R is labelled "ث". The meeting of $A\widehat{T}C$ and $B\widehat{G}$ is labelled "ح" (although there is already such a letter) and its meeting with the circle is labelled "ر". Within the circle and above $L\widehat{M}$, between DH and AR, is توران [Persian name for the land of the Turks]; between LM and NS, on the left, is شام [Syria]; between NS and QF, on the right, is ايران [Iran]; between QF and BG, only slightly left of centre, is روم [Byzantium]. These geographical names in III 5, 6 and 7 are written in a different hand at a later time and perhaps by a Persian or Turk, because they are written without the Arabic article, *al-*. Cf. also the name *Tūrān*, above, and the expression *ṭaraf-i Hind wa-Sind*, below in the figure for III 6. In the margin are short notes, beginning طرف الشرق [end of the East].

In **N** the diagram is obscured in our copy by another diagram, on a small separate sheet, pasted on top of it, leaving visible only the leftmost quarter or third. The obtrusive diagram is labelled "15" and underneath is الدوائر [circles]. It is the figure for II 14 (q.v.).

6. Copied from **Cz**. **P** is very similar, but the curve on arc *DHTKZE* is shallower and *KO* and *BG* are lower down.

In **A**, which in general is again similar, the diagram and the text have ض "*Ź*" *in corr.*; there is no letter at the other end of the arc (in our diagram *Ď*). There are again geographical inscriptions within the circle (cf. III 5). On the right, between *SQ* and *FC*, is فارس [Persia]; between *FC* and *DŹ* is شمال [north]; between *DŹ* and *RO* is جنوب [south]; between *RO* and *BG* is روم [Byzantium]. On the left, between *FC* and *DŹ*, is اسكندرية [Alexandria]; and between *DŹ* and *RO* is شام [Syria]. As in III 5, there are short notes in the margin. These begin: طرف هند وسند [end of Hind and Sind].

In **N** "*K*", "*P*" and "*Ď*" are missing. "ص" appears for "ظ".

7. Based on **CzP**, though in **Cz** ⊙*AD* touches ⊙*MTN* – at *O*, which is no longer on *PK̂F*; and in **P** "*F*" "*S*" "*M*" are lost in the binding, and *PK̂Y* and *DK̂R* are continued to *B̂G*.

In **A** there is an extra "ح" apparently labelling *D*, although "*D*" is also there. "*D*" (i.e. د, for ذ) is nearer *O* than *F*. *PK̂Y* is drawn as a straight line and is continued by a straight line, at an angle to it, reaching *B̂G*. *DK̂R* is continued to *B̂G*, apparently by a straight line. There are some geographical inscriptions inside the diagram (cf. III 5 and 6). On the right, between *MN* and *SQ*, is برو [?]; between *SQ* and *FC* is فارس [Persia], between *FC* and *BG* is يمن, below *BG* is سمر . On the left side, above *MN*, upper right ايران [Iran] and lower left توران ; between *MN* and *SQ* عرب [Arabs], between *SQ* and *FC* is جنوب , between *FC* and *BG* is شام [Syria] and شمال [north], below *BG*, near the bottom, is بيت المقدس [Jerusalem]. To the left of the diagram can be recognized طرف شمال [end of the north] and several isolated letters.

In **N** *PK̂Y* and *DK̂R* are continued to meet *B̂G* at points labelled ر and ر [sic]. There is an arc, drawn and crossed out, between the intersection of *DK̂R* and *B̂G* and a point on *LĤ*. The genuine ر is written as ن , although ن is also written in its proper place.

8. Copied from **CzPAN** by accepting the unusual translation of Ω into ض and ⱶ (in our edition: *Ď*) and of Ψ into ذ and *I* (in our edition: *Ť*).

In **A** "*T́*" is missing. There is an extra ص , written just above the "ص" (for ض or *D́*), near the top of the figure.

In **N** ن is written for ر , although the genuine "ن" is in the right place. "*H*" is missing.

9. The three constituent figures are copied from **CzP**. In **P** "*H*" is hard to read in all three constituent figures – perhaps there is a connection with the presence of an extra "*H*" in the top right of the diagram.

AN have the constituent figures on the same page. **P** labels them "prima", etc., and **AH** with الأول etc. In this case **H** puts the first figure on the left (though on the right for III 10).

In **A** each constituent figure is a L. to R. mirror image, but "*B*" is retained on the left. Figs. (i) and (iii) are drawn and lettered as we have drawn them, but Fig. (ii) is quite different. The arcs from A or from arc DE are, in order from L. to R.: $A\widehat{K}S$ $P\widehat{Q}$ $A\widehat{T}N$ $R\widehat{O}$ $C\widehat{F}$ $A\widehat{H}M$ $A\widehat{R}L$.

The figure in **N** is badly drawn, with the letters in (ii) and (iii) so crowded that it is not always certain to which points they refer. In Fig. (i) the arcs from A or arc DE are, in order from R. to L., $A\widehat{T}S$ $F\widehat{R}$ $A\widehat{H}R$ $A\widehat{Q}M$ $A\widehat{L}C$ $A\widehat{R}Z$. The similar series of arcs on Fig. (ii) is: $A\widehat{K}S$ $A\widehat{C}Q$ $A\widehat{F}R$ $A\widehat{T}M$ $A\widehat{H}S$ $R\widehat{L}$ $A\widehat{N}B$. On Fig. (iii): $A\widehat{K}Q$ $T\widehat{R}$ $A\widehat{L}Y$ $A\widehat{H}P$ $A\widehat{M}R$ $A\widehat{X}S$ $A\widehat{Q}L$ $A\widehat{N}B$.

10. Copied from **CzP**. **P** has Fig. (ii) first, then (i) and then (iii), labelled "secunda", etc. The figures are labelled الأول etc. in **A** and الأولى etc. in **N**.

In **A** the constituent figures are all rotated anticlockwise, Fig. (ii) by about 45°. In both **A** and **N** there is an additional arc in (ii), $L\widehat{S}$, with S on $\widehat{BE}$.

In Fig. (i): at F there are two letters, ف and ط. The arcs between $\widehat{ED}$ and $\widehat{EB}$ (with or without A) are, from L. to R., $L\widehat{B}$ $M\widehat{S}$ $A\widehat{Z}Q$ $N\widehat{T}$ $A\widehat{H}K$.

In **N**, Fig. (ii), "*B*" is cut off on the L.H.S. of the page. Most of (iii) is rotated 45° anticlockwise, $\widehat{BE}$ about 90°.

11. Copied from **CzPA**, though **Cz** has line ATK further to the right.

In **N** A is cut off in the top margin. " ن " and " ز " are swopped and " ح " is written for " ج ". H does not lie on $A\widehat{T}K$, but it remains the intersection of $N\widehat{S}$ and $D\widehat{Z}$. $A\widehat{T}K$ is drawn over point C. HQ is missing. A line was drawn, and then crossed out, between F and the intersection of $A\widehat{T}K$ and $N\widehat{H}S$.

12. Copied from **P**. In **Cz** Z is on the intersection of $\odot MB$ and $\odot HEM$, but this is geometrically impossible, for $\odot EKZ$ must touch a circle equal and parallel to $\odot EFH$ at a point opposite to point E. Further, **Cz** draws $L\widehat{K}S$ through A, but this is not specified by the text.

A, which is rotated 90° anticlockwise, also has Z on the intersection of $\odot$s EMH and MB; and **A**'s $L\widehat{K}S$ goes through A.

N does not place Z (which is wrongly labelled "E", despite the presence of the genuine "E") where **CzA** do, but it does have $L\widehat{K}S$ running through A. "H" is displaced from its post and now labels the intersection, on the right, of $\odot$s AG and LEM. " ر " stands in the place of " ن ". Q is on the outer circle LEM. "F" has the dots and shape of ق (C). $G\widehat{K}D$ goes through F.

13. Based on **CzP**, though in **Cz** $A\widehat{Z}G$ is curved in the opposite sense, and $Z\widehat{E}H$ $T\widehat{E}K$ appear in **P** as straight lines.

The figure in **A** is a mirror-image reflected about AD. " م " appears to label the same point as " ز ". There is another, smaller diagram in the margin, to the left of the one just described, with the lettering as we have it; $G\widehat{K}D$ and $B\widehat{H}D$ are curved in the opposite sense to ours and $A\widehat{E}D$ is very nearly a straight line.
In **N** the figure is rotated 180°. "D" is cut off on the L.H.S. $D\widehat{E}A$ is very erratically drawn. At the end of the treatise there is another diagram, likewise rotated through 180°. Top left of the diagram there is the following note, in a hand other than that of the text:

وذلك فى يوم [...] صحرة النهار [This is in the day of ...]. On the right of the diagram, apparently in the same hand as the previous note:

هذا الشكل ههنا معاد وهو الشكل المتقدم على الشكل الأخير ومن شرطه أن

يكون كرى فإن فى الأول ...

[This figure is a repeat: it is the figure preceding the last figure and from its condition is that it is spherical, because in the first ...]. The rest is cut off at the bottom of the page. On the top right of the diagram is another short note, in our copy illegible, in a different hand.

14. Based on **CzPA**, though **Cz** has $S\widehat{N}K$ curved in the opposite sense, and in **P** Z and T are cut off on the L.H.S. In **N** G is cut off at the bottom of the page, and "N" replaces "Z" (despite the presence of the real "N").

Lemmas and Notes

The first diagram in **A**, which serves for both **A1** and **A3**, appears only once, within the text for **A1**. The second diagram, which is for **A2** and **A4**, appears three times: next to the first diagram, on the right; at the end of **A2**; and at the end of the text.

The diagram of the Latin lemma "Quod proportio ... " is taken from **PR**, though **R** has "D" for "B". There is no diagram in **M**.

For the lemma for II 11 in **Kg** etc. there is no diagram in **KgFiPsPt**. Our figure is based on **OBZ**. Of these, **Z** puts B and O at the bottom of the small circle, **O** to the left and B distinctly to the right.

Mathematical Summary

Abbreviations and Symbols

In this section enunciations (and definitions) are rendered by full translations, but, to make the flow of the argument easier to follow, the proofs are divided into mathematical lines. Such a line is, wherever possible, rendered by a single line of print; further lines, when necessary, are indented.

Proofs are indented to distinguish them from the enunciation and the principal construction. Certain routines are further indented. Typical of these routines is the proof of the congruence of two right triangles when two sides, not including the right angle, are correspondingly equal[1].

To write each mathematical line in as few printed lines as possible, abbreviations have often been used. First, the symbol common in English textbooks, $\therefore$ (therefore) has been adopted. Other symbols in this category are $\parallel$ (is parallel to), $\perp$ (is perpendicular to), $\perp^{\mathrm{r}}$ (perpendicular, i.e. noun or adjective), $\odot$ (circle), $\triangle$ (triangle), $\equiv$ (is congruent to, i.e. of triangles), $\sim$ (is similar to, i.e. of arcs, angles or triangles), $\divideontimes$ (contradiction).

$\widehat{AB}$ means arc AB; $\overline{AB}$ means straight line AB. When the meaning is obvious, the curved or straight line over the letters is sometimes omitted (so: AB simply). $A\hat{B}G$ means angle ABG; $\hat{D}$ means angle D.

Secondly, articles and prepositions are sometimes omitted for brevity. Thus "plane $\odot ABG$" means "the plane of circle ABG". Further abbreviations are also used, e.g. diam. (diameter), rad. (radius), etc.

Finally, two special abbreviations are introduced here: si.sq.gt.$\odot$ (the side of a square inscribed in a great circle) and arc.si.sq.gt.$\odot$ (the arc subtended by the side of the square inscribed in a great circle).

Square brackets, [ ], include material added by the editors. References to the *Sphaerica* are made simply by the Book number and Proposition number (e.g. I 6). References to Euclid's *Elements* (in Heiberg's numbering) are distinguished by "Euclid" (e.g. Euclid XI 18). In these citations "Def." means definition.

---

[1]See note 10 to Proposition I 1.

"[enunciation]" at the end of a proposition means that the enunciation is repeated there.

In a proof, only the mathematical steps are indicated, not the exact meaning. Typically, construction is indicated by verbs in the imperative, irrespective of the part of the verb in the Arabic. The marginal notes in the Arabic are translated in the footnotes.

Book I

[Definitions][1]
1[2]. (I) A sphere is a corporeal figure surrounded by one surface [such that] all the straight lines drawn from one of the points inside it and meeting the surface are equal to each other.

2. (II) The centre of the sphere is that point.

3. (III) An axis of the sphere is some straight line passing through the centre and ending at both sides on the surface of the sphere, when the line is fixed and the sphere is rotated about it.

4. (IV) The two poles of the sphere are the two ends of the axis.

5. (V) The thing that is called in the sphere the pole of a circle is a point on the surface of the sphere [such that] all the straight lines drawn from it to the circumference of the circle are equal to each other.

---

[1]The Arabic numbers of the definitions have been added by the editors. The Roman numerals refer to those in Ver Eecke's translation of the Greek text, which are the same as Heiberg's (in Arabic numerals) in his edition. There are no such numbers in Czinczenheim's edition. Since Theodosius' definitions, in both Greek and Arabic, show signs of influence from Euclid's *Elements*, a comparison with Euclid is outlined in the following footnotes.

[2]Definitions 1-4 are the definitions of a sphere, its centre, axes and poles. Euclid's definition of a sphere is different: the figure produced by rotating a semicircle about its diameter. Euclid's definition of axis and centre of a sphere are related to this definition of a sphere. Theodosius' definition of an axis of a sphere is the same as Euclid's definition of a diameter of a sphere (see the discussion in Ver Eecke, p. 1 – some Greek manuscripts have διάμετρος instead of ἄξων in Def. 3) together with the addition, "when the line is fixed and the sphere is rotated about it", an obscure allusion, no doubt, to Euclid's definition of axis (Def. 15). Ver Eecke doubts the authenticity of this addition.

6[3]. It is said: in a sphere the distance of circles from its centre is equal when the perpendiculars drawn from the centre of the sphere to the planes of the circles are equal to each other.

7. The circle that is further away is the one on which a longer perpendicular falls.

8[4]. It is said: a plane is inclined to another plane when some point is marked on the common section of the two planes and from it in each of the two planes a straight line is drawn, standing at right angles on the common section, so that the two lines drawn contain an acute angle.

9. The inclination is the angle that these two lines contain.

10. (VI) It is said: the inclination of a plane to a plane is as the inclination of another plane to another plane when the straight lines drawn from the common section of the planes at right angles [to the common section] in each of the planes from the same point contain equal angles.

11. Those whose angles are smaller have the greater inclination[5].

1. When a spherical surface is cut by any plane, the resulting section is the circumference of a circle.

Let a spherical surface be cut by a plane to make a section: line ABG
I say: ABG is the circumference of a circle.

If the cutting plane passes through the centre of the sphere, line ABG is the circumference of a circle, since straight lines from the centre to line ABG are equal[1] [Def. sphere]

---

[3]Definitions 6 and 7, on the distance of circles in a sphere from the centre, are not in Greek. They and most of the following definitions are also not in the redaction by Ibn Abī al-Shukr. See Lorch [1996], 171, and Czinczenheim, 200.

[4]Definitions 8–11 are on the inclination of planes, of which Def. 10 is in Greek, though enclosed in square brackets by Czinczemheim, but the rest are not. As in Euclid XI Def. 6–7, they refer the inclination of planes to the angle made at a point on the common section of the planes, one in each plane and both perpendicular to the common section. Theodosius' definition of the inclination of one plane to another (Def. 8–9 together) is essentially the same as Euclid's (Euclid XI Def. 6), except he first defines being inclined (Def. 8) and then (Def. 9) the amount of inclination. Theodosius' Def. 10, on the similarity of the inclination of planes is essentially the same as Euclid Def. 7, except that he describes again the angle used to define inclination instead of referring to his Def. 8–9.

[5]No explanation is offered for this convention.

[1]since ... equal: the Greek has: since straight lines from the centre of the sphere to the surface are equal and ABG is in the surface.

Since the matter is so[2] it is clear that the centres of the sphere
and circle are the same

If the cutting plane does not pass through the centre of the sphere, let
the centre of the sphere be D

From D draw $DE \perp$ plane through ABG[3], meeting it at E

Draw[4] EB, EG; join DB, DG

Since D is the centre of the sphere[5], DB[6] $= DG$

$$\therefore \ DB^2 = DG^2 \ ^7$$

But $DE^2 + EB^2 = DB^2$, since $D\hat{E}B$ is right[8]

and $DE^2 + EG^2 = DG^2$, since $D\hat{E}G$ is right[9]

$$\therefore \ DE^2 + EB^2 = DE^2 + EG^2$$

subtract the common DE^2: there remains $EB^2 = EG^2$

$$\therefore \ BE = GE \ ^{10}$$

Similarly, all straight lines from E to line ABG are equal

$\therefore \ ABG$ is the circumference of a circle with centre E.

Corollary: When a perpendicular is drawn from the centre of a sphere
to one of the circles on the sphere, it falls on the centre of the circle[11].
Q.E.D.

2. To find the centre of a given sphere.

Let the given sphere be cut by some plane: the section arising is a circle
[I 1]; let this circle be AB

---

[2]Since ... so: the Greek has, since the plane is supposed through the centre of
the sphere.

[3]plane through ABG: in Greek "the cutting plane"; Euclid XI 2 *supra* **A**.

[4]From here on, in this proposition, the Arabic-Latin tradition has different letters from the Greek: B for Greek A and G for Greek B.

[5]Since D is the centre of the sphere: not in Greek.

[6]quocumque modo cadat *add.* Latin.

[7]This line is not in Greek.

[8]since $D\hat{E}B$ is right: not in Greek, though "$D\hat{E}B$" is in Arabic here rendered
in the Greek style as "the angle that lines DE EB contain". Against EB is, in **A**
supra, Euclid I 47.

[9]since $D\hat{E}G$ is right: not in Greek. See previous note.

[10]This follows from $DB = DG$ by the congruence of $\triangle$s BDE and GDE. But
Euclid does not give the case of two sides and a non-included right angle.

[11]The corollary in Greek runs: when there is a circle in a sphere, the perpendicular
drawn from the centre to it [the circle] will fall on the centre [of the circle].

If the plane passes through the the centre of the sphere, the centre of
    the sphere and the centre of the circle are the same; and we have
    found the centre of a known circle [Euclid III 1][1]

If the plane does not pass through the centre [of the sphere], let the
    centre of $\odot AB$[2] be G

From G draw a line perpendicular to plane $\odot AB$[3]: GD; produce it in
    both directions to meet the surface of the sphere at D E

Bisect DE at Z

I say: Z is the centre of the sphere.

> If it were not so, and the centre [of the sphere] could be another
>   point, let it be H
>
> From H draw a $\perp^{\mathrm{r}}$ to the plane of the $\odot$, meeting it at T[4]
>
> But a perpendicular from the centre of the sphere to a $\odot$ on the
>   sphere goes through its centre[5] [I 1 Cor.]: $\therefore$ T is the centre
>   of $\odot[AB]$
>
> But G is also its centre ※
>
> And if the $\perp^{\mathrm{r}}$ falls on G, then from the same point two $\perp^{\mathrm{rs}}$ are
>   drawn from one point to one plane[6] ※[Euclid XI 13][7]
>
> $\therefore$ H is not the centre of the sphere
>
> Similarly centre of the sphere cannot be a point other than Z
>
> $\therefore$ Z is the centre of the sphere.

Corollary: When there is a circle on a sphere and a straight line is
drawn from the centre of the circle perpendicular to the plane of the
circle, the centre of the sphere is on the line. Q.E.D.

3. When a sphere touches a plane without being cut by it, it [the
sphere] touches it [the plane] in only one point[1].

---

[1]If ... circle: is not in Greek; Euclid III 1 *supra* **A**. Latin has (lines 7–8) in addition: Ergo iam manifestum est nobis quomodo reperiamus centrum spere. Perhaps
this was originally meant as an improvement of the previous sentence.

[2]In the Greek text $\odot AB$ is $\odot ABG$. In the Arabic-Latin tradition the subsequent
diagram letters are one earlier than the Greek in the alphabet: it has G for Greek
D, D for E, E for Z, Z for H, H for T and T for K.

[3]Euclid XI 12 *supra* **A**.

[4]Euclid XI 11 *supra* **A**.

[5]But ... centre: this sentence is not in Greek.

[6]And if ... plane: not in Greek.

[7]Euclid XI 18 *supra* **A**.

[1]Here and in the proof the Arabic speaks of a sphere touching a plane, but in
the Latin a plane touches a sphere.

If it were possible, let the sphere touch the plane, without its cutting it, in more than one point: at A and B

Let the centre of the sphere be G; join AG BG

> Draw plane through AG GB producing sections: with the sphere a circle [I 1] and with the [original] plane a straight line; let the circle be DAB and the straight line be $\overline{EABZ}$
>
> Since the [original] plane does not cut the sphere, $\overline{EABZ}$ does not cut $\odot DAB$
>
> Since A and B are marked, however they fall, on the circumference, $\overline{AB}$ falls within the circle DAB[2]
>
> But it also falls outside ⁕

$\therefore$ [enunciation]

4. If a sphere touches a plane without being cut by it, the straight line joining the centre and the point of contact is perpendicular to the tangent plane.

Let a sphere touch a plane at one point [I 3], A, without being cut by it

Let the centre of the sphere be B; join AB

I say: $AB \perp$ plane.

> For if a plane is drawn through AB, it produces [1] in the surface of the sphere $\odot AGD$ [I 1] and [2] in the plane a straight line EAZ[1]
>
> Let another plane pass through AB, producing [1] in the surface of the sphere $\odot AT$ and [2] in the plane $\overline{KAL}$
>
> Since the [first] plane touches the sphere, $\overline{EAZ}$ is also tangent to $\odot ADG$
>
> Since $\overline{EAZ}$ touches $\odot ADG$ at A, and $\overline{AB}$ is drawn from A to the centre of the circle,
>
> $\therefore$ $AB \perp EAZ$[2]
>
> B is centre of $\odot AGD$ [Euclid III 19], since plane $\odot AGD$ passes through $\overline{BA}$, which is drawn from the centre of the sphere[3]
>
> Similarly $BA \perp \overline{KAL}$

---

[2]Euclid III 2 *supra* **A**.

[1]Euclid XI 3 *supra* **A**.

[2]Euclid III 7 *supra* **A**

[3]B ... the centre of the sphere: as Ver Eecke notes, this sentence is useless and is probably interpolated. Heiberg and Czinczenheim put it between square brackets.

And since $\overline{BA}$ is $\perp^{\mathrm{r}}$ at the intersection of *EZ* and *KL*,
  AB $\perp$ plane that passes through them [Euclid XI 4], and the
  plane that passes through[4] *EZ KL* touches the sphere[5]. Q.E.D.

5. When a sphere touches a plane without being cut by it, and at the
place of contact a line is drawn standing on it at right angles, the centre
of the sphere is on that perpendicular line.

Let a sphere touch a plane at *A* without being cut by it
From *A* draw a $\perp^{\mathrm{r}}$[1] to the plane: *AB*
I say: The centre of the sphere is on *AB*.

  If it is not, and something else were possible: let the centre of the
    sphere be *G*; join *GA*
  Since a sphere touches a plane at *A*[2] without being cut by it
    and *GA* is drawn from the centre of the sphere to the place of
    contact, *GA* $\perp$ plane [I 4]
  But *BA* also $\perp$ plane: $\therefore$ from the same point on one plane two
    straight lines are drawn at right angles on the same side, *AB*
    AG $\divideontimes$ [Euclid XI 13][3]
  $\therefore$ *G* is not the centre of the sphere
  Similarly it is clear that it is not possible that the centre is an-
    other point, not on *BA*
  $\therefore$ the centre is on *BA*. Q.E.D.

6. Circles on a sphere passing through the centre are its greatest [i.e.
"great"] circles; and, of the remaining circles, those whose distances
from the centre are equal are equal, and those whose distances from
the centre are greater are smaller.

Let there be in the sphere $\odot$s *AB GD EZ*. Let *GD* pass through the
    centre of the sphere
First let the distance of $\odot$s *AB EZ* be equal

---

[4]Euclid XI 4 *supra* **A**.

[5]*Note in* **A**. Because if it [*m.*] cuts it [*f.*], it must cut it in more than one place;
and then the plane will be cutting the sphere. We have already supposed it as
touching it. This is a contradiction. The Greek has additionally: Therefore the
straight line *AB* is perpendicular to the plane tangent to the sphere.

[1]Euclid XI 12 *supra* **A**.

[2]In Latin the plane touches the sphere.

[3]Euclid XI 13 *supra* **A**.

[1] *I say:* The greatest of these circles is $\odot GD$, and $\odot$s AB EZ are equal.

For we make the centre of the sphere H
∴ it is the centre of $\odot GD$
From H draw ⊥ʳs, HT HK, to planes of $\odot$s AB EZ,[1] meeting the planes at T K
∴ T K are the centres of $\odot$s AB EZ [I 1 Cor.]
From T K H draw straight lines to the circumferences of $\odot$s AB GD EZ: lines TL KN HM
Join HL HN
Since HT ⊥ plane $\odot AB$, it produces right angles with all lines drawn from its end in the plane $\odot AB$ [Euclid XI Def. 3]
From its end has been drawn TL, in plane $\odot AB$
∴ $L\hat{T}H$ is right
Similarly $H\hat{K}N$ is right
Since $L\hat{T}H$ is right, $L\hat{T}H >$[2] $L\hat{H}T$
∴ $LH > LT$ [Euclid I 17][3]
And $LH = HM$ since H is the centre of the sphere and HL HM are radii[4] of the sphere
∴ $HM > LT$
And HM is rad. $\odot GD$ and TL is rad. $\odot AB$: ∴ $\odot GD > \odot AB$
Similarly $[\odot GD] > \odot EZ$
∴ $\odot GD$ is the greatest of the circles in the sphere.

[2] *I say also:* $\odot AB = \odot EZ$.

For their distance from the centre is equal, $HT = HK$
Since H is the centre of the sphere, $HL = HN$
∴ $HL^2 = HN^2$
But $LT^2 + TH^2 = HL^2$ and $NK^2 + HK^2 = HN^2$[5]
∴ $LT^2 + TH^2 = HK^2 + KN^2$
And $TH^2 = HK^2$: there remains $TL^2 = KN^2$
∴ $LT = KN$[6]

---

[1]Euclid XI 11 *supra* **A**.

[2]Euclid I 32 *supra* **A**.

[3]Euclid I 18 *supra* **A**.

[4]What is translated as "is the radius of" is in Arabic "goes out from the centre of the circle to its circumference". This may be considered a technical term for radius in Arabic, equivalent to the Greek ἡ ἐκ τοῦ κέντρου, "the [line] from the centre [to the circumference]". In the passage here (which is enclosed by Czinczenheim in square brackets) the Greek expresses the idea more fully.

[5]Euclid I 47 *supra* **A**.

[6]This is clear from $HL = HN$. See note 10 on Prop. 1.

TL is rad. $\odot AB$; KN is rad. $\odot EZ$
$\therefore$ rad. $\odot AB$ = rad. $\odot EZ$
$\therefore$ $\odot AB = \odot EZ$.

[3] Again, let distance $\odot AB$ from centre > distance $\odot EZ$ from it
I say: $\odot AB < \odot EZ$.

We make just the same construction as before

Since distance $\odot AB$ from the centre of the sphere > distance
$\odot EZ$ from it, $\therefore$ $HT > HK$
Since $HL = HN$,
$$HL^2 = HN^2$$
But $HT^2 + TL^2 = HL^2$ and $HK^2 + KN^2 = HN^2$
$\therefore$ $LT^2 + TH^2 = HK^2 + KN^2$
And $TH^2 > HK^2$
$\therefore$ there remains $LT^2 < NK^2$
$\therefore$ $LT < KN$ [7]
And TL is the radius of $\odot AB$ and KN is the radius of $\odot EZ$
$\therefore$ $\odot AB < \odot EZ$.

$\therefore$ [enunciation] Q.E.D.

7. When there is a circle on a sphere and the centre of the sphere and
the centre of the circle are joined by a line, then the joining line is
perpendicular to the plane of the circle.

Let the circle in the sphere be $ABGD$; let the centre of the sphere be
E and the centre of the circle be Z; join EZ
I say: $EZ \perp \odot ABGD$.

From the centre of the circle draw $AZG\ BZD$; join $EB\ ED$
Since $ZB = ZD$ and ZE is common,
$BZ\ ZE = DZ\ ZE$ respectively [in $\triangle$s ZBE and ZDE]
And base BE = base DE (for E is the centre of the sphere and
$B\ D$ are on the surface of the sphere)
$\therefore$ $B\hat{Z}E = D\hat{Z}E$ [Euclid I 8][1]
When a straight line stands on a straight line and makes the
angles on the two sides equal, then each of the equal angles
is right [Euclid I Def. 10], and the standing line is said to be

---

[7]This follows from $HT > HK$ by reasoning exactly parallel to that proving LT
$= KN$ from $HT = HK$ in the second section of this proposition.
[1]Euclid I 8 *supra* **A**.

perpendicular to the line on which it stands

∴ each of $B\hat{Z}E$ and $D\hat{Z}E$ is right and $EZ \perp BD$

Similarly $[EZ] \perp AG$

Since $\overline{EZ} \perp$ the common section of the intersecting lines $AG\ BD$,

∴ it also $\perp$ plane that passes through $AG\ BD$[2]

And the plane that passes through $AG\ BD$ is plane $\odot ABGD$

∴ $EZ \perp$ [plane] $ABGD$. Q.E.D.

8. When there is a circle on a sphere and a perpendicular to it is drawn from the centre of the sphere and is produced in both directions, it will fall on the circle's poles.

Let the circle be ABG; let the centre of the sphere be D[1]

From D draw DE perpendicular to plane $\odot ABG$, meeting the plane of the circle at E,

∴ E is the centre of $\odot ABG$ [I 1 Cor.]

Produce DE in both directions, meeting surface of the sphere at $Z\ H$

I say: $Z\ H$ are the poles of $\odot ABG$.

Draw $AEG\ BET$

Join $AZ\ ZG\ AH\ HG\ BZ\ ZT\ BH\ HT$

Since $ZE \perp \odot ABG$ and [so] makes right angles with all straight lines drawn from its end in $\odot ABG$ [Euclid XI Def. 3],

∴ each of $Z\hat{E}A\ Z\hat{E}G\ Z\hat{E}B\ Z\hat{E}T$ is right

Also, since $AE = EG$, and EZ is common [to $\triangle$s $AEZ\ GEZ$] and at right angles [to AE and EG],

∴ base $AZ =$ base ZG [Euclid I 4][2]

Similarly the remaining lines from Z to $A\hat{B}G$ are equal

∴ Z is pole $\odot ABG$

Similarly H is pole $\odot ABG$

∴ $Z\ H$ are the poles of $\odot ABG$.

∴ [enunciation]

9[1]. If there is a circle on a sphere and one of its poles and the centre are joined by a straight line, then the line is perpendicular to the circle.

---

[2]Euclid XI 4 *supra* **A**.

[1]Here and in the other two occurrences D is replaced by K in Gerard's Latin – with the exception of MS **O**, which has D.

[2]Euclid I 4 *supra* **A**.

[1]This proposition, which, like Prop. 8, is a converse of Prop. 7, is not in Greek.

The proof of this proposition is similar to the proof of the proposition
  that is before it[2].

10[1]. When there is a circle on a sphere and a line perpendicular to it
is drawn to it from one of its poles, [the line] falls on the centre of the
circle. If it is produced on the other side, it falls on the other of the
two poles of the circle.

Let the circle on the sphere be $\odot ABG$[2]

From one of its poles, D, draw $DE \perp$, to [the circle] to meet the plane
  of the circle at E

Produce DE to meet the surface of the sphere on the other side at Z

I say: E is the centre of the $\odot ABG$, and Z is the other of the two poles
  of $\odot ABG$.

From E draw EA EB; join AD DB AZ ZB

Since $DE \perp \odot ABG$, it produces right angles with all straight
  lines in plane $\odot ABG$ drawn from the end of [DE] in plane
  $\odot ABG$ [Euclid XI Def. 3]

And AE EB emerge from its end in plane $\odot ABG$

  $\therefore$ each of $D\hat{E}A$ $D\hat{E}B$ is right

So since $AD = DB$,

  $\therefore AD^2 = DB^2$

  But $AD^2 = DE^2 + EA^2$ and $DB^2 = DE^2 + EB^2$

  $\therefore AE^2 + ED^2 = BE^2 + DE^2$

  $\therefore$, by subtracting the common DE^2,

    there remains $AE^2 = EB^2$

$\therefore AE = EB$[3]

---

[2]The proofs of both Prop. 8 and Prop. 9 rely on the congruence of $\triangle$s ZEA and
ZEG. See figure for Prop. 8. In **N** and **P** the diagram for Prop. 8 is repeated for
Prop. 9. In **A** there is, beside Fig. 8 in the margin, the following note: صورة
الشكل التاسع وبرهانه مثل صورة الشكل الثامن وبرهانه ولذلك لم تتصور صورته (The
diagram of the ninth proposition and its proof are similar to the diagram of the
eighth proposition and its proof; and therefore its diagram is not drawn.) Since the
radii are equal and since the line drawn from the pole, also to the circumference of
the circle, are equal *add.* **A**. And its diagram is like its diagram. Q.E.D. *add.* **H**.

[1]From here on all the Greek proposition-numbers are smaller by one than the
numbers in this edition. Thus our Prop. 10 is Prop. 9 in Greek.

[2]In MSS **PAN**, and in the MS followed by Czinczenheim for the figures, G is
put on line DB, at the intersection with the sphere. Although this is misleading,
we have copied this in our figure.

[3]This follows from $D\hat{E}A = D\hat{E}B$ and $AD = DB$, since $\triangle$s DAE and DBE are
therefore congruent. See n. 10 on Prop. 1.

Similarly all the lines drawn from E to ABG are equal

$\therefore$ E is cen. $\odot ABG$

I say also: Z is the other of the two poles of $\odot ABG$.

Since $EA = EB$ and ZE is common [to $\triangle$s EAZ EBZ] and $\perp$ [EA EB],

$\therefore$ base $AZ =$ base ZB [Euclid I 4]

Similarly all the straight lines drawn from Z to the circumference of ABG are equal

$\therefore$ Z is the other of the two poles of $\odot ABG$

We have proved that E is the centre of $\odot ABG$

$\therefore$ E is the centre of $\odot ABG$ and Z is the other of the two poles of $\odot ABG$. Q.E.D.

11. When there is a circle in a sphere, the straight line that passes through its poles is perpendicular to it and passes through its centre and through the centre of the sphere.

Let the circle in a sphere be $\odot ABGD$; let its poles be E Z; join EZ

I say: $EZ \perp \odot ABGD$ and passes through its centre and the centre of the sphere.

In plane $\odot ABGD$ let it pass through H

From H draw AHG BHD[1]

Join BE ED BZ ZD

Since $EB = ED$ and EZ is common [to $\triangle$s EBZ DEZ],

$\therefore$ BE $EZ = DE$ EZ resp.

And base $BZ =$ base ZD

$\therefore$ $B\hat{E}Z = D\hat{E}Z$ [Euclid I 8]

Also, since $BE = DE$ and EH is common [to $\triangle$s BEH DEH],

$\therefore$ BE $EH = DE$ EH resp.

And $B\hat{E}H = D\hat{E}H$

$\therefore$ base $BH =$ base DH and $\triangle BEH \equiv \triangle EDH$ [Euclid I 4];

and all the angles $=$ all the angles that the equal sides subtend

$\therefore$ $D\hat{H}E = B\hat{H}E$

When a straight line stands on a straight line, making the two angles equal on the two sides, then each of the equal angles is right [Euclid I Def. 10][2]

$\therefore$ $EH \perp DB$

---

[1] Instead of AHG BHD **A** has: AH GH, and let AH be on a straight line with HG; from point H let lines HB HD also be drawn; and let HB be on a straight line with HD.

[2] When … right: not in Greek.

Similarly $EH \perp AG$

$\therefore [EH]^3 \perp$ plane thru $BD\ AG$ [Euclid XI 4], i.e. plane $\odot ABGD$

$\therefore EHZ \perp \odot ABGD.^4$

I say also: It passes through the centre of the circle and through the centre of the sphere[5].

For $\odot ABGD$ is in a sphere and the perpendicular EH is drawn to it from one of its poles, E, meeting its plane at H

$\therefore H$ is the centre of $\odot ABGD$ [I 10].

I say: It passes through the centre of the sphere.

$\odot ABGD$ is in a sphere and a perpendicular line EHZ is drawn from its centre to the plane of the circle

$\therefore$ the centre of the sphere is on EHZ [I 2 Cor.]

$\therefore EZ$ passes through the centre of the sphere.

$\therefore EHZ \perp \odot ABGD$ and passes through its centre and the centre of the sphere. Q.E.D.

12. Great circles on a sphere bisect each other.

Let there be two great circles on a sphere, $AB\ GD$, intersecting at $E\ Z$

I say: Circles $AB\ GD$ bisect each other.

Mark their centre H; this point is the centre of the sphere

Join $EH\ HZ$

Since $E\ H\ Z$ are in plane $\odot AB$ and also in plane $\odot GD$,

$E\ H\ Z$ are in the planes of both $\odot$s $AB\ GD$

$\therefore E\ H\ Z$ are on their common section

The common section of any two planes is a straight line

$\therefore$ line EHZ is straight

Since H is the centre of $\odot AB$,

EHZ is a diameter of it

$\therefore$ each of lines $EAZ\ EBZ$ is the arc of a semicircle

Since H is also the centre of $\odot GD$,

EHZ is diameter of it

$\therefore$ each of $EGZ\ EDZ$ is the arc of a semicircle

$\therefore \odot$s $AB\ GD$ bisect each other. Q.E.D.

---

[3]**H** *add.:* and line EH is perpendicular to lines $BD\ AG$; and when a line is perpendicular to two lines in one plane, then it is perpendicular to the plane that passes through them.

[4]$\therefore$ line EH passes through the centre of $\odot ABGD$ *add.* **H**.

[5]and … sphere: not in Greek.

13. Those of the circles on a sphere that bisect each other are great circles.

Let ⊙s *AB GD* in a sphere bisect each other at *E Z*
I say: ⊙s *AB GD* are great.
      Join their common section, *EZ*
      ∴ *EZ* is diameter of ⊙s *AB GD*
      Bisect *EZ* at *H*
      ∴ *H* is the centre of ⊙s *AB GD*
I say: It is also the centre of the sphere.
      Through *H* let *HT* ⊥ plane ⊙*GD* and *HK* ⊥ plane ⊙*AB*
      Since *GD* is on a sphere and a line, *HT*, is drawn from its centre
         at right angles to the plane of the circle, the centre of the
         sphere is on line *HT* [I 2 Cor.]
      Similarly it is also on line *HK*
      ∴ the centre of the sphere is on the common section of lines *HT*
         HK
      The common section of the two is *H*
      ∴ *H* is the centre of the sphere
      Circles that pass through the centre of the sphere are great [Def.]
      ∴ ⊙s *AB GD* are great. Q.E.D.

14. When a great circle in a sphere cuts another circle, of the circles on the sphere, at right angles, it bisects it and passes through its poles.

Let great ⊙*ABGD* cut another of the circles on the sphere, ⊙*EBZD*,
      at right angles
I say: It bisects it and passes through its poles.
      Join their common section, line *BD*
      Let centre ⊙*ABGD* be *H*; it is also the centre of the sphere [I
         Def.]
      From *H* draw *HT* ⊥ *BD*; and produce it in both directions to
         meet the surface of the sphere at *A G*
      Since each of the planes stands on the other at right angles, i.e.
         plane ⊙*ABGD* and plane ⊙*EBZD*, and line *TA* is erected at
         right angles on their common section, *BD*, and it is in one of
         the planes, i.e. plane ⊙*ABGD*,
      ∴ *AG* ⊥ plane ⊙*EBZD* [Euclid XI 4]
      Since ⊙*EBZD* is in a sphere and from the centre of the sphere
        ⊥ʳ *HT* is drawn to it, and it meets plane ⊙*EBZD* at *T*,
      ∴ *T* is the centre of ⊙*EBZD* [I 1 Cor.]

∴ each of $\widehat{BED}$ $\widehat{BZD}$ is a semicircle
∴ $\odot ABGD$ bisects $\odot EBZD$.

I say: It also passes through its poles.

That is because $\odot EBZD$ is on a sphere and $\perp^{\mathrm{r}}$ HT is drawn
to it from the centre of the sphere and is produced in both
directions, meeting the sphere at A G
When there is a circle in a sphere and a $\perp^{\mathrm{r}}$ is drawn to it from
the centre of the sphere, and it is produced in both directions,
it falls on its poles[1] [I 8]
∴ A G are the poles of the circle
∴ $\odot ABGD$ passes through poles $\odot EBZD$.

But it has already bisected it
∴ $\odot ABGD$ bisects $\odot EBZD$ and passes through its poles. Q.E.D.

15. When there is a great circle in a sphere and it bisects some non-
great circle, of those that are in the sphere, then it cuts it at right
angles and passes through its poles.

Let the great circle in the sphere be $ABGD$ and let it bisect some
non-great circle, of those that are on the sphere, $\odot EBZD$
I say: It cuts it at right angles and passes through its poles.

Join their common section: BD
Since $\odot ABGD$ bisects $\odot EBZD$, each of $\widehat{BED}$ $\widehat{BZD}$ = $\frac{1}{2}\odot$[1]
Bisect BD at T
∴ T is the centre of $\odot EBZD$
Let centre of $\odot ABGD$ be H; it is also the centre of the sphere
Join HT and produce it in both directions, meeting the surface
of the sphere at A G
Since $\odot EBZD$ is in a sphere and line HT joins its centre and the
centre of the sphere, $HT \perp \odot EBZD$ [I 7]
∴ all planes through $HT \perp \odot EBZD$ [Euclid XI 18]
One of the planes passing through HT is $\odot ABGD$
∴ $\odot ABGD \perp \odot EBZD$
∴ $\odot ABGD$ cuts $\odot EBZD$ at right angles

I say: It passes through its poles.

---

[1]When ... poles: not in Greek.
[1]Greek has here, after $\frac{1}{2}\odot$, "∴ BD is a diam. of $\odot BZDE$", which is missing
from the Arabic.

For since *EBZD* is on a sphere and ⊥ʳ*HT* is drawn to it from
the centre of the sphere and it is produced in both directions
to meet the surface of the sphere at *A G*, *A G* are the poles of
⊙*EBZD* [I 8]
∴ ⊙*ABGD* passes through poles ⊙*EBZD*

And it already cuts it at right angles
∴ ⊙*ABGD* cuts ⊙*EBZD* at right angles and passes through its poles.
Q.E.D.

16. When a great circle in a sphere cuts some circle, of those that are
in the sphere, and passes through its poles, then it bisects it at right
angles.

Let great ⊙*ABGD*, which is on the sphere, cut some circle ⊙*EBZD*, of
the circles on the the sphere, and go through its poles
I say: It bisects it at right angles.

Let poles ⊙*EBZD* be *A G*
A G are on ⊙*ABGD*, for *ABGD* passes through poles ⊙*EBZD*;
join *AG*
∴ ⊙*EBZD* is on a sphere and a straight line is drawn in the
sphere passing through its two poles: line *AG*
When there is a circle on the sphere, the straight line passing
through its poles ⊥ the circle; and it passes through its centre
and through the centre of the sphere [I 11]
∴ *AG* ⊥ ⊙*EBZD*
∴ all planes passing through *AG* ⊥ ⊙*EBZD* [Euclid XI 18]
One of the planes passing through *AG* is ⊙*ABGD*
∴ ⊙*ABGD* cuts ⊙*EBZD* at right angles,
∴ it also bisects it [I 14]

And it already cuts it at right angles
∴ ⊙*ABGD* bisects ⊙*EBZD* at right angles. Q.E.D.

17. When a great circle is in a sphere, then the line drawn from its
pole to its circumference is equal to the side of the square drawn in
the great circle.

Let the great circle in the sphere be ⊙*ABGD*
I say: A straight line from its pole to its circumference = si.sq.gt.⊙ .
Draw two diameters of circle *ABGD* cutting each other at right angles:
AG BD

Since $\odot ABGD$ is great, its centre and the centre of the sphere are the
      same [I 6]: E[1]
$\therefore$ E is the centre of the sphere and the centre of $\odot ABGD$
Erect from E on plane $\odot ABGD$ a perpendicular to the $\odot$: EZ. Let it
      meet the surface of the sphere at Z
$\therefore$ Z is pole $\odot ABGD$ [I 8]
Join ZA AB: AB is the side of the square drawn in $\odot ABGD$; and ZA
      goes from the pole to the circumference of the circle
I say: $ZA = AB$.

    Since $ZE \perp \odot ABGD$, it produces right angles with all straight
            lines from its end in plane $\odot ABGD$ [Euclid XI Def. 3]
    $\therefore$ $ZE \perp$ each of AE EB EG ED
    Since E is the centre of the sphere, $EB = EZ$
    And EA is common [in $\triangle$s EBA EZA]
        $\therefore$ EB $EA = EA$ EZ respectively[2]
    And right angles $B\hat{E}A = A\hat{E}Z$
        $\therefore$ base $BA =$ base AZ [Euclid I 4]
    ZA is from pole $\odot ABGD$ to its circumference and AB is si.sq.gt.$\odot$
            $ABGD$
    $\therefore$ line from pole $\odot ABGD$ to its circumference $=$ si.sq.gt.$\odot$.
            Q.E.D.

18. When there is a circle in a sphere and a line from its pole to its
circumference is equal to the side of a square drawn in a great circle,
then the circle is great also.

Let $\odot ABG$ be in a sphere and let its pole be D. Let line DG, drawn
      from its pole to its circumference, be equal to si.sq.gt.$\odot$
I say: $\odot ABG$ is great.

    Draw the plane passing through DG and the centre of the sphere,
            producing a section that forms a great circle in the surface of
            the sphere [I 6]: $\odot BDGE$
    Let the common section of [$\odot BDGE$] and $\odot ABG$ be BG
    Join DB – and so $\overline{DB} = \overline{DG}$ [from definition of pole]
    Since $DB = DG$ and GD is si.sq.gt.$\odot$,
        $\therefore$ BD is also si.sq.gt.$\odot$
        $\therefore$ each of $\hat{BD}$ $\hat{DG} = \frac{1}{4}\odot$

---

[1]Since ... E: not in Greek.
   [2]$\therefore$ ... resp.: so also in Greek. Logically, "EB $EA = EZ$ EA resp." would be
better.

$\therefore B\widehat{D}G = \frac{1}{2}\odot$, and so BG is the diameter of $\odot DEG$

Since great $\odot DEG$ is in a sphere and cuts one of the circles in
the sphere, $\odot ABG$, and passes through its poles,
it also bisects it [I 16][1]

$\therefore \odot$s ABG DEG bisect each other

And $\odot$s bisecting each other in a sphere are great [I 13]

$\therefore \odot ABG$ is great also. Q.E.D.[2]

19. How to find a line equal to the diameter of a given circle in a sphere.

Let the given circle in a sphere be $\odot ABG$. We want to draw a line
equal to its diameter[1].

Mark on ABG three points, however they fall: A B G

From three straight lines construct $\triangle DEZ$ so that $DE = AB$, $DZ =$
AG and also $EZ = BG$

Imagine: AG GB BA are joined

From E Z on DE DZ draw two lines at right angles: EH ZH[2]

Draw DH

I say: DH = diam. $\odot ABG$.

Imagine diam. $\odot AT$; join AT GT

Since AB $BG = DE$ EZ respectively [in $\triangle$s ABG DEZ], and
bases $AG = DZ$,

$\therefore A\widehat{B}G = D\widehat{E}Z$ [Euclid I 8]

But $A\widehat{B}G = A\widehat{T}G$, since they are on the same section, $\overparen{AG}$, of
the circle

---

[1]Since DB ... bisects it. **H** has: which is the side of the square that is drawn
in $\odot ABG$ and line BD is the side of the square drawn in $\odot ABG$; because $\odot EBDG$
[passes] through the pole of $\odot ABG$, $\therefore$ it bisects it at a right angle; $\therefore$ $B\widehat{A}G$ is
a semicircle; $\therefore$ line BG is a diameter of $\odot BAG$; and since each of lines BD DG
is the side drawn in it, the square of BG is equal to the square[s] of BD DG;
$\therefore$ $B[\widehat{D]}G$ is right; $\therefore$ $\overparen{BDG}$ is a semicircle.

[2]Since $DB = DG$ ... Q.E.D. These lines are different in Greek: $DG = DB$; and
each of DG DB = si.sq.gt.$\odot$. $\therefore$ BDG is the arc of a $\frac{1}{2}\odot$ $\therefore$ BG is diam. $\odot DE$.
And since D is pole $\odot ABG$, $\odot DBEG$ cuts $\odot ABG$ through the poles. Since in a
sphere great $\odot DE$ cuts a circle on the sphere, ABG, through the poles, it bisects
it at right angles. Their common section is BG: $\therefore$ BG is diameter of $\odot ABG$. But
it is also a diameter of the sphere. $\therefore$ ABG is a great circle.

[1]Theodosius copies onto a separate plane a triangle inscribed in the circle and
constructs the diameter on that; and we "imagine" the diameter etc. of the orig-
inal circle. It may be that he is thinking of the practical difficulties of making
constructions within the sphere. Similar remarks apply to the next proposition.

[2]I.e. $D\widehat{E}H = D\widehat{Z}H$ = right.

And $D\hat{E}Z = D\hat{H}Z$, since $D\ E\ H\ Z$ are concyclic[3]
∴ $A\hat{T}G = D\hat{H}Z$
And right angles $D\hat{Z}H = A\hat{G}T$
    ∴ $AGT\ DHZ$ are two triangles, $A\hat{T}G$ and $A\hat{G}T$ of one of
    them $= D\hat{H}Z\ D\hat{Z}H$ of the other, respectively, and side AG of
    one of them subtending one of the equal angles $=$ side DZ, its
    counterpart
∴ all sides $=$ all sides, respectively [Euclid I 26]
∴ $AT = DH$ and AT is a diameter of $\odot ABG$
∴ DH is a diameter of $\odot ABG$. Q.E.F.

20. How to find a line equal to the diameter of a given sphere.

Imagine the sphere for which we want to draw a line equal to its dia-
    meter
On the surface of the sphere mark points $A\ B$, however they fall
About pole A and with distance AB draw $\odot BGD$
It is possible to draw a line equal to the diameter of $\odot DBG$ [I 19]: ZH
From the three straight lines, of which two $=$ two lines from the pole
    to [the circumference of] the circle and one $=$ diam. $\odot [DBG]$,
    construct $\triangle EZH$: each of $ZE\ EH =$ line from pole A to the
    circumference of $\odot DBG$, and $ZH =$ the diameter [of $\odot DBG$]
From $Z\ H$ on $EZ\ EH$ draw two lines at right angles, $ZT\ HT$ [i.e. $E\hat{Z}T$
    $= E\hat{H}T =$ right]; join ET
I say: $ET =$ diameter of the sphere.

    Imagine the diameter of the sphere $= AK$. Through AK let a
        plane pass, producing a section, a great circle, i.e. $\odot ABD$
    Join $AB\ BD\ AD\ DK$
    Since $AB\ BD = EZ\ ZH$ respectively [in $\triangle$s $ABD\ EZH$], and
        bases $AD = EH$,
        ∴ $A\hat{B}D = E\hat{Z}H$ [Euclid I 8]
    But $A\hat{B}D = A\hat{K}D$ and $E\hat{Z}H = E\hat{T}H$ [since $E\ Z\ T\ H$ concyclic]
        ∴ $A\hat{K}D = E\hat{T}H$
    And right angles $A\hat{D}K = E\hat{H}T$
        ∴ in $\triangle$s $AKD\ ETH$, $A\hat{D}K\ D\hat{K}A = E\hat{T}H\ T\hat{H}E$ respectively[1]

---

[3]Since D ... concyclic: not in Greek. The points are concyclic because $D\hat{E}H =$
$D\hat{Z}H =$ right.

[1]$E\hat{T}H\ T\hat{H}E$: probably this should read "$T\hat{H}E\ E\hat{T}H$", as indicated by the
Greek.

and side AD of one, that subtending one of the equal angles,
= side EH, its counterpart of the other
∴ all the sides = all the sides respectively [Euclid I 26]
∴ $AK = ET$, and AK is diameter of the sphere
∴ ET is diameter of the given sphere. Q.E.D.

21. How to draw a great circle through two given points on the surface
of a sphere.

Let the given points on the surface of the sphere be A and B. We want
to draw a great circle through them
If the two points are on a diameter of the sphere, it is clear that infinite
great circles may be drawn through A and B
If A B are not on a diameter of the sphere, draw $\odot EGD$ about pole A
and with distance = si.sq.gt.$\odot$
∴ $\odot EGD$ is great, since a straight line from the pole of a circle
to its circumference = si.sq.gt.$\odot$ [I 18]
Draw also $\odot EZH$ about B and with dist. of si.sq.gt.$\odot$
$\odot EZH$ is great, since a straight line from its pole to its circumference = si.sq.gt.$\odot$
Join E with A B: EA EB
Each of EA EB = si.sq.gt.$\odot$
∴ $EA = EB$
∴ the circle about pole E and with distance EB also passes
through A, since $EA = EB$
Let it be drawn: $\odot ABT$
ABT is great, since a straight line from its pole to its circumference = si.sq.gt.$\odot$
∴ great circle ABT is drawn and passes through the given
points A B on the surface of the sphere. Q.E.F.

22^1. How to find the pole of a given circle on a sphere.

---

[1] After this proposition two more follow, though enclosed in square brackets, in
Heiberg and Czinczenheim; Nizze also considers them spurious. The first of these is
the theorem that if a line in a sphere through the centre bisects a line not passing
through the centre, it also cuts it at right angles, and conversely, reference being
made to the proof ἐν τοῖς κυκλικοῖς [Euclid III 18]. The second proposition is a
repetition of Prop. 7.

Let the given circle on a sphere be $\odot ABG$. We want to find its pole
On the circumference we mark point A, however it falls
Cut off equal arcs $AD\ AE$
Bisect the remaining arc $D\widehat{B}E$ at Z
$\odot ABG$ either is a great circle or it is not.

[i] Let it not be great

> Draw a great circle through the given points $Z\ A$: $\odot ZAT$ [I 21]
> Since $D\widehat{A} = A\widehat{E}$ and $D\widehat{Z} = Z\widehat{E}$, $\therefore$ total $A\widehat{D}Z = $ total $A\widehat{E}Z$
> $\therefore$ $\odot ZAT$ bisects $\odot ABG$
> $\therefore$ great circle AZT on the sphere bisects one of the non-great
> circles on the sphere
> > $\therefore$ it cuts it at right angles and passes through its poles [I 15]
> $\therefore$ $\odot ZAT$ cuts $\odot ABG$ at right angles and goes through its poles
> Bisect $Z\widehat{H}A^{2}$ at H: $\therefore$ H is pole $\odot ABG$.

[ii] Let $\odot ABG$ be great

> Similarly [as above] $A\widehat{D}Z = A\widehat{E}Z$
> Let $A\widehat{D}Z^{3}$ be bisected at G
> > $\therefore$ each of $A\widehat{G}\ G\widehat{Z} = \frac{1}{4}\odot$
> and the circle drawn about pole G with distance GZ also passes
> through A, since A is opposite Z
> Let it be drawn: $\odot ZAT$
> $\odot ZAT$ is great, since a line from its pole to its circumference $=$
> si.sq.gt.$\odot$ [since $AG = GZ = \frac{1}{4}\odot$] and G is pole $\odot ZAT$
> $\therefore$ $\odot ABG$ cuts $\odot ZAT$ and passes through its poles
> $\therefore$ great $\odot ABG$ is on a sphere cutting another circle, of the
> circles on the sphere, $\odot ZAT$, and passes through its poles
> > $\therefore$ it bisects it at right angles [I 15]
> $\therefore$ $\odot ABG \perp \odot ZAT$
> $\therefore$ $\odot ZAT \perp \odot ABG$
> $\therefore$ great $\odot ATZ$ is on a sphere, cutting another circle, ABG, at
> right angles
> > $\therefore$ it bisects it and passes through its poles [I 14]
> $\therefore$ $\odot ATZ$ bisects $\odot ABG$ and passes through its poles
> Bisect $Z\widehat{H}A^{4}$ at H
> > $\therefore$ H is pole $\odot ABG$. Q.E.D.

$^{2}Z\widehat{H}A$: the Greek has ZA. H has not yet been defined.
$^{3}A\widehat{D}Z$: the Greek MSS have $A\widehat{Z}$ or $A\widehat{G}Z$.
^{4}Again, Greek has ZA.

Book II

[Def.] It is said that circles on a sphere touch each other when the common section of their planes touches both circles.

1. The poles of parallel circles on the sphere are the same.

Let the parallel circles *ABG DEZ* be on the sphere
I say: The poles of each of ⊙s *ABG DEZ* are the poles of the other.
> Let the poles of ⊙*ABG* be *H T*; join *HT*
> When there is a circle on a sphere, the line that passes through
> its poles ⊥ it [the circle] and passes through its centre and
> through the centre of the sphere [I 11]
> ∴ *HT* ⊥ ⊙*ABG* and goes through its centre and the centre of
> the sphere
> Since *HT* ⊥ ⊙*ABG* and ⊙*ABG* ∥ ⊙*DEZ*, *HT* ⊥ ⊙*DEZ*[1]
> And since ⊙*DEZ* is on a sphere and ⊥[r] *HT* is drawn to it from
> the centre of the sphere and is produced on both sides and
> meets the surface of the sphere at *H T*,
> ∴ *H T* are poles ⊙*DEZ* [I 8]
> And they are also poles ⊙*ABG*
> > ∴ the poles of each of ⊙s *ABG DEZ* are the poles of the
> > other. Q.E.D.

2. Circles on a sphere about common poles are parallel.

Let ⊙s *ABG DEZ* be on a sphere on poles *H T*
I say: ⊙*ABG* ∥ ⊙*DEZ*.
> Join *HT*
> Since *ABG* is on a sphere and a line is drawn passing through
> its poles, *HT*,
> > ∴ *HT* ⊥ ⊙*ABG* [I 11]
> Similarly [HT] ⊥ *DEZ*
> And planes on which the same line falls and is perpendicular to
> them, when produced, do not meet [Euclid XI 14][1]

---

[1] A converse of Euclid XI 14. It also follows from Euclid XI 16 – see Heath [1956], III 301.

[1] when … meet: sc. are parallel. The expression is in Theodosius' Greek. It is not in the enunciation of XI 14 Heiberg's Greek of the *Elements*, though it is in the body of the theorem. But the expression is in the enunciation in the Latin texts of the *Elements* by Adelard of Bath, Robert of Chester (attrib.), Gerard of Cremona and Campanus. Euclid's definition of parallel planes is (XI Def. 8): those which do not meet.

∴ when planes ⊙s *ABG DEZ* are produced, they do not meet:
  ∴ ⊙*ABG* ∥ ⊙*DEZ*. Q.E.D.

3. When there are two circles on a sphere cutting the circumference of any of the great circles from among the circles that are on it at one and the same point and the two [circles]'s poles are on that circle, then the [two circles] touch.

Let ⊙s *ABG DEG* cut the circumference of great ⊙*AGE* at the same point, *G*; and let their poles be on ⊙*AGE*
I say: ⊙s *ABG DEG* touch.

> Let the common section of ⊙*AGE* and ⊙*ABG* be line *AG*
> Let the common section of ⊙*AGE* and ⊙*GDE* be line *GE*
> Let the common section of ⊙*ABG* and ⊙*GDE* be line *ZGH*
> Great ⊙*AGE* cuts another ⊙, *ABG*, and goes through its poles
>   ∴ it bisects it at right angles [I 16]
> ∴ line *AG* is a diameter of ⊙*ABG*
> Similarly *GE* is a diameter of ⊙*GDE*
> Since ⊙*AGE* stands at right angles to ⊙s *ABG GDE*,
>   each of ⊙*ABG* and ⊙*GDE* ⊥ ⊙*AGE*
> ∴ their [*sc.* ⊙s *ABG GDE*'s] common section ⊥ plane ⊙*AGE*
>   [Euclid XI 19]
> For when two planes stand on one plane at right angles, their
>   common section also ⊥ that plane itself [Euclid XI 19]
> ∴ it ⊥ all the straight lines that emerge from its end in plane
>   ⊙*AGE* [Euclid XI Def. 3]
> Now *AG GE* emerge from its end in plane ⊙*AGE*
> ∴ *ZH* ⊥ *AG*, *GE*
> Since from the end of diam. ⊙*ABG* line *ZH* is drawn at right
>   angles,
>   it will touch ⊙*ABG* at *G* [Euclid III 16 Por.]
> Similarly *ZH* touches ⊙*GDE* also at *G*
> And the circles that are said to touch each other in a sphere are
>   those the common section of whose planes touch them together
>   [II Def.]
> And *ZG* touches the two circles at *G* together
> ∴ ⊙s *ABG GDE* touch each other. Q.E.D.

4. When two circles on a sphere touch each other, the great circle passing through their poles also goes through their point of contact.

Let ⊙s *ABG GDE* on a sphere touch at *G*

Let *Z* be a pole of ⊙*ABG* and *H* a pole of ⊙*GDE*

I say: The great circle passing through poles *Z H* passes also through *G*.

> If it were possible, let [the great circle through poles *Z H*] not
> pass through [*G*]: let it be, e.g., ⊙*ZBH*
>
> About pole *H* and with distance *HB* draw ⊙*BKT*
>
> ∴ ⊙*GDE* ∥ ⊙*BKT* – for they are about the same pole [II 2]
>
> Since ⊙s *ABG BKT* are on a sphere cutting the circumference
> of some great circle , line *ZBH*, at *B*, and the poles are on that
> circle
>
> ∴ ⊙s *ABG BKT* touch each other [II 3]
>
> And they cut each other ※
>
> For it is not possible that the great through *Z H* does not pass
> through *G*
>
> ∴ [enunciation for ⊙s *ABG GDE*]. Q.E.D.

5. When two circles on a sphere touch, then the great circle passing through the poles of one of the two circles and the place of contact passes also through the poles of the other circle.

Let ⊙s *ABG GDE* on a sphere touch at *G*

Let pole ⊙*ABG* be *Z* and pole ⊙*GDE* be *H*

I say: The great circle passing through *Z G* also passes through *H*.

> If that were not so, and something else were possible, let it be
> drawn and let it be, e.g., ⊙*ZGT*
>
> Let another great ⊙ be drawn passing through poles *Z H* – so
> that it passes through *G* [II 4]
>
> Since each of ⊙s *ZGH ZGT* is great, each bisects the other [I 12]
>
> ∴ each of arcs *ZKG ZLG* is a semicircle
>
> ∴ *ZG* is a diameter of the sphere, since it is diameter of great
> circles *ZGH ZGT*
>
> But it also emerges from pole ⊙*ABG* ※
>
> ∴ the great circle passing through *Z G* passes also through *H*.
> Q.E.D.

6. When a great circle in a sphere touches another circle, of the circles on the sphere, it touches another circle equal to it and parallel to it.

Let great circle ABG on a sphere touch another circle in the sphere, GD, at G

I say: $\odot ABG$ touches another circle, equal and parallel to $\odot GD$.

Mark E, pole $\odot GD$ [I 22]

Draw a great circle passing through $G\ E$ [I 21], $\odot GEDBZH$

Cut off from it $\widehat{BZ} = \widehat{GE}$

Draw, about pole Z and with distance ZB, $\odot BH$

Since $\odot$s $ABG\ GD$ touch on a sphere and a circle is drawn on the sphere passing through E, pole $\odot GD$, and through the place of contact, $\odot GEDBZH$,

$\odot GEDBZH$ passes also through the poles of $\odot ABG$ [II 5]

And since $\odot$s $ABG\ BH$ on a sphere cut the circumference of another great $\odot$, at the same point, B, and their poles are on the circle,

$\therefore$ $\odot$s $ABG\ BH$ touch each other [II 3]

And since $\widehat{GE} = \widehat{BZ}$ and $\widehat{EB}$ is common, total $\widehat{GEB}$ = total $\widehat{EZ}$

And $\widehat{GEB}$ is a semicircle [I 12]: $\therefore$ EZ is a semicircle

$\therefore$ E is opposite Z

And E is pole $\odot GD$

Also, since $\widehat{EZ}$ is a semicircle and Z is pole $\odot BH$,

E is also pole $\odot BH$

$\therefore$ the $\odot$s $GD\ BH$ are about the same poles

and circles about the same poles are parallel [II 2],

$\therefore$ $\odot GD \parallel \odot BH$

And since $GE = BZ$, $\odot GD = \odot BH$; and it also $\parallel [BH]$,

$\therefore$ [enunciation with diagram letters] Q.E.D.

7. When there are two equal and parallel circles on a sphere, then a great circle that touches one of them touches the other also.

Let there be two equal, parallel circles on a sphere: $AB\ GD$

I say: A great circle that touches $\odot AB$ also touches $\odot GD$.

If it is possible that it is not so, let great $\odot AE$ touch $\odot AB$ at A and let it not touch $\odot GD$

Since great $\odot AE$, which is on the sphere, touches some circle, of the circles that are on the sphere, $\odot AB$, then it touches another circle, equal and parallel to $\odot AB$ [II 6]

Let it touch $\odot EZ$

$\therefore$ $\odot AB$ = and $\parallel \odot EZ$

And $\odot AB$ already = and $\parallel \odot GD$

∴ on the same sphere three ⊙s are equal and parallel ✳ [I 6]
∴ it is not possible that a great circle which touches AB does
    not touch ⊙GD
∴ it touches it. Q.E.D.

8. If a great circle on a sphere inclines to another circle, of the circles
on the sphere, it touches two equal circles parallel to the other circle
mentioned before.

Let ABG be the great circle on the sphere be inclining on one of the
circles that are on the sphere, ⊙BD, i.e. it does not pass through the
poles of ⊙BD[1]
I say: ⊙ABG touches two equal circles parallel to ⊙BD.
Since ⊙ABG is inclined to ⊙BD, pole ⊙BD is not on ⊙ABG

Mark pole ⊙BD: E
Draw a great circle through E and through poles ⊙ABG: ⊙AEH
About pole E and with distance EA draw ⊙AZ
∴ ⊙AZ ∥ ⊙BD – this is because the two are about the same
    poles [II 1]
Since ⊙s ABG AZ, which are on the sphere, cut the circumference
    of a great circle, from the circles on the sphere, ⊙$AEZH$[2], at
    the same point, A, and the poles of the two are on it,
    the two circles touch [II 3]
∴ ⊙ABG touches ⊙AZ
Since great ⊙ABG is on a sphere and touches some circle, of the
    circles on the sphere,
    ∴ it touches another circle equal and parallel to ⊙AZ [II 6]
∴ let it touch ⊙HG
Since AZ = and ∥ ⊙GH and ⊙AZ ∥ ⊙BD,
    ∴ ⊙GH ∥ ⊙BD
∴ ⊙ABG touches two equal ⊙s parallel to ⊙BD. Q.E.D.

9. When there are two intersecting circles on a sphere and a great circle
is drawn passing through their poles, then it divides the sections cut
off from the circles into halves.

Let ⊙s $ZAEB$ $ZGED$ on the sphere intersect at E Z

---

[1]i.e. ... ⊙BD: Czinczenheim puts "i.e. ... poles" in square brackets; Heiberg
says it seems to be interpolated; "of ⊙BD" is apparently an explanatory addition
in Arabic.
[2]⊙$AEZH$: not in Greek.

Draw a great $\odot$ passing through their poles: $\odot AGBD$

I say: $\odot AGBD$ bisects the sections cut off from the circles,

   i.e. $\overset{\frown}{ZA} = \overset{\frown}{AE}$, $\overset{\frown}{ZB} = \overset{\frown}{BE}$, $\overset{\frown}{ZG} = \overset{\frown}{GE}$, $\overset{\frown}{ZD} = \overset{\frown}{DE}$.

   Let the common section of $\odot$s $AGBD$ $ZAEB$ be AB

   Let the common section of $\odot$s $AGBD$ $ZGED$ be GD

   Join ZH HE[1]

   Since Z H E are in plane $\odot ZAEB$ and also in plane $\odot ZGED$,
      they are on the common section of the planes of the first two
      circles

   The common section of any [two] planes is a straight line
      $\therefore$ ZH joins HE in a straight line

   $\odot AGBD$ is a great circle on a sphere and cuts another circle on
      the sphere, $\odot ZAEB$, and passes through its poles
      $\therefore$ it bisects it at right angles [I 16]

   $\therefore$ AB is a diameter of $\odot ZAEB$

   Similarly GD is a diameter of $\odot ZDEG$

   Since $\odot AGBD \perp$ both $\odot$s $ZAEB$ $ZDEG$,
      $\therefore$ each of $\odot$s $ZAEB$ $ZDEG \perp \odot AGBD$

   When [the planes of] two intersecting $\odot$s $\perp$ some plane, the com-
      mon section of the two also $\perp$ that plane [Euclid XI 19 – of
      planes in general],
      The common section of $\odot$s $ZAEB$ $ZDEG \perp$ plane $AGBD$; and
      their common section is ZHE

   $\therefore$ $ZHE \perp \odot AGBD$

   $\therefore$ it forms a right angle with all straight lines emerging from a
      point of it in plane $\odot AGBD$ [Euclid XI Def. 3]

   Now AB GD, which are in plane $\odot AGBD$, emerge from H (in
      line ZHE)

   $\therefore$ $ZHE \perp$ both AB GD and both AB $GD \perp ZHE$

   Since a line is drawn in [plane] $\odot ZAEB$ passing through the
      centre, line AB, cutting another line not passing through the
      centre, ZHE, at right angles, $\therefore$ it bisects it [Euclid III 3]

   $\therefore$ $ZH = HE$

   HA is common to [ZH HE, of $\triangle$s ZHA EHA] and $\perp$ them
      $[\therefore \ \overline{ZA} = \overline{AE}]$

   $\therefore$ $\overset{\frown}{ZA} = \overset{\frown}{AE}$

   Similarly $ZB = BE$, $ZG = GE$, $ZD = DE$

$\therefore$ $\odot AGBD$ divides the sections that are cut off from the two circles
in half. Q.E.D.

---

[1] H must be the intersection of AB and GD.

10. If there are parallel circles on a sphere and great circles are drawn passing through their poles, then the arcs of the parallel circles which are between the great circles are similar and the arcs of the great circles which are between the parallel circles are equal.

Let there be two parallel circles on a sphere: *ABGD EZHT*; let *K* be
    their pole
Draw two great circles passing through their poles: *AEHG BZTD*
I say: The arcs of the parallel circles between the great circles are
    similar,
    i.e. $\overarc{BG} \sim \overarc{ZH}$, $\overarc{GD} \sim \overarc{HT}$, $\overarc{DA} \sim \overarc{TE}$, $\overarc{AB} \sim \overarc{EZ}$.
I say also: The arcs of the great circles between the parallel circles are
    equal,
    i.e. $\overarc{ZB} = \overarc{HG} = \overarc{TD} = \overarc{EA}$.

    Let common section of $\odot ABGD$ and $\odot AEHG$ be AG
    Let common section of $\odot BZTD$ and $\odot ABGD$ be BD
    Let common section of $\odot EZHT$ and $\odot ZKT$ be ZT
    Let common section of $\odot EZHT$ and $\odot EKH$ be HE
    The great $\odot AEHG$, of the circles on the sphere, cuts some circle,
        of the circles on the sphere, $\odot ABGD$, and passes through its
        poles
        ∴ it bisects it at right angles [I 16]
    ∴ AG is a diameter of $\odot ABGD$
    Similarly, BD is a diameter of $\odot ABGD$
    ∴ L is the centre of $\odot ABGD$
    Also, since great $\odot AEHG$, of the circles on the sphere, cuts some,
        of the circles on the sphere, $\odot EZHT$, and passes through its
        poles,
        ∴ it bisects it at right angles [I 16]
    ∴ EH is a diameter of $\odot EZHT$
    Similarly ZT is also a diameter of $\odot EZHT$
    ∴ M is the centre of $\odot EZHT$
    Since planes of $\odot$s $ABGD$ $\odot EZHT$ are parallel and plane $\odot BZTD$
        cuts them,
        their two common sections are also parallel [Euclid XI 16]
    ∴ $BD \parallel TZ$
    Similarly $AG \parallel EH$
    ∴ BL LG, which touch, $\parallel$ ZM MH, which touch; and they are
        not lines in one plane
        ∴ they contain equal angles [Euclid XI 10]

∴ $Z\hat{M}H = B\hat{L}G$; and they are on the two centres; $Z\hat{M}H$ has
  base $\overset{\frown}{ZH}$ and $B\hat{L}G$ has base $\overset{\frown}{BG}$
∴ $BG \sim ZH$
Similarly $GD \sim HT$, $AD \sim ET$, $AB \sim EZ$
∴ [first part of enunciation]

I say also: The arcs of the great circles between the parallel circles are
  equal.

For K is pole $\odot ABGD$: ∴ $\overset{\frown}{KA} = \overset{\frown}{KB} = \overset{\frown}{KG} = \overset{\frown}{KD}$
Also, since K is pole $\odot EZHT$, $\overset{\frown}{KE} = \overset{\frown}{KZ} = \overset{\frown}{KH} = \overset{\frown}{KT}$
∴ the remaining arcs $\overset{\frown}{EA} = \overset{\frown}{ZB} = \overset{\frown}{HG} = \overset{\frown}{TD}$
∴ [second part of enunciation] Q.E.D.

11. When equal sections of circles are constructed on the diameters of
equal circles, standing on them at right angles, and then equal arcs are
cut off from them, starting at[1] the ends of the diameters, those arcs
being less than half of the sections, and when, from the points that are
created in the place of cutting off, equal straight lines are drawn to the
circumferences of the first circles, then they cut off equal arcs from the
first circles starting at the ends of the diameters which I mentioned.

On two of the diameters of the equal $\odot$s ABG DEZ let two equal
  sections of a circle be constructed, standing on them at right
  angles, sections AHG DTZ
Cut off from them equal arcs from the ends of the diameters, i.e. at A
  D $\overset{\frown}{AH}$ $\overset{\frown}{DT}$, less than the halves of AHG DTZ
To the circumferences of $\odot$s ABG DEZ draw two equal straight lines
  HB TE
I say: $\overset{\frown}{AB} = \overset{\frown}{DE}$.

From H T draw perpendiculars to the planes of $\odot$s ABG DEZ
  ∴ they fall on their common sections[2], on AG DZ
Let the perpendiculars be HK TL
Let the centres of $\odot$s ABG DEZ be M N
Join KB MB LE NE

---

[1]at = مما يلى *hic et saepius.*

[2]Ver Eecke gives Euclid XI 38, in the numeration of Peyrard, in support of this
statement. This proposition was regarded by Heiberg and Heath as interpolated.
See Heiberg's edition of the *Elements*, IV, p. 354-5 ("Uulgo XI prop. 38") and
Heath's translation [1956], III 360. Either Theodosius had the proposition to hand
or he regarded it as obvious. See Note to II 11 in **M** etc., below, p. 426, n. 1.

Since $HK \perp$ plane $\odot ABG$, it $\perp$ all lines that meet it and are in plane $\odot ABG$ – it makes right angles with them [Euclid XI Def. 3]

$\therefore$ $H\hat{K}B$ is right

Similarly $T\hat{L}E$ is right

Since sections AHG DTZ are equal, and AH DT, which are cut off, are also equal, and $\perp^{rs}$ HK TL are drawn,

$\therefore$ $AK = DL$ and $HK = TL^3$

Since $BH = TE$ [in $\triangle$s BHK ETL],

$$BH^2 = TE^2$$

But $HK^2 + KB^2 = BH^2$ and $TL^2 + LE^2 = TE^2$

$\therefore$ $HK^2 + KB^2 = TL^2 + LE^2$

But $HK^2 = TL^2$

There remains $KB^2 = LE^2$

$\therefore$ $KB = LE^4$

Since $AM = DN$ and $AK = DL$, $\therefore$ the remaining $KM = LN$

And $BM = EN$

$\therefore$ KM $MB = LN$ NE, respectively [in $\triangle$s KMB LNE]; and bases $KB = LE$

$\therefore$ $K\hat{M}B = L\hat{N}E$ [Euclid I 8]

$\therefore$ $\hat{AB} = \hat{DE}$ [Euclid IV 26]

Similarly[5], when, on the diameters of equal circles, equal sections of circles are constructed, standing on them at right angles, and when equal arcs are cut off from them at the ends of the diameters, less than their halves, and equal arcs are cut off from the first circles on the same side, and from those ends straight lines are drawn joining the points created and the cutting-off places, then those lines are equal.

On the equal $\odot$s ABG DEZ let there be constructed on diameters AG DZ, from among their diameters, two equal sections of circles, standing on them at right angles, sections AHG DTZ

Cut off, starting from the ends of the diameters, A D, two equal arcs, AH DT, < halves of the first sections AHG DTZ

Cut off from the circles equal arcs AB DE, from the ends of the diameters, on the same side

---

[3]For a proof, by considering the plane figures $AHGMK$ $DTZNL$, see Ver Eecke, p. 46, n. 1.

[4]This follows from $BH = TE$, for in $\triangle$s BHK and ETN we have also $HK = TL$ and $\hat{K} = \hat{L} =$ right. See the tenth note on Prop. I 1.

[5]The rest of this proposition is counted as Proposition 12 in Greek.

Join HB TE
I say: $\overline{HB} = \overline{TE}$.

    From H T draw perpendiculars to the planes of $\odot$s ABG DEZ: they will fall on AG DZ, which are the common sections of the planes[6], HK TL

    Let centres of the circles be M N

    Join KM BM LE EN

    Since $A\widehat{B} = D\widehat{E}$, $\therefore$ $A\widehat{M}B = D\widehat{N}E$

    Since sections $AHG = DTZ$, and the arcs cut off $AH = DT$, and $\perp^{\text{rs}}$ HK TL are drawn,

    $\therefore$ $\perp^{\text{rs}}$ $AK = DL$[7] and $HK = TL$

    Since $AM = DN$ and $AK = DL$, there remains $KM = LN$

    And $BM = EN$: $\therefore$ KM $MB = LN$ NE, respectively [in $\triangle$s KMB LNE], and $K\widehat{M}B = L\widehat{N}E$

    $\therefore$ base $KB =$ base LE [Euclid I 4]

    And $HK \perp$ plane $\odot ABG$

    $\therefore$ with all lines which touch it and are in plane ABG it makes right angle [Euclid XI Def. 3]

    KB touches it: $\therefore$ $H\widehat{K}B$ is right

    Similarly TLE is right

    Since $HK = TL$ and $KB = LE$, HK $KB = TL$ LE respectively [in $\triangle$s HKB TLE]; they enclose right angles

    $\therefore$ bases $HB = TE$ [Euclid I 4]. Q.E.D.

12^1. To draw on a sphere a great circle, of the circles of [the sphere], that touches a known circle, and touches it at a known point.

Let there be on a sphere a known circle $<$ a great circle: $\odot AB$
Let the known point on its circumference be B
[*Problem:*] To draw a great circle touching the known $\odot AB$ and passing through B.

    Let G be pole $\odot AB$ [I 22].

    Draw a great circle through G B: $\odot GBD$ [I 21].

    Cut off from it an arc equal to the arc.si.sq.gt.$\odot$: $\widehat{BD}$.

    $\widehat{GB}^2 \neq \frac{1}{4}\odot$, because the line drawn from pole $\odot AB$ to its circumference $\neq$ si.sq.gt.$\odot$

---

[6]See the second note on this proposition.

[7]See note to corresponding statement of the first part of the proposition.

[1]This proposition is Prop. 14 in Greek.

[2]In Arabic, **A** has "arc" and **N** has "side"; **H** omits the term. In Latin, MSS **KgOBPsVa** have "arcus" and **PRVMFiZ** have "latus".

That is because $\odot AB$ would then be great [I 18] and that is
not so

$\therefore\ \widehat{BG} \neq \frac{1}{4}\odot$, but $< \frac{1}{4}\odot^3$

On pole D and with distance DB draw $\odot EBZ$

$\therefore\ \odot EBZ$ is a great $\odot$. That is because the line drawn from its
pole to its circumference $=$ si.sq.gt.$\odot$ [I 18]

And $\odot$s AB EBZ are in a sphere and cut the circumference of
another, great, circle, of the circles in the sphere, GBD, at the
same point, B, and their poles are on it

$\therefore\ $ one of the two circles touches the other [II 3]

$\therefore\ \odot AB$ touches $\odot EBZ$

$\therefore\ $ a great circle is drawn, EBZ, through the known point B and
touching $\odot AB$ at B. Q.E.D.

13. When there are parallel circles on a sphere and when two great circles are drawn in that sphere touching one of those circles and cutting
the remaining circles, then the arcs of the parallel circles between the
halves of the great circles, [i.e. arcs] that do not meet, are similar; and
the arcs of the great circles between the parallel circles are equal.

Let there be in a sphere parallel $\odot$s $ABGD$ $EZHT$ KL

Draw in the sphere two great $\odot$s $AEKHGO^1$ $BZLTDO$, touching one
of those circles, LK, at L and K;

they cut $\odot$s $ABGD$ $EZHT$

I say: The arcs of the parallel circles between the halves of the great
circles that do not meet are similar; and the arcs of the two great
circles between the parallel circles are equal.

We can recognize the arcs that are between the semicircles that
do not meet from what I shall describe,

i.e. that, since great circles in a sphere bisect each other, $R\widehat{KA}O$

$^3GB \ldots < \frac{1}{4}\odot$ is not in the Greek text. A similar argument is given in Scholium
207 (Czinczenheim, p. 406 [Greek] and p. 872 [French]); cf. also Ver Eecke, p. 52.

^{1}In this proposition some of the Arabic and Latin diagram letters have an unusual correspondence with the Greek. First, س (for Greek Ξ) and ش (for Greek
T) are interchanged in our Arabic manuscipts, **ANH**, but not in **P** (according to
the usual representation of Arabic letters by Gerard). The representations of three
other letters may be indicated by the following table:

Gk.	Ar.	Lat.
Π	ث	P
Φ	ف	F
Υ	ت	Y

If ث and ف were interchanged in Arabic (or Π and Φ in Greek), then a later
interchange of ت and ث in Arabic would explain Gerard's lettering.

is a semicircle

$\therefore \ K\widehat{A}O < \frac{1}{2}\odot$

Put $K\widehat{AO}F = \frac{1}{2}\odot^2$

Since $R\widehat{B}O^3$ is also $\frac{1}{2}\odot$, $\therefore \ LR\widehat{B}O^4 > \frac{1}{2}\odot$

Let $LR\widehat{B}Y^5 = \frac{1}{2}\odot$

$\therefore$ the semicircle, drawn from K towards A, $K\widehat{AO}F$, does not meet the semicircle drawn from L towards B, $LR\widehat{B}Y$

Similarly $K\widehat{R}F$, which is a semicircle, does not meet the semicircle drawn from L towards K, $LT\widehat{DO}Y^6$

$\therefore$ the arcs of the parallel circles between the halves of the great circles that do not meet are $\widehat{KL} \ \widehat{EZ} \ \widehat{AB} \ \widehat{HT} \ \widehat{GD}$

I say: $\widehat{KL} \ \widehat{EZ} \ \widehat{AB}^7$ are similar.

I say also: Arcs of great circles between the parallel circles are equal,
i.e. the four $AE \ ZB \ HG \ TD$ are equal,
and the four $KE \ KH \ ZL \ LT$ are equal.

Mark pole of parallel $\odot$s: M

Draw [arcs of] great $\odot$s passing through M and each of $K \ L$: $\odot$s $MKSN \ MLQP$

Since there are two circles on the sphere, $AEKHG \ KL$, touching at K, and a great circle is drawn through the poles of one of the two $\odot$s, KL, and through the place of contact of the other, $\odot MKSN$,

$\therefore \ \odot MKSN$ also passes through the poles of $AEKHG$ and stands at right angles to it [I 15]

Similarly $\odot MLQP$ passes through poles $\odot BZLTDO$ and $\perp$ it

In the equal circles, $AEKHGO \ BZLTDO$, on the diameters drawn from $K \ L$ two equal circular sections, $LM \ MK$, have been constructed, at right angles to them, and the segments connected with these two to complete two semicircles[8]; from them the two equal arcs, $KM \ ML$, are cut off, and they $<$ the halves of the portions made; and $\overline{MA} = \overline{MD}$, for they are both drawn

---

[2]This line is different in Greek.

[3]Latin has RYO. But Y has not yet been defined. Perhaps Gerard worked from a manuscript that had رس with false dots or no dots under the middle letter.

[4]Gerard has $LRYO$.

[5]Gerard has $LZBY$.

[6]drawn ... $LT\widehat{DO}Y$: Latin has LDY.

[7]$\widehat{HT}$ and $\widehat{GD}$ add. **H**.

[8]to complete two semicircles: not in Greek.

from pole $\odot ABD$ to its circumference[9]

$\therefore$ they cut off equal arcs [II 11]: $\widehat{AK} = \widehat{LD}$

Similarly $EK = LT$

Since $\odot$s $ABGD$ $AEKHGO$ are in a sphere, cutting each other, and a great circle, $MKSN$ is drawn through their poles, $\odot MKSN$ bisects the sections cut off [II 9]

$\therefore$ $A\widehat{E}K = K\widehat{H}G$ and $\widehat{AN} = \widehat{NG}$

Similarly $\widehat{BL} = \widehat{LD}$ and $\widehat{BP} = \widehat{PD}$

And since $A\widehat{E}K = L\widehat{T}D$ [proved above] and $AEKG = 2A\widehat{E}K$ and $D\widehat{T}LB = 2L\widehat{T}D$, $\therefore$ $AK\widehat{H}G = D\widehat{T}LB$,

And the $\odot$s are equal, for they are great,

$\therefore$ $\overline{AG} = \overline{DB}$

$\therefore$ $AN\widehat{B}G = B\widehat{P}D$, since the straight lines subtending them are equal, and they belong to the same circle[10]

$\widehat{AN} = \frac{1}{2}AN\widehat{B}G$ and $\widehat{BP} = \frac{1}{2}\widehat{BD}$

$\therefore$ $\widehat{AN} = \widehat{BP}$

Add common arc BN: $\therefore$ total $A\widehat{N}B = N\widehat{B}P$

And they belong to same circle: $\therefore$ $A\widehat{N}B \sim N\widehat{B}P$

But $N\widehat{B}P \sim \widehat{KL}$. For when there are in a sphere parallel circles, and great circles are drawn through their poles, then the arcs of the parallels that are between the great circles are similar [II 10]

And the two arcs of the parallel circles between MN and MP, which are from the great circles passing through their poles, are $\widehat{KL}$, $\widehat{NP}$

And $A\widehat{N}B \sim \widehat{KL}$

And for this reason also $\widehat{KL} \sim \widehat{EZ}$

$\therefore$ $\widehat{EZ} \sim \widehat{AB}$, since they have the same form, for they are between $\odot$s AEG and DZB

$\therefore$ $\widehat{EZ} \sim \widehat{KL}$[11]

Similarly $G\widehat{P}D \sim H\widehat{Q}T$; and this arc $\sim \widehat{EZ}$, for $H\widehat{Q}T \sim \widehat{KL}$[12]

$\therefore$ the arcs of the two parallels that are between the halves of great circles that do not meet are similar.

I also say: Arcs of great circles between parallel circles are equal, i.e. the four arcs AEK KHG BZL LTD are equal,

---

[9]for ... circumference: not in Greek.

[10]since ... same circle: not in Greek.

[11]since they ... $\widehat{KL}$: not in Greek.

[12]and this arc ... $\widehat{KL}$: not in Greek.

from which the four arcs *EK KH ZL LT* are equal.
For great circle *KN* bisects sections *EKH ESH*, which were cut off;
and sections *ZLT ZQT* are similarly divided

∴ $\widehat{EK} = \widehat{KH}$
But clearly $\widehat{EK} = \widehat{LT}$
∴ $\widehat{KH} = \widehat{TL}$
   And $\widehat{TL} = \widehat{LZ}$
∴ $\widehat{LZ} = \widehat{KH}$
∴ the four arcs *EK KH ZL LT* are equal
   and the remaining four arcs *AE BZ GH DT* are equal
∴ arcs of parallel circles between the halves of the great circles
   that do not meet are similar[13] and the arcs of the great circles
   between the parallel circles are equal. Q.E.D.

14[1] When there is in a sphere a known circle smaller than a great circle
and there is a known point on the sphere between the aforementioned
circle and the circle equal and parallel to it, to draw a great circle
passing through the known point and touching the non-great circle.

Let the known circle in the sphere smaller than a great circle be ⊙*AB*,
   and let the known point on the surface of the sphere between
   ⊙*AB* and the equal and parallel circle be *G*
To draw a great circle passing through *G* and touching ⊙*AB*
   Mark pole ⊙*AB*: *D* [I 22]
   About *D* with distance *DG* draw ⊙*GEZH*
   Through *D G* draw great ⊙*DBGT* [I 20]
   From it cut off an arc equal to the arc subtended by the si.sq.gt.⊙:
   $\widehat{BT}$
   About pole *T* with distance *TB* draw ⊙*EBH*
   ∴ ⊙*EBH* is a great ⊙, for the line from pole to circumference
   = si.sq.gt.⊙ [I 18]
   It touches ⊙*AB*. For they cut the circumference of great ⊙*DBGT*
   at one point, *B*, and their poles are on that circumference
   Draw two great circles through *D* and through *E H*: ⊙s *DMEK*
   DNHL
   [From them] cut off $\widehat{EK}$ $\widehat{HL}$, [each] = $\widehat{GT}$

---

13 ∴ arcs ... similar: not in Greek.
   [1]This proposition is Prop. 15 in Greek. From here to the end of the book the
proposition numbers in the Greek text are greater by one than in the Arabo-Latin
tradition.

Since in the sphere $\odot$s *EBH ZEGH* cut each other and a great $\odot$ is drawn through their poles, $\odot DBGT$, bisecting the sections cut off [II 9],

$\therefore$ $\widehat{EG} = \widehat{GH}$, $\widehat{EB} = \widehat{BH}$

Since $\widehat{DE} = \widehat{DG} = \widehat{DH}$ (since they are drawn from *D*, the common pole of the two circles) and [since] $\widehat{DM} = \widehat{DB} = \widehat{DN}$, the remaining arcs $\widehat{ME} = \widehat{BG} = \widehat{NH}$

And $\widehat{EK} = \widehat{GT} = \widehat{HL}$: $\therefore$ $\widehat{MK} = \widehat{BT} = \widehat{NL}$

And $\widehat{BT} = $ arc.si.sq.gt.$\odot$

$\therefore$ $\widehat{MK}$ and $\widehat{NL}$ also $= $ arc.si.sq.gt.$\odot$

Since $\odot DBGT$ is great in a sphere and cuts one of the circles in the sphere, *ZEGH*, and passes through its poles, it bisects it at right angles [II 16]

$\therefore$ $\odot DBGT \perp \odot ZEGH$

Similarly $\odot DNHL \perp \odot ZEGH$, $\odot DMEK \perp \odot ZEGH$

Draw $\overline{LN}$ $\overline{LG}$ $\overline{TE}$

On the diameters of $\odot ZEGH$ drawn from *G H* two equal sections of circles are constructed, standing on them at right angles, *GT HL*, and what is connected to them, and each of $\widehat{GT}$ $\widehat{HL} < \frac{1}{2}\odot$, and $\widehat{EG} = \widehat{GH}$

$\therefore$ $\overline{TE} = \overline{LG}$ [II 11, part 2]

And si.sq.gt.$\odot = TE$: $\therefore$ $LG = $ si.sq.gt.$\odot$

And $\overline{LN} = $ si.sq.gt.$\odot$: $\therefore$ $\overline{LG} = \overline{LN}$

$\therefore$ the circle about pole *L* with distance *LG* passes through *N*: $\odot GNS$

This circle is a great circle. For a line drawn from pole to circumference is si.sq.gt.$\odot$ [I 18]

Since $\odot$s *AB GNS* in a sphere cut the circumference of a circle at one point, *N*, and their poles are on the circle,

$\therefore$ the circles touch [II 3]

$\therefore$ $\odot GNS$ touches $\odot AB$

Similarly the circle about *K* with distance *KG* goes through *M*, for if we join $\overline{GK}$ $\overline{TH}$, they are equal

And $\overline{TH}$ is the side of a square [i.e. si.sq.gt.$\odot$]. For it is drawn from pole of great $\odot EBH$ to its circumference

$\overline{GK}$ is also the side of a square [i.e. si.sq.gt.$\odot$] and similarly $\overline{KM}$

$\therefore$ $\overline{KM} = \overline{KG}$

$\therefore$ the circle about pole *K* with distance *KM* passes through *G*: $\odot GMQ$

It touches *AB*

If someone says: The arc which is cut off, i.e. $\widehat{BG}$, = arc.si.sq.gt.$\odot$, we explain that to him in the following way:

That when each of $\widehat{DE}$ $\widehat{DH}$ = $\widehat{DG}$, and each of $\widehat{DM}$ $\widehat{DN}$ = $\widehat{DB}$,
$\therefore$ the remaining arc $\widehat{BG}$ = each of $\widehat{NH}$ $\widehat{ME}$

And $\widehat{BG}$ is subtended by the side of the square [inscribed in a great circle]
$\therefore$ each of $\widehat{NH}$ $\widehat{ME}$ is subtended by the side of a square [i.e. si.sq.gt.$\odot$]

So since $\widehat{NH}$ is subtended by the side of a square and similarly $\widehat{GH}$ is subtended by a side of a square [i.e. si.sq.gt.$\odot$],
$\therefore$ $\overline{NH} = \overline{GH}$

$\therefore$ the circle about pole H with distance HG passes through N and similarly the circle about pole E with distance EG passes through M

$\therefore$ what we wanted to construct is of two kinds[2]. Q.E.D.[3]

15. Great circles in a sphere that cut off between them similar arcs from parallel circles either pass through the poles of the parallel circles or touch the same one of the parallel circles.

Let there be in a sphere two great circles, *AHG BTD*, cutting off between them similar arcs from the parallel $\odot$s *ABGD EZHT*:
$\widehat{AB} \sim \widehat{EZ}$, $\widehat{BG} \sim \widehat{ZH}$, $\widehat{GD} \sim \widehat{HT}$, $\widehat{AD} \sim \widehat{ET}$

I say: $\odot$s *AHG BTD* either pass through the poles of the parallel circles or they touch the same parallel circle.

For $\odot AHG$ either passes through the poles of the parallel circles or it does not.

[1] Let it pass through the poles of the parallel circles: see first diagram
I say: $\odot BTD$ also passes through poles of the parallel circles,
i.e. K is pole of parallel $\odot$s *ABGD EZHT*.

If it were not so, let L be pole of these two parallel circles
Draw great $\odot$ through L and Z: $\odot LZM$

---

[2] على ضربين. Gerard has "secundum duas ... figuras", presumably having read
على صورتين

[3]Here **H** adds: rather, it is of three kinds, but the construction when the arc of the side of the square passes through (?) point G, it falls short on one side; he therefore made it two kinds. The Greek text also has an extra passage here, on the case in which $BG > \frac{1}{4}\odot$, but Czinczenheim encloses it in square brackets. The text she edits here is different from Heiberg's.

$\therefore\ \widehat{ABM} \sim \widehat{EZ}$ [II 10]

And $\widehat{EZ} \sim \widehat{AB}$ [hypothesis]

$\therefore\ \widehat{AB} \sim \widehat{MA}$; and it is on the same circle: ✳

$\therefore$ L is not pole of the two parallel ⊙s

Similarly it is impossible that their pole is a point other than K

$\therefore$ K is the pole of the parallel circles

$\therefore$ ⊙s AHG BTD pass through the poles of the two parallel circles

[2] Let ⊙AEG[1] not pass through the poles of the parallel circles:
   it either touches ⊙$EZHT$ or it is inclined to it

First let it touch it, at E, as in second diagram

I say: ⊙ZB also touches it.

For, if it is possible, let it not touch it

Through Z draw a great circle touching ⊙EZH[2]: ⊙ZK[3]

Let $\frac{1}{2}$⊙ZK not meet $\frac{1}{2}$⊙EA

$\therefore\ \widehat{AK} \sim \widehat{EZ}$ [II 13]

$\widehat{EZ} \sim \widehat{AB}$ [hypothesis]

$\therefore\ \widehat{AK} \sim \widehat{AB}$, and it is on the same circle : ✳

$\therefore$ it is not possible that ⊙BZ does not touch ⊙EZH

$\therefore$ it touches it

[3] Let ⊙AHG be inclined to the parallel circles, as in third diagram

It touches two equal circles parallel to ⊙s $ABGD$ $EZHT$ [II 8]

I say: ⊙$BZTD$ [also] touches the two [equal parallel ⊙s].

If possible, let ⊙$AEHG$ touch one of these two [equal and] parallel circles, MLS, at L

Let ⊙$BZTD$ not touch it, if possible

Draw a great circle, passing through Z, between ⊙LMS and the circle equal and parallel to it, touching ⊙LMS at M: ⊙$NZMQ$

$\widehat{ABN} \sim \widehat{EZ}$ [II 13]

And $\widehat{EZ} \sim \widehat{AB}$

$\therefore\ \widehat{ABN} \sim \widehat{AB}$, and it belongs to the same circle ✳

---

[1]In Greek: AHG. H is not shown in the diagram reproduced by Czinczenheim and also not in **AN**.

[2]EZΘ in Greek.

[3]For K Greek has Γ in this and the following three occurrences. **A** and **N** vary between K and L; **H** and Gerard have L in all four places.

∴ it is not possible that ⊙ZTD does not touch it also
∴ it does touch it
∴ ⊙s $AEHG$ $BZTD$ touch one and the same circle from the parallel circles on the sphere. Q.E.D.

16. Parallel circles in a sphere that cut off from a great circle, starting from the greatest of the parallel circles, equal arcs are equal. Circles cutting off greater arcs are smaller.

Let there be two parallel circles, AB GD, in a sphere, cutting off from great ⊙$ABGD$ equal arcs, BZ ZD, starting from the greatest of the parallel circles, EZ

I say: ⊙AB = ⊙GD.

Let the common section of ⊙AB and ⊙$ABGD$ be line AB
Let the common section of ⊙EZ and ⊙$ABGD$ be line EZ
Let the common section of ⊙GD and ⊙$ABGD$ be line GD

Since the parallel planes ETZ GKD are cut by some plane, plane ⊙$ABGD$, their common sections are parallel [Euclid XI 16]
∴ $EZ \parallel GD$
Similarly $AB \parallel GD$ and EZ
Since in ⊙$ABGD$ two parallel lines, EZ GD, are drawn,
∴ $\widehat{DZ} = \widehat{EG}$. For if we join ED, the alternate angles [$Z\widehat{E}D$ $E\widehat{D}G$] are equal [Euclid I 29] and in equal circles equal angles stand on equal arcs [Euclid III 29]
∴ $\widehat{EG} = \widehat{ZD}$
Similarly $\widehat{BZ} = \widehat{AE}$
But $\widehat{BZ} = \widehat{ZD}$ [hypothesis]: ∴ $\widehat{AE} = \widehat{EG}$
∴ $\widehat{AE}$ $\widehat{BZ} = \widehat{EG}$ $\widehat{ZD}$ together
Since total arcs $\widehat{EALBZ} = \widehat{EGMDZ}$, for ⊙s ETZ $ABGD$ are great [and so bisect each other, I 12]
and $\widehat{AE}$ $\widehat{BZ} = \widehat{EG}$ $\widehat{ZD}$ together,
∴ remainders $\widehat{ALB} = \widehat{GMD}$
They are on the same circle: ∴ $\overline{AB} = \overline{GD}$ [Euclid III 29]

⊙$ABGD$ either cuts ⊙s AHB GKD and passes through their poles or cuts them and does not pass through their poles

[1] First let it cut them and pass through their poles
∴ it bisects them [I 16]
∴ AB is the diameter of ⊙AHB and GD is the diameter of ⊙GKD
And $AB = GD$ [above]: ∴ ⊙AHB = ⊙GKD

[2] Let ⊙*ABGD* cut ⊙s *AHB GKD*, not passing through their poles

Mark pole of the two parallel circles: *N* [I 22]

Draw great circle through *N* and through one of the poles of ⊙*ABGD*: ⊙*LNTMS*

Cut off $\widehat{MS} = \widehat{LN}$

Since $\widehat{LN} = \widehat{MS}$ and $\widehat{NKM}$ is common, total arcs $\widehat{LKM} = \widehat{NKMS}$

But $\widehat{LKM} = \frac{1}{2}\odot$: ∴ *NKMS* is also $\frac{1}{2}\odot$

∴ *N* is opposite *S*

And *N* is pole of the parallel circles: ∴ *S* is the other pole of the parallel circles

Since ⊙s *ABGD GKD* are on a sphere, cutting each other, and great ⊙*LTKS* is drawn through their poles,

∴ ⊙*LTKS* bisects the sections that were cut off from the circles [II 9]

∴ $\widehat{GM} = \widehat{MD}$

∴ $\widehat{GMD} = 2\widehat{MD}$

Similarly $\widehat{ALB} = 2\widehat{AL}$

And $\widehat{GMD} = \widehat{ALB}$ [above]: ∴ $\widehat{MD} = \widehat{AL}$

Since on diameter LM of ⊙*ABGD* two equal sections of a circle, *LN MS*[1] (together with the section connected with these completing the $\frac{1}{2}\odot$)[2], are constructed, standing on it at right angles, and two equal arcs, *LN MS*, are cut off from them which $< \frac{1}{2}$ of the two and equal arcs, *AL DM*, are cut off from the first circle,

∴ $\overline{NA} = \overline{SD}$ [II 11]

$\overline{NA}$ goes from pole ⊙*AHB* to its circumference

and $\overline{SD}$ goes from pole ⊙*GKD* to its circumference

∴ line from pole ⊙*AHB* to its circumference = line from pole ⊙*GKD* to its circumference

But circles in which the lines from pole to circumference are equal are equal[3]

---

[1]*LN MS*: Gerard has *LTM NMS*. Greek has only one section, *LTMS*. Ver Eecke's translation has two sections, *LTKM* and *MS*.

[2](together ... $\frac{1}{2}\odot$): not in Greek.

[3]Extra passage in **A**: Because these lines cut off from the semicircle equal arcs passing through its poles, so the line joining the two common points of the circumference of the circle passing through the two poles and each of the circumferences of the two circles is parallel to the diameter of the sphere. So the two perpendiculars going from these two points to the diameter of the sphere are equal: they are the two lines going from the centre of each of the two circles to both their circumferences.

$\therefore \odot AHB = \odot GKD.$

Now let $\widehat{DZ} > \widehat{ZB}$
I say: $\odot GKD < \odot AHB.$

For $\widehat{DZ} > \widehat{ZB}$: cut off from $\widehat{DZ}$ an arc which $= \widehat{ZB}$: $\widehat{ZQ}$
Draw $\odot$ parallel to $\odot ETZ$ through Q: $\odot QFC$
$\therefore \ \odot QFC = \odot AHB$ [part I of this prop.], for $\widehat{ZQ} = \widehat{ZB}$
And $\odot CFQ > \odot GKD$, for $\odot CFQ$ is nearer the centre of the
sphere than $\odot GKD$ [I 6]
$\therefore \ \odot AHB > \odot GKD$
$\therefore \ \odot GKD < \odot AHB.$ Q.E.D.

17. Equal and parallel circles on a sphere cut off from any great circle
equal arcs starting from the greatest of the parallel circles. Circles that
are greater cut off smaller arcs.

Let there be in a sphere two equal and parallel circles, AB GD
From great $\odot ABGD$ let them cut off $\widehat{ZB}$ $\widehat{ZD}$ beginning from the great-
est $[EZ]$ of the parallel circles
I say: $\widehat{ZB} = \widehat{ZD}.$

For if $\widehat{ZB} \neq \widehat{ZD}$, $\odot AB \neq \odot GD$ [II 16]
But it is equal to it: $\therefore \ \widehat{BZ} = \widehat{ZD}$

Now let $\odot AB > \odot GD$
I say: $\widehat{BZ} < \widehat{ZD}.$

For if $\widehat{BZ} \not< \widehat{ZD}$, $\odot AB \not> \odot GD$ [II 16]
But it is greater: $\therefore \ \widehat{BZ} < \widehat{ZD}$
$\therefore$ [enunciation] Q.E.D.

18. When on a sphere there is a great circle which cuts parallel circles
in the sphere and does not pass through their poles, it divides them
into unequal parts – except the greatest of the parallel circles. As for
the sections that are cut off in one of the hemispheres, those between
the greatest of the parallels and the visible pole, are each greater than
a semicircle; and as for the remaining sections which are in that half
of the sphere, each of them is less than a semicircle and the reciprocal
sections of the parallel, equal circles are equal to each other.

Let $ABGD$ be a great circle on a sphere, cutting parallel circles, AD
      EZ BG, but not passing through their poles

Let the greatest of the parallel circle be EZ

I say: $\odot ABGD$ divides these circles into unequal parts, except $\odot EZ$,
      the greatest of the parallel circles;

      each of the sections cut off in one of the hemispheres between
      $\odot EZ$ and the visible pole $> \frac{1}{2}\odot$; each of the remaining sections
      $< \frac{1}{2}\odot$; and the reciprocal sections of the parallel equal circles are
      equal.

      Let the visible pole of the parallel circles be H

      Draw a great circle through E H: $\odot TEH$

      $\odot HET$, when completed, passes through Z [1]

      Let it be drawn: $\odot HNZK$

      Complete $\odot BG$, so that it reaches T K

      Since great circle $TEHNZK$ on a sphere cuts parallel $\odot$s (from
      among the circles in the sphere), i.e. $AMND$ EZ $TBLGK$, and
      passes through their poles,
        $\therefore$ it bisects them at right angles [I 16][2]

      $\therefore$ each of sections MN EZ $TBLGK$ is a semicircle

      Since section MN is $\frac{1}{2}\odot$, section $AMND > \frac{1}{2}\odot$

      Similarly all sections between $\odot EZ$ and pole $H > \frac{1}{2}\odot$

      Also, since section $TLGK$ is $\frac{1}{2}\odot$, section $BG < \frac{1}{2}\odot$

      Similarly all the sections between $\odot EZ$ and the invisible pole in
      this hemisphere $< \frac{1}{2}\odot$.

Now let $\odot AD =$ and $\parallel \odot BG$

I say: The alternate sections of $\odot$s AD BG are equal.

      For since $\odot AD =$ and $\parallel \odot BG$, $\overset{\frown}{EA} = \overset{\frown}{EB}$ and $\overset{\frown}{DZ} = \overset{\frown}{ZG}$ [II 17]
      $\therefore$ $\overset{\frown}{AE} + \overset{\frown}{DZ} = \overset{\frown}{EB} + \overset{\frown}{ZG}$

      And $\overset{\frown}{EA} + \overset{\frown}{AD} + \overset{\frown}{DZ} = \overset{\frown}{EB} + \overset{\frown}{BG} + \overset{\frown}{GZ}$, since each of $EADZ$
      $EBGZ$ is a semicircle – that is since $\odot$s $ABGD$ EZ are great[3]
      [II 12]

      $\therefore$ the remaining arcs $\overset{\frown}{AD} = \overset{\frown}{BG}$

      And $\overset{\frown}{AD}$ $\overset{\frown}{BG}$ are of the same circle: $\therefore$ $\overline{AD} = \overline{BG}$

---

   [1]**A** *add.:* because it bisects $\odot EZ$; and $\overset{\frown}{EZ}$ is $\frac{1}{2}\odot EZ$.

   [2]For great circles cut each other into two halves and point ... cutting the circle
on which they are into two halves *marg.* **N**.

   [3]since each ... great: put into square brackets, and placed after "the remaining
arcs $\overset{\frown}{AD} = \overset{\frown}{BG}$" of the next line, by Czinczenheim.

And $\overline{AD}$ subtends $A\widehat{M}D$ and $\overline{BG}$ subtends $B\widehat{L}G$[4]

Equal straight lines in equal circles cut off equal arcs, greater cut off greater, and smaller smaller [Euclid III 28]

∴ the greater arc of $\odot AMND$ = the greater arc of $\odot BLG$; and the smaller arc of $\odot AHD$ = the smaller arc of $\odot BLG$

And section $AMD > \frac{1}{2}\odot$ and section $BLG < \frac{1}{2}\odot$

∴ the reciprocal sections of the equal and parallel circles are equal to each other. Q.E.D.

19. When in a sphere a great circle cuts some of the parallel circles, of the circles in the sphere, but not through their poles, of the arcs that are cut off in one of the hemispheres that which is nearer the visible pole is greater than the arc of that circle similar to the [arc] cut off from [the circle] that is further away from that pole.

Let there be in a sphere great $\odot AEZB$[1] cutting some of the parallel circles and not passing through their poles

Let the parallel circles be AB GD EZ

I say: Of the arcs cut off in the two hemispheres the one that is nearer the pole > the arc of that circle similar to the arc cut off in the circle further from the visible pole,

i.e. $\widehat{AB}$ > the arc of its circle that $\sim \widehat{GD}$, and $\widehat{GD}$ > arc of its circle which $\sim \widehat{EZ}$.

Let the visible pole of the parallel circles be H

Draw great circle through H and G: $\odot HLGT$[2]

Draw great circle through H and D: $\odot HMDK$

$\odot$s $HLGT$ $HMDK$ cut off between them similar arcs [II 10]:

∴ $\widehat{LM} \sim \widehat{GD}$

∴ $A\widehat{LM}B$ > the arc of its circle that $\sim GD$[3]

Similarly $\widehat{GD}$ > the arc of its circle that $\sim \widehat{EZ}$ [as is clear] when we draw great circles through H and E Z

---

[4]And $\overline{AD}$ ... $B\widehat{L}G$ is put into square brackets by Czinczenheim.

[1]Greek has $ABDG$.

[2]The proof in Greek refers to a diagram (Ver Eecke, p. 66, note 1, the first of the two diagrams) that is similar to ours, but has T between H and G and K between H and Z. In fact T K label, repectively, points L M – letters "L" and "M" do not appear on the diagram. Accordingly, the proof sometimes has different letters (never L or M).

[3]of $\odot$ $ALMB$ and it is arc LM *add.* **A** (not in Greek).

It is possible to show that without drawing these two circles, but by restricting ourselves to completing ⊙*EZ*, as in the previous proposition.[4]

20. When on equal spheres there are great circles inclined to other great circles, then any circle whose pole happens to be higher has a greater inclination over its companion[1]. As for the circles the distance of whose poles from the planes of the circles on which they stand is equal, their inclination is equal.

Let there be in equal spheres two great circles, *BKD ZLT*, inclined to great circles *ABGD EZHT*
Let pole ⊙*BKD* be *M*, and pole ⊙*ZLT* be *N*
Let pole *M* be higher than pole *N*
I say: Inclination ⊙*BKD* on ⊙*ABGD* > inclination ⊙*ZLT* on ⊙*EZHT*.
Draw a great circle through *M* and a pole of *ABGD*: ⊙*AKMG* [I 21]
Draw another great circle through *N* and a pole of ⊙*EZHT*: ⊙*ELNH* [I 21]
∴ they pass through poles of ⊙*BKD* and ⊙*ZLT*, bisecting them at right angles [I 21][2]
Let common section of ⊙*ABGD* and ⊙*BKD* be *BD*
Let common section of ⊙*ABGD* and ⊙*AKMG* be *AG*
Let common section of ⊙*BKD* and ⊙*AKMG* be *KS*
Let common section of ⊙*EZHT* and ⊙*ZLT* be *ZQT*
Let common section of ⊙*EZHT* and ⊙*ELNH* be *EH*
Let common section of ⊙*ZLT* and ⊙*ELNH* be *LQ*

Since ⊙*AKMG* is in a sphere and it cuts some of the great circles in the sphere – ⊙s *ABGD* and *BKD* – and it goes through their poles, then it bisects them at right angles [I 16]
∴ ⊙*AKMG* ⊥ both ⊙s *ABGD BKD*
∴ each of ⊙s *ABGD BKD* ⊥ ⊙*AKMG*
When two planes cut each other and ⊥ another plane, then their common section also ⊥ the plane [Euclid XI 19]

---

[4]when we draw ... proposition: in Greek in square brackets.
[1]I.e. over the circle to which it is inclined. "over its companion" is omitted in Greek. Here there is an extra passage in **A**: "with his words 'the pole of the circle is higher' he means: when the perpendicular falling from the pole of the inclined circle onto the plane of the circle on which it is inclined is longer, and, when the two perpendiculars are equal, the two inclinations are equal".
[2]This line, which is superfluous (see below), is not in the Greek.

$\therefore$ common section $\odot ABG$ and $\odot BKD \perp \odot AKMG$, and their common section is line BD

$\therefore$ $BD \perp \odot AKMG$

$\therefore$ it produces right angles with all the straight lines passing through its end and lying in plane $AKMG$ [Euclid XI Def. 3]

Each of KS SA passes through its end and is in plane $\odot AKMG$:

$\therefore$ $BD^3 \perp$ each of KS SA

Since planes $ABGD$ BKD cut each other and KS SA are drawn at right angles to BD, which is the common section,

and KS is in plane $\odot BKD$ and SA is in plane of $\odot ABGD$,

$\therefore$ $K\hat{S}A =$ inclination of plane BKD to plane $ABGD$

Similarly $L\hat{Q}E =$ inclination of plane ZLT to plane $EZHT$

I say: $K\hat{S}A < L\hat{Q}E$.

Since M is higher than N,

$\perp^r$ from M to plane $\odot ABGD > \perp^r$ from N to plane $\odot EZHT$

And $\perp^r$ from M to plane $\odot ABGD$ falls on the common section of $\odot$s $AKMG$ and $ABGD$, i.e. on AG, since planes $AKMG$ and $ABGD$ are mutually perpendicular,

and $\perp^r$ from N to plane $\odot EZHT$ falls on EH

So $\perp^r$ from M to $\overline{AG} > \perp^r$ from N to $\overline{EH}$

Since sections $AKMG$ $ELNH$ are sections of equal circles and M N are marked, however they fall,

and $\perp^r$ from M to $AG > \perp^r$ from N to EH,

$\therefore$ $\widehat{MG} > \widehat{NH}$

and $\widehat{MK} = \widehat{NL}$, each $=$ arc.si.sq.gt.$\odot$ (that is since they are drawn from one of the poles of $\odot$s BKD ZLT)[4]

$\therefore$ total arc $\widehat{KMG} >$ total arc $\widehat{LNH}$,

since total arc $\widehat{AKMG} =$ total arc $\widehat{ELNH}$ and $\widehat{KMG} > \widehat{LNH}$,

$\therefore$ there remains $\widehat{AK} < \widehat{EL}$

And $K\hat{S}A$ has base AK and $L\hat{Q}E$ has base LE, and these two angles are at the centres of the circles[5]

$\therefore$ $K\hat{S}A < L\hat{Q}E$

But $K\hat{S}A$ is inclination of plane $\odot BKD$ to plane $\odot ABGD$,

and $L\hat{Q}E$ is inclination of plane $\odot ZLT$ to plane $\odot EZHT$

$\therefore$ inclination of $\odot BKD$ to $\odot ABGD >$ inclination of $\odot ZLT$ to $\odot EZHT$

So $\odot BKD$ is more inclined to $\odot ABGD$ than $\odot ZLT$ to $\odot EZHT$

---

[3] BS in Greek.

[4] (that ... $\odot ZLT$): not in Greek. **A** *add.*: to its circumference.

[5] and these ... circles: not in Greek.

Now let the distance of the poles of ⊙s *BKD ZLT* from the planes on
    which they stand be equal,
      i.e. the $\perp^{\mathrm{r}}$ from *M* to plane ⊙*ABGD* = $\perp^{\mathrm{r}}$ from *N* to plane
      ⊙*EZHT*
I say: Inclination of ⊙s *BKD ZLT* to ⊙s *ABGD EZHT* is equal,
      i.e. $K\hat{S}A = L\hat{Q}E$.
When we do these things, we explain that $K\hat{S}A$ is inclination of plane
    ⊙*BKD* to plane ⊙*ABGD* and $L\hat{Q}E$ is inclination of plane ⊙*ZLT*
    to plane ⊙*EZHT*
I say: $K\hat{S}A = L\hat{Q}E$.

    Since the $\perp^{\mathrm{rs}}$ from *M N* to the planes of ⊙s *ABGD EZHT* are
      equal
      and the $\perp^{\mathrm{rs}}$ from *M N* to planes ⊙s *ABGD EZHT* fall on lines
      AG EH,
      the $\perp^{\mathrm{rs}}$ from *M N* to lines *AG EH* are equal
And since $A\widehat{KM}G$ $EL\widehat{N}H$ are sections of two equal circles, and
    M N have been marked on them, however they fall,
    and the $\perp^{\mathrm{r}}$ from *M* to *AG* = $\perp^{\mathrm{r}}$ from *N* to *EH*,
    ∴ $\widehat{MG} = \widehat{NH}$ and $\widehat{KM} = \widehat{NL}$, that is since each is subtended
    by si.sq.gt.⊙
    ∴ total $K\widehat{M}G$ = total $L\widehat{N}H$ and total $A\widehat{KM}G$ = total $EL\widehat{N}H$
    ∴ remainders $\widehat{AK} = \widehat{EL}$
And $K\hat{S}A$ has base *AK* and $L\hat{Q}E$ has base *EL*
    ∴ $K\hat{S}A = L\hat{Q}E$
And $K\hat{S}A$ is inclination of plane ⊙*BKD* to plane ⊙*ABGD*,
    and $L\hat{Q}E$ is inclination of plane ⊙*ZLT* to plane ⊙*EZHT*
    ∴ inclination of ⊙*BKD* to ⊙*ABGD* = inclination of ⊙*ZLT*
    to plane ⊙*EZHT*
∴ inclination of ⊙s *BKD ZLT* to ⊙s *ABGD EZHT* are similar.
We know: It is said that the inclination of a plane to a plane
    is similar to the inclination of another plane to another plane
    when the straight lines going from the common sections of
    the planes at right angles in each of the planes contain equal
    angles[6]. Q.E.D.

---

[6]Czinczenheim puts this last sentence in square brackets. It is almost identical to
the penultimate definition of Book I, which is not in the Greek text. It is equivalent
to Euclid XI Def. 7 combined with Def. 6.

21^1. When a great circle in a sphere touches one of the non-great circles that are in the sphere and cuts another circle parallel to that one, from among the circles that are between the centre of the sphere and the circle that the first circle touches, and [when] the pole of the great circle is between the two parallel circles, and great circles are drawn touching the greater of the two parallel circles, then these circles are inclined to the great circle and the highest of these circles is the circle whose point of contact is in the middle of the greater of the two sections of that circle and the lowest [is] the circle whose point of contact is at the middle of the smaller of the two sections of the circle. As for the other circles, those of them [for which] the distance of the point of contact from one of the mid-points of the two sections, whichever it is, is equal have a similar inclination and the circle whose point of contact is further away from the middle of the greater section has a greater inclination than the circle whose point of contact is nearer; and the poles of the great circles are also on one circle parallel to the two circles that we have mentioned [this circle being] smaller than the circle that is touched by the first circle.

Let great $\odot ABG$ be in a sphere, touching some non-great circle AD at
    A and cutting another circle parallel to this circle from among
    the $\odot$s between the centre of the sphere and $\odot AD$: $\odot EZHT$
Let pole $\odot ABG$ be between $\odot$s AD and $EZHT$, the greater of the two
    parallel circles[2]
Draw great circles touching $\odot EZHT$: $\odot$s MNS BZG QFC PT RO
Let $\odot BZG$ touch $\odot EZHT$ at Z, the middle of the greater of the two
    sections of $\odot EZHT$, section EZH
Let $\odot PT$ touch it in the middle of the smaller of the two sections,
    at T
Let the distance of place of contact of $\odot$s MNS QFC from the point of
    one of the two halves, which ever, be equal[3], however it comes.

I say: $\odot$s MNS BZG QFC PT RO are inclined to $\odot ABG$,
    the highest of them is $\odot BZG$ and the lowest PT,
    and $\odot$s MNS QFC have a similar inclination,
    $\odot RO$ is more inclined to $\odot ABG$ than $\odot QFC$,
    the poles of $\odot$s MNS BZG QFC RO PT are on one circle par-

---

[1]As Ver Eecke notes, there are several places in this proposition where proof is lacking.

[2]Greek *add.*: it is point K.

[3]and the distance of the place of contact of $\odot RO$ with it from point Z is further from the point of contact of $\odot$s MNS QFC with it *add.* **A**.

allel to ⊙s *AD* and *EZHT*, which is smaller than the first circle
touched by ⊙*ABG*.

Mark one of the poles of the parallel ⊙s *AD EZHT*: *L*

Draw a great circle through *A* and *L*: ⊙*AL*

Since ⊙s *ABG AD* are in a sphere, one touching the other, and
 a great circle is drawn through the pole of one of them and the
 place of contact, ⊙*AL*,
  ⊙*AL* passes through the pole of ⊙*ABG* also [II 5]
∴ it is at right angles to [⊙*ABG*] [I 16]

Let *K* be pole of ⊙*ABG*
∴ ⊙*AL*, when completed, also passes through *K*, say *ALK*

⊙s *ABG EZHT* cut each other in a sphere and a great circle has
 been drawn passing through their poles, ⊙*ALK*,
  ∴ ⊙*ALK* bisects the sections cut off from them [II 9]

The mid-point of section *EZH* is *Z*

The mid-point of section *ETH* is *T*

∴ ⊙*ALK*, when completed, also passes through *Z* and *T*:
 ⊙*TALKZ*

Since *K* is pole ⊙*ABG* and ⊙*ABG* is a great circle,
 the line subtending $\widehat{AK}$ is the si.sq.gt.⊙
∴ $A\widehat{K}Z >$ arc.si.sq.gt.⊙

And since ⊙*EZHT* < a great circle (since it is between the centre
 of the sphere and ⊙*AD*), and its pole is *L*,
  ∴ *LZ* < arc.si.sq.gt.⊙

And since $ALZ^4 >$ arc.si.sq.gt.⊙ and $L\widehat{Z} <$ arc.si.sq.gt.⊙,
  ∴ when an arc that = arc.si.sq.gt.⊙ is cut off from $A\widehat{L}Z$ at
 Z, its other end falls between *A* and *L*

Cut off arc = mentioned arc [i.e. arc.si.sq.gt.⊙]: $\widehat{YZ}$

Draw about pole *L* and with distance *LY* ⊙*YXT́Ź*

[⊙*YXT́Ź*] ∥ ⊙s *AD EZHT*

Draw great circles through *L* and [severally] through *N F O*: ⊙s
 NLŹ FLX OLT́

And since $\widehat{NL} = \widehat{LZ}$ (for they are drawn from pole ⊙*EZHT* to
 its circumference),
 and $\widehat{LY} = \widehat{LZ}$ (for they are drawn from pole ⊙*T́ŹX* to its
 circumference),
  ∴ total $N\widehat{L}Ź = Z\widehat{L}Y$

And $Z\widehat{L}Y =$ arc.si.sq.gt.⊙,
  ∴ $N\widehat{L}Ź =$ arc.si.sq.gt.⊙

---

[4]Greek has *AKZ*.

Similarly each of $\widehat{XLF}$ $\widehat{ULT}$[5] $\widehat{TLO}$ = arc.si.sq.gt.$\odot$

Since $\odot$s *MNS* *EZHT* are in a sphere, touching each other, and a great circle is drawn passing through the poles of one of them [*EZHT*] and through the place of contact, i.e. $\odot NL\acute{Z}$, $\odot NL\acute{Z}$ passes though the poles of $\odot MNS$ [II 5] and is at right angles to it [I 16]

Since $\odot MNS$ is a great circle, an arc from pole to circumference = arc.si.sq.gt.$\odot$, and $NL\acute{Z}$ = arc.si.sq.gt.$\odot$

∴ line from N to $\acute{Z}$ = line from circumference of $\odot MNS$ to its pole

∴ $\acute{Z}$ is pole $\odot MNS$

Similarly Y is pole $\odot BZG$,

X is pole $\odot QFC$

$\acute{T}$ is pole $\odot RO$

U is pole $\odot PT$

∴ poles of $\odot$s *MNS* *BZG* *QFC* *RO* *PT* are on $\odot YX\acute{T}$, which is parallel $\odot$s *AD* and *EZHT* and $< \odot AD$

I say: $\odot$s *MNS* *BZG* *QFC* *RO* *PT* are inclined to $\odot ABG$;
the one with the highest elevation is $\odot BZG$;
the one with the greatest depression is *PT*;
$\odot$s *MNS* *QFC* are of similar inclination;
$\odot RO$ has a greater inclination to $\odot ABG$ than has $\odot QFC$.

Since $\widehat{NZ} = \widehat{FZ}$ and they are on same circle, $\widehat{NZ} \sim \widehat{FZ}$

But $\widehat{NZ} \sim \widehat{UD}$ and $\widehat{ZF} \sim \widehat{UG}$

∴ $\widehat{UD} \sim \widehat{UG}$[6]

They are on same circle: ∴ $\widehat{UD} = \widehat{UG}$

But $\widehat{DU} = \widehat{YZ}$, for they are opposite, between two arcs of great circles that pass through their poles, and $\widehat{UG} = \widehat{YX}$

∴ $\widehat{YZ} = \widehat{YX}$

In $\odot YX\acute{T}\acute{Z}$ a section of a circle is constructed on diameter YU, at right angles to it: section UKZ and what is joined to this section

From [UKZ] an arc is cut off which $< \frac{1}{2}$ total section: $\widehat{UK}$

Two equal arcs have been cut off from the first circle: YX $Y\acute{Z}$,

∴ $\overline{KX} = \overline{K\acute{Z}}$ [II 11]

∴ the circle drawn about pole K with distance KX passes through $\acute{Z}$

---

[5] $\widehat{ULT}$: *om.* **H** and Latin.

[6] ∴ $\widehat{UD} \sim \widehat{UG}$: not in Greek.

Let it pass, like $X\acute{Z}$

∴ $\odot X\acute{Z} \parallel \odot ABG$, for they are on the same poles, for K is pole $\odot ABG$

Since $\odot X\acute{Z} \parallel \odot ABG$, $\perp^{\mathrm{r}}$ from X to plane $\odot ABG = \perp^{\mathrm{r}}$ from $\acute{Z}$ to plane $\odot ABG$

Similarly it also $= \perp^{\mathrm{r}}$ from I to plane $\odot ABG$[7]

And $\perp^{\mathrm{r}}$ from I[8] to plane $\odot ABG > \perp^{\mathrm{r}}$ from Y to plane $\odot ABG$

and $\perp^{\mathrm{r}}$ from $\acute{Z}$ to plane $\odot ABG > \perp^{\mathrm{r}}$ from Y to plane ABG (and similarly $\perp^{\mathrm{r}}$ from X)[9] (for each $= \perp^{\mathrm{r}}$ from I)[10]

∴ $\acute{Z}$ is higher than Y

And $\acute{Z}$ is pole $\odot MNS$ and Y is pole $\odot BZG$,

∴ pole $\odot MNS$ is higher than pole $\odot BZG$

And circles whose poles are higher have greater inclination to the planes on which they are [II 20]

∴ $\odot MNS$ is more inclined to $\odot ABG$ than $\odot BZG$

∴ $\odot BZG$ is more elevated than $\odot MNS$

Similarly $\odot BZG$ is more elevated than all the $\odot$s touching $\odot EZHT$

∴ $\odot BZG$ is the most elevated of all these circles.

I say: $\odot PT$ has the greatest depression.

For $\perp^{\mathrm{r}}$ from U to plane $\odot ABG > \perp^{\mathrm{r}}$ from $\acute{T}$ to plane $\odot ABG$, so U is higher than $\acute{T}$

And U is pole $\odot PT$ and $\acute{T}$ is pole $\odot RO$

∴ pole $\odot PT$ is higher than pole $\odot RO$

Circles whose poles are higher have greater inclination to the plane on which they are [II 20]

∴ $\odot PT$ is more inclined to $\odot ABG$ than $\odot RO$

∴ $\odot PT$ is more depressed than $\odot RO$

Similarly [$\odot PT$] is the lowest of all these circles that touch $\odot EZHT$

∴ $\odot PT$ is lower than all these circles

Since $\perp^{\mathrm{r}}$ from $\acute{Z}$ to plane $\odot ABG = \perp^{\mathrm{r}}$ from X to plane $\odot ABG$, distance of $\acute{Z}$ and X from plane $\odot ABG$ is equal

And $\acute{Z}$ is pole $\odot MNS$ and X is pole $\odot QFC$

∴ distance of poles of $\odot$s MNS QFC from plane $\odot ABG$ is equal

---

[7]Ver Eecke calls this line "inutile" and says it was probably interpolated. It is deleted by Heiberg and is included by Czinczenheim in square brackets.

[8]I: X in Greek. We may note that I is not defined.

[9]and similarly ... X: not in Greek. It must be understood as appended to the previous line.

[10](for ... I): apparently another interpolation (Ver Eecke). It is deleted by Heiberg; in Czinczenheim it is in square brackets.

Circles whose poles have an equal distance from the planes on
which they are inclined have a similar inclination [II 20][11]
Since $\perp^{\mathrm{r}}$ from $\acute{T}$ to plane $\odot ABG > \perp^{\mathrm{r}}$ from X to plane $\odot ABG$,
  $\therefore$ $\acute{T}$ is higher than X
And $\acute{T}$ is pole $\odot RO$ and X is pole $\odot QFC$
  $\therefore$ pole $\odot RO$ is higher than pole $\odot QFC$
Circles whose poles are higher have a greater inclination to the
planes on which they are [II 20]
  $\therefore$ $\odot RO$ is more inclined to $\odot ABG$ than $\odot QFC$
  $\therefore$ $\odot$s MNS BZG QFC RO PT are inclined to $\odot ABG$,
  and the most elevated is $\odot BZG$,
  and the most depressed is $\odot PT$,
  and $\odot$s MNS QFC have similar inclinations,
  and $\odot RO$ is more inclined to $\odot ABG$ than $\odot QFC$,
  and also their poles are [all] on one of the parallel circles, which
  is smaller than $\odot AD$. Q.E.D.

22. When these things are as we have described, and the arcs drawn
between the "knot" places, i.e. between the places of contact of the
circles, and their cutting of the first circle, are equal, then the great
circles are similar in inclination.

Let the arcs drawn from the two knots, N F, i.e. from the place[s] of
  contact, to the places of intersection of $\odot ABG$ and $\odot$s $MN\acute{Y}$ [1]
  QFC [2], i.e. $\widehat{MN}$ $\widehat{FC}$, be equal.
I say: The inclinations of $\odot MN\acute{Y}$ and $\odot QFC$ to $\odot ABG$ are similar.
  Mark pole of parallel $\odot$s AD $EZHT$: L [I 22]
  Draw a great circle through A L: $\odot TALZP$
  It passes through K, pole $\odot ABG$
  Draw great circles, each through L, and through N F: $\odot$s LNB
  LFG
  Since $\odot$s $EZHT$ $MN\acute{Y}$ are in a sphere touching each other, and
  a great circle is drawn passing through the pole of one and the
  place of contact, $\odot LNB$,
    $\therefore$ $\odot LNB$ also passes through poles of $\odot MN\acute{Y}$ [II 5] and at
  right angles [I 16]

---

[11]Circles ... inclination: in square brackets in Czinczenheim. $\therefore$ inclination of
$\odot$s MNS QFC to $\odot ABG$ is equal *add.* **A**.

[1]We have, as usual, kept to Gerard's letters. In this proposition he renders the
Arabic ر (for Greek Σ) by S and the Arabic س (for Greek Ξ) by Y. We represent
these letters in Gerard's text by Ś and Ý.

[2]i.e. from the places ... QFC: not in Greek.

Similarly $\odot LFG$ also passes through poles $\odot QFC$ and at right
   angles
Since there are constructed in equal circles[3], on their diameters,
   drawn from N F, two equal sections of circles standing on
   them at right angles, i.e. NL FL, together with the sections
   connected with them, and from them two equal arcs, NL FL,
   are cut off and they are less than half of the two arcs, and two
   equal arcs, MN FC, are cut off from the first circles,
   $\quad \therefore \ \overline{LM} = \overline{LC}$ [II 11]
$\therefore$ the circle about pole L and with distance LM passes through
   C also
Let it pass and be, say, $M\acute{Y}QC$, which $\parallel \odot$s AD $EZHT$ (since
   they have the same poles) [II 2]
Since $\odot$s ABG $M\acute{Y}QC$ are in a sphere, cutting each other and a
   great circle, $\odot TDKZP$, is drawn through their poles,
   $\quad \therefore \ \odot TDKZP$ bisects the sections cut off from the circles
$\therefore \ \widehat{MP} = \widehat{PC}$
Also, since $\odot$s $MN\acute{Y}$ $M\acute{Y}PC$ intersect and a great circle, $\odot LNB$,
   is drawn through their poles, $\odot LNB$ bisects the sections cut
   off from the $\odot$s [II 9]
$\therefore \ \widehat{MN} = \widehat{N\acute{Y}}$ and $\widehat{M\acute{Y}} = \widehat{S\acute{Y}}$
Similarly $\widehat{QF} = \widehat{FC}$ and $\widehat{QO} = \widehat{OC}$
Since $\widehat{MN} = \widehat{FC}$ [by hypothesis] and $\widehat{MN\acute{Y}} = 2\widehat{MN}$ and $\widehat{QFC} =$
   $2\,\widehat{FC}$,
   $\quad \therefore \ \widehat{MN\acute{Y}} = \widehat{QFC}$
And the two circles are equal
   $\quad \therefore$ line subtending $\widehat{MN\acute{Y}} =$ line subtending $\widehat{QFC}$
But line subtending $\widehat{MN\acute{Y}}$ also subtends $\widehat{MS\acute{Y}}$ and the line sub-
   tending $\widehat{QFC}$ also subtends $\widehat{QOC}$; and the arcs $\widehat{MS\acute{Y}}$ and
   $\widehat{QOC}$ are of the same circle,
   $\quad \therefore \ \widehat{MS\acute{Y}} = \widehat{QOC}$
And $\widehat{MS} = \frac{1}{2}\widehat{MS\acute{Y}}$ and $\widehat{CO} = \frac{1}{2}\widehat{QOC}$,
   $\quad \therefore \ \widehat{MS} = \widehat{CO}$
And total arcs $\widehat{MS\acute{Y}P} = \widehat{PQOC}$,
   $\quad \therefore$ remaining arcs $\widehat{S\acute{Y}P} = \widehat{PQO}$
And they are of the same circle: $\therefore \ \widehat{S\acute{Y}P} \sim \widehat{PQO}$[4]

---

[3] $MN\acute{Y}$ QFC are specified in Greek.

[4] $\therefore \ \widehat{S\acute{Y}P} \sim \widehat{PQO}$: not in Latin.

But $\widehat{S\hat{Y}P} \sim \widehat{NZ}$ and $\widehat{P\hat{Q}O} \sim \widehat{ZF}$ [II 10],

   $\therefore$ $\widehat{NZ} \sim \widehat{ZF}$[5]

They are on the same circle: $\therefore$ $\widehat{NZ} = \widehat{ZF}$

$\therefore$ distance of $\odot$s $MN\acute{Y}$ QFC from the midpoint of one of the
   two arcs into which $\odot EZHT$ is divided are equal

And the circles whose distance from the midpoint of one of these
   two arcs – whichever it is – is equal have a similar inclination[6]
   [II 21]

   $\therefore$ inclination of $\odot$s $MN\acute{Y}$ QFC to $\odot ABG$ is similar. Q.E.D.

Book III

1. When in a circle a straight line is drawn, dividing the circle into
two unequal parts, and a section of a circle not greater than its half is
constructed on it, standing on the circle at right angles, and the arc of
the section constructed on the line is divided into unequal parts, then
the line subtending the smaller arc is the shortest of all the straight
lines drawn from the point at which the arc was divided to the greater
arc of the first circle. Similarly, if the line drawn is a diameter of the
circle and the other things which belong to the section which is not
greater than a semicircle[1] [and which is] constructed on the line are as
they were, then the aforementioned line that is drawn is the shortest
of all the straight lines drawn from that same point and meeting the
circumference of the first circle; the greatest of them is the line that
subtends the greatest arc.

In $\odot ABGD$ draw a straight line BD dividing the circle into two unequal
   parts; and let $\widehat{B\hat{G}D} > \widehat{B\hat{A}D}$

On BD construct a section of a circle, $\not> \frac{1}{2}\odot$, standing on $\odot ABGD$ at
   right angles: section BED

Divide $\widehat{BED}$ into unequal parts at E; let $\widehat{BE} < \widehat{ED}$

Draw EB

I say: BE is the shortest of all the straight lines drawn from E to
   $\widehat{B\hat{G}D}$.

---

[5] $\therefore$ $\widehat{NZ} \sim \widehat{ZF}$: not in Latin.
[6] to $\odot ABG$ *add.* **A**
[1] which is ... semicircle: not in Greek.

Let $EZ \perp$ plane $\odot ABGD$

$\therefore$ it falls on the common section of planes $ABGD$ BED, i.e. BD,
since section $BED \perp \odot ABGD$ [Euclid XI Def. 4]

Mark centre of $\odot ABGD$: H

Draw ZH and produce it in both directions to T and K

From E draw EL to $[L,$ anywhere on$]$ $\widehat{BGD}$; join ZL

$EZ \perp$ plane $\odot ABGD$, and so produces right angles with all
straight lines drawn from its end and lying in plane $\odot ABGD$
[Euclid XI Def. 3]

And each of ZB ZL, which are in plane $\odot ABGD$, are drawn from
the end of line EZ,

$\therefore$ each of $B\widehat{Z}E$ $L\widehat{Z}E$ is right

Since $ZB < ZL$ [Euclid III 7],

$ZB^2 < ZL^2$

$\therefore$, with EZ^2 common, $EZ^2 + ZB^2 < EZ^2 + ZL^2$

But $BE^2 = EZ^2 + ZB^2$ and $LE^2 = LZ^2 + ZE^2$

$\therefore$ $BE^2 < LE^2$

$\therefore$ $EB < EL$

Similarly $[EB] <$ all straight lines drawn from E and meeting $\widehat{BGD}$

So BE is shorter than all the straight lines from E and meeting $\widehat{BGD}$

I say: Of the straight lines drawn from E [to $\widehat{BGK}$] between points
K B the line nearer $[EB]$ is always shorter than the one further
from it.

Join GE, ZG

Since $LZ < ZG$ [Euclid III 7],

$\therefore$ $ZL^2 < ZG^2$

$\therefore$, with ZE^2 common, $ZL^2 + ZE^2 < EZ^2 + ZG^2$

But $LZ^2 + ZE^2 = LE^2$ and $GZ^2 + ZE^2 = EG^2$

$\therefore$ $LE^2 < EG^2$

$\therefore$ $LE < EG$

Similarly, of the straight lines from E [to a point] between B and K,
those that are nearer EB are shorter than those further from it

Join EK ED

I say: EK is the longest of all the straight lines drawn from E meeting
$B\widehat{K}D^2$,

and ED is shorter than all straight lines from E [to] between D
and K.

---

[2]For BKD Greek has KD.

Since $KZ > ZG$ [Euclid III 7],

$\therefore$ $KZ^2 > ZG^2$

$\therefore$, with ZE^2 common, $KZ^2 + ZE^2$, i.e. EK^2, $> EZ^2 + ZG^2$, i.e. EG^2

$\therefore$ $EK > EG$

Similarly $EK >$ all straight lines from E to $\widehat{KD}$[3]

$\therefore$ EK is the longest of all the straight lines from E to $\widehat{BKD}$

I say also: $ED <$ all straight lines from E [to] between K and D.

Draw another straight line, EM; join MZ

Since $DZ < ZM$ [Euclid III 7],

$\therefore$ $DZ^2 < ZM^2$

$\therefore$, with ZE^2 common, $EZ^2 + ZD^2$, which $= ED^2$, $< EZ^2 + ZM^2$, which $= EM^2$

$\therefore$ $DE < ME$

Similarly ED is shorter than all straight lines from E to $\widehat{KD}$ between K and D

$\therefore$ ED is shorter than all the straight lines drawn from E and meeting KD between K and D

And of the lines drawn between K and D, those that are near [to ED] are shorter than those that are far from it.

Since $ED > EB$ (since $\widehat{DE} > \widehat{EB}$), $\overline{EB} \ll$ all straight lines from E to $\widehat{KD}$

It is shorter than all the straight lines drawn to $\widehat{KB}$ [above]

$\therefore$ line BE is shorter than all the straight lines drawn from E to $\widehat{BKD}$[4]

Let line BD be the diameter of the circle and let the other things be as before

I say: $EB <$ all straight lines from E to circumference $\odot ABGD$; ED is the longest of them.

Since everything is as described, $\widehat{DE} > \widehat{EB}$ and $\perp^{\mathrm{r}}$ EZ is drawn,

$\therefore$ $DZ > ZB$

And BD is a diameter $\odot ABGD$

$\therefore$ centre of $\odot[ABGD]$ is on line ZD

$\therefore$ $ZD > ZG$ and $ZG > ZB$ [both Euclid III 7]

---

[3] KD: Greek has BGD, though some Greek manuscripts have BKD, **A** BLD.

[4] Since $ED > EB$ ... $\widehat{BKD}$: not in Greek.

$\therefore$ $ZD^2 > GZ^2 >$ and $GZ^2 > ZB^2$

$\therefore$, with ZE^2 common, $DZ^2 + ZE^2$, which $= DE^2$, $>$
$GZ^2 + ZE^2$, which $= GE^2$
and $GZ^2 + ZE^2$, which $= GE^2$, $> BZ^2 + ZE^2$, which
$= BE^2$

$\therefore$ $DE > EG$ and $EG > EB$

Similarly ED is the longest of all the lines drawn from E meeting
the circumference of $\odot ABGD$, and EB is the shortest of them[5].
Q.E.D.

2. When a straight line is drawn in a circle, and from it a section not
less than a semicircle is cut off, and a section of a circle not greater than
a semicircle is drawn on it inclined to the section that is not greater
than a semicircle, and [if] the arc of the section [so] constructed is
divided into two unequal parts, then the line subtending the smaller
arc is less than all the straight lines going from the point of division
to the arc of the section that is not less than a semicircle.

In $\odot ABGD$ draw $\overline{AG}$, cutting from the circle a section $ABG \not< \frac{1}{2}\odot$
On line AG construct circular section AEG, not greater than a semi-
circle and inclined to section ADG, which $\not> \frac{1}{2}\odot$
Divide $A\widehat{E}G$ into unequal parts at E, and let $\widehat{GE} > \widehat{EA}$
Join $\overline{EA}$
I say: $EA <$ all straight lines from E to $A\widehat{B}G$.

From E draw the $\perp^{\mathrm{r}}$ to plane $\odot ABGD$: it falls between $\overline{AG}$ and
$A\widehat{D}G$, since section AEG is inclined to section ADG
Draw it and let it be EZ, meeting plane $\odot[ABGD]$ at Z
Mark centre of $\odot ABGD$: its centre is either on $\overline{AG}$ or between
$\overline{AG}$ and $A\widehat{B}G$, since we have laid down section ABG as not
less than a semicircle
First let it be between $\overline{AG}$ and $A\widehat{B}G$: let it be H
Draw ZH, and produce it in both directions, to D B
From E to $A\widehat{B}G$ draw $\overline{ET}$; draw AZ ZT
Since $EZ \perp$ plane $\odot ABGD$, it makes right angles with all lines
meeting it and being in plane $\odot ABGD$ [Euclid XI Def. 3]
Since each of AZ ZT, which are in plane $\odot ABGD$, meets EZ,
$\therefore$ each of $A\widehat{Z}E$ $T\widehat{Z}E$ is a right angle
Since $AZ < ZT$ [Euclid III 7],

---

[5]This last mathematical line is repeated in both Arabic and Greek.

$$AZ^2 < ZT^2$$

$\therefore$, with ZE^2 common, $AZ^2 + ZE^2$, which $= AE^2$, $<$
    $TZ^2 + ZE^2$, which $= TE^2$

$\therefore$ $AE < TE$

Similarly $[AE] <$ all straight lines from E to $A\widehat{T}B$

$\therefore$ AE is shorter than all straight lines drawn from E to $A\widehat{T}B$ between
    A and B

Similarly, of the straight lines going from E to $A\widehat{T}B$ between points A
    and B, those that are nearer to $[AE] <$ those that are further off

Draw line BE

I say: BE is the longest of all the lines drawn from E to $A\widehat{B}G$.

Since $BZ > ZT$ [Euclid III 7],

$\therefore$ $BZ^2 > ZT^2$

$\therefore$, with ZE^2 common, $EZ^2 + ZB^2$, which $= EB^2$, $>$
    $EZ^2 + ZT^2$, which $= ET^2$

$\therefore$ $BE > ET$

Similarly $[EB] >$ all straight lines from E to $A\widehat{B}G$

$\therefore$ EB is longer than all the lines from E to $A\widehat{B}G$

Draw EG

I say: EG is the shortest of all lines from E to $B\widehat{G}$ between B and G.

Draw another straight line EK, and also ZK ZG

Since $ZG < ZK$ [Euclid III 7],

$\therefore$ $ZG^2 < ZK^2$

$\therefore$, with ZE common, $ZG^2 + ZE^2$, which $= EG^2$, $<$
    $KZ^2 + ZE^2$, which $= EK^2$

$\therefore$ $GE < EK$

Similarly $[EG]$ is shorter than all straight lines drawn from E to
    $B\widehat{K}G$ between B and G

$\therefore$ EG is the shortest of all lines from E to $B\widehat{K}G$ between B and G[1]

Similarly, of the straight lines drawn from E to $B\widehat{G}$ between B and G,
    what is nearer to $[EG]$ is shorter than what is further

Since AE is the shortest straight line from E to $A\widehat{T}B$ and EG is the
    shortest of the straight lines from E to $B\widehat{K}G$ and $AE < EG$
    (since arc $<$ arc),

$\therefore$ $AE \ll$ the lines drawn from E to $B\widehat{G}$

---

[1] $\therefore$ EG ... G: not in Greek.

∴ EA is the shortest of all straight lines from E to $A\widehat{B}G$[2]

Similarly, if section ABG is a semicircle, then AE is the shortest of all straight lines from E to $A\widehat{B}G$. Q.E.D.

3. When two great circles intersect on a sphere, and from each two consecutive equal arcs are cut off on the two sides of one of the points of intersection, then the straight lines joining the end-points of the arcs on the same side are equal.

Let there be two great circles on a sphere intersecting at E: ⊙s AB GD

From each cut off two consecutive equal arcs on the two sides of E:
    $A\widehat{E}$ $E\widehat{B}$, $G\widehat{E}$ $E\widehat{D}$[1];
    let $A\widehat{E} = E\widehat{B}$ and $G\widehat{E} = E\widehat{D}$[2]

Join GA BD

I say: $GA = BD$.

> For the circle with pole E and distance EA also passes through B;
>
> and it either passes through G or not
>
> [1] First let it pass through G, as in Fig. 1
>
> ∴ it passes also through D, since $GE = ED$
>
> Draw ⊙$AGBD$
>
> Let common section of ⊙s $AGBD$ AEB be $\overline{AB}$
>
> Let common section of ⊙s $AGBD$ GED be $\overline{GD}$
>
> Since great ⊙AEB cuts one of the ⊙s in the sphere, ⊙$AGBD$, and passes through its poles, it bisects it at right angles [I 16]
>
> ∴ AB is a diameter of ⊙$AGBD$
>
> Similarly GD is a diameter of ⊙$AGBD$
>
> ∴ Z is centre of ⊙$AGBD$[3]
>
> ∴ the four lines ZA ZB ZG ZD are equal
>
> Since ZA $ZG = ZB$ ZD respectively and $A\hat{Z}G = D\hat{Z}B$ (vertically opposite)[4],
>
> ∴ base $AG =$ base DB [Euclid I 4]

---

[2]Since AE ... $A\widehat{B}G$: not in Greek, but cf. Scholium 344. See Czinczenheim, p. 420 (Greek) and p. 886 (French).

[1]AE ... ED: not in Greek.

[2]That is because arc $[E]G$ either is equal to arc EB or is unequal. If it is equal, the circle passes [through] G; if it is not equal, it does not pass *marg.* **N**.

[3]∴ Z ... $AGBD$: not in Greek.

[4](vertically opposite): not in Greek.

[2] Again, let the circle drawn about pole E with distance EA
not pass through G, but fall at a distance from it, as in Fig. 2
So it passes through B and falls at a distance from D; construct
it: $\odot AHBT$
Complete $\odot GED$ and let it meet $\odot AHBT$ at H and T
Let common section $\odot$s $AHBT$ AEB be AB,
  let common section $\odot$s $AHBT$ HET be HT
As before, Z is centre of $\odot AHBT$; and each of AEB $HET \perp$
  $\odot AHBT$
From G D draw GK and $DL \perp \odot AHBT$
Join AK LB
Since $\overset{\frown}{EH} = \overset{\frown}{ET}$ (since pole $\odot AHBT$ is E) and $\overset{\frown}{GE} = \overset{\frown}{ED}$,
  $\therefore$ remaining arcs $\overset{\frown}{GH} = \overset{\frown}{DT}$
Since $\overset{\frown}{HET}$ is a section of a circle, and equal arcs HG DT are
  cut off from it, and $\perp$ ^{rs} KG DL are drawn,
  $\therefore$ $\perp^r$ $KG = \perp^r DL$ and $HK = TL$[5]
And total arcs $HZ = ZT$,
  $\therefore$ remaining arcs $ZL = KZ$
And $AZ = ZB$: $\therefore$ $AK = $ and $\|$[6] LB [in $\triangle$s AZK BZK; Euclid
  I 4]
And since $AK = LB$ and $KG = DL$,
  AK $KG = BL$ LD resp. and $G\hat{K}A = D\hat{L}B$ (since each is right),
  $\therefore$ base $AG = $ base DB [in $\triangle$s GKA DLB; Euclid I 4]. Q.E.D.

4. When two great circles on a sphere intersect, and from one of them
two consecutive equal arcs are cut off on the two sides of one of the
two points of intersection, and two parallel planes are drawn passing
through the resulting points, one of them meeting the common section
of the planes of the two circles outside the surface of the sphere, on
the side of the point mentioned, and each of the equal arcs is greater
than each of the two arcs cut off from the other great circle by the
two planes drawn from that very point, then the arc between the point
at which the two great circles intersect and the plane not meeting the
common section is greater than the arc between that point and the
plane of the circle meeting the common section.

---

[5]Ver Eecke justifies this by applying Euclid I 26 to $\triangle$s GHK and DTL, for $\overline{HG}$
$= \overline{DT}$ (since $\overset{\frown}{HG} = \overset{\frown}{DT}$), $G\hat{H}K = G\hat{T}L$ and $\hat{K} = \hat{L} = $ right.
[6]and $\|$: not in Greek.

Let two great ⊙s *AEB GED* intersect at *E*

Let two consecutive equal arcs, *AE EB*, on either side of *E*, be cut off
    from ⊙*AEB*

Let two parallel planes be drawn through *A* and *B*: planes *AD GB*

Let plane *AD* meet the common section of planes *AEB GED* outside
    the surface of the sphere on the *E*-side [see below]

Let each of equal arcs $\widehat{AE}$ $\widehat{EB}$ > each of arcs *GE ED*

I say: $\widehat{GE} > \widehat{ED}$.

> For the circle about pole *E* with distance *EA* passes through *B*
>   and falls further from *G D* (since $\widehat{AE}$ $\widehat{EB} > \widehat{GE}$ $\widehat{ED}$)
> Let it be drawn; and let it be ⊙*AHBZ*
> Complete the two circles; let ⊙*GED* meet ⊙*AHBZ* at *H Z*[1];
>   let ⊙*AD* meet ⊙*AHBZ* at *T*; let ⊙*BG* meet ⊙*AHBZ* at *K*
> Let common section ⊙s *AEB AHBZ* be *AB*
> Let common section ⊙s *HEZ AHBZ* be *HZ*
> Let common section ⊙s *ADT AHBZ* be *AT*
> Let common section ⊙s *KGB AHBZ* be *KB*
> Let common section ⊙s *HEZ AEB* be *EL*
> Let common section ⊙s *HEZ ADT* be *MD*
> Let common section ⊙s *KGB HEZ* be *GN*
> Plane *AD* meets the common section of planes *HEZ AEB*, i.e.
>   *EL*, outside the surface of the sphere on the *E*-side: at *S*
> ∴ *S* is in plane *ADT*
> But it is also in plane *HEZ*; and *D M* are in planes *ADT HEZ*
>   ∴ *MD* meets *LE* outside the surface of the sphere, on the
>   *E*-side: at *S*; so let them meet on it
> Great ⊙*AEB* in the sphere cuts ⊙*AHBZ* of the circles in the
>   sphere and passes through its poles
>   ∴ it bisects it at right angles [I 16]
> ∴ *AB* is a diameter of ⊙*AHBZ*
> Similarly *HZ* is a diameter of *AHBZ*
> ∴ *L* is the centre of ⊙[*AHBZ*]
> Since parallel planes *KGB ADT* are cut by plane *AHBZ*, their
>   two common sections are parallel,
> ∴ *KB* ∥ *AT*
> Also, since parallel planes *KGB ADT* are cut by plane *HEZ*,
>   their two common sections are parallel [Euclid XI 16]
> ∴ *GN* ∥ *DM*

---
[1]let ⊙*GED* ... *Z*: not in Greek.

Since each of planes AEB HEZ $\perp$ plane $AHBZ$, their common
  section also $\perp$ $AHBZ$ [Euclid XI 19]
Their common section is EL: $\therefore$ $EL \perp$ plane $AHBZ$
$\therefore$ it produces right angles with all the straight lines that meet
  it in plane $AHBZ$ [Euclid XI Def. 3]
Each of AB HZ, which are in plane $AHBZ$, meets EL
  $\therefore$ $EL \perp AB$ and HZ
Since $S\hat{L}N$, exterior angle to $\triangle SLM$, > interior angle $S\hat{M}L$, opposite it, and $S\hat{L}N$ is right,
  $\therefore$ $S\hat{M}L$ is acute
$\therefore$ $S\hat{M}Z$ is obtuse
Since $GN \parallel DM$, and HZ falls on them,
  $\therefore$ $G\hat{N}H = S\hat{M}L$
And $S\hat{M}L$ is acute
  $\therefore$ $G\hat{N}H$ is acute
Since $AT \parallel KNB$ and lines AB MN are drawn between them,
  $\therefore$ $AL = LB$
$\therefore$ $NL = LM$ [in $\triangle$s BLN ALM; Euclid I 26]
And total lines $HL = LZ$
  $\therefore$ after subtraction $HN = MZ$
Since HEZ is the section of a circle, and two equal lines, HN MZ,
  are cut off from its chord[2], and $GN \parallel DM$, and $G\hat{N}H$ is acute
  and $D\hat{M}Z$ is obtuse,
  $\therefore$ $\widehat{HG} < \widehat{DZ}$[3]
$\therefore$ total arcs $\widehat{HE} = \widehat{EZ}$; and $\widehat{HG} < \widehat{DZ}$
  $\therefore$ remaining arcs $\widehat{GE} > \widehat{ED}$. Q.E.D.

5. When the pole of parallel circles in a sphere is on the circumference
of a great circle of its circles [*sc.* from among the sphere's circles],
and this circle is cut by two great circles at right angles, one from the
parallel circles and one inclined to the parallel circles, and from the
inclined circle two consecutive equal arcs are cut off on the same side
of the greatest of the parallel circles, and parallel circles are drawn
through the points generated, then they cut off from the first great
circle unequal arcs between them, and of these arcs the [arc] that are
nearer to the greatest of the parallel circles are greater than the arc
that is further from it.

---

[2]from its chord: not in Greek.
[3]See Ver Eecke, p. 93.

Let the pole of the parallel circles, A, be on the circumference of great
  circle ABG
Let this circle be cut at right angles by two great circles, BZG DZE,
  one of them (BZG) belonging to the parallel circles and the other
  (DZE) inclined to them
Let consecutive equal arcs, $\overset{\frown}{KT}$ $\overset{\frown}{TH}$, be cut off the inclined circle, on
  the same side of great $\odot BZG$ (of the parallel circles)
Let parallel circles be drawn through K T H: $\odot$s QKF NTS LHM
I say: From the first great $\odot ABG$ they cut off unequal arcs, and the
  arc of the parallel circles nearer to the greatest of the parallel
  circles > the arc that is further away.
I say: $\overset{\frown}{QN} > \overset{\frown}{NL}$.

Let a great circle be drawn through A T: $\odot ATC$
And since A is pole $\odot QKF$, $\overset{\frown}{ANQ} = \overset{\frown}{ATC}$
And since A is pole $\odot NTS$, $\overset{\frown}{ALN} = \overset{\frown}{ART}$
$\therefore$ remaining arcs $\overset{\frown}{NQ} = \overset{\frown}{TC}$
Similarly $\overset{\frown}{NL} = \overset{\frown}{RT}$
  $\therefore$ $\overset{\frown}{NQ} = \overset{\frown}{TC}$ and $\overset{\frown}{LN} = \overset{\frown}{RT}$
Great circle ATC in a sphere cuts a circle, of the circles in the
  sphere, QCF[1], and passes through its poles
  $\therefore$ it bisects it at right angles [I 16]
$\therefore$ $\odot ATC \perp \odot QCF$

On the diameter of $\odot QCF$ drawn from C a section of a circle
  is constructed standing on $\odot QCF$ at right angles: section CT
  and what is connected with it
Cut off from it an arc $< \frac{1}{2}$ section: $\overset{\frown}{TC}$
  $\therefore$ $\overline{CT}$ is the shortest of all straight lines from T to $\odot QFC$
  [III 1]
$\therefore$ $\overline{CT} < \overline{TK}$
And $\odot$s DE AC[2] are equal, for they are great
  $\therefore$ $\overset{\frown}{TC} < \overset{\frown}{TK}$
Similarly $\overset{\frown}{TR} < \overset{\frown}{TH}$, since a section of a circle is constructed on
  diameter LHM, standing on it at right angles: section RT with
  what is connected with it
And $\overset{\frown}{RT}$ is cut off $< \frac{1}{2}$ section constructed
Also, since $\overline{KT} = \overline{TH}$, so each of $\overline{KT}$ $\overline{TH}$ > each of $\overline{CT}$ $\overline{TR}$

---

[1]Here and in the following Greek has K for C in this combination. Of course,
the same arc is meant.

[2]DE AC: not in Greek.

Since $\odot BZG \parallel \odot LHM$ and $\odot BZG$ meets the common section of [planes] $\odot$s HTK ATC inside, i.e. at the centre of the sphere,
∴ plane LHM meets the common section of $\odot$s HTK ATC outside the surface of the sphere, on the T-side[3]

And since great $\odot$s HTK RTC intersect, and two consecutive equal arcs, KT TH, have been cut from $\odot HTK$ on each side of T

and [since] two parallel planes, LHM QCF, are constructed through H K, and of them plane LHM meets the common section of planes HTK RTC outside the surface of the sphere on the T-side and each of the equal arcs $\widehat{KT}$ $\widehat{TH}$ > each of $\widehat{CT}$ $\widehat{TR}$ [III 4],

∴ $\widehat{CT} > \widehat{TR}$

But $\widehat{CT} = \widehat{QN}$ and $\widehat{TR} = \widehat{NL}$

∴ $\widehat{QN} > \widehat{NL}$. Q.E.D.

6. When the pole of the parallel circles in a sphere is on the circumference of one of the great circles and this circle is cut at right angles by two great circles, and one of the[se] circles is a parallel circle and the other is inclined to the parallel circles, and consecutive equal arcs are cut off from the inclined circle on the same side of the circle that is the greatest of the parallel circles, and great circles are drawn passing through the resulting points and through the pole, then they cut off from the great[est] of the parallel circles unequal arcs and the arc near the first great circle is always greater than those that are further from it.

On line ABG, the circumference of the great circle, let there be the pole of the parallel circles: A

Let $\odot ABG$ be cut at right angles by two great circles: BZG DZE

Let $\odot BZG$ be the greatest of the parallel circles and $\odot DZE$ inclined to the parallel circles

Let two consecutive equal arcs, KT TH, be cut off from $\odot DZE$ on the same side of $\odot BZG$, which is the greatest of the parallel circles

Draw great circles passing through A and severally through H K T: $\odot$s AHL ATM AKN

I say: $\widehat{LM} > \widehat{MN}$.

Draw parallel circles passing through H T K: $\odot$s SHQ FTC RKO

---

[3]See Ver Eecke's note 8 on p. 85.

$\therefore$ $\widehat{RF} > \widehat{FS}$ because of what we explained earlier [III 5]
But $\widehat{RF} = \widehat{PT}$ and $\widehat{FS} = \widehat{TY}$
$\therefore$ $\widehat{PT} > \widehat{TY}$
Assume $\widehat{TX} = \widehat{YT}$
And $\widehat{HT} = \widehat{TK}$
$\therefore$ $\overline{HY} = \overline{XK}$ [III 3; cf. Euclid I 4]
Draw the circle parallel to the first circles and through X: $\odot X\acute{T}\acute{D}$[1]
Since great $\odot A\acute{T}KN$ in a sphere cuts one of the circles in the sphere, $X\acute{T}\acute{D}$, and passes through its poles, it bisects it at right angles [I 16]
$\therefore$ $\odot A\acute{T}KN \perp \odot X\acute{T}\acute{D}$
Since the parallel planes BZG $X\acute{T}\acute{D}$ are cut by plane $A\acute{T}KN$, their common sections are parallel [Euclid XI 16]
$\therefore$ common section of planes $A\acute{T}KN$ BZG $\parallel$ the common section of planes $A\acute{T}KN$ $X\acute{T}\acute{D}$
And the common section of planes BZG $A\acute{T}KN$ is the diameter of $\odot A\acute{T}KN$ from N
$\therefore$ the common section of planes $A\acute{T}KN$ $X\acute{T}\acute{D}$ $\parallel$ the diameter $\odot A\acute{T}KN$ from N
In $\odot A\acute{T}KN$ a line is drawn, the common section of $\odot$s $A\acute{T}KN$ $X\acute{T}\acute{D}$, dividing the circle into unequal parts, and $\parallel$ the diameter of $\odot A\acute{T}KN$ from N
A section of a circle is constructed which $\perp$ $A\acute{T}KN$, i.e. section $X\acute{T}$ and the section joined to it
The arc of the standing section is divided into unequal parts at X,
and $\widehat{X\acute{T}} < \frac{1}{2}$ the section made
$\therefore$ $\overline{X\acute{T}}$ is the shortest of all lines from X to $\widehat{\acute{T}KN}$ [III 1]
$\therefore$ $\overline{X\acute{T}} < \overline{XK}$ and $\overline{XK} > \overline{X\acute{T}}$
And $\overline{XK} = \overline{HY}$ [above]
$\therefore$ $\overline{HY} > \overline{X\acute{T}}$
Since $\odot X\acute{T}\acute{D}$ is nearer the centre of the sphere than $\odot SHQ$, $\odot X\acute{T}\acute{D} > \odot SHQ$ [I 6]
Since $\odot$s SHQ $X\acute{T}\acute{D}$ are not equal and $\odot SHQ$ is the smaller, and $\overline{HY}$ is drawn in $\odot SHQ$ and $\overline{X\acute{T}}$ is drawn in $\odot X\acute{T}\acute{D}$,

[1] In this proposition some of the rare letters are unusually represented:

Gk.	Ar.	Lat.	edn.
Ψ	ذ	*I*	$\acute{T}$
Ω	ض	*Fi*	$\acute{D}$

and since $\overline{HY} > \overline{X\acute{T}}$,

∴ $H\widehat{Y}$ > the arc of its circle which ~ $X\widetilde{T}$

But $H\widehat{Y} \sim L\widehat{M}$ and $X\widetilde{T} \sim M\widehat{N}$

∴ $L\widehat{M}$ > the arc of its circle which ~ $M\widehat{N}$

They both belong to the same circle

∴ $L\widehat{M} > M\widehat{N}$. Q.E.D.

7. When there is a great circle in a sphere touching one of the parallel circles and there is another great circle inclined to the parallel circles, touching two circles greater than the two circles that the first circle touches, and the places of contact are also on the first great circle, and consecutive equal arcs are cut off from the inclined circle on the same side of the great circle (from among the parallel circles), and parallel circles are drawn passing through the created points, then they cut off between them unequal arcs from the first great circle, and the arc near the greatest of the parallel circles is greater than the arc that is further from it.

Let great $\odot ABG$ in a sphere touch one of the circles in a sphere, $\odot AD$, at A

Let another great circle, inclined to the parallel circles, $\odot EZH$, touch two circles greater than the two circles touched by $\odot ABG$

Let the places of contact also be on $\odot ABG$: E H

Let the greatest of the parallel circles be $\odot BZG$

From the circle inclined to the parallel circles, $\odot EZH$, let two consecutive equal arcs be cut off, LK KT, on one side of the greatest of the parallel circles

Let parallel circles be drawn through T K L: MTN SKQ FLC

I say: $FS > MS$.

Draw a great $\odot$ through K and touching $\odot AD$: $\odot RKD$ [II 14]

∴ the semicircle $[\frac{1}{2}\odot ABG]$ drawn from A on the B-side does not meet the semicircle $[\frac{1}{2}\odot RKD]$ drawn from D to the R-side

Mark the pole of the parallel circles: P [I 22; II 1]

Draw a great circle through P K: $\odot PKY$ [I 21]

Great $\odot PKY$ is in a sphere cutting one of the circles in the sphere, $\odot FLC$, and passes through its poles

∴ it bisects it at right angles [I 16]

∴ $\odot PKY \perp \odot FLC$

∴ on the diameter of $\odot FLC$, emerging from Y, a section of a circle is constructed standing on it at right angles: section PY and what is connected to it

The arc of the section constructed is divided into unequal divisions at K, and $\widehat{KY}$ is the smaller of the divisions

∴ $\overline{KY}$ < all straight lines drawn from K to the circumference of $\odot FLC$

and the line near it always < the one further off [III 1]

∴ $\overline{KR} < \overline{KL}$

[... [1]] $> \overline{KR}$

And $\odot$s DR ELH are equal, for they are great,

∴ $\widehat{KL} > \widehat{KR}$

Similarly $\widehat{TK} > \widehat{K\acute{T}}$[2]

And $\widehat{TK} = \widehat{KL}$

∴ $\widehat{TK}\ \widehat{KL} > \widehat{K\acute{T}}\ \widehat{KR}$

Since $\odot BZG \parallel \odot MTN$ and $\odot BZG$ meets the common section of $\odot$s TKL $\acute{T}KR$ inside the surface of the sphere,

∴ $\odot MTN$ will meet the common section of $\odot$s TKL $\acute{T}KR$ outside the surface of the sphere on the K-side

But great $\odot$s TKL $\acute{T}KR$ in the sphere intersect at K;

and from one of them two consecutive equal arcs are cut off, TK KL, on the two sides of the point at which [the circles] intersect;

through T L parallel planes, FLC MTN, pass, and of these plane MTN meets the common section of planes TKL $\acute{T}KR$ outside the surface of the sphere on the K-side;

and each of $\widehat{TK}\ \widehat{KL}$ > each of $\widehat{RK}\ \widehat{K\acute{T}}$

∴ $\widehat{RK} > \widehat{K\acute{T}}$ [III 4]

But $\widehat{RK} = \widehat{FS}$ and $\widehat{K\acute{T}} = \widehat{MS}$

∴ $\widehat{FS} > \widehat{MS}$. Q.E.D.

8. When a great circle is in a sphere touching one of the parallel circles, and another great circle is on it, inclined to the parallel circles and touching two circles greater than those which the first circle touches,

---

[1] *lac.* in all MSS **ANH**. Gerard has the full text: Ergo linea que coniungit inter punctum K et punctum L.

[2] In this proposition Gerard renders ذ by I (our $\acute{T}$) – cf. note 1 to III 6. But in line 51, for the combination $\acute{T}KR$ only **B** has IKR, all the others having DKR. Again, in line 57 four manuscripts have DKR for $\acute{T}KR$.

and the place of meeting [المماسة] is also on the the first great circle
and two equal arcs are cut off from the inclined circle, on the same
side of the circle which is the greatest of the parallel circles, and great
circles are drawn passing through the points created and cutting off
similar arcs from the parallel circles between them, then they cut off
unequal arcs from the circle which is the greatest of the parallel circles
between them; and of these the arc which is near the first great circle
is greater than the one that is far from it.

Let great $\odot ABG$ be on a sphere and let it touch one of the parallel
    circles on the sphere, AD, at A
Let another great circle be inclined to the parallel circles: $\odot EZG$,
    touching two circles which $>$ the parallel circles touched by the
    first circle, $\odot ABG$
Let places of meeting [مواضع المماسة] on $\odot ABG$ be E G
Let greatest parallel circle be $\odot BZ$
From the inclined circle EZG let two equal arcs be cut off consecutively,
    $\widehat{HT}$ $\widehat{TK}$, on the same side of $\odot BZ$
Draw great circles through H T K: DHL MTN SKQ
    meeting [and tangent to] $\odot AD$ at D M S [II 14] and cutting off
    from the parallel circles between them similar arcs [II 13]
I say: LN > NQ.

    Draw parallel circles through H T K: FHC RT OPK
    $\therefore$ $\widehat{RO} > \widehat{RF}$ [III 7]
    But $\widehat{RO} = \widehat{TP}$ and $\widehat{RF} = \widehat{TC}$ [II 13]
    $\therefore$ $\widehat{PT} > \widehat{TC}$
    Let $\widehat{TY} = \widehat{TC}$
    Now $\widehat{HT} = \widehat{TK}$ [by hypothesis]: $\therefore$ $\overline{HC} = \overline{YK}$ [III 3]
    Let a circle be drawn parallel to any of $\odot$s FHC RT OPK BZ,
        passing through Y: $\odot XY\acute{T}^1$
    Mark the pole of the parallel circles: $\acute{D}$ [I 22; II 1]
    Draw the great circle through $\acute{D}$ Q: $\odot \acute{D}Q$ [I 21]
      $\therefore$ great $\odot \acute{D}Q$ in the sphere cuts one of the circles in the sphere,
        $\odot BZ$, and passes through its pole
        $\therefore$ it bisects it at right angles [I 16]
    $\therefore$ $\odot \acute{D}Q \perp \odot BZ$
    $\therefore$ $\odot SQ$ is inclined to $\odot BZ$ on the A-E-B-side
    $\therefore$ $\odot BZ$ is inclined to $\odot SQ$ on the S-side

---

[1]In this proposition, as in III 6, Gerard puts I (written as $\acute{T}$ in the edition) for
Arabic ذ (Greek Ψ) and Fi ($\acute{D}$ in the edition) for Arabic ض (Greek Ω).

And $\odot BZ \parallel \odot XY\acute{T}$

$\therefore \odot XY\acute{T}$ is inclined to $\odot SQ$ on the S-side

Since two parallel planes, $BZ \ XY\acute{T}$, are cut by plane SQ,

   $\therefore$ their common sections are parallel [Euclid XI 16]

$\therefore$ the common sections of planes $SQ \ XY\acute{T} \parallel$ the common sections of planes $BZ \ SQ$

The common section of planes $BZ \ SQ$ is the diameter of $\odot SQ$ drawn from Q

$\therefore$ the common section of planes $SQ \ XY\acute{T} \parallel$ is the diameter of $\odot SQ$ drawn from Q

In $\odot SQ$ a line – the common section of planes $SQ \ XY\acute{T}$ – is drawn, dividing the circle into two unequal parts,

   for it $\parallel$ diameter of $\odot SQ$, and a section of a circle has been constructed, section $Y\acute{T}$, together with what is connected with it, which is inclined to the section which $\ngtr \frac{1}{2}\odot$

The arc of the standing section is divided into unequal parts at Y, and $Y\tilde{T} <$ half section constructed[2]

   $\therefore \ \overline{Y\acute{T}}$ is the shortest of all the straight lines from Y to the arc that $\nless \frac{1}{2}\odot$ [III 2]

$\therefore \ \overline{Y\acute{T}} < \overline{YK}$

And $\overline{YK} = \overline{HC}$, as above, $\therefore \ \overline{Y\acute{T}} < \overline{HC}$

$\therefore \ \overline{HC} > \overline{Y\acute{T}}$

And since $\odot XY\acute{T}$ is nearer the centre of the sphere than $\odot FHC$, $\odot XY\acute{T} > \odot FHC$ [I 6]

And since $\odot XY\acute{T} \neq \odot FHC$ and $\odot FHC$ is the smaller, and in $\odot FHC$ is drawn $\overline{HC}$,

   and in $\odot XY\acute{T}$ is drawn $\overline{Y\acute{T}}$,

   and the line drawn in the smaller circle $>$ line drawn in the greater $\odot$, because $\overline{HC} > \overline{Y\acute{T}}$,

   $\therefore \ \overset{\frown}{HC} >$ arc of its circle that $\sim Y\tilde{T}$

But $\overset{\frown}{HC} \sim \overset{\frown}{LN}$ and $\overset{\frown}{Y\tilde{T}} \sim \overset{\frown}{NQ}$

$\therefore \ \overset{\frown}{LN} >$ arc of its circle that $\sim \overset{\frown}{NQ}$

They are both of the same circle

   $\therefore \ \overset{\frown}{LN} > \overset{\frown}{NQ}$. Q.E.D.

9. When the pole of the parallel circles is on the circumference of a great circle and two great circles cut this circle at right angles, one of them being one of the parallel circles and the other being inclined to

---

[2] See the long footnote 1 by Ver Eecke on p. 104.

the parallel circles, and [when] from the inclining circle two equal, but not consecutive, arcs are cut off on the same side of the circle which is the greatest of the parallel circles [and when] then circles are drawn through the resulting points and through the pole, then they cut off unequal arcs from the greatest of the parallel circles between them. An arc starting from the first great circle is always greater than those that are further from it.

Let pole of the parallel circles be on the circumference of $\odot ABG$: A
Let two great $\odot$s, DEG BE, cut $\odot ABG$ at right angles
    Let $\odot BE$ be one of the parallel circles, and $\odot DEG$ be inclined to the parallel circles
From $\odot DEG$ cut off two equal arcs, ZH TK, not consecutive, on the same side of the greatest of the parallel circles
Through Z H T K and through pole A draw great $\odot$s: AZL AHM ATN AKS
I say: $\widehat{LM} > \widehat{NS}$.
For $\widehat{HT}$ either is commensurable with $\widehat{ZH}$ $\widehat{TK}$ or it is not
[1] In Fig. 1 let $\widehat{HT}$ be commensurable with $\widehat{ZH}$ $\widehat{TK}$

    Let $\widehat{ZH}$ $\widehat{HT}$ $\widehat{TK}$ be divided by that quantity which is common to them: [e.g.] at Q F C R
    Draw great circles through Q F C R and through pole A: $\odot$s QO FP CY RX
    Since $\widehat{ZQ}$ $\widehat{QH}$ $\widehat{HF}$ $\widehat{FC}$ $\widehat{CT}$ $\widehat{TR}$ $\widehat{RK}$ are equal consecutive arcs,
    $\therefore$ $\widehat{LO}$ $\widehat{OM}$ $\widehat{MP}$ $\widehat{PY}$ $\widehat{YN}$ $\widehat{NX}$ $\widehat{XS}$ are unequal consecutive arcs, the greatest being $\widehat{LO}$ [and decreasing] after that according to the order [III 6]
    $\therefore$ $\widehat{LO} > \widehat{NX}$ and $\widehat{OM} > \widehat{XS}$
    $\therefore$ total arcs $\widehat{LM} > \widehat{NS}$

[2] Now let $\widehat{HT}$ be incommensurable with $\widehat{ZH}$ $\widehat{TK}$
I say: $\widehat{LM}$ again $> \widehat{NS}$. For if $\widehat{LM} \not> \widehat{NS}$, it either $<$ or $=$ it.
    [2a] If possible, let $\widehat{LM} < \widehat{NS}$, as in Fig. 2
    Let $\widehat{LM} = \widehat{NQ}$
    Draw a great $\odot$ passing through pole A and point Q: $\odot QF$[1]
    Since there are three arcs, $\widehat{KT}$ $\widehat{TF}$ $\widehat{HT}$, we mark an arc, $\widehat{TC}$, which $> \widehat{TF}$ and $< \widehat{TK}$ and commensurable with $\widehat{HT}$

---

[1]Let $\widehat{TC}$ be assumed [to be] greater than $\widehat{TF}$ and smaller than $\widehat{TK}$ and commensurable [? وقسمه] with $\widehat{HT}$ add. **N** (not in Greek or Latin; it seems to represent another version of what is coming next).

Let $\widehat{HR} = \widehat{TC}$

Let two great circles be drawn through R C and through pole
  A: ⊙s RO $C\acute{Y}$[2]

Since $\widehat{RH} = \widehat{TC}$ and $\widehat{HT}$ is commensurable with $\widehat{RH}$ $\widehat{TC}$ and
  $\widehat{MO} > \widehat{N\acute{Y}}$ [by part 1]

[But] $\widehat{LM} > \widehat{OM}$[3]: ∴ $\widehat{LM} \gg \widehat{NQ}$

But it was also equal to it ※
  ∴ $\widehat{LM} \not< \widehat{NS}$

[2b] *I say:* $[\widehat{LM}] \neq \widehat{NS}$ [see Fig. 3].

For, if it were possible, let it be equal to it, as in Fig. 3
Bisect $\widehat{ZH}$ $\widehat{TK}$ at Q F
Draw great circles through Q F and through pole A: ⊙s QC FR
Since $\widehat{ZQ} = \widehat{QH}$, ∴ $\widehat{LC} > \widehat{CM}$
∴ $\widehat{LM} > 2\widehat{MC}$
Also, since $\widehat{TF} = \widehat{FK}$, ∴ $\widehat{NR} > \widehat{RS}$
∴ $\widehat{SR} < \widehat{NR}$
∴ $\widehat{NS} < 2\widehat{NR}$
Since $\widehat{LM} = \widehat{NS}$ and $\widehat{LM} > 2\widehat{MC}$ and $\widehat{NS} < 2\widehat{NR}$, ∴ $\widehat{CM} <$
  $\widehat{NR}$
Already supposed: $\widehat{HQ} = \widehat{TF}$ ※, as is clear from Fig. 2[4]
∴ $\widehat{LM} \neq \widehat{NS}$

And it $\not< $ it: ∴ $\widehat{LM} > \widehat{NS}$. Q.E.D.

10. When the pole of the parallel circles is on the circumference of
a great circle, and two great circles cut this circle at right angles,
one of them being from the parallel circles and the other inclined to
the parallel circles, and two points are marked on the inclining circle,
however they fall, on the same side as the great circle from among
the parallel circles, and great circles are drawn passing through the
resulting points and through the pole, then the ratio of the arc of the
greatest of the parallel circles that falls between the first great circle
and the great circle drawn afterwards and so passes through the poles

---

[2]Here the letter $\acute{Y}$ appears in the Greek as Υ. The Arabic has the undotted
letter ‍و‍, rendered in our edition as ‍ڡ‍ (see the table of correspondence of the
diagram letters in the Introduction above). Gerard evidently read this letter as ‍ٮ‍
and rendered it, according to his system, as Y; in our edition we have written this
letter as $\acute{Y}$.

[3]and $\widehat{NY} > \widehat{NQ}$ *add.* **NH** and Gerard (not in Greek).

[4]I.e. where it is proved that if $\widehat{HQ} = \widehat{TF}$ then $\widehat{CM} = \widehat{NR}$.

to the arc of the inclining circle that falls between these two circles is
as the ratio of the arc of the greatest of the parallel circles that pass
through the pole of the parallel circles and through the marked points
to some arc which is less than the arc of the inclining circle between
the points that are marked.

Let the pole of the parallel circles, A, be on the circumference of great
    $\odot ABG$
Let two great circles cut $\odot ABG$ at right angles: $\odot$s DEG BE,
    $\odot BE$ being of the parallel circles, $\odot DEG$ inclined to them
Let two points, as they come, be marked on $\odot DEG$ on the same side
    of great $\odot BE$ of the parallel $\odot$s: Z H
Through Z H and through pole A draw great circles: AZT AHK
I say: $\overarc{BT} : \overarc{DZ} = \overarc{TK} :$ (an arc $< ZH$).
For $\overarc{ZH}$ is either commensurable with $\overarc{ZD}$ or not

[1] Let it be commensurable, as in Fig. 1
    Divide $\overarc{ZD}$ $\overarc{ZH}$ by the [common] magnitude, at points L M N
    Draw great circles through L M N and through pole A: $\odot$s LS
        MQ NF
    Since arcs DL LM MZ ZN NH are joined consecutively [and]
        are equal,
        $\therefore$ of arcs BS SQ QT TF FK, each $>$[1] the next [III 6], with
        $\overarc{BS}$ the greatest
    Since arcs BS SQ QT TF FK are consecutive, each $>$ the next,
        and arcs DL LM MZ ZN NH are consecutive [and] equal
        and number of arcs BS SQ QT = number of arcs DL LM MZ
        and number of arcs TF FK = number of arcs ZN NH,
        $\therefore$ $\overarc{BT} : \overarc{DZ} > \overarc{TK} : \overarc{ZH}$
    For since $\overarc{BS} > \overarc{TF}$ and $\overarc{DL} = \overarc{ZN}$ [etc.]
        and when there are [two] unequal quantities, the ratio of the
        greater quantity : a given quantity $>$ the ratio of the smaller
        quantity : [the given quantity] [Euclid V 8] and Σ antecedents
        : Σ consequents $> \Sigma$ antecedents : Σ consequents[2]
    If we make $\overarc{BT} : \overarc{DZ} = \overarc{TK} :$ some arc, then the arc $< \overarc{ZH}$

[2] Or $\overarc{HZ}$ is not commensurable with $\overarc{DZ}$
I say: $\overarc{BT} : \overarc{DZ} = \overarc{TK} :$ (arc $< \overarc{ZH}$).
For if that is not so, $[\overarc{BT}] : [\overarc{DZ}]$ either $= \overarc{TK} :$ (arc which $> \overarc{ZH}$) or
    $= [\overarc{TK}] : \overarc{ZH}$

---

[1]Here **N** and Latin have "$<$".
[2]For ... Σ consequents: not in Greek. Cf. Ver Eecke's note 3 on p. 109.

[2a] First, if possible let it = $\widehat{TK}$: (arc which > $\widehat{ZH}$, so $\widehat{ZL}$ in Fig. 2)

There are three[3] arcs, $\widehat{LZ}$ $\widehat{ZH}$ $\widehat{ZD}$. Cut off[4] another arc < $\widehat{LZ}$ and > $\widehat{ZH}$, commensurable with $\widehat{ZD}$. Let it be $\widehat{ZM}$

Draw great circle passing through M and pole A: $\odot MN$

Since $\widehat{ZM}$ is commensurable with $\widehat{DZ}$, ∴ $\widehat{BT}$: $\widehat{DZ}$ = $\widehat{TN}$: (arc which < $\widehat{ZM}$) [by part 1]

And $\widehat{BT}$: $\widehat{DZ}$ = $\widehat{TK}$: $\widehat{ZL}$

∴ $\widehat{TK}$: $\widehat{LZ}$ = $\widehat{TN}$: (arc which < $\widehat{ZM}$)

And $\widehat{NT}$ > $\widehat{TK}$: ∴ (the arc which < $\widehat{ZM}$) > $\widehat{ZL}$

But [$\widehat{ZM}$] < [$\widehat{ZL}$] ⁂

∴ $\widehat{BT}$: $\widehat{ZD}$ ≠ $\widehat{TK}$: (arc which > $\widehat{ZH}$)

[2b] *I say:* [$\widehat{BT}$] : [$\widehat{ZD}$] ≠ $\widehat{TK}$: $\widehat{ZH}$.

If it were possible, let $\widehat{BT}$: $\widehat{ZD}$ = $\widehat{TK}$: $\widehat{ZH}$ (see Fig. 3)

Bisect $\widehat{DZ}$ $\widehat{ZH}$ at L M

Draw two great circles passing through pole A and one of L M: LN MS

Since $\widehat{DL}$ = $\widehat{LZ}$, ∴ $\widehat{BN}$ > $\widehat{NT}$ [III 6]

∴ $\widehat{BT}$ > $2\widehat{TN}$

Similarly $\widehat{KT}$ < $2\widehat{TS}$

Since $\widehat{BT}$ > $2\widehat{TN}$ and $\widehat{TK}$ < $2\widehat{TS}$[5], ∴ $\widehat{BT}$: $\widehat{TK}$ > $\widehat{NT}$: $\widehat{TS}$

∴ $\widehat{NT}$: $\widehat{TS}$ < $\widehat{BT}$: $\widehat{TK}$

And $\widehat{BT}$: $\widehat{TK}$ = $\widehat{DZ}$: $\widehat{ZH}$ [by hypothesis and *alternando*]

∴ $\widehat{NT}$: $\widehat{TS}$ < $\widehat{DZ}$: $\widehat{ZH}$

And $\widehat{DZ}$: $\widehat{ZH}$ = $\widehat{LZ}$: $\widehat{ZM}$

∴ $\widehat{NT}$: $\widehat{TS}$ < $\widehat{LZ}$: $\widehat{ZM}$

∴, *alternando*, $\widehat{NT}$: $\widehat{LZ}$ < $\widehat{TS}$: $\widehat{ZM}$

If we make $\widehat{NT}$: $\widehat{LZ}$ = $\widehat{TS}$: some arc, this arc > $\widehat{ZM}$

It is clear from Fig. 2 [6] that this is ⁂

∴ $\widehat{BT}$: $\widehat{DZ}$ ≠ $\widehat{TK}$: $\widehat{ZH}$

But [$\widehat{BT}$] : [$\widehat{DZ}$] ≠ $\widehat{TK}$: (arc which > $\widehat{ZH}$)

∴ $\widehat{BT}$: $\widehat{DZ}$ = $\widehat{TK}$: (arc which < $\widehat{ZH}$). Q.E.D.

11. When the pole of the parallel circles is on the circumference of a great circle and the circle is cut by two great circles at right angles,

---

[3] three arcs $\widehat{LZ}$ $\widehat{ZH}$ $\widehat{ZD}$: arcus *LZ ZH* Latin.

[4] Cut off, Latin: *ponemus*.

[5] $2\widehat{TS}$. Greek has additionally: $\widehat{BT}$: $\widehat{TN}$ > $\widehat{KT}$: $\widehat{TS}$ and *alternando*.

[6] from Fig. 2: not in Greek

one being [the greatest] of the parallel circles and the other inclined to
the parallel circles, and another great circle passes through the poles
of the parallel circles, cutting the inclined circle between the greatest
of the parallel circles and the circle of the parallel circles touched by
the inclined circle, then the ratio of the diameter of the sphere to the
diameter of the [parallel] circle touched by the inclined circle is greater
than the ratio of the arc of the greatest of the parallel circles falling
between the first great circle and the circle following it and passing
through the poles of the parallel circles to the arc of the inclined circle
falling between the same circles.

Let the pole of the parallel circles be on great $\odot ABG$: A
Let $\odot ABG$ be cut by two great circles at right angles: $\odot$s BEG DEZ
Let $\odot BEG$ be the greatest of the parallel circles and let $\odot DEZ$ be
    inclined to the parallel circles
Let another great $\odot AHK$ cut $\odot DEZ$, passing through the poles of the
    parallel circles, [at H] between $\odot BEG$ and the [parallel] circle
    touched by $\odot DEZ$, i.e $\odot DLM$
I say: Diameter sphere : diameter $\odot DLM > \overset{\frown}{BT} : \overset{\frown}{DH}$.

    Draw the parallel circle through H: $\odot NHS$
    Join the common sections of these planes [i.e. of the planes of
        the mentioned great circles and the parallels through B H D]:
        AK DZ BG NS DM TQ HF HC QH
    So great $\odot ABG$ in a sphere cuts some of the circles in the sphere,
        DLM NHS BEG, and passes through their poles
        $\therefore$ it bisects them at right angles [I 16]
    [$\because$] $\overline{DM}$ $\overline{NS}$ $\overline{BG}$ are diameters of $\odot$s DLM NHS BEG
    And $\odot ABG \perp \odot$s DLM NHS BEG
    And since $\odot$s DLM NHS BEG are parallel in a sphere
        and a straight line, AK, is drawn passing through their poles,
        $\therefore$ line $AK \perp \odot$s DLM NHS BEG and passes through their
        centres and through the centre of the sphere [I 11]
    $\therefore$ R F Q are centres of $\odot$s DLM NHS BEG
    And since the parallel planes DLM NHS BEG are cut by plane
        ABG, their common sections are parallel [Euclid XI 16]
        $\therefore$ $\overline{DM}$ $\overline{NS}$ $\overline{BG}$ are parallel
    Also, since the parallel planes NHS BEG are cut by plane AHK,
        $\therefore$ their common sections are parallel [Euclid XI 16]
    Since $\overline{HF} \parallel \overline{TQ}$ and $\overline{NF}$ meets $\overline{FH}$, and these two $\parallel \overline{BQ}$ $\overline{QT}$,
        which meet, and these lines are not in the same plane,
        $\therefore$ they contain equal angles, i.e. $N\hat{F}H = B\hat{Q}T$ [Euclid XI 10]

Since $\odot$s *NHS DEZ* stand at right angles to $\odot ABG$, the common sections of $\odot$s *NHS DEZ* also $\perp \odot ABG$ [Euclid XI 19]

Their common section is *HC*

$\therefore$ *HC* $\perp \odot ABG$ and [$\because$] produces right angles with all straight lines which emerge from [*C*] in plane $\odot ABG$ [Euclid XI Def. 3]

And each of *FC QC*, which are in plane $\odot ABG$, is drawn from [the end of] $\overline{HC}$

$\therefore$ each of $H\hat{C}F$ $H\hat{C}Q$ is right

Since $AK \perp NS$[1], $C\hat{F}Q$ is right

Since $C\hat{F}Q$ is right, $F\hat{Q}C$ is acute [by considering right $\triangle FQC$]

$\therefore$ $QC > CF$ [QC is hypotenuse in right $\triangle FQC$; Euclid I 19]

Let $CO = FC$; draw *HO*

Since $FC = CO$ and *HC* is common [in $\triangle$s *FCH* and *OCH*],

$\therefore$ lines FC $CH = CO$ CH resp.

And right $F\hat{C}H =$ right $O\hat{C}H$

$\therefore$ base $HO =$ base HF and $\triangle FCH \equiv \triangle OCH$, the remaining angles are respectively equal to the corresponding angles, equal angles being subtended by equal sides [Euclid I 4]

$\therefore$ $H\hat{F}C = H\hat{O}C$

But $H\hat{F}C = T\hat{Q}B$: $\therefore$ $H\hat{O}C = T\hat{Q}B$

Since *HQC* is a triangle and the angle at *C* is right, and *HO* is drawn in it,

$\therefore$ $QC : CO > C\hat{O}H : C\hat{Q}H$[2]

And $CF = CO$ and $C\hat{O}H = T\hat{Q}B$

$\therefore$ $QC : CF > B\hat{Q}T : C\hat{Q}H$

But $CQ : CF = QD : DR$ [Euclid VI 4] and $= DZ : DM$

And $B\hat{Q}T : C\hat{Q}H = \overset{\frown}{BT} : \overset{\frown}{DH}$,

$ZD : DM > \overset{\frown}{BT} : \overset{\frown}{DH}$

And *DZ* is the diameter of the sphere and *DM* is the diameter of $\odot DLM$

$\therefore$ the diameter of the sphere : the diameter of $\odot DLM > \overset{\frown}{BT} : \overset{\frown}{DH}$. Q.E.D.

12. When two great circles are in a sphere touching the same circle of the parallel circles, cutting off similar arcs between them from the

---

[1] *NS*. Latin has: superficiem *NS*.

[2] See the section on lemmas to III 11 at the end of this chapter.

parallel circles, and another great circle is inclined to the parallel circles, touching two circles greater than the two circles that the first two [great] circles touch, and cuts the two circles that touch the same circle of the parallel circles between the greatest of the parallel circles and the circle that the first two circles touch, then the ratio of double the diameter of the sphere to the diameter of the circle which the inclining circle touches is greater than the ratio of the arc of the greatest of the parallel circles that falls between the two circles that touch the same circle to the arc of the inclining circle that falls between these same circles.

In a sphere let great circles AB GD touch the same circle of the parallel circles, $\odot AG$, at A G

From the parallel circles let them cut off, between them, similar arcs [II 13]

Let another great circle, $\odot EZ$, inclined to the parallel circles, touch two circles greater than the two circles touched by $\odot$s AB GD

Let [$\odot EZ$] cut $\odot$s AB GD between the greatest of the parallel circles and $\odot AG$, which is touched by $\odot$s AB GD

Let the greatest of the parallel circles be $\odot MBZ$

Let the circle touched by $\odot EZ$ from among the parallel circles be $\odot EH$

I say: 2×diameter of the sphere : diameter $\odot EH > \widehat{BD} : \widehat{TK}$.

    Let the pole of the parallel circles be L

    Draw great circles through L and each of E T K: *EMHL LTN LKS*

    Draw the parallel circle through K: $\odot QK$

    Let great circle QTF pass through T and touch $\odot EH$, at F [II 14]

    Since $\odot$ $QK \parallel \odot EFH$, and two great circles are drawn – *ETKZ QTF* – touching $\odot EFH$ at E F, and a great circle is drawn through pole L and through T – great circle *LTC*,

    $\therefore$ $\widehat{QC} = \widehat{CK}$ [1]

    $\therefore$ $\widehat{RC} < \widehat{CK}$

    $\therefore$ $\widehat{RK} < 2\widehat{KC}$

    But $\widehat{RK} \sim \widehat{BD}$ and $\widehat{KC} \sim \widehat{NS}$

    $\therefore$ $\widehat{BD} < 2\widehat{NS}$

    Since diameter of sphere : diameter $\odot EH > \widehat{MN} : \widehat{ET}$ [III 11]

    and $\widehat{MN} : \widehat{ET} > \widehat{NS} : \widehat{TK}$ [III 10],

    $\therefore$ diameter of the sphere : diameter $\odot EFH > \widehat{NS} : \widehat{TK}$

---

[1] See Ver Eecke, p. 116, note 4, for explanation.

$\therefore$, doubling the antecedents, 2×diameter of the sphere : diam.
$\odot EFH > 2\widehat{NS} : \widehat{TK}$
And $2\widehat{NS} : \widehat{TK} > \widehat{BD} : \widehat{TK}$, for $2\widehat{NS} > \widehat{BD}$
$\therefore$ 2×diam. of the sphere : diam. $\odot EFH \gg \widehat{BD} : \widehat{TK}$. Q.E.D.

13. When in a sphere there are parallel circles cutting off from some great circle equal arcs starting from the greatest of the parallel circles, and great circles are drawn passing through the resulting points and either passing through the poles of the parallel circles or touching the same circle of the parallel circles, then they cut off, betweeen them, equal arcs from the greatest of the parallel circles.

Let there be two parallel circles on a sphere, AB GD, and let them cut off two equal arcs, AE ED, from $\odot AED$ starting from $\odot ZEH$, the greatest of the parallel circles
Draw great circles through A E D: $\odot$s AZG TEK BHD, either passing through the pole of the parallel circles or touching the same parallel circle
I say: $\widehat{ZE} = \widehat{EH}$.

For when in a sphere there are two parallel circles, AB GD, cutting off equal arcs $\widehat{AE}$ $\widehat{ED}$ from a great circle, AED, starting from the greatest of the parallel circles, ZH,
[then] $\odot AB = \odot GD$ [II 16]
And since two parallel and equal circles, AB GD, cut off from great circle KT the two arcs TE EK starting from the greatest of the parallel circles, ZH,
[then] $TE = EK$ [II 17]
And $AE = ED$: $\therefore$ $\overline{AT} = \overline{KD}$ [III 3]
$\therefore$ chord($\widehat{AT}$) = chord($\widehat{KD}$)[1]
And the circles are equal: $\therefore$ $\widehat{AT} \sim \widehat{KD}$
But $\widehat{AT} \sim \widehat{ZE}$ and $\widehat{KD} \sim \widehat{EH}$
$\therefore$ $\widehat{ZE} \sim \widehat{EH}$
They are on the same circle: $\therefore$ $\widehat{ZE} = \widehat{EH}$. Q.E.D.

14. When a great circle touches one of the parallel circles in a sphere, and another great circle, inclined to the parallel circles, touches circles greater than those touched by the first [great] circle, then the two great

---

[1] $\therefore$ chord($\widehat{AT}$) = chord($\widehat{KD}$): Greek has $\widehat{AT} = \widehat{KD}$. According to Ver Eecke, Theodosius is appealing to Euclid III 28.

circles cut off, between them, dissimilar arcs from the parallel circles: whichever of the arcs that is near[er] one of the poles (whichever it is) is greater than the arc of its circle similar to one of them which is farther.

Let there be a great circle, $\odot ABG$, in a sphere, touching one of the parallel circles in the sphere, $\odot ADS$, at A

Let another great $\odot BEG$, inclined to the parallel circles, touch circles greater than the circles touched by the the first circle ABG

I say: The two circles ABG BEG cut off from the parallel circles, between them, dissimilar arcs; and whichever of them is near[er] one of the two poles is greater than the arc of its circle similar to what is far[ther] off.

Mark E K on the inclined circle, BG, however they fall

Draw two circles on E K which $\parallel \odot ADS$: $\odot ZEH$ TKL

I say: $\overset{\frown}{EH} >$ arc of its circle which $\sim \overset{\frown}{KL}$

and $\overset{\frown}{TK} >$ arc of its circle which $\sim \overset{\frown}{ZE}$.

Draw two great circles passing through E K touching $\odot ADS$: $\odot$s DEM SNK [II 12]

$\therefore$ the semicircle going out from D on the M-side does not meet the semicircle from A on the Z-T-side

and the semicircle from S on the K-side does not meet the semicircle from A on the L-side

Since the two semicircles AL SK do not meet

and between them, from among the parallel circles are the two arcs $\overset{\frown}{NH}$ $\overset{\frown}{KL}$,

$\therefore$ $\overset{\frown}{NH} \sim \overset{\frown}{KL}$ [II 13]

Hence also $\overset{\frown}{ZE} \sim \overset{\frown}{TM}$

And since $\overset{\frown}{NH} \sim \overset{\frown}{KL}$,

$\overset{\frown}{EH} >$ arc of its circle which $\sim \overset{\frown}{KL}$

and similarly $\overset{\frown}{TK}$ also $>$ arc of its circle which $\sim \overset{\frown}{ZE}$. Q.E.D.

Lemmas to III 11

Note. In III 11[1] Theodosius states an inequality without proof. In the letters of Fig. **A**2, for any triangle ABG with GD drawn to a point D on AB:

---

[1] Line 48-50 (Arabic) and 67-68 (Latin). Only the diagrams for the Arabic lemmas are given here (in translation). For the other diagrams the reader is referred to the Latin text above.

$$AB : BD > B\hat{D}G : B\hat{A}G$$

The inequality is derived from the fact that a sector is less than a triangle that contains it, but greater than one that it contains; a ratio of sectors is reduced to a ratio of angles and the corresponding ratio of triangles is reduced to a ratio of lengths.

A1 and **A**3 have a relatively complicated figure and argument, but the proof is essentially the same as outlined above. **A**2, with its simpler figure and argument, has again the same type of proof. Containing this lemma are several other texts, all mathematically equivalent, but differing in details such as diagram letters – e.g. the "Qusṭā" translation in Florence and Cambridge has a proof in which $\hat{A}$ is the right angle – and the order of the lines of reasoning: a scholion to the Greek text[2]; an appendix in **H** (in a different hand to the text); an isolated theorem in MS Aya Sofya 4830, f. 102r, "Lemma omitted in Theodosius' Book on Spheres"; and the redactions of the *Spherics* by al-Ṭūsī and Ibn Abī [al-]Shukr[3]. Of the same family but carrying a slightly different diagram is the proof in **P**, f. 18r *marg.*, and other manuscripts.

There are other statements of the lemma in Latin, such as MS Oxford, Bodleian Library, Digby 168, 126vb, and Paris BnF, lat. 7377B (**Px**), 48v-49r. These two are both in collections of notes to various mathematical texts. The enunciation of the Theodosius lemma, which is elaborate, is very similar in these two manuscripts.

A1

This is the proposition that we mentioned at the end of the book.
[Let there be] $\triangle ABG$ with $\hat{B}$ right. Draw GD as it comes [and D
    between A and B]
I say: $AB : BD > B\hat{D}G : B\hat{A}G$.
This being assumed, $AD : DB$ on the common section[1] $> D\hat{G}A : B\hat{A}G$
On $\triangle ADG$ draw $\odot AG$
Draw DH to H so that it $\parallel BG$ and produce it in a straight line to K
Join $GK\ AK$
I say: $AH : HG > A\hat{D} : D\hat{G}$.

---

[2]Czinczenheim, no. 450: p. 435 (Greek) and 901 (French).

[3]For a detailed treatment of this lemma, both in material related to Theodosius and elsewhere, see Knorr [1985] and [1986].

[1]This phrase appears tobe a corrupt form of a phrase for *separando*.

About K and with distance GK draw $G\overset{\frown}{Z}T$

From the assumption, $AH : HG >$ sector TZK : sector ZGK

It is like that because of what we shall mention:

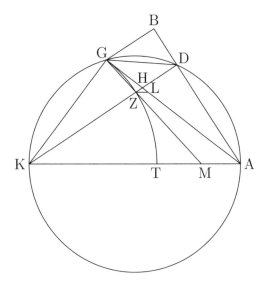

We join GZ and produce it in a straight line to M

$\therefore$ sector TZK : sector $GZK = Z\hat{K}T : G\hat{K}Z = D\hat{G}A : D\hat{A}G$

And $\triangle KZM : \triangle KGZ >$ sector : sector

$\therefore MZ : ZG >$ sector : sector [Euclid VI 1]

Draw $ZL \parallel AK$

$\therefore AL : [LG] >$ sector [: sector] [Euclid VI 2]

And $AH : HG > AL : LG^2$

$\therefore AH : HG >$ sector : sector

Composition of the proof: We do what we have done and there results:

[1] sector : sector $< AH : HG$;

[2] $AH : HG = AD : BD$ [since $\triangle AHD \sim \triangle AGB$ and Euclid VI 2];

[3] sector : sector $= Z\hat{K}T : G\hat{K}Z = D\hat{G}A : D\hat{A}G$ (which $= B\hat{A}G)^3$ [Euclid III 27]

$\therefore AD : DB > D\hat{G}A : D\hat{A}G$

$\therefore$, *componendo*, $AB : BD > B\hat{D}G : B\hat{A}G$. Q.E.D.[4]

---

[2]It is assumed that L lies between H and A.

[3]Perhaps "(which $= B\hat{A}G$)" is displaced from the next line.

[4]There follows (lines 24–29) an obscure fragment on ratios of chords. It appears to refer to the diagram for Lemma **A**1.

A2

[Let there be] △ABG with $\hat{B}$ right
Draw line GD, however it falls
I say: $AB : BD > B\hat{D}G : B\hat{A}G$.

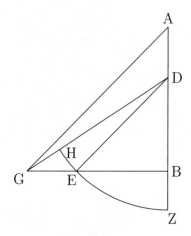

Proof: From D draw $DE \parallel AG$
Clearly, $DE > DB$ [Euclid I 18] and $< DG$
Make D a centre; draw a circle with distance DE: it falls inside
   △DBG and outside. Let it be ZEH
Since $ED \parallel GA$, $AD : DB = GE : EB$ [Euclid VI 2]
And $GE : EB = $ △$DEG : $ △BDE [Euclid VI 1]
∴ $AD : DB = $ △$DEG : $ △DBE
And △$DGE : $ △$DBE > $ △$DGE : $ sector DZE, since sector DZE
   $> $ △BDE
∴ $AD : DB > $ △$DEG:$ sector DEZ
And △$DEG > $ sector DEH
   ∴ $AD : DB \gg $ sector $DEH : $ sector DZE
∴, *componendo*, $AB : BD > $ sector $DZH : $ sector DZE
But sector $DZH : $ sector $DZE = B\hat{D}H : B\hat{D}E$
And $B\hat{D}E = B\hat{A}G:$ ∴ $AB : BD > B\hat{D}G : B\hat{A}G$. Q.E.D.

A3[1]

[Let there be] △ABG with $\hat{B}$ right and GD drawn as it comes
I say: $AB : BD > B\hat{D}G : B\hat{A}G$ and, *separando*, $AD : DB > D\hat{G}A :$
   $D\hat{A}G$.

---

[1]See diagram for Lemma **A**1

Proof: Draw circle about $\triangle ADG$

Draw $DHK \parallel BG$

Draw $AK\ KG$

With distance KG draw $G\widehat{Z}T$

Join GZ and produce it to M

Draw $LZ \parallel AM$

Since sector $KTZ < \triangle\ MZK$ and $\triangle KZG <$ sector KZG,

 $\therefore\ \triangle MZK : \triangle KZG$, i.e. $MZ : ZG$ [Euclid VI 1],

  i.e. $AL : GL$ [Euclid VI 2], $>$ sector $TZK :$ sector KZG

But $AH : HG > AL : LG$

$\therefore\ AH : HG$, i.e. $AD : DB$ [Euclid VI 2] $\gg$ sector $TZK :$ sector KZG,

 i.e. $T\widehat{K}Z : Z\widehat{K}G$, i.e. $\widehat{AD} : \widehat{DG}$, i.e. $A\widehat{G}D : D\widehat{A}G$ [Euclid III 26 and 27]

$\therefore$, *componendo*, $AB : DB > T\widehat{K}G : G\widehat{K}Z$, i.e. $B\widehat{D}G : D\widehat{A}G$ [Euclid I 32]. Q.E.D.

A4[2]

If DE is drawn parallel to GA and if a circle is constructed on centre D with distance DE, BG is cut and BD is cut when produced in a straight line towards Z; and let it be ZEH, etc. in this sense.

H

Lemma that Theodosius needed in the 11th proposition of the third book of this treatise, by Thābit ibn Qurra

Thābit said:

Let there be $\triangle ABG$ [*lit.* a triangle on which are $A\ B\ G$], with $\widehat{B} \not< \widehat{A}$

From $\widehat{A}$ draw AD to BG, however it falls, inside the triangle

I say: $GB : BD > A\widehat{D}B : A\widehat{G}B$.

 Proof: From line GD draw a line that $\parallel AG$, meeting AB at E

 Make D a centre and describe a circle through E, meeting AD at H

 Produce GB till it meets it at Z

 $\therefore$ sector $DEH :$ sector $DEZ = H\widehat{D}E : E\widehat{D}Z$

---

[2]This fragment of a proposition refers to the diagram for Lemma **A2**.

∴ $\triangle ADE$: sector $EDZ > H\hat{D}E$: $E\hat{D}Z$

∴ $\triangle ADE$: $\triangle EDB \gg A\hat{D}E$: $E\hat{D}B$

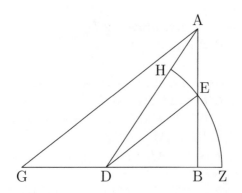

∴, *componendo*, $\triangle ADB$: $\triangle EDB > A\hat{D}B$: $E\hat{D}B$

And $E\hat{D}B = A\hat{G}B$ and $\triangle ADB$: $\triangle EDB = AB$: EB

∴ AB : $EB > A\hat{D}B$: $\hat{G}$

But AB : $BE = GB$: BD [Euclid VI 2]

  ∴ GB : $BD > A\hat{D}B$: $\hat{G}$. Q.E.D.

P

That CQ : $CO > C\hat{O}H$: $C\hat{Q}H$ is proved [as follows][1]:

From Q draw[2] $QF \parallel OH$, meeting CH at F

With centre Q and distance QH draw $\odot GHB$

Join CG

Since sector BHQ : sector $HQG = B\hat{Q}H$: $G\hat{Q}H$

  and sector BHQ : $\triangle HQC >$ sector BHQ : sector HQG [since $\triangle HQC <$ sector BHQ],

  ∴ sector BHQ : $\triangle CHQ > H\hat{Q}B$: $H\hat{Q}G$

∴, *componendo*, surface $BHCQ$: $\triangle CHQ > B\hat{Q}C$: $G\hat{Q}H$

This is because we cut off something [X, say] from sector BQH

  such that [X] : $\triangle HQC = B\hat{Q}H$: $H\hat{Q}G$

  ∴ $\triangle QFC$: $\triangle HQC \gg B\hat{Q}C$: $H\hat{Q}C$[3]

---

[1] The lettering is curious. The figure of $\triangle QHC$ and point O are taken, with its letters, from the figure for III 11, but the rest of the figure is labelled with B, G, F, which all have a different meaning in III 11 proper and its diagram.

[2] The text has a superfluous CQ before the word for "parallel".

[3] Since $\triangle QFC >$ surface $BHCQ$. The previous line is irrelevant.

But $\triangle FQC : \triangle HQC = FC : CH$
  and $FC : CH = CQ : CO$, since $OH \parallel QF$ [Euclid VI 2]
  $\therefore\ CQ : OC > G\hat{Q}B : G\hat{Q}H$
But $G\hat{Q}B = C\hat{O}H$ [Euclid I 29]
  $\therefore\ CQ : OC > C\hat{O}H : C\hat{Q}H$. Q.E.D.

Note to II 11 in **Kg** etc.

Theodosius proposed, in the *Spherics* II 11, perpendicular HK falling on the common section of planes $AGH\ ABG$ is within triangle ABG and perpendicular TL falling inside triangle DEZ. But it is possible that the sections erected on circles $ABG\ DEZ$, which are $AHG\ DTE$ are so large and the arcs that are cut off from them, starting from the extremities of the diameters, are so small that the perpendiculars drawn to them do not fall as Theodosius showed, i.e. inside circles ABG DEZ, but fall outside them on the common sections if produced, and the perpendiculars $KH\ TL$ fall outside, as is clear below, and diameters $GA\ ZD$ meet them at $K\ L$.

I say: If $HB = TE$, $\widehat{AB} = \widehat{DE}$ and conversely.

  Proof: Since section AHG = section DTZ and $\widehat{DT} = \widehat{AH}$
    $\therefore$ the remaining arcs $\widehat{HG} = \widehat{TZ}$
  $\therefore\ \overline{DT} = \overline{AH},\ \overline{HG} = \overline{TZ}$ and $DZ = AG$: $\therefore\ \triangle TDZ \equiv \triangle AHG$
    [Euclid I 8], [so that] corresponding angles are equal
  $\therefore\ T\hat{D}Z = H\hat{A}G$
  $\therefore\ T\hat{D}L = H\hat{A}K$
  And $T\hat{L}D$ is right and also $A\hat{K}H$, since HK and TL are perpendiculars, [$\therefore$] $LD = KA$ [Euclid I 26]
  But $DN = MA$: $\therefore\ LN = KM$
  Also $HB^2 = TE^2$ [given], $HB^2 = HK^2 + KB^2$, $TE^2 = TL^2 + LE^2$
  And $HK^2 = TL^2$: $\therefore\ KB = LE$
  But $KM = LN$, $MB = NE$ and base KB = base LE [in $\triangle$s KMB LNE]
    $\therefore\ K\hat{M}B = L\hat{N}E$ [Euclid I 8]
  They are at the centres of equal circles: $\therefore\ \widehat{AB} = \widehat{DE}$.

Note to II 11 in **M** *etc.*

These propositions are necessary in Theodosius' *Sphaerica* II 11:

[1]¹ The perpendicular from H to plane ABG falls on the common section of circles $ABG\ AHG$.

For draw from some point in the plane of $\odot ABG$ a perpendicular to the plane of $\odot ABG$. So a plane drawn from [*sc.* through] the perpendicular is perpendicular to the plane of $\odot ABG$. Therefore the two planes – i.e. [1] drawn from the perpendicular and [2] plane $\odot AHG$ – intersect and are perpendicular to plane $\odot ABG$

∴ their common section ⊥ plane $\odot ABG$ [Euclid XI 19], which also falls on AG

From any marked point in AG draw $KH \parallel$ this common section

∴ $KH \perp$ plane $\odot ABG$, for, if there are two parallel lines and one ⊥ a plane, then the other also ⊥ that plane

∴ draw a perpendicular from H [to AG] that falls outside AG: when the feet of the two perpendiculars are drawn, a triangle is made with two right angles ✳

[2]² Equal sectors $AHG\ DTZ$ of equal circles erected on diameters of equal circles can be semicircles, less than semicircles, or greater³. When they are semicircles or less, perpendiculars fall inside the arcs; but when they are greater, they sometimes fall outside the arcs on lines produced rectilinearly joining the diameters.

When perpendiculars are drawn from $H\ T$, falling outside the arcs, and $AH\ HG\ DT\ TZ$ are drawn,

$G\hat{A}H\ Z\hat{D}T$ will be obtuse, since they fall in sectors less than semicircles [Euclid III 31]

Since $A\hat{H} = D\hat{T}$ and $A\hat{G} = D\hat{Z}$, total arcs $H\hat{A}G = T\hat{D}Z$

And $ZT = GH$: ∴ $GH^2 = ZT^2$

But $GH^2 = GA^2 + AH^2 + 2GA \cdot [AK]^4$ [Euclid II 12]⁵

---

¹For the first part, see note 2 to II 11 in the mathematical summary.

²See the diagram for the note in **Kg** etc. (above, p. 324). Lines AB and DE must be supplied.

³A later note in this series, on the spherical trigonometry of Jābir b. Aflaḥ, is concerned with whether certain arcs are quadrants, greater than quadrants or less. See Lorch [1995], item VII, 12–20.

⁴Here AK is written for "the line that separates the perpendicular from the obtuse angle". The square brackets, here and in the following, indicate this and similar interpretations.

⁵For further use of theorems of this type in the notes, see Lorch [1995], item VII, 35.

and $ZT^2 = ZD^2 + DT^2 + 2ZD \cdot [DL]$ [Euclid II 12]
and GA^2 and $AH^2 = ZD^2$ and DT^2
$\therefore\ 2GA \cdot [AK] = 2ZD \cdot [DL]$
$\therefore\ GA \cdot [AK] = ZD \cdot [DL]$
But $GA = ZD$: $\therefore\ [AK] = [DL]$ (Euclid VI 14)

[3] Let centre of $\odot ABG$ be M and the centre of $\odot DEZ$ be N.

Draw $HB\ TE\ MB\ NE\ BA\ ED$
Join $B\ E$ to the two feet of the perpendiculars $[BK\ EL]$
Since $\overline{HA} = \overline{TD}$, $\therefore\ AH^2 = TD^2$
But the angle that the perpendicular $[sc.\ HK]$ makes with $[AK]$
= the angle that the other perpendicular makes with $[DL]$,
since each $[H\hat{K}G\ T\hat{L}Z]$ is right;
and $[AK] = [DL]$
$\therefore\ \perp^{\mathrm{r}}\ [sc.\ HK] = \perp^{\mathrm{r}}\ [sc.\ TL]$ [Euclid I 47]
And $HB = TE$ [hypothesis]: $\therefore\ H[K] = E[L]$
And since $M[K] = N[L]$ and $MB = NE$ and $B[K] = E[L]$,
$\therefore\ \hat{M} = \hat{N}\ [sc.\ B\hat{M}K = E\hat{N}L]$ [Euclid I 8]
But angle : angle = arc : arc, $\therefore\ BA = ED$

[4] *And I say:* If $AB\ AH = DE\ DT$ respectively, then $\overline{HB} = \overline{TE}$.

Since $B\hat{A}G = Z\hat{D}E$ [Euclid III 27], the angle made by BA with
$[AK]$ = the angle made by ED with $[DL]$ [i.e. $B\hat{A}K = E\hat{G}L$]
And BA and $[AK] = DE$ and $[DL]$
$\therefore\ B[K] = E[L]$ [Euclid I 4]
And $\perp^{\mathrm{r}}\ [sc.\ HK] = \perp^{\mathrm{r}}\ [sc.\ TL]$ and $[\hat{K}] = [\hat{L}]$ [= right]
$\therefore\ HB = TE$ [Euclid I 4]. Q.E.D.

Bibliography

J. L. Berggren, "The Relation of Greek Spherics to Early Greek Astronomy", in A. Bowen, *Science and Philosophy in Classical Greece*, N.Y. 1991, 227–248.

A. Björnbo, "Über zwei mathematische Handschriften aus dem vierzehnten Jahrhundert", *Bibliotheca Mathematica*, 3. F., 3 (1902) 63–75.

I. Bulmer-Thomas, "Theodosius of Bithynia", *Dictionary of Scientific Biography*, XIII, New York 1976, 319a–321a.

C. Burnett, "The Coherence of the Arabic-Latin Translation Program in Toledo in the Twelfth Century", *Science in Context* 14 (2001) 249–288.

Busard [1968]: H. L. L. Busard (ed.), *The Translation of the* Elements *of Euclid from the Arabic into Latin by Hermann of Carinthia (?), books I–VI*, Leiden 1968 (repr. from *Janus* LIV, 1967, 1-140).

Busard [1977]: H. L. L. Busard (ed.), *The Translation of the Elements of Euclid from the Arabic into Latin by Hermann of Carinthia (?), Books VII–XII*, Amsterdam 1977 (Math. Centre Tracts 84).

Busard [1983, 1]: H. L. L. Busard (ed.), *The First Latin Translation of Euclid's* Elements *Commonly Ascribed to Adelard of Bath*, Toronto 1983.

Busard [1883, 2]: H. L. L. Busard (ed.), *The Latin translation of the Arabic version of Euclid's* Elements *commonly ascribed to Gerard of Cremona*, Leiden 1983.

Busard [2005]: H. L. L. Busard, *Campanus of Novara and Euclid's* Elements, 2 vols., Stuttgart 2005.

H. L. L. Busard and M. Folkerts (eds.), *Robert of Chester's (?) Redaction of Euclid's* Elements, *the so-called Adelard II Version*, 2 vols., Basel–Boston–Berlin 1992.

Cat. BN: *Bibliothèque Nationale. Catalogue général des manuscrits latins*, vol. V (3278 à 3535), Paris 1966.

Cat. Paris: *Catalogus codicum manuscriptorum bibliothecae regiae.* Pars tertia, tomus quartus, Paris 1744.

C. Czinczenheim (ed.), *Édition, traduction et commentaire des* Sphériques *de Théodose*, 2 vols., Lille 2000. Thèse pour l'obtention du grade de Docteur de l'Université Paris IV.

J. Daly and C. Ermatinger, "Mathematics in the Codices Ottoboniani Latini, Part II", *Manuscripta* 9 (1965) 12–29.

Euclid: *vide* Busard, Busard-Folkerts, Heiberg, Tummers.

R. Fecht (ed.), *Theodosii de habitationibus liber, De diebus et noctibus libri duo*, Abhandlungen der Gesellschaft der Wissenschaften zu Göttingen, phil.-hist. Klasse, N.F. XIX, 4 (1927).

A. L. Gabriel, *A Summary Catalogue of Microfilms of One Thousand Scientific Manuscripts in the Ambrosian Library*, Notre Dame 1968.

Ḥājjī Khalīfa, *Keşf-el-Zunun*, ed. Ş. Yaltkaya and K. R. Bilge, I, Istanbul 1971.

Heath [1921]: Sir Thomas Heath, *A History of Greek Mathematics*, 2 vols., Oxford 1921.

Heath [1956]: Sir Thomas Heath (transl.), *The Thirteen Books of Euclid's Elements*, 2nd edn., Cambridge 1925, repr. N.Y. 1956.

Heiberg [1883–88]: J. L. Heiberg (ed.), *Euclidis Elementa*, 5 vols., Leipzig 1883–1888.

Heiberg [1927]: J. L. Heiberg (ed.), Theodosius Tripolites, *Sphaerica*, Abhandlungen der Gesellschaft der Wissenschaften zu Göttingen, phil.-hist. Klasse, N.F. XIX, 3 (1927).

Ibn al-Nadīm, *Kitâb al-Fihrist*, ed. G. Flügel and J. Roediger, 2 vols., Leipzig 1871–1872; repr. Beirut: Maktabat Khayyāṭ, s.d.

Ibn al-Qifṭī, *Ta'rīḫ al-ḥukamā'*, ed. A. Müller and J. L. Lippert, Leipzig 1903; repr. Baghdad: Maktabat al-Muthannā, and Cairo: Mu'assasat al-Khānjī, s.d.

Knorr [1985]: W. R. Knorr, "Ancient Versions of Two Trigonometrical Lemmas", *Classical Quarterly* 35 (ii) (1985) 362–391.

Knorr [1986]: W. R. Knorr, "The Medieval Tradition of a Greek Mathematical Lemma", *Zeitschrift für Geschichte der Arabisch-Islamischen Wissenschaften* 3 (1986) 230–261 + 6 pp. illustrations.

Kunitzsch [1974]: P. Kunitzsch, *Der Almagest. Die Syntaxis Mathematica des Claudius Ptolemäus in arabisch-lateinischer Überlieferung*, Wiesbaden 1974.

Kunitzsch [1991-92]: P. Kunitzsch, "Letters in Geometrical Diagrams, Greek – Arabic – Latin", *Zeitschrift für Geschichte der Arabisch-Islamischen Wissenschaften* 7 (1991–92) 1–20.

Lorch [1995]: R. Lorch, *Arabic Mathematical Sciences. Instruments, Texts, Transmission*, Ashgate (Variorum), Aldershot 1995.

Lorch [1996]: R. Lorch, "The Transmission of Theodosius' *Sphaerica*", in *Mathematische Probleme im Mittelalter. Der lateinische und arabische Sprachbereich*, ed. M. Folkerts, Wiesbaden 1996.

Lorch [2001]: Thābit ibn Qurra, *On the Sector-Figure and Related Texts*, ed. R. Lorch, Frankfurt am Main 2001; repr. Augsburg 2008.

G. D. Macray, *Catalogi codicum manuscriptorum bibliothecae Bodleianae*, Pars nona: *Codices a viro clarissimo Kenelm Digby ... donatos, complectens*, Oxford 1883.

T. J. Martin, *The Arabic Translation of Theodosius's* Sphaerica, unpublished dissertation, University of St. Andrews, 1975.

J. Millás Vallicrosa, *Las traducciones orientales en los manuscritos de la Biblioteca Catedral de Toledo*, Madrid 1942.

E. Nizze (ed.), *Theodosii Tripolitae sphaericorum libros tres ...*, Berlin 1852.

B. Nogara, *Codices Vaticani Latini*, III, Rome 1912.

N. Sidoli and T. Kusuba, "Naṣīr al-Dīn al-Ṭūsī's revision of Theodosius's *Spherics*", *Suhayl* 8 (2008) 9–46.

N. Sidoli and K. Saito, "The Role of Geometrical Construction in Theodosius's *Spherics*", *Archive for History of Exact Sciences* 63 (2009) 581–609.

M. Steinschneider, "Schriften der Araber in hebräischen Handschriften. Ein Beitrag zur arabischen Bibliographie", *Zeitschrift der Deutschen Morgenländischen Gesellschaft* 47 (1893) 335–384.

Theodosius: *vide* Czinczenheim, Fecht, Heiberg, Nizze, Ver Eecke.

P. M. J. E. Tummers (ed.), *The Latin Translation of Anaritius' Commentary on Euclid's Elements of Geometry, Books I–IV*, Artistarium, supplementa IX, Nijmegen 1994.

P. Ver Eecke (transl.), *Les sphériques de Théodose de Tripoli*, Bruges 1927.

W. Wisłocki, *Katalog Biblioteki Uniwersytetu Jagiellońskiego*, vol. I, Cracow 1877–81.

J. Young and P. H. Aitken, *A Catalogue of the Manuscripts in the Library of the Hunterian Museum in the University of Glasgow*, Glasgow 1908.

K. Ziegler, "Theodosius 5)", *Paulys Real-Encyclopädie der classischen Altertumswissenschaft*, 2. Reihe (R–Z), vol. V.A.2, Stuttgart 1934, col. 1930–1935.

BOETHIUS
Texte und Abhandlungen zur Geschichte der Mathematik und der Naturwissenschaften

Begründet von Joseph Ehrenfried Hofmann, Friedrich Klemm und Bernhard Sticker.
Herausgegeben von Menso Folkerts.

Franz Steiner Verlag   ISSN 0523–8226

ISBN 978-3-515-07975-4

46. Yvonne Dold-Samplonius / Joseph W. Dauben / Menso Folkerts / Benno van Dalen (Hg.)
From China to Paris
2000 Years Transmission of Mathematical Ideas
2002. X, 470 S., kt.
ISBN 978-3-515-08223-5

47. Ulf Hashagen
Walther von Dyck (1856–1934)
Mathematik, Technik und Wissenschafts-organisation an der TH München
2003. XV, 802 S., geb.
ISBN 978-3-515-08359-1

48. Rudolf Seising / Menso Folkerts / Ulf Hashagen (Hg.)
Form, Zahl, Ordnung
Studien zur Wissenschafts- und Technikgeschichte. Ivo Schneider zum 65. Geburtstag
2004. XI, 926 S., geb.
ISBN 978-3-515-08525-0

49. Michael Weichenhan
„Ergo perit coelum …"
Die Supernova des Jahres 1572 und die Überwindung der aristotelischen Kosmologie
2004. 688 S., geb.
ISBN 978-3-515-08374-4

50. Friedrich Steinle
Explorative Experimente
Ampère, Faraday und die Ursprünge der Elektrodynamik
2005. 450 S. mit zahlr. Abb., geb.
ISBN 978-3-515-08185-6

51. Hubertus Lambertus Ludovicus Busard
Campanus of Novara and Euclid's Elements
2005. 2 Bde. mit zus. XII, 768 S. mit zahlreichen Diagr. und Tab., geb
ISBN 978-3-515-08645-5

52. Richard Lorch (Hg.)
Al-Farghani. On the Astrolabe
Arabic Text Edited with Translation and Commentary
2005. VIII, 447 S. mit zahlr. Diagr. und Tab., geb.
ISBN 978-3-515-08713-1

53. Christian Tapp
Kardinalität und Kardinäle
Wissenschaftshistorische Aufarbeitung der Korrespondenz zwischen Georg Cantor und katholischen Theologen seiner Zeit
2005. 607 S. mit 30 Abb., geb.
ISBN 978-3-515-08620

54. Rudolf Seising
Die Fuzzifizierung der Systeme
Die Entstehung der Fuzzy Set Theorie und ihrer ersten Anwendungen.
Ihre Entwicklung bis in die 70er Jahre des 20. Jahrhunderts
2005. XIX, 395 S. mit 139 Abb., geb.
ISBN 978-3-515-08768-1

55. Harald Siebert
Die große kosmologische Kontroverse
Rekonstruktionsversuche anhand des *Itinerarium exstaticum* von Athanasius Kircher SJ (1602–1680)
2006. 383 S. mit 13 Abb., geb.
ISBN 978-3-515-08731-5

56. David A. King
Astrolabes and Angels, Epigrams and Enigmas
From Regiomontanus' Acrostic for Cardinal Bessarion to Piero della Francesca's Flagellation of Christ
2007. XI, 348 S. mit zahlr. z.T. farb. Abb. und CD-ROM, geb.
ISBN 978-3-515-09061-2

57. *in Vorbereitung*

58. Hartmut Hecht / Regina Mikosch / Ingo Schwarz / Harald Siebert / Romy Werther (Hg.)
Kosmos und Zahl
Beiträge zur Mathematik- und Astronomie-geschichte, zu Alexander von Humboldt und Leibniz
510 S. mit zahlr. Abb., geb.
ISBN 978-3-515-09176-3

59. Horst Kranz / Walter Oberschelp
Mechanisches Memorieren und Chiffrieren um 1430
Johannes Fontanas *Tractatus de instrumentis artis memorie*
2009. 167 S. mit 33 Abb., geb.
ISBN 978-3-515-09296-8